Boolean Functions in Coding Theory and Cryptography

Translations of

MATHEMATICAL MONOGRAPHS

Volume 241

Boolean Functions in Coding Theory and Cryptography

O. A. Logachev

A. A. Salnikov

V. V. Yashchenko

American Mathematical Society
Providence, Rhode Island

О. А. Логачев, А. А. Сальников, В. В. Ященко

БУЛЕВЫ ФУНКЦИИ В ТЕОРИИ КОДИРОВАНИЯ И КРИПТОГРАФИИ

М.: МЦНМО, 2004

This work was originally published in Russian by Издательство МЦНМО under the title "Булевы функции в теории кодирования и криптографии" © 2004. The present translation was created under license for the American Mathematical Society and is published by permission.

Translated by Svetla Nikova

2000 *Mathematics Subject Classification.* Primary 94–02; Secondary 94A60, 94C10.

For additional information and updates on this book, visit
www.ams.org/bookpages/mmono-241

Library of Congress Cataloging-in-Publication Data

Logachev, Oleg A.

[Bulevy funktsii v teori kodirovaniia i kriptologii. English]

Boolean functions in coding theory and cryptography / O.A. Logachev, A.A. Salnikov, V.V. Yashchenko ; translated by Svetla Nikova.

p. cm. — (Translations of mathematical monographs ; v. 241)

Includes bibliographical references and index.

ISBN 978-0-8218-4680-3 (alk. paper)

1. Coding theory. 2. Cryptography. 3. Algebra, Boolean. I. Sal′nikov, A. A. (Aleksei Aleksandrovich) II. IAshchenko, V. V. III. Title.

QA268.L6413 2011

003′.54—dc23

2011035308

∞ The paper used in this book is acid-free and falls within the guidelines
established to ensure permanence and durability.
Visit the AMS home page at `http://www.ams.org/`

10 9 8 7 6 5 4 3 2 1 17 16 15 14 13 12

Contents

Foreword

For the last 10 years there have been practically no books in Russian which have the word "cryptography" in the title. Nowadays many people already know that cryptography is the science which studies ciphers, and that only cryptography gives the most reliable tools for ensuring the security of information technology. However, there are not many specialists in this area, because in order to fully understand cryptography it is necessary to have knowledge in many scientific branches such as mathematics, physics, communication theory, and cybernetics. Thus, at present, cryptography (the theoretical branch of cryptology) becomes a university science. A detailed discussion of this issue has been held during the two conferences at Moscow State University (MGU): "Moscow University and development of cryptography in Russia" (October 17–18, 2002) and "Mathematics and information technology security" (October 23–24, 3003).

Institute for Problems of Information Security, a new division of MGU, publishes a series of fundamental books on scientific and methodological problems of information security, including those parts of cryptology that are already included in the university mathematical curriculum.

The book by O. A. Logachev, A. A. Salnikov, and V. V. Yashchenko "Boolean functions in coding theory and cryptology" belongs to this series. It is written by mathematicians-cryptographers for mathematicians and presents in a systematic way certain results in one branch of cryptology: application of Boolean functions in the analysis and design of ciphers.

The book is recommended to readers with basic university knowledge, namely students and graduate students in mathematics, research mathematicians, and cryptographers.

Rector of MGU, Academician V. A. Sadovnichii
February, 2004

Preface

The notion of Boolean function was introduced in the second half of the 19th century in connection with investigations in mathematical logic and foundations of mathematics. Boolean functions are named after George Boole (1815–1864), an English mathematician, one of the founders of mathematical logic. In the first half of the 20th century Boolean functions attain fundamental importance in the foundations of mathematics. However, for a long time Boolean functions have not been used in applications.

This situation changed drastically in the middle of the 20th century, when the intensive development of communication technology, instrument-building, and computer technology required the creation of an adequate mathematical apparatus. In this period, applied parts of mathematics such as the theory of finite functional systems, information theory, coding theory, and finally mathematical cryptography have been developed. The practice showed the fruitfulness of the application of Boolean functions to the problems of analysis and synthesis of discrete devices for processing and transformation of information.

The concept of cryptography that has been established in the scientific literature includes a range of scientific areas, each of them having its own subject of investigations and using specific mathematical techniques. Some researchers do abstract investigations "with cryptographic motifs" in the area of computational complexity theory; others are busy constructing and analyzing algorithms for particular cryptographic systems. In many cryptographic areas, Boolean function techniques are often used while formulating and solving various problems. This applies mainly to traditional cryptographic systems with a secret key. The title of the book "Boolean functions in coding theory and cryptography" reflects the relation between many cryptographic problems and encoding and decoding problems for Reed–Muller codes.

In this book, for the first time in Russian, we present cryptographic aspects using Boolean functions techniques. The only exceptions are questions related to complexity theory and solving systems of Boolean functions. In this book both classical and recent results are presented.

To understand the material, university courses of linear algebra, group theory, finite fields theory and polynomials, combinatorics and discrete mathematics will suffice. A knowledge of basics of probability theory is also assumed.

The book is based on courses given by the authors in MGU for students of Mechanics–Mathematics and Computational Mathematics and Cybernetics Departments who major in "Information security". Recent results obtained by the authors in the framework of the scheduled work of the MGU Laboratory on Mathematical Problems of Cryptography are also included in the book.

The book consists of nine chapters.

Chapter 1 is preliminary. It contains basic notions and results of algebra used in the book. In Chapter 2, basic notions and theorems of Boolean function theory are proved. In Chapter 3, problems of Boolean function classification under different groups of transformations are considered. Chapter 4 presents basics of coding theory. In Chapter 5, properties of Boolean functions are considered from the point of view of coding theory. In Chapter 6, properties of maximum-nonlinear functions are studied. Chapter 7 investigates the correlation immunity property of a function. In Chapter 8, various cryptographic characteristics of Boolean functions and mappings are considered in detail. Chapter 9 contains elements of cryptanalysis.

To avoid making the book too large, some of the results are presented as problems. Some of the problems included in the book are still open; they may be a basis for future research.

All items in the text are numbered consecutively within chapters: definitions, theorems, examples, etc. Thus, for example, Definition 1.121 refers to item 121 in Chapter 1 (which turns out to be a definition). The mathematical expressions and figures have similar but independent numbering.

The authors will accept with gratitude any comments on the book. They could be sent to the internet site `http://www.cryptography.ru`.

The authors express their gratitude to Mikhail Vladimirovich Stepanov for his support during the work on the book.

Notation

$\mathbb{N}$ — the set of natural numbers $(1, 2, 3, \ldots)$;
$\mathbb{Z}$ — the ring of integers $(\ldots, -2, -1, 0, 1, 2, \ldots)$;
$\mathbb{Z}_n$ — the ring of residues modulo $n \in \mathbb{N}$;
$\mathbb{R}$ — the field of real numbers;
$\mathbb{Q}$ — the field of rational numbers;
$\mathbb{C}$ — the field of complex numbers;
$\#A$ — cardinality of a set A;
$A \times B$ — Cartesian product of sets A and B;
A^n — nth Cartesian power of a set A $(n \in \mathbb{N})$;
$\mathcal{P}\{\cdot\}$ — probability of the event in the brackets;
$E[\cdot]$ — mathematical expectation of the random variable in the brackets;
B^A — set of all maps from a set A to a set B;
Π_n — set of minimal representatives of cyclotomic classes;
$\varphi^{-1}(b)$ — complete preimage of $b \in B$ under a map $\varphi \in B^A$;
$\mathbb{F}$ — finite field;
$\mathbb{F}_q$ — finite field of q elements ($q = p^n$, p is a prime number, $n \in \mathbb{N}$);
$\mathbb{F}_q^*$ — multiplicative group of invertible elements of the field $\mathbb{F}_q$;
$\mathbb{F}_q[x, y, \ldots, z]$ — ring of polynomials in variables $x, y, \ldots, z$ over the field $\mathbb{F}_q$;
$\mathrm{Tr}_{q^m/q}(\alpha)$ — relative trace of an element $\alpha \in \mathbb{F}_{q^m}$ over the field $\mathbb{F}_q$;
$\mathrm{Tr}_m(\alpha)$ — absolute trace of an element $\alpha \in \mathbb{F}_{p^m}$ over the field $\mathbb{F}_p$ (p is prime);
$\deg P(x, y, \ldots, z)$ — degree of a polynomial $P(x, y, \ldots, z)$;
V — linear vector space;
$V_{n,q}$ — vector space of columns of height $n \in \mathbb{N}$ with coordinates in the field $\mathbb{F}_q$;
V_n — vector space of columns of height $n \in \mathbb{N}$ with coordinates in the field $\mathbb{F}_2$ (n-dimensional Boolean space);
$\preccurlyeq$ — partial ordering relation of vectors from V_n;
$\dim V$ — dimension of a vector space V;
$\top$ — operator for matrix transposition;
$\boldsymbol{v} = (v^{(1)}, \ldots, v^{(n)})^\top$ — column vector in n-dimensional vector space in coordinate notation (in a fixed basis);
$\mathrm{wt}(\boldsymbol{v})$ — Hamming weight of a vector $\boldsymbol{v}$;
$\mathrm{dist}(\boldsymbol{v}, \boldsymbol{u}) = \mathrm{wt}(\boldsymbol{v} - \boldsymbol{u})$ — Hamming distance between vectors $\boldsymbol{v}$ and $\boldsymbol{u}$;
$\mathrm{dist}(A, B) = \min_{\boldsymbol{v} \in A, \boldsymbol{u} \in B} \{\mathrm{dist}(\boldsymbol{v}, \boldsymbol{u})\}$ — Hamming distance between sets of vectors $A \subseteq V$ and $B \subseteq V$;
$\oplus$ — coordinatewise addition of vectors of the same dimension over the field $\mathbb{F}_2$;

$\mathfrak{S}_n$ — symmetric (permutation) group of order $n \in \mathbb{N}$;
$\mathfrak{S}_V$ — symmetric permutation group acting on elements of a space V;
$\mathfrak{N}_n$ — group of translations acting on the vector space $V_{n,q}$ of dimension $n \in \mathbb{N}$;
$\mathfrak{D}_n$ — Jevons group acting on the vector space $V_{n,q}$ of dimension $n \in \mathbb{N}$;
$\langle A \rangle$ — group generated by a set A;
$\mathfrak{GL}(V)$ — full linear group acting on a vector space V;
$\mathfrak{GA}(V)$ — full affine group acting on a vector space V;
r-vector — vector of dimension $r \in \mathbb{N}$;
r-subset — subset of cardinality $r \in \mathbb{N}$;
$(m \times n)$ matrix — matrix with $m \in \mathbb{N}$ rows and $n \in \mathbb{N}$ columns;
$\operatorname{rank} M$ — rank of a matrix M;
$\det M$ — determinant of a square matrix M;
$\otimes$ — tensor product;
$a \mid b$ — a divides b ($a, b \in \mathbb{Z}$);
$\lfloor x \rfloor$ — largest integer less than or equal to $x \in \mathbb{R}$;
$\lceil x \rceil$ — smallest integer greater than or equal to $x \in \mathbb{R}$;
$\varnothing$ — empty set;
gcd — greatest common divisor;
lcm — least common multiple;
$\boldsymbol{i} = \sqrt{-1} \in \mathbb{C}$ — imaginary unit;
$T_n = \left\{ \exp\left\{2\pi \boldsymbol{i} \frac{k}{n}\right\} \mid k \in \mathbb{Z}_n \right\}$ — group of nth roots of unity;
$T = \{x \in \mathbb{C} \mid |x| = 1\}$ — multiplicative group of complex numbers of absolute value 1;
$\mathcal{F}_n$ — set of all Boolean functions of $n \in \mathbb{N}$ variables;
$\mathcal{F}_n(S)$ — set of partially defined Boolean functions of $0.5n \in \mathbb{N}$ variables with defining set $S \subseteq V_n$;
$\mathcal{F}_{n,m}$ — set of all Boolean functions from V_n to V_m ($n, m \in \mathbb{N}$);
$\exp f(\boldsymbol{x}) = (-1)^{f(\boldsymbol{x})}$ — function on V_n with values in $\{-1, 1\}$;
$\mathcal{L}_n$ — set of all linear Boolean functions of $n \in \mathbb{N}$ variables;
$\mathcal{L}_{n,m}$ — set of all linear Boolean mappings from V_n to V_m ($n, m \in \mathbb{N}$);
$\mathcal{A}_n$ — set of all affine Boolean functions of $n \in \mathbb{N}$ variables;
$\mathcal{A}_{n,m}$ — set of all affine Boolean mappings from V_n to V_m ($n, m \in \mathbb{N}$);
$\mathcal{B}_n$ — set of all Boolean bent functions (maximum-nonlinear functions) of $n \in \mathbb{N}$ variables;
$\mathcal{B}_n(S)$ — set of partially defined bent functions of $n \in \mathbb{N}$ variables with defining set $S \subseteq V_n$;
$\mathcal{S}_n$ — set of all symmetric Boolean functions of $n \in \mathbb{N}$ variables;
$\langle \boldsymbol{x}, \boldsymbol{y} \rangle = \sum_j x^{(j)} y^{(j)}$ — scalar product of vectors $\boldsymbol{x}$ and $\boldsymbol{y}$;
$\boldsymbol{x} \cdot \boldsymbol{y} = (x^{(1)} y^{(1)}, \ldots, x^{(n)} y^{(n)})^\top$ — product of vectors $\boldsymbol{x}$ and $\boldsymbol{y}$;
$W_f(\boldsymbol{\alpha}) = \sum_{\boldsymbol{x} \in V_n} (-1)^{f(\boldsymbol{x}) \oplus \langle \boldsymbol{x}, \boldsymbol{\alpha} \rangle}$ — Walsh–Hadamard transformation of a Boolean function $f \in \mathcal{F}_n$ ($\boldsymbol{\alpha} \in V_n$);
$W_f(\boldsymbol{a}, s) = \sum_{\boldsymbol{x} \in \mathbb{F}_{2^n}} (-1)^{f(\boldsymbol{x}) \oplus Tr(\boldsymbol{a}\boldsymbol{x}^s)}$ — extended Walsh–Hadamard transformation of a Boolean function $f \in \mathcal{F}_n$ ($\boldsymbol{\alpha} \in V_n$, $s \in \Pi_n$);

$W_f^D(\boldsymbol{\alpha}) = \sum_{\boldsymbol{x}\in D}(-1)^{f(\boldsymbol{x})\oplus\langle\boldsymbol{x},\boldsymbol{\alpha}\rangle}$ — partial Walsh–Hadamard transformation of a Boolean function $f \in \mathcal{F}_n$ with respect to the set $D \subseteq V_n$ ($\boldsymbol{\alpha} \in V_n$);

$\overline{W}_f(\boldsymbol{\alpha}) = \sum_{\boldsymbol{x}\in V_n} f(\boldsymbol{x})(-1)^{\langle\boldsymbol{x},\boldsymbol{\alpha}\rangle}$ — Fourier transform of a Boolean function $f \in \mathcal{F}_n$ ($\boldsymbol{\alpha} \in V_n$);

$\overline{W}_f^D(\boldsymbol{\alpha}) = \sum_{\boldsymbol{x}\in D} f(\boldsymbol{x})(-1)^{\langle\boldsymbol{x},\boldsymbol{\alpha}\rangle}$ — partial Fourier transform of a Boolean function $f \in \mathcal{F}_n$ with respect to a set $D \subseteq V_n$ ($\boldsymbol{\alpha} \in V_n$);

N_f — nonlinearity of a Boolean function $f \in \mathcal{F}_n$;

GN_f — generalized nonlinearity of a Boolean function $f \in \mathcal{F}_n$;

$D_{\boldsymbol{u}}\Phi$ — derivative of a Boolean mapping $\Phi \subset \mathcal{F}_{n,m}$ in the direction $\boldsymbol{u} \in V_n$;

$\Delta_f(\boldsymbol{\alpha}) = \sum_{\boldsymbol{x}\in V_n}(-1)^{f(\boldsymbol{x}\oplus\boldsymbol{\alpha})\oplus f(\boldsymbol{x})}$ — autocorrelation of a Boolean function $f \in \mathcal{F}_n$ with shift $\boldsymbol{\alpha} \in V_n$;

$\Delta_f = \max_{\substack{\boldsymbol{\alpha}\in V_n\\ \boldsymbol{\alpha}\neq\boldsymbol{0}}} |\Delta_f(\boldsymbol{\alpha})|$, $\sigma_f = \sum_{\substack{\boldsymbol{\alpha}\in V_n\\ \boldsymbol{\alpha}\neq\boldsymbol{0}}} \Delta_f^2(\boldsymbol{\alpha})$ — numerical measures of the global avalanche characteristics of a Boolean function $f \in \mathcal{F}_n$;

$\mathrm{ill}(F)$ — linearity index of a Boolean mapping $F \in \mathcal{F}_{n,m}$;

$\widetilde{f}$ — dual of a bent function $f \in \mathcal{B}_n$;

$\mathcal{J}_{\mathfrak{G}}(f)$ — moment group of a function f in a group $\mathfrak{G}$;

L_F — subspace of linearity (subspace of linear structures) of a Boolean mapping $F \in \mathcal{F}_{n,m}$;

PC_F — set of directions (vectors) for which a mapping $F \in \mathcal{F}_{n,m}$ satisfies the propagation criterion;

$\mu l(F)$ — maximum element from the difference table of a Boolean mapping $F \in \mathcal{F}_{n,m}$;

$RM(r, n)$ — binary Reed–Muller code of order $r \in \mathbb{N}$ and length 2^n ($n \in \mathbb{N}$);

$RM^*(r, n)$ — binary punctured Reed–Muller code of order $r \in \mathbb{N}$ and length $2^n - 1$ ($n \in \mathbb{N}$);

$C_j(f)$ — number of code words in the code $\boldsymbol{C}$ that are at distance j from f ($0 \leqslant j \leqslant r_{\boldsymbol{C}}$);

$\mathrm{Aut}(\boldsymbol{C})$ — automorphism group of a code $\boldsymbol{C}$;

$\boldsymbol{C}^{\perp}$ — dual code of a code $\boldsymbol{C}$;

$d_{\boldsymbol{C}}$ — minimum distance of a code $\boldsymbol{C}$;

$k_{\boldsymbol{C}}$ — dimension of a linear code $\boldsymbol{C}$;

$r_{\boldsymbol{C}}$ — covering radius of a code $\boldsymbol{C}$;

$R_{\boldsymbol{C}}$ — rate of a code $\boldsymbol{C}$;

$\mathcal{W}_{\boldsymbol{C}}(x, y)$ — weight function of a code $\boldsymbol{C}$;

$[n, k, d]$-code — linear code of length $n \in \mathbb{N}$, dimension $k \in \mathbb{N}$, and minimum distance $d \in \mathbb{N}$;

$N_{\max}(n, m)$ — maximum possible nonlinearity of an m-resilient Boolean function on $\mathcal{F}_n$;

$\rho(\boldsymbol{x}, r)$ — ball centered at $\boldsymbol{x} \in V_n$ of radius $r \in \{0, 1, \ldots, n\}$;

μ — Möbius function;

δ — Dirac δ-function;

I_M — indicator function of a set M;

E_n — identity $(n \times n)$ matrix;

$H_n = \left(\begin{smallmatrix}1 & 1\\ 1 & -1\end{smallmatrix}\right)^{[n]}$ — Sylvester–Hadamard matrix of order $n \in \mathbb{N}$; $[n]$ is Kronecker (tensor) power;

NW_f — number of binary vectors for which the Walsh–Hadamard coefficients of a function f are nonzero;

$N\Delta_f$ — number of binary vectors for which the values of the autocorrelation function f are nonzero;

$\mathrm{con}_J^{\boldsymbol{a}}$ — operation of fixing part of variables of a set of functions; it is given by a set of indices $J = \{j_1, \ldots, j_l\}$, $1 \leq j_1 < \cdots < j_l \leq n$, and binary vectors $\boldsymbol{a}^\top = (a^{(1)}, \ldots, a^{(l)})^\top \in V_l$;

pr_i — projection operator of a Boolean mapping to the ith coordinate;

$\mathrm{per}(\boldsymbol{u})$ — period of a sequence $\boldsymbol{u}$;

$\mathrm{loc}(\boldsymbol{c})$ — set of locators of a row $\boldsymbol{c}$;

$\mathrm{char}_{\boldsymbol{u}}(\lambda)$ — characteristic polynomial of a recurrent sequence $\boldsymbol{u}$.

CHAPTER 1

Arithmetics of Finite Fields and Polynomials

In this chapter we present the results from the theory of finite fields and polynomials that are necessary for reading the book. In order to understand the material in this chapter, basic knowledge of set theory is required.

1.1. Basic Algebra

One of the most important notions in algebra is the notion of algebraic systems. In general, an algebraic system is a set endowed with operations and predicates which satisfy certain conditions. In the sequel we will only consider algebraic systems with an empty set of predicates. Examples of such algebraic systems are groups, rings, and fields. Those are the algebraic systems most often used in this book when describing and investigating constructions from coding theory and cryptography.

Let S be a set and $S \times S$ the set of all ordered pairs (s, t), where $s, t \in S$. Then an arbitrary mapping from $S \times S$ into S is called a *(binary) operation on the set* S.

The term "group" is a common name for algebraic systems with one binary operation satisfying certain conditions. We will denote groups by capital letters and elements of groups, by small letters.

DEFINITION 1.1. *A group* $(\mathfrak{G}, *)$ is a set $\mathfrak{G}$ with a binary operation $*$ such that the following three conditions are satisfied.

1. The operation $*$ is associative, i.e., for any $\mathfrak{a}$, $\mathfrak{b}$, $\mathfrak{c} \in \mathfrak{G}$ the following equality holds:

$$\mathfrak{a} * (\mathfrak{b} * \mathfrak{c}) = (\mathfrak{a} * \mathfrak{b}) * \mathfrak{c}.$$

2. In $\mathfrak{G}$ there exists an *identity element* $\mathfrak{e}$ such that for any $\mathfrak{a} \in \mathfrak{G}$ the following equalities hold:

$$\mathfrak{a} * \mathfrak{e} = \mathfrak{e} * \mathfrak{a} = \mathfrak{a}.$$

3. For any element $\mathfrak{a} \in \mathfrak{G}$ there exists an *inverse element* $\mathfrak{a}^{-1} \in \mathfrak{G}$ such that

$$\mathfrak{a} * \mathfrak{a}^{-1} = \mathfrak{a}^{-1} * \mathfrak{a} = \mathfrak{e}.$$

If the group also satisfies the condition

4. For any $\mathfrak{a}, \mathfrak{b} \in \mathfrak{G}$,

$$\mathfrak{a} * \mathfrak{b} = \mathfrak{b} * \mathfrak{a},$$

then the group is called *abelian* (or *commutative*).

We will denote the group $(\mathfrak{G}, *)$ simply by $\mathfrak{G}$. It is easy to show that the identity element of the group $\mathfrak{G}$, as well as the inverse element, is unique. For any $\mathfrak{a}, \mathfrak{b} \in \mathfrak{G}$, the following equality holds:

$$(\mathfrak{a} * \mathfrak{b})^{-1} = \mathfrak{b}^{-1} * \mathfrak{a}^{-1}.$$

For the sake of simplicity the multiplication $\cdot$ is often used to denote the group operation, and instead of $\mathfrak{a} * \mathfrak{b}$ we will write $\mathfrak{a} \cdot \mathfrak{b}$ or simply $\mathfrak{ab}$ (we will say that this element is the product of the elements $\mathfrak{a}$ and $\mathfrak{b}$). Note that it is not always the case that the group operation is multiplication. Some times it is convenient to use the additive notation $\mathfrak{a} + \mathfrak{b}$ instead of $\mathfrak{a} * \mathfrak{b}$ (we will call this element the sum of the elements $\mathfrak{a}$ and $\mathfrak{b}$), 0 instead of $\mathfrak{e}$ (this element is called *zero element of the group* $\mathfrak{G}$), and $-\mathfrak{a}$ instead of $\mathfrak{a}^{-1}$. As a rule, the additive notation is used for abelian groups.

EXAMPLE 1.2. Let $\mathfrak{G}$ be the set of integers with the usual addition $+$. It is known that this operation is commutative. It is easy to see that $\mathfrak{G}$ is a commutative group with identity element 0 in which the inverse element of $\mathfrak{a}$ is $-\mathfrak{a}$. This group is usually denoted by $\mathbb{Z}$.

DEFINITION 1.3. A (multiplicative) group $\mathfrak{G}$ is called *cyclic* if there exists an element $\mathfrak{a} \in \mathfrak{G}$ such that any element $\mathfrak{b} \in \mathfrak{G}$ is a power of $\mathfrak{a}$, i.e., there exists an integer k such that $\mathfrak{b} = \mathfrak{a}^k$. Such an element $\mathfrak{a}$ is called a *generator* of the group G.

For a cyclic group $\mathfrak{G}$ with a generator $\mathfrak{a}$ the following notation is used: $\mathfrak{G} = \langle \mathfrak{a} \rangle$. For additive groups, we talk about multiples of an element instead of powers. From the definition and the commutativity of the additive group $\mathbb{Z}$ it follows that any cyclic group is commutative. In a cyclic group there may be more than one generator. For example, the additive group $\mathbb{Z}$ has generators 1 and -1.

DEFINITION 1.4. An *equivalence relation* on a set S is the subset Γ of the set $S \times S$ of ordered pairs (s, t), $s, t \in S$, that has the following properties:
1. $(s, s) \in \Gamma$ for any $s \in S$ (reflectivity).
2. If $(s, t) \in \Gamma$, then $(t, s) \in \Gamma$ (symmetry).
3. If $(s, t) \in \Gamma$ and $(t, u) \in \Gamma$, then $(s, u) \in \Gamma$ (transitivity).

Two elements $s, t \in S$ are called *equivalent* if $(s, t) \in \Gamma$. Any equivalence relation on the set S generates a partition of S into disjoint subsets of equivalent elements. All elements of the set S equivalent to a certain fixed element $s \in S$ form the equivalence class of the element s, which is denoted by

$$\{s\}_\Gamma = \{t \in S \mid (s, t) \in \Gamma\}.$$

Note that $\{s\}_\Gamma = \{t\}_\Gamma$ if and only if s and t are equivalent, i.e., $(s, t) \in \Gamma$. All the equivalence classes yield a partition of S.

PROBLEM 1.5. Let $a, b \in \mathbb{Z}$ and $n \in \mathbb{N}$. We will say that a is congruent to b modulo n, $a \equiv b \pmod{n}$, if the difference $a - b$ is divisible by n, i.e., if $a = b + kn$ for some $k \in \mathbb{Z}$. Prove that
1. The relation $\equiv$ is an equivalence relation.
2. The set $\mathbb{Z}_n$ of equivalence classes with the operation $\{\mathfrak{a}\}_\equiv + \{\mathfrak{b}\}_\equiv = \{\mathfrak{a} + \mathfrak{b}\}_\equiv$ is a cyclic group (this group is called the *group of residue classes* modulo n).

DEFINITION 1.6. A group is called *finite* (resp. *infinite*) if it contains a finite (resp. infinite) number of elements. The number of its elements is called the *order of the group*.

We will use $\#\mathfrak{G}$ to denote the order of a finite group $\mathfrak{G}$.

EXAMPLE 1.7. Let $T_n = \{\exp 2\pi \boldsymbol{i}\frac{k}{n} \mid k \in \mathbb{Z}_n\}$ be the set of nth (complex) roots of unity ($\boldsymbol{i} = \sqrt{-1}$). Endowed with multiplication of complex numbers as the group operation,

$$\left(\exp 2\pi \boldsymbol{i}\frac{k_1}{n}\right) \cdot \left(\exp 2\pi \boldsymbol{i}\frac{k_2}{n}\right) = \exp 2\pi \boldsymbol{i}\frac{k_1 + k_2}{n},$$

$k_1, k_2 \in \mathbb{Z}$, the set T_n is the multiplicative group of the nth roots of unity. Moreover $\#T_n = n$.

DEFINITION 1.8. A subset $\mathfrak{H} \subseteq \mathfrak{G}$ of a group $\mathfrak{G}$ is called a *subgroup* of this group ($\mathfrak{H} \leq \mathfrak{G}$) if the subset $\mathfrak{H}$ is a group with respect to the same operation as $\mathfrak{G}$. The subgroups of $\mathfrak{G}$ different from the trivial subgroups $\{\mathfrak{e}\}$ and $\mathfrak{G}$ are called *nontrivial subgroups.*

DEFINITION 1.9. The subgroup of a group $\mathfrak{G}$ that consists of all powers of an element $\mathfrak{g} \in \mathfrak{G}$ is called the *subgroup generated by the element* $\mathfrak{g}$ and is denoted by $\langle \mathfrak{g} \rangle$. If $\langle \mathfrak{g} \rangle$ is a finite cyclic group, then its order $\#\langle \mathfrak{g} \rangle$ is called the *order of the element* $\mathfrak{g}$. Otherwise the element $\mathfrak{g}$ is called an *element of infinite order*.

Thus, the order of an element $\mathfrak{g}$ is equal to the smallest natural number k for which $\mathfrak{g}^k = \mathfrak{e}$. It is easy to show that any integer m such that $\mathfrak{g}^m = \mathfrak{e}$ is divisible by k.

If N is a nonempty subset of a group $\mathfrak{G}$, then the subgroup $\mathfrak{H}$ of the group $\mathfrak{G}$ that consists of all finite products of powers of the elements of N is called the *subgroup generated by the set* N and is denoted by $\mathfrak{H} = \langle N \rangle$. Then N is called the *generating set of the subgroup* $\mathfrak{H}$.

PROBLEM 1.10. Let $\mathfrak{H}$ be a subgroup of a group $\mathfrak{G}$. Show that the relation $\Gamma_{\mathfrak{H}}$ on $\mathfrak{G}$ defined by the condition

$$(\mathfrak{g}_1, \mathfrak{g}_2) \in \Gamma_{\mathfrak{H}} \iff \mathfrak{g}_1 = \mathfrak{g}_2\mathfrak{h}$$

for some $\mathfrak{h} \in \mathfrak{H}$ is an equivalence relation.

The equivalence classes of the equivalence relation $\Gamma_{\mathfrak{H}}$ are called the *left cosets* of the subgroup $\mathfrak{H}$ in the group $\mathfrak{G}$ and are denoted by

$$\mathfrak{g}\mathfrak{H} = \{\mathfrak{g}\mathfrak{h} \mid \mathfrak{h} \in \mathfrak{H}\}$$

(or $\mathfrak{g} + \mathfrak{H} = \{\mathfrak{g} + \mathfrak{h} \mid \mathfrak{h} \in \mathfrak{H}\}$ in the additive notation), where $\mathfrak{g}$ is a fixed element of the group $\mathfrak{G}$. Similarly one defines the partition of the group $\mathfrak{G}$ into the *right cosets* of the subgroup $\mathfrak{H}$; right cosets have the form $\mathfrak{H}\mathfrak{g} = \{\mathfrak{h}\mathfrak{g} \mid \mathfrak{h} \in \mathfrak{H}\}$. If the group $\mathfrak{G}$ is abelian, then obviously the left cosets of the subgroup $\mathfrak{H}$ in $\mathfrak{G}$ coincide with the right cosets.

PROBLEM 1.11. If $\mathfrak{H}$ is a finite subgroup of a group $\mathfrak{G}$, then any (left or right) coset of the subgroup $\mathfrak{H}$ in $\mathfrak{G}$ contains as many elements as the subgroup $\mathfrak{H}$ itself.

DEFINITION 1.12. If a subgroup $\mathfrak{H}$ of a group $\mathfrak{G}$ is such that the set of cosets of $\mathfrak{H}$ in $\mathfrak{G}$ is finite, then the number of these cosets (left or right) is called the *index of the subgroup* $\mathfrak{H}$ in the group $\mathfrak{G}$ and is denoted by $(\mathfrak{G} : \mathfrak{H})$.

PROBLEM 1.13. Show that the order of a finite group $\mathfrak{G}$ is equal to the product of the order of any of its subgroups $\mathfrak{H}$ and the index $(\mathfrak{G} : \mathfrak{H})$ of this subgroup in the group $\mathfrak{G}$. In particular, the order of an element $\mathfrak{g} \in \mathfrak{G}$ divides the order of the group $\mathfrak{G}$.

DEFINITION 1.14. The smallest natural number m (if it exists) for which the equality $\mathfrak{g}^m = \mathfrak{e}$ holds for all $\mathfrak{g} \in \mathfrak{G}$ is called the *exponent of the group* $\mathfrak{G}$ and is denoted by $\exp \mathfrak{G}$. The exponent of a finite group $\mathfrak{G}$ divides the order of the group, i.e., $\exp \mathfrak{G} \mid \#\mathfrak{G}$.

The subgroups and orders of elements of cyclic groups are described in the following theorem.

THEOREM 1.15. *The following statements hold.*

1. *Any subgroup of a cyclic group is also cyclic.*

2. *In a finite cyclic group* $\langle \mathfrak{g} \rangle$ *of order* m, *the element* $\mathfrak{g}^k$ *generates a subgroup of order* $\frac{m}{\gcd(k,m)}$ *(where* $\gcd(k,m)$ *is the greatest common divisor of the numbers* k *and* m*).*

3. *If* $\langle \mathfrak{g} \rangle$ *is a finite cyclic group of order* m *and* d *is a positive divisor of* m, *then* $\langle \mathfrak{g} \rangle$ *contains a unique subgroup of index* d. *If* l *is a positive divisor of* m, *then the group* $\langle \mathfrak{g} \rangle$ *contains a unique subgroup of order* l.

4. *If* $\langle \mathfrak{g} \rangle$ *is a finite cyclic group of order* m *and* l *is a positive divisor of* m, *then the group* $\langle \mathfrak{g} \rangle$ *contains exactly* $\varphi(l)$ *elements of order* l *(here* $\varphi(l)$ *is Euler's function of* l, *i.e., the number of integers* k *such that* $1 \le k \le l$ *and* k *is coprime with* l*).*

5. *A finite cyclic group* $\langle \mathfrak{g} \rangle$ *of order* m *contains exactly* $\varphi(m)$ *generators. All generators of the group* $\langle \mathfrak{g} \rangle$ *are of the form* $\mathfrak{g}^r$, *where* $\gcd(r,m) = 1$.

PROBLEM 1.16. Prove Theorem 1.15.

DEFINITION 1.17. A mapping $\psi\colon \mathfrak{G} \to \mathfrak{H}$ from a group $(\mathfrak{G}, *)$ to a group $(\mathfrak{H}, \cdot)$ is called a *homomorphism* from $\mathfrak{G}$ to $\mathfrak{H}$ if it respects the operations in the groups, i.e., for all $\mathfrak{g}_1, \mathfrak{g}_2 \in \mathfrak{G}$,

$$\psi(\mathfrak{g}_1 * \mathfrak{g}_2) = \psi(\mathfrak{g}_1) \cdot \psi(\mathfrak{g}_2).$$

If in addition the homomorphism ψ is a mapping onto $\mathfrak{H}$ (i.e., if ψ is surjective), then ψ is called an *epimorphism* and in this case $\mathfrak{H}$ is called a *homomorphic image of the group* $\mathfrak{G}$. A homomorphism of a group $\mathfrak{G}$ to $\mathfrak{G}$ is called an *endomorphism* of this group. If ψ is a one-to-one homomorphism of a group $\mathfrak{G}$ to a group $\mathfrak{H}$, then it is called an *isomorphism*, and in this case we say that the groups $\mathfrak{G}$ and $\mathfrak{H}$ are *isomorphic* (notation: $\mathfrak{G} \cong \mathfrak{H}$). An isomorphism of a group $\mathfrak{G}$ to $\mathfrak{G}$ is called an *automorphism* of this group.

If $\psi\colon \mathfrak{G} \to \mathfrak{H}$ is a homomorphism of a group $(\mathfrak{G}, *)$ to a group $(\mathfrak{H}, \cdot)$, then from the identity $\mathfrak{a} * \mathfrak{e} = \mathfrak{a}$ in the group $(\mathfrak{G}, *)$ it follows that $\psi(\mathfrak{a}) \cdot \psi(\mathfrak{e}) = \psi(\mathfrak{a})$. Thus, the image of the identity element $\psi(\mathfrak{e}) = \mathfrak{e}'$ is the identity element of the group $\mathfrak{H}$. Besides, from the identity $\mathfrak{g} * \mathfrak{g}^{-1} = \mathfrak{e}$ in the group $(\mathfrak{G}, *)$ it follows that $\psi(\mathfrak{g}^{-1}) = (\psi(\mathfrak{g}))^{-1}$.

EXAMPLE 1.18. Let the mapping ψ of the group $\mathbb{Z}$ to the group $\mathbb{Z}_n$, $n \in \mathbb{N}$, be given by the equality $\psi(z) = \{z\}_{\equiv}$. Then ψ is an epimorphism of the group $\mathbb{Z}$ onto the group $\mathbb{Z}_n$.

DEFINITION 1.19. Let $\mathfrak{G}$ be a finite abelian group and $\mathfrak{T}$ the multiplicative group of complex numbers of modulus one. A homomorphism χ from the group $\mathfrak{G}$ to the group $\mathfrak{T}$ is called a *character of the group* $\mathfrak{G}$.

Let χ be a character of the group $\mathfrak{G}$. Then for any $\mathfrak{g} \in \mathfrak{G}$ the following equalities hold:

$$(\chi(\mathfrak{g}))^{\exp \mathfrak{G}} = \chi((\mathfrak{g})^{\exp \mathfrak{G}}) = \chi(\mathfrak{e}) = 1.$$

Thus, the values of the character χ are the roots of unity of order $\exp \mathfrak{G}$. Besides, $\chi(\mathfrak{g}^{-1}) = (\chi(\mathfrak{g}))^{-1} = \overline{\chi(\mathfrak{g})}$ for any $\mathfrak{g} \in \mathfrak{G}$ (here the bar denotes complex conjugation). Among the characters of the group $\mathfrak{G}$ there is a *trivial character* χ_0, defined by the rule $\chi_0(\mathfrak{g}) = 1$ for all $\mathfrak{g} \in \mathfrak{G}$. Other characters are called *nontrivial.*

PROBLEM 1.20. Prove that the set $\mathfrak{G}^\wedge$ of all characters of an abelian group $\mathfrak{G}$ is an abelian group with respect to multiplication that is defined as follows:

$$(\chi_1\chi_2)(g) = \chi_1(\mathfrak{g})\chi_2(\mathfrak{g}), \quad \text{where} \quad \chi_1, \chi_2 \in \mathfrak{G}^\wedge, \ \mathfrak{g} \in \mathfrak{G}.$$

Since the values of the characters can only be roots of unity of order $\exp \mathfrak{G}$, the group $\mathfrak{G}^\wedge$ is a finite group.

PROBLEM 1.21. Let $\mathfrak{G}$ be a finite cyclic group of order n with generator $\mathfrak{g}$. Prove that $\mathfrak{G}^\wedge$ consists of the mappings χ_j, $0 \le j \le n-1$, of the form

$$\chi_j(\mathfrak{g}^k) = e^{2\pi \boldsymbol{i} \frac{jk}{n}} = \cos 2\pi \frac{jk}{n} + \boldsymbol{i} \sin 2\pi \frac{jk}{n}.$$

PROBLEM 1.22. Prove that for any two different elements $\mathfrak{g}_1$, $\mathfrak{g}_2$ of an abelian group $\mathfrak{G}$ there exists a character $\chi \in \mathfrak{G}^\wedge$ of the group $\mathfrak{G}$ such that $\chi(\mathfrak{g}_1) \ne \chi(\mathfrak{g}_2)$ (such a character *distinguishes* between $\mathfrak{g}_1$ and $\mathfrak{g}_2$).

THEOREM 1.23. *If χ is a nontrivial character of a finite abelian group $\mathfrak{G}$, then*

$$\sum_{\mathfrak{g}\in\mathfrak{G}} \chi(\mathfrak{g}) = 0. \tag{1.1}$$

If $\mathfrak{g} \in \mathfrak{G}$ and $\mathfrak{g} \ne \mathfrak{e}$, then

$$\sum_{\chi\in\mathfrak{G}^\wedge} \chi(\mathfrak{g}) = 0. \tag{1.2}$$

PROOF. Since χ is a nontrivial character of the group $\mathfrak{G}$, there exists an element $\mathfrak{g}_1 \in \mathfrak{G}$ such that $\chi(\mathfrak{g}_1) \ne 1$. For such an element $\mathfrak{g}_1$ the following chain of equalities holds:

$$\chi(\mathfrak{g}_1) \sum_{\mathfrak{g}\in\mathfrak{G}} \chi(\mathfrak{g}) = \sum_{\mathfrak{g}\in\mathfrak{G}} \chi(\mathfrak{g}_1\mathfrak{g}) = \sum_{\mathfrak{g}\in\mathfrak{G}} \chi(\mathfrak{g})$$

(if an element $\mathfrak{g}$ runs over the whole group $\mathfrak{G}$, the product $\mathfrak{g}_1\mathfrak{g}$ also runs over the whole group $\mathfrak{G}$). Thus,

$$(\chi(\mathfrak{g}_1) - 1) \sum_{\mathfrak{g}\in\mathfrak{G}} \chi(\mathfrak{g}) = 0,$$

whence (1.1) follows.

For the second statement of the theorem, observe that the function on $\mathfrak{G}^\wedge$ assigning to a character χ the number $\widehat{\mathfrak{g}}(\chi) = \chi(\mathfrak{g})$, is a character of the finite abelian group $\mathfrak{G}^\wedge$. This character is nontrivial because, according to the results of Problem 1.22, there exists a character $\chi \in \mathfrak{G}^\wedge$ such that $\chi(\mathfrak{g}) \ne \chi(\mathfrak{e}) = 1$. To obtain (1.2) it remains to apply (1.1) to the group $\mathfrak{G}^\wedge$. □

THEOREM 1.24. *The number of different characters of a finite abelian group $\mathfrak{G}$ is equal to the order of this group:*

$$\#\mathfrak{G}^\wedge = \#\mathfrak{G}.$$

PROOF. The theorem results from the equalities

$$\#\mathfrak{G}^\wedge = \sum_{\mathfrak{g}\in\mathfrak{G}} \sum_{\chi\in\mathfrak{G}^\wedge} \chi(\mathfrak{G}) = \sum_{\chi\in\mathfrak{G}^\wedge} \sum_{\mathfrak{g}\in\mathfrak{G}} \chi(\mathfrak{G}) = \#\mathfrak{G},$$

where the first equality follows from (1.2) and the fact that $\chi(\mathfrak{e}) = 1$, while the second equality follows from (1.1) and the fact that $\chi_0(\mathfrak{g}) = 1$. □

PROBLEM 1.25. Prove that the group of characters $\mathfrak{G}^\wedge$ of a finite abelian group $\mathfrak{G}$ is isomorphic to $\mathfrak{G}$ itself. (*Hint*: Apply a theorem about finite abelian groups that in its simplest form says that any such group is isomorphic to the direct product of finite cyclic groups. Prove first that the group of characters of a finite cyclic group is isomorphic to this same group.)

It is easy to check that the set of all automorphisms of a group is also a group with composition as the operation (or, in other words, with consecutive application of automorphisms).

DEFINITION 1.26. The automorphism $\psi_{\mathfrak{g}}$, $\mathfrak{g} \in \mathfrak{G}$, of a group $\mathfrak{G}$ given by the formula $\psi_{\mathfrak{g}}(\mathfrak{h}) = \mathfrak{g}\mathfrak{h}\mathfrak{g}^{-1}$ for all $\mathfrak{h} \in \mathfrak{G}$ is called an *internal automorphism* of the group $\mathfrak{G}$.

The elements $\mathfrak{h}$ and $\mathfrak{g}\mathfrak{h}\mathfrak{g}^{-1}$ are called *conjugate*. If S is a nonempty subset of the group $\mathfrak{G}$, the set $\mathfrak{g}S\mathfrak{g}^{-1} = \{\mathfrak{g}\mathfrak{h}\mathfrak{g}^{-1} \mid \mathfrak{h} \in S\}$ is called *conjugate* with S.

DEFINITION 1.27. The *kernel of a homomorphism* $\psi\colon \mathfrak{G} \to \mathfrak{H}$ from a group $\mathfrak{G}$ to a group $\mathfrak{H}$ is the set

$$\ker\psi = \{\mathfrak{g} \in \mathfrak{G} \mid \psi(\mathfrak{g}) = \mathfrak{e}'\},$$

where $\mathfrak{e}'$ is the identity element of the group $\mathfrak{H}$.

DEFINITION 1.28. A subgroup $\mathfrak{H}$ of the group $\mathfrak{G}$ is called *normal* if $\mathfrak{g}\mathfrak{h}\mathfrak{g}^{-1} \in \mathfrak{H}$ for all $\mathfrak{g} \in \mathfrak{G}$, $\mathfrak{h} \in \mathfrak{H}$.

PROBLEM 1.29. Prove that the kernel of any homomorphism $\ker\psi$ is a normal subgroup.

PROBLEM 1.30. Prove that a subgroup $\mathfrak{H}$ of a group $\mathfrak{G}$ is normal if and only if for any element $\mathfrak{g} \in \mathfrak{G}$ the left coset $\mathfrak{g}\mathfrak{H}$ coincides with the right coset $\mathfrak{H}\mathfrak{g}$.

PROBLEM 1.31. Prove that if $\mathfrak{H}$ is a normal subgroup of a group $\mathfrak{G}$, then the set of (left) cosets of the group $\mathfrak{H}$ in $\mathfrak{G}$ is a group with respect to the following operation:

$$(\mathfrak{g}_1\mathfrak{H})(\mathfrak{g}_2\mathfrak{H}) = (\mathfrak{g}_1\mathfrak{g}_2)\mathfrak{H}.$$

DEFINITION 1.32. Let $\mathfrak{H}$ be a normal subgroup of a group $\mathfrak{G}$. Then the group formed by the (left) cosets of the group $\mathfrak{H}$ in $\mathfrak{G}$ with the operation introduced in Problem 1.31 is called the *quotient group* of the group $\mathfrak{G}$ modulo $\mathfrak{H}$ and is denoted by $\mathfrak{G}/\mathfrak{H}$.

If the quotient group $\mathfrak{G}/\mathfrak{H}$ is finite, its order coincides with the index $(\mathfrak{G} : \mathfrak{H})$ of the subgroup $\mathfrak{H}$ in $\mathfrak{G}$. In particular, for a finite group $\mathfrak{G}$ the following equalities hold:

$$\#\mathfrak{G}/\mathfrak{H} = (\mathfrak{G} : \mathfrak{H}) = \frac{\#\mathfrak{G}}{\#\mathfrak{H}}.$$

It is easy to see that any normal subgroup of a group $\mathfrak{G}$ determines a homomorphism of this group, and vice versa. Let $\psi\colon \mathfrak{G} \to \mathfrak{G}_1 = \psi(\mathfrak{G})$ be a homomorphism from a group $\mathfrak{G}$ to a group $\mathfrak{G}_1$. The kernel $\ker\psi$ is a normal subgroup of the group $\mathfrak{G}$, and the group $\mathfrak{G}_1$ is isomorphic to the quotient group $\mathfrak{G}/\ker\psi$. Conversely, if $\mathfrak{H}$ is a normal subgroup of a group $\mathfrak{G}$, then the mapping $\psi_{\mathfrak{H}}:\mathfrak{G}\to\mathfrak{G}/\mathfrak{H}$, defined by the conditions $\psi_{\mathfrak{H}}(\mathfrak{g})=\mathfrak{g}\mathfrak{H}$ for any $\mathfrak{g}\in\mathfrak{G}$, is a homomorphism of the group $\mathfrak{G}$ onto the group $\mathfrak{G}/\mathfrak{H}$, where $\ker\psi_{\mathfrak{H}}=\mathfrak{H}$.

Let S be a nonempty subset of a group $\mathfrak{G}$. The *normalizer* of the group $\mathfrak{G}$ is the set

$$\mathfrak{N}(S)=\{\mathfrak{g}\in\mathfrak{G}\mid \mathfrak{g}S\mathfrak{g}^{-1}=S\}.$$

If $S=\{\mathfrak{g}\}$, then $\mathfrak{N}(\mathfrak{g})$ is called the normalizer of the element $\mathfrak{g}$ in the group $\mathfrak{G}$.

PROBLEM 1.33. Prove that $\mathfrak{N}(S)$ is a subgroup of the group $\mathfrak{G}$.

The *center* of a group $\mathfrak{G}$ is the set

$$\mathfrak{C}=\{\mathfrak{c}\in\mathfrak{G}\mid \mathfrak{c}\mathfrak{g}=\mathfrak{g}\mathfrak{c}\text{ for all }\mathfrak{g}\in\mathfrak{G}\}.$$

PROBLEM 1.34. Prove that $\mathfrak{C}$ is a normal subgroup of $\mathfrak{G}$.

DEFINITION 1.35. A *ring* $(R,+,\cdot)$ is a set R with two binary operations denoted by the symbols $+$ and $\cdot$ and such that

1. R is an abelian group with operation $+$.
2. The operation $\cdot$ is associative, i.e., for all $a,b,c\in R$ one has

$$(a\cdot b)\cdot c=a\cdot(b\cdot c).$$

3. Distributive laws are satisfied, i.e., for all $a,b,c\in R$ the following equalities hold:

$$a\cdot(b+c)=a\cdot b+a\cdot c,\quad (b+c)\cdot a=b\cdot a+c\cdot a.$$

(Note that the operations $+$ and $\cdot$ do not necessarily have to be the usual addition and multiplication.) For the sake of brevity, the ring $(R,+,\cdot)$ is usually denoted by only one letter R. The identity element of the additive group of a ring R is called the *zero element* (or *zero*) of the ring R and is denoted by 0, and the inverse of an element a from the same group is denoted by $-a$. Instead of $a+(-b)$ one writes $a-b$, and instead of $a\cdot b$ one writes simply ab. From the definition of a ring it follows that $a\cdot 0=0\cdot a=0$ for all $a\in R$, as well as $(-a)b=a(-b)=-ab$ for all $a,b\in R$.

DEFINITION 1.36. 1. A ring R is called a *ring with identity* if it has a multiplicative identity, i.e., if there exists an element $e\in R$ such that $ae=ea=a$ for all $a\in R$.

2. A ring R is called commutative if the operation $\cdot$ is commutative.

3. A ring R is called a *domain* if it is a commutative ring with identity $e\neq 0$ and if the equality $ab=0$ implies that $a=0$ or $b=0$. If $ab=0$ and $a\neq 0$, $b\neq 0$, then a and b are called *zerodivisors*. Thus, a domain is a commutative ring with identity and without zerodivisors.

4. A ring R is called a *division ring* or a *skew field* if $R\neq\{0\}$ and the nonzero elements of R form a group with the operation $\cdot$.

5. A commutative division ring is called a *field*.

In a field, which we will denote by $\mathbb{F}$, the two operations are called addition and multiplication. It has two special elements 0 and e, where $0\neq e$. A field $\mathbb{F}$ is an abelian group with respect to addition; the identity element of this group is 0. The elements of $\mathbb{F}$ different from 0 form an abelian group with respect to multiplication with the identity element e. Further on we will denote the multiplicative identity element of a field by 1. "Addition" and "multiplication" in a field are connected by the distributive law.

We will illustrate the notions of ring and field in the following example.

EXAMPLE 1.37. 1. The integers form a domain but not a field.

2. Even numbers form a commutative ring without identity.

3. The mappings $\psi\colon R \to R$ (where R is a ring) form a ring if the sum and product of the mappings are defined pointwise:

$$(\psi_1 + \psi_2)(x) = \psi_1(x) + \psi_2(x), \quad (\psi_1 \cdot \psi_2)(x) = \psi_1(x) \cdot \psi_2(x)$$

for all $x \in R$.

4. The set of all $(n \times n)$ matrices $(n \in \mathbb{N})$ with entries from a given ring R form a ring.

5. The set $\mathbb{Z}_n$ for $n = p$ prime and with operations $\{a\}_\equiv + \{b\}_\equiv = \{a+b\}_\equiv$, $\{a\}_\equiv \cdot \{b\}_\equiv = \{a \cdot b\}_\equiv$, $a, b \in \mathbb{Z}$ is a field $\mathbb{Z}_p = \mathbb{F}_p$ of residues modulo p (it is easy to see that addition and multiplication of classes of congruent elements are well defined). This follows immediately from the fact that the equality $\{a\}_\equiv \cdot \{b\}_\equiv = \{a \cdot b\}_\equiv = \{0\}_\equiv$ holds if and only if $ab = kp$ for some integer k. But since p is prime, it divides at least one of the factors a or b. Therefore, either $\{a\}_\equiv = \{0\}_\equiv$ or $\{b\}_\equiv = \{0\}_\equiv$, i.e., $\mathbb{Z}_p$ does not have a zerodivisor. Besides, for any $a \in \mathbb{Z}$, $a_\equiv \neq \{0\}_\equiv$, and p there exist (see for example [**404**]) integers u and v such that $ua + pv = 1$. Thus, $\{u\}_\equiv \{a\}_\equiv = \{1\}_\equiv$ and $\{a\}_\equiv^{-1} = \{u\}_\equiv$. Therefore, $\mathbb{Z}_p$ is a field.

DEFINITION 1.38. If R is a ring, a subset S of the set R is called a *subring* if it is closed under the ring operations of R; a subring is a ring with respect to these operations.

DEFINITION 1.39. A subset J of a ring R is called a *(two-sided) ideal* of the ring if it is a subring of the ring R and if for all $a \in J$ and $r \in R$ one has $ar \in J$ and $ra \in J$.

EXAMPLE 1.40. 1. Suppose that R is a commutative ring with identity, $a \in R$, and put $J = \{ra \mid r \in R\}$. Then J is an ideal of the ring R.

2. Let R be a commutative ring. Then the smallest ideal which contains a given element $a \in R$ is the ideal $(a) = \{ra + na \mid r \in R,\ n \in \mathbb{Z}\}$. If the ring R has an identity element, then $(a) = \{ra \mid r \in R\}$.

DEFINITION 1.41. Let R be a commutative ring. An ideal J of the ring R is called *principal* if there exists an element $a \in R$ such that $J = (a)$. In this case, the ideal J is also called the principal ideal generated by the element a.

Since ideals are normal subgroups of the additive group of a ring, any ideal J of the ring R defines a partition of the set R into cosets of the additive subgroup J; they are called residue classes of the ideal J in the ring R. The residue class containing an element $a \in R$ will be denoted by $\{a\} = a + J$; it consists of all elements of the ring R of the form $a + c$, where $c \in J$. Elements $a, b \in R$ belonging to the same residue class modulo J (i.e., such that $a - b \in J$) will be called *congruent* modulo J; this will be denoted by $a \equiv b \pmod{J}$ (cf. Problem 1.5). It is easy to check that if $a \equiv b \pmod{J}$, then $a + r \equiv b + r \pmod{J}$, $ar \equiv br \pmod{J}$, $ra \equiv rb \pmod{J}$ and $na \equiv nb \pmod{J}$ for any $r \in R$ and $n \in \mathbb{Z}$. If, in addition, $r \equiv s \pmod{J}$, then $a + r \equiv b + s \pmod{J}$ and $ar \equiv bs \pmod{J}$. It is easy to show by direct check that the set of residue classes of a ring R modulo an ideal J is a ring with respect to the operations "+" and " · " defined by the equalities

$$(a+J)+(b+J)=(a+b)+J, \tag{1.3}$$

$$(a+J)(b+J)=ab+J. \tag{1.4}$$

DEFINITION 1.42. The ring of residue classes of a ring R modulo an ideal J with the operations (1.3) and (1.4) is called the *quotient ring* of the ring R modulo the ideal J and is denoted by R/J.

EXAMPLE 1.43. Let $(n) = \{kn \mid k \in \mathbb{Z}\}$ be the ideal of the ring of integers consisting of multiples of n. Then the quotient ring $\mathbb{Z}/(n)$ consists of residue classes

$$0+(n),\ 1+(n),\ \dots,\ n-1+(n).$$

If $n = p$ is prime (see Example 1.37(5)), then $\mathbb{Z}/(p) = \mathbb{Z}_p = \mathbb{F}_p$ is a field.

Note that not all the properties of the initial ring are automatically preserved in the quotient ring. For example, the property that there is no zerodivisor is not always preserved; say, the ring $\mathbb{Z}/(n)$ does not have this property if n is not prime.

The notion of group homomorphism allows a generalization for the case of rings. A mapping $\psi\colon R \to S$ of a ring R to a ring S is called a homomorphism if for any $a, b \in R$ one has

$$\psi(a+b)=\psi(a)+\psi(b), \quad \psi(ab)=\psi(a)\psi(b).$$

Thus, a homomorphism $\psi\colon R \to S$ respects the operations + and · in the rings R and S and induces the corresponding homomorphism of the additive group of the ring R to the additive group of the ring S. The set

$$\ker\psi = \{a \in R \mid \psi(a) = 0 \in S\}$$

is called the *kernel of the homomorphism* ψ. Other notions, such as isomorphism, etc., are defined as in Definition 1.17.

It ψ is a homomorphism of a ring R onto a ring S, then $\ker\psi$ is an ideal of the ring R, and S is isomorphic to the quotient ring $R/\ker\psi$. Conversely, if J is an ideal of a ring R, then the mapping $\psi_J\colon R \to R/J$ defined by the equality $\psi_J(a) = a + J$ for all $a \in R$, is a homomorphism of R onto R/J with kernel J.

Let R be a commutative ring with identity. An element $a \in R$ is called a *divisor* of the element $b \in R$ if there exists an element $c \in R$ such that $b = ac$. The divisors of unity are called *invertible* elements. Elements $a, b \in R$ are called *associates* it there exists an invertible element $\varepsilon \in R$ such that $a = b\varepsilon$. An element $c \in R$ is called *prime* if it is not invertible and does not have divisors other than invertible elements and its associates.

DEFINITION 1.44. 1. An ideal $P \neq R$ of a ring R is called *prime* if for any $a, b \in R$ the inclusion $ab \in P$ holds if and only if either $a \in P$ or $b \in P$.

2. An ideal $M \neq R$ of the ring R is called *maximal* if for any ideal J of the ring R the inclusion $M \subseteq J$ holds if and only if either $J = M$ or $J = R$.

3. A ring R is called a *principal ideal domain*, if it is a domain and each ideal J of R is principal; i.e., there exists an element $a \in R$ such that $J = (a) = \{ra \mid r \in R\}$.

THEOREM 1.45. *Let R be a commutative ring with identity.*

1. *An ideal M of the ring R is maximal if and only if the quotient ring R/M is a field.*

2. *An ideal P of the ring R is prime if and only if the quotient ring R/P is a domain.*

3. *Any maximal ideal is prime.*

4. *If R is a principal ideal domain, then the quotient ring $R/(c)$ is a field if and only if c is a prime element of the ring R.*

PROOF. 1. Let M be a maximal ideal of the ring R. Then for $a \in R$, $a \notin M$, the set $J = \{ar + m \mid r \in R,\ m \in M\}$ is an ideal of the ring R which contains M and is different from M; hence, $J = R$. In particular, there exist elements $r \in R$ and $m \in M$ such that $ar + m = 1$, where 1 is the multiplicative identity of the ring R. This means that if $a + M \neq 0 + M$, i.e., the residue class $a + M$ is a nonzero element of the quotient ring, then it has a multiplicative inverse since $(a + M)(r + M) = ar + M = (1 - m) + M = 1 + M$. Hence, R/M is field.

Conversely, let R/M be a field and let J be an ideal of the ring R such that $J \supseteq M$, $J \neq M$. Then for $a \in M \subseteq J$ the residue class $a + M$ has a multiplicative inverse, so $(a + M)(r + M) = 1 + M$ for some $r \in R$. This means that $ar + m = 1$ for some $m \in M$. Since J is an ideal, $1 \in J$, whence $(1) = R \subseteq J$ and $J = R$. Thus, M is a maximal ideal of the ring R.

2. Let P be a prime ideal of the ring R. Then the quotient ring R/P is a commutative ring with identity $1 + P \neq 0 + P$. If $(a + P)(b + P) = 0 + P$, then $ab \in P$. Since P is a prime ideal, either $a \in P$ or $b \in P$, i.e., either $a + P = 0 + P$ or $b + P = 0 + P$. Thus, the quotient ring R/P does not have zerodivisors, i.e., it is a domain. If we apply the same argument in reverse order, we will immediately obtain the converse statement.

3. This statement follows from 1 and 2 since any field is a domain.

4. Suppose that $c \in R$. If c is an invertible element, then $(c) = R$ and the quotient ring $R/(c)$ consists of one element, so it cannot be a field. If c is not invertible and not prime, then c has a divisor $a \in R$ which is not associated with the element c and not invertible. Observe that $a \neq 0$: if $a = 0$, then $c = 0$ and a would be associated with the element c. Suppose that $c = ab$, where $b \in R$. We claim that $a \notin (c)$. Indeed, otherwise we have $a = cd = abd$, where $d \in R$, i.e., $a(1 - bd) = 0$. Since $a \neq 0$, one has $bd = 1$. Thus, b is an invertible element, which contradicts the fact that a is not an associate of c. Hence, $(c) \subseteq (a) \subseteq R$, where all inclusions are proper and therefore the quotient ring $R/(c)$ cannot be a field as in item 1. The remaining case is: c is a prime element of the ring R. Then $(c) \neq R$ as c is not invertible. If $J \supseteq (c)$ is an ideal of the ring R, then $J = (a)$ for some $a \in R$, as R is a principal ideal domain. Therefore, $c \in (a)$ and a is a divisor of the element c. Thus, a is either invertible or associated with the element c, so either $J = R$ or $J = (c)$. This shows that c is a maximal ideal of the ring R. Taking into account part 1 of the theorem, we conclude that the quotient ring $R/(c)$ is a field. □

Let $\mathbb{F}$ be a field. A subset K of the field $\mathbb{F}$ that is a field itself with respect to the operations of the field $\mathbb{F}$ is called a *subfield*. In this case the field $\mathbb{F}$ is called an *extension* of the filed K. If $K \neq \mathbb{F}$, then K is called a *proper subfield* of the field $\mathbb{F}$.

The identity of any subfield K of the field $\mathbb{F}$ is the identity of the field $\mathbb{F}$. Indeed, let e' be the identity of the field $\mathbb{F}$, and let e be the identity of the subfield K. As $e \in \mathbb{F}$, one has $e' \cdot e = e \cdot e = e$ (since e is the identity of K). As $e \neq 0$, the equality $e' \cdot e = e \cdot e$ implies $e' = e$.

DEFINITION 1.46. A field that does not contain proper subfields is called *prime*.

EXAMPLE 1.47. 1. A field $\mathbb{F}_p = \mathbb{Z}/(p)$, with p prime, is a prime field. This follows from the fact that the field $\mathbb{F}_p$ has no proper additive subgroups.

2. The field $\mathbb{Q}$ of rational numbers is a prime field. Indeed, let $\mathbb{F}$ be a subfield of the field $\mathbb{Q}$. Then $1 \in \mathbb{F}$ and therefore $\mathbb{F}$ contains any integer $n \in \mathbb{Z}$. If $n \neq 0$, then $n^{-1} = \frac{1}{n} \in \mathbb{F}$. Therefore, $\frac{m}{n} \in \mathbb{F}$, whence $\mathbb{F} = \mathbb{Q}$.

PROBLEM 1.48. Prove that the intersection of any two subfields of a field $\mathbb{F}$ is also a subfield of $\mathbb{F}$.

LEMMA 1.49. *Any field has a unique prime subfield.*

PROOF. Let $\mathbb{F}$ be a given field. We consider the set of all subfields of it (this set is not empty as it contains the field $\mathbb{F}$ itself). Let $\mathbb{F}_0$ be the intersection of all subfields of the field $\mathbb{F}$. Then $\mathbb{F}_0$ is a subfield of the field $\mathbb{F}$ according to Problem 1.48. Moreover, $\mathbb{F}_0$ is a prime subfield of the field $\mathbb{F}$. Indeed, if $\mathbb{F}_1$ is a subfield of the field $\mathbb{F}_0$, then at the same time it is also a subfield of the field $\mathbb{F}$, and thus contains $\mathbb{F}_0$, so that $\mathbb{F}_0 = \mathbb{F}_1$. The uniqueness of the prime field follows directly from the way it is constructed. $\square$

THEOREM 1.50. *Let $\mathbb{F}$ be a prime field and e the identity of the field $\mathbb{F}$. Then one of the following conditions holds.*

1. *If n is a nonzero integer and e is the identity, then $ne \neq 0$ (by ne we mean the sum $e + \cdots + e$ of n terms for $n > 0$; $(-n)e = -(ne)$). In this case the field $\mathbb{F}$ is isomorphic to the field of rational numbers $\mathbb{Q}$.*

2. *There exists an integer $n \neq 0$ for which $ne = 0$. Then the smallest natural number p with this property is prime. In this case the field $\mathbb{F}$ is isomorphic to the field $\mathbb{F}_p$.*

PROOF. 1. Let the element $ne = e + \cdots + e$ be nonzero for any $n \in \mathbb{Z}$, $n \neq 0$. Then the elements of the form ne in $\mathbb{F}$ constitute a ring isomorphic to $\mathbb{Z}$. As $\mathbb{F}$ is a field, the element $(ne)^{-1}$ for $n \neq 0$ also belongs to the field $\mathbb{F}$, as well as $(me)(ne)^{-1} = \frac{me}{ne} \in \mathbb{F}$ for any m and n, $n \neq 0$. It is easy to see that these elements form the subfield $\mathbb{F}_0$ in $\mathbb{F}$ isomorphic to $\mathbb{Q}$. Since $\mathbb{F}$ is a prime field, it follows that $\mathbb{F} = \mathbb{F}_0$ and $\mathbb{F}$ is isomorphic to $\mathbb{Q}$.

2. Suppose there exists an integer $n \neq 0$ such that $ne = 0$. Among all such numbers we choose the smallest natural n_0. The elements $0, e, 2e, \ldots, (n_0 - 1)e$ form a subring R in $\mathbb{F}$ isomorphic to $\mathbb{Z}/(n_0)$. Suppose that n_0 is not prime, i.e., there are natural numbers k and m, $k < n_0$, $m < n_0$, such that $n_0 = k \cdot m$. Then $0 = n_0 e = (ke) \cdot (me)$, and since there are no zerodivisors in the field $\mathbb{F}$, we have $ke = 0$ or $me = 0$, which in any case contradicts the minimality of n_0. Therefore, $n_0 = p$ is a prime and the ring R is isomorphic to the field $\mathbb{Z}/(p)$. Since $\mathbb{F}$ is a prime field, $\mathbb{F} = R$ and $\mathbb{F}$ is isomorphic to the field $\mathbb{Z}/(p) = \mathbb{F}_p$. $\square$

DEFINITION 1.51. The smallest natural number p that satisfies the condition $pe = 0$ is called the *characteristic* of the field $\mathbb{F}$ and is denoted by $\operatorname{char}\mathbb{F}$. If such a number does not exist, we say that the characteristic of the field is equal to zero.

Obviously, the characteristic of any field coincides with the characteristic of its prime subfield, because the definition of the characteristic only depends on the identity of the field and the identity is contained in the prime subfield too.

Let $\mathbb{F}$ be an arbitrary field. We will call a *polynomial* over the field $\mathbb{F}$ an expression

$$a(\lambda) = \sum_{i=0}^{\infty} a_i\lambda^i = a_0 + a_1\lambda + \cdots + a_n\lambda^n + \cdots,$$

where n is a nonnegative integer, the coefficients a_i, $i = 0, 1, \ldots,$ are elements of the field $\mathbb{F}$, and only a finite number of them are different from zero. Here, λ is a formal symbol which does not belong to the field $\mathbb{F}$. It is called a *variable* (or an *unknown*) over $\mathbb{F}$. For convenience, we will omit the term $a_i\lambda^i$ if $a_i = 0$. Therefore, a polynomial $a(\lambda)$ can be represented as $a(\lambda) = \sum_{i=0}^{n} a_i\lambda^i$, where a_n is the last (with respect to the index i) nonzero coefficient. From now on, when we compare two polynomials $a(\lambda)$ and $b(\lambda)$ over $\mathbb{F}$, we can assume that both contain the same powers of the variable λ. Polynomials $a(\lambda) = \sum_{i=0}^{n} a_i\lambda^i$ and $b(\lambda) = \sum_{i=0}^{n} b_i\lambda^i$ over $\mathbb{F}$ are considered to be equal if and only if $a_i = b_i$ for $0 \le i \le n$. We define the sum of the polynomials $a(\lambda)$ and $b(\lambda)$ by the equality

$$a(\lambda) + b(\lambda) = \sum_{i=0}^{n} (a_i + b_i)\lambda^i,$$

and the product of the polynomials

$$a(\lambda) = \sum_{i=0}^{m} a_i\lambda^i \quad \text{and} \quad b(\lambda) = \sum_{i=0}^{n} b_i\lambda^i$$

by the equality

$$a(\lambda)b(\lambda) = \sum_{k=0}^{m+n} c_k\lambda^k, \quad \text{where} \quad c_k = \sum_{\substack{i+j=k,\\ 0\le i\le m,\ 0\le j\le n}} a_ib_j.$$

It is easy to see that the set of polynomials with these operations is a commutative ring with identity and without zerodivisors. If it is clear from the context which the variable is, we will write a instead of $a(\lambda)$.

DEFINITION 1.52. The ring formed by the polynomials over the field $\mathbb{F}$ with the above-introduced operations is called the *ring of polynomials* over $\mathbb{F}$ and is denoted $\mathbb{F}[\lambda]$.

The zero element of the field $\mathbb{F}[\lambda]$ is a polynomial with all zero coefficients. It is called the zero polynomial and is denoted by 0.

DEFINITION 1.53. If $a(\lambda) = \sum_{i=0}^{n} a_i\lambda^i \in \mathbb{F}[\lambda]$, $a_n \ne 0$, then a_n is called the *leading coefficient of the polynomial*, a_0 is called its *constant term* and n its *degree*. The latter is denoted by $\deg a(\lambda) = \deg a$. One sets $\deg 0 = -1$.

Polynomials of degree ≤ 0 are called *constant polynomials* (or *constants*).

If the leading coefficient of a polynomial $a(\lambda)$ is equal to 1, then the polynomial is called *monic* (or *unitary*).

PROBLEM 1.54. Suppose that $a(\lambda)$, $b(\lambda) \in \mathbb{F}[\lambda]$. Prove that

$$\deg\,(a(\lambda) + b(\lambda)) \le \max(\deg a(\lambda), \deg b(\lambda)),$$
$$\deg\,(a(\lambda) \cdot b(\lambda)) = \deg\,(a(\lambda) + \deg b(\lambda)).$$

We will say that a polynomial $a(\lambda) \in \mathbb{F}[\lambda]$ divides a polynomial $b(\lambda) \in \mathbb{F}[\lambda]$ if there exists a polynomial $c(\lambda) \in \mathbb{F}[\lambda]$ such that $b(\lambda) = a(\lambda)c(\lambda)$. In this case we will say that $a(\lambda)$ is a divisor of $b(\lambda)$, and that $b(\lambda)$ is divisible by $a(\lambda)$ (or is a multiple of $a(\lambda)$). The invertible elements in the field $\mathbb{F}[\lambda]$ are divisors of the constant polynomial 1, and therefore they are just nonzero constant polynomials. In the ring $\mathbb{F}[\lambda]$, division with remainder is possible. To wit, let $b(\lambda) \neq 0$ be a polynomial in $\mathbb{F}[\lambda]$. Then for any $a(\lambda) \in \mathbb{F}[\lambda]$ there exist polynomials $c(\lambda), d(\lambda) \in \mathbb{F}[\lambda]$ such that

$$a(\lambda) = b(\lambda)c(\lambda) + d(\lambda), \quad \deg d(\lambda) < \deg b(\lambda).$$

THEOREM 1.55. *The ring $\mathbb{F}[\lambda]$ of polynomials over a field $\mathbb{F}$ is a principal ideal domain. In other words, for any nonzero ideal $J \neq (0) \subset \mathbb{F}[\lambda]$ there exists a uniquely determined monic polynomial $a(\lambda) \in \mathbb{F}[\lambda]$ such that $J = (a(\lambda))$.*

PROOF. Let $J \neq (0)$ be an ideal of the ring $\mathbb{F}[\lambda]$. Denote by $b(\lambda)$ a nonzero polynomial of minimum degree contained in J. If b_n is the leading coefficient of the polynomial $b(\lambda)$, put $a(\lambda) = b_n^{-1}b(\lambda)$. Then $a(\lambda)$ is a monic polynomial contained in J. If $c(\lambda)$ is an arbitrary polynomial in J, then, applying division with remainder, we find $q(\lambda)$, $d(\lambda) \in \mathbb{F}[\lambda]$ such that $c(\lambda) = q(\lambda)a(\lambda) + d(\lambda)$, $\deg d(\lambda) < \deg a(\lambda) = \deg b(\lambda)$. As J is an ideal, $d(\lambda) = c(\lambda) - q(\lambda)a(\lambda) \in J$; since the degree of $b(\lambda) \in J$ is minimal, $d(\lambda) = 0$. Thus, $c(\lambda)$ is divisible by $a(\lambda)$, and so $J = (a(\lambda))$.

If $a_1(\lambda) \in \mathbb{F}[\lambda]$ is another monic polynomial such that $J = (a_1(\lambda))$, then $a(\lambda) = c_1(\lambda)a_1(\lambda)$ and $a_1(\lambda) = c_2(\lambda)a(\lambda)$, where $c_1(\lambda), c_2(\lambda) \in \mathbb{F}[\lambda]$. Hence, $a(\lambda) = c_1(\lambda)c_2(\lambda)a(\lambda)$, so $c_1(\lambda)c_2(\lambda) = 1$, i.e., $c_1(\lambda)$ and $c_2(\lambda)$ are constant polynomials. Since both $a(\lambda)$ and $a_1(\lambda)$ are monic, we have $a(\lambda) = a_1(\lambda)$. □

DEFINITION 1.56. If $a(\lambda), b(\lambda) \in \mathbb{F}[\lambda]$ are nonzero polynomials, then the *greatest common divisor of polynomials* $a(\lambda)$ and $b(\lambda)$ is a monic polynomial $d(\lambda) \in \mathbb{F}[\lambda]$ having the following properties:

1) $d(\lambda)$ divides both $a(\lambda)$ and $b(\lambda)$;

2) any polynomial that divides each of the polynomials $a(\lambda)$ and $b(\lambda)$, also divides $d(\lambda)$.

The greatest common divisor of polynomials $a(\lambda)$ and $b(\lambda)$ is denoted by $\gcd(a(\lambda), b(\lambda))$.

PROBLEM 1.57. Prove the existence and uniqueness of the polynomial $d(\lambda)$ from Definition 1.56. (*Hint*: The set of polynomials of the form $\{a(\lambda)q(\lambda) + b(\lambda)c(\lambda) \mid q(\lambda), c(\lambda) \in \mathbb{F}[\lambda]\}$ is an ideal of the ring $\mathbb{F}[\lambda]$.)

PROBLEM 1.58. Put $\gcd(a(\lambda), b(\lambda)) = d(\lambda)$, where $a(\lambda), b(\lambda), d(\lambda) \in \mathbb{F}[\lambda]$. Prove that there exist polynomials $u(\lambda), v(\lambda) \in \mathbb{F}[\lambda]$, such that

$$d(\lambda) = a(\lambda)u(\lambda) + b(\lambda)v(\lambda).$$

Prove this statement for an arbitrary finite number of polynomials.

Note that the notion of greatest common divisor can be generalized to an arbitrary finite collection of polynomials. The greatest common divisor of two polynomials $a(\lambda), b(\lambda) \in \mathbb{F}[\lambda]$ can be found by using the *Euclidean algorithm*. Suppose without loss of generality that the nonzero polynomial $b(\lambda)$ does not divide the polynomial $a(\lambda)$. Then, applying the algorithm for division with remainder several times, we obtain

$$\begin{aligned}
a(\lambda) &= q_1(\lambda)b(\lambda) + r_1(\lambda), && 0 \le \deg r_1(\lambda) < \deg b(\lambda),\\
b(\lambda) &= q_2(\lambda)r_1(\lambda) + r_2(\lambda), && 0 \le \deg r_2(\lambda) < \deg r_1(\lambda),\\
r_1(\lambda) &= q_3(\lambda)r_2(\lambda) + r_3(\lambda), && 0 \le \deg r_3(\lambda) < \deg r_2(\lambda),\\
&\cdots\cdots\cdots && \cdots\cdots\cdots\\
r_{s-2}(\lambda) &= q_s(\lambda)r_{s-1}(\lambda) + r_s(\lambda), && 0 \le \deg r_s(\lambda) < \deg r_{s-1}(\lambda),\\
r_{s-1}(\lambda) &= q_{s+1}(\lambda)r_s(\lambda).
\end{aligned}$$

Here $q_1(\lambda), \dots, q_{s+1}(\lambda)$ and $r_1(\lambda), \dots, r_s(\lambda)$ are polynomials from $\mathbb{F}[\lambda]$. As the degree $\deg b(\lambda)$ is finite, the procedure will terminate after a finite number of steps. If the leading coefficient of the last nonzero remainder $r_s(\lambda) = r_{s0}+r_{s1}\lambda+\cdots+r_{sk}\lambda^k$ is equal to r_{sk}, then $\gcd(a(\lambda), b(\lambda)) = r_{sk}^{-1}r_s(\lambda)$.

DEFINITION 1.59. Suppose that $a(\lambda), b(\lambda) \in \mathbb{F}[\lambda]$ are not zero polynomials. The *least common multiple of the polynomials* $a(\lambda)$ and $b(\lambda)$ is a monic polynomial $c(\lambda) \in \mathbb{F}[\lambda]$ having the following properties:

1) $c(\lambda)$ is divisible by both $a(\lambda)$ and $b(\lambda)$;

2) any polynomial that is divisible by $a(\lambda)$ and $b(\lambda)$, is divisible by $c(\lambda)$.

The least common multiple of the polynomials $a(\lambda)$ and $b(\lambda)$ is denoted by $\mathrm{lcd}(a(\lambda), b(\lambda))$.

PROBLEM 1.60. Prove the existence and uniqueness of the polynomial $c(\lambda)$ from Definition 1.59. (*Hint*: Consider the intersection of the principal ideals $(a(\lambda))$ and $(b(\lambda))$.)

Prove the existence and uniqueness of the least common divisor for an arbitrary finite collection of polynomials.

DEFINITION 1.61. A polynomial $a(\lambda) \in \mathbb{F}[\lambda]$ is called *irreducible* over the field $\mathbb{F}$ (or in the ring $\mathbb{F}[\lambda]$) if its degree is positive and the equality $a(\lambda) = b(\lambda) \cdot c(\lambda)$, $b(\lambda)$, $c(\lambda) \in \mathbb{F}[\lambda]$, holds only when either $b(\lambda)$ or $c(\lambda)$ is a constant polynomial.

If a polynomial of positive degree from $\mathbb{F}[\lambda]$ is not irreducible over $\mathbb{F}$, it is called *reducible* over the field $\mathbb{F}$ (or in the ring $\mathbb{F}[\lambda]$).

PROBLEM 1.62. Prove that if an irreducible polynomial $a(\lambda) \in \mathbb{F}[\lambda]$ divides the product $a_1(\lambda) \cdots a_m(\lambda)$ of polynomials from $\mathbb{F}[\lambda]$, then at least one of the factors $a_j(\lambda)$, $j = 1, \dots, m$, is divisible by $a(\lambda)$.

THEOREM 1.63. *Any polynomial of positive degree* $a(\lambda) = a_0+a_1\lambda+\cdots+a_n\lambda^n$, $a_n \ne 0$ *from* $\mathbb{F}[\lambda]$ *can be represented as the product*

$$a(\lambda) = a_n b_1^{t_1}(\lambda) \cdots b_k^{t_k}(\lambda), \tag{1.5}$$

where $a_n \in \mathbb{F}$, $b_1(\lambda), \dots, b_k(\lambda)$ *are different irreducible monic polynomials from* $\mathbb{F}[\lambda]$, *and* $t_1, \dots, t_k$ *are natural numbers. This factorization is unique up to the order of factors.*

PROOF. We apply induction on the degree of the polynomial $a(\lambda)$. The case $\deg a(\lambda) = 1$ (induction basis) is trivial since any polynomial of degree one is irreducible over $\mathbb{F}$. Assume that the theorem holds for all polynomials from $\mathbb{F}[\lambda]$ of degree less than n.

Suppose that $\deg a(\lambda) = n$ (induction step). If $a(\lambda)$ is irreducible over $\mathbb{F}$, then $a(\lambda) = a_n(a_n^{-1}a(\lambda))$ is the factorization we are looking for since $a_n^{-1}a(\lambda)$ is an irreducible (over $\mathbb{F}$) monic polynomial from $\mathbb{F}[\lambda]$. If $a(\lambda)$ is reducible over $\mathbb{F}$, it can be represented as $a(\lambda) = b(\lambda)c(\lambda)$, where $b(\lambda), c(\lambda) \in \mathbb{F}[\lambda]$, $1 \le \deg b(\lambda) < n$, $1 \le \deg c(\lambda) < n$. By the induction hypothesis, any polynomials $b(\lambda)$ and $c(\lambda)$ can be represented as in (1.5), whence the polynomial $a(\lambda)$ can also be represented as in (1.5).

Suppose now that a polynomial $a(\lambda)$ has two factorizations as in (1.5). If we compare leading coefficients of these factorizations, we see that the constant factors are both equal to a_n:

$$a(\lambda) = a_n b_1^{t_1}(\lambda) \cdots b_k^{t_k}(\lambda) = a_n d_1^{l_1}(\lambda) \cdots d_r^{l_r}(\lambda). \tag{1.6}$$

The irreducible polynomial $b_1(\lambda)$ divides the right-hand side of (1.6). According to the result of Problem 1.62, it must divide one of the polynomials $d_j(\lambda)$, $1 \le j \le r$. Since the polynomial $d_j(\lambda)$ is also irreducible, the equality $d_j(\lambda) = c(\lambda)b_1(\lambda)$ holds, where $c(\lambda) = c_0$ is a constant polynomial. Since $d_j(\lambda)$ and $b_1(\lambda)$ are both monic, $d_j(\lambda) = b_1(\lambda)$ and we can cancel $d_j(\lambda)$ and $b_1(\lambda)$ in (1.6). Apply the same argument to the resulting equality. After a finite number of steps we find that the two factorizations (1.6) coincide up to the order of factors. □

We will call the factorization (1.5) the *canonical factorization of the polynomial* $a(\lambda)$ in the ring $\mathbb{F}[\lambda]$. An important question about polynomials in $\mathbb{F}[\lambda]$ is whether a given polynomial is reducible over the field $\mathbb{F}$ or not.

LEMMA 1.64. *Suppose that $a(\lambda) \in \mathbb{F}[\lambda]$. For the quotient ring $\mathbb{F}[\lambda]/(a(\lambda))$ to be a field, it is necessary and sufficient that the polynomial $a(\lambda)$ be irreducible over $\mathbb{F}$.*

PROOF. Since irreducible polynomials over $\mathbb{F}$ are elements of the ring $\mathbb{F}[\lambda]$, the assertion immediately follows from Theorems 1.45(4) and 1.55. □

The quotient ring $\mathbb{F}[\lambda]/(a(\lambda))$, where $a(\lambda) \ne 0$ is a polynomial in $\mathbb{F}[\lambda]$, consists of the residue classes $\{b(\lambda)\} = b(\lambda) + (a(\lambda))$, where $b(\lambda) \in \mathbb{F}[\lambda]$ and the ring operations are defined by (1.3) and (1.4). Two residue classes $b(\lambda) + (a(\lambda))$ and $c(\lambda) + (a(\lambda))$ coincide if and only if $b(\lambda) \equiv c(\lambda) \pmod{a(\lambda)}$, i.e., when the polynomial $b(\lambda) - c(\lambda)$ is divisible by $a(\lambda)$. This is equivalent to the fact that the remainders of division of $b(\lambda)$ and $c(\lambda)$ by $a(\lambda)$ are the same. A residue class $b(\lambda) + (a(\lambda))$ contains only one polynomial $r(\lambda) \in \mathbb{F}[\lambda]$ for which $\deg r(\lambda) < \deg a(\lambda)$. This polynomial is the remainder of division of $b(\lambda)$ by $a(\lambda)$. The uniqueness of the polynomial $r(\lambda)$ follows from the fact that if there exists another polynomial $r_1(\lambda) \in b(\lambda) + (a(\lambda))$ such that $\deg r_1(\lambda) < \deg a(\lambda)$, then the difference $r_1(\lambda) - r(\lambda)$ must be divisible by $a(\lambda)$; since $\deg(r_1(\lambda) - r(\lambda)) < \deg a(\lambda)$, this is possible only if $r_1(\lambda) = r(\lambda)$. Thus, the elements of the quotient ring $\mathbb{F}[\lambda]/(a(\lambda))$ can be described as follows: they are the residue classes $r(\lambda) + (a(\lambda))$, where $r(\lambda)$ runs over all polynomials from $\mathbb{F}[\lambda]$ of degree less than $\deg a(\lambda)$.

Suppose that $a(\lambda) \in \mathbb{F}[\lambda]$. If we replace the variable λ in the polynomial $a(\lambda)$ by an arbitrary element of the field $\mathbb{F}$, we transform this polynomial into a well-defined element of the field $\mathbb{F}$. If $a(\lambda) = a_0 + a_1\lambda + \cdots + a_n\lambda^n \in \mathbb{F}[\lambda]$ and $\alpha \in \mathbb{F}$, then substituting α for λ we obtain an element $a(\alpha) = a_0 + a_1\alpha + \cdots + a_n\alpha^n \in \mathbb{F}$. This element is called the value of the polynomial $a(\lambda)$ at the point $\lambda = \alpha$. If in the ring $\mathbb{F}[\lambda]$ there exists a polynomial identity, then replacing λ by an arbitrary fixed element $\alpha \in \mathbb{F}$ in this identity, we obtain a polynomial identity in the field $\mathbb{F}$. The mapping $\alpha \mapsto a(\alpha) \in \mathbb{F}$, $\alpha \in \mathbb{F}$, is called a mapping of the field into itself defined by the polynomial $a(\lambda)$.

DEFINITION 1.65. An element $\alpha \in \mathbb{F}$ is called a *root of the polynomial* $a(\lambda) \in \mathbb{F}[\lambda]$ if $a(\alpha) = 0$.

PROBLEM 1.66. Prove that an element $\alpha \in \mathbb{F}$ is a root of a polynomial $a(\lambda) \in \mathbb{F}[\lambda]$ if and only if the polynomial $\lambda - \alpha$ divides $a(\lambda)$.

DEFINITION 1.67. Let $\alpha \in \mathbb{F}$ be a root of a polynomial $a(\lambda) \in \mathbb{F}[\lambda]$. The *multiplicity of the root* α is the natural number k such that $a(\lambda)$ is divisible by $(\lambda - \alpha)^k$ but not by $(\lambda - \alpha)^{k+1}$. If $k = 1$, the root α is called *simple*, and if $k > 1$, it is called *multiple*.

PROBLEM 1.68. Suppose that $a(\lambda) \in \mathbb{F}[\lambda]$ and $\deg a(\lambda) = n \geq 0$. Prove that the polynomial $a(\lambda)$ has at most n roots.

DEFINITION 1.69. The *derivative of a polynomial* $a(\lambda) = a_0 + a_1\lambda + \cdots + a_n\lambda^n \in \mathbb{F}[\lambda]$ is the polynomial $a'(\lambda) = a_1 + 2a_2\lambda + \cdots + na_n\lambda^{n-1} \in \mathbb{F}[\lambda]$ (in this formula, mb, where $b \in F$ and m is a natural number, means $b + \cdots + b$ (m times).

PROBLEM 1.70. Prove that a root $\alpha \in \mathbb{F}$ of a polynomial $a(\lambda) \in \mathbb{F}[\lambda]$ is a multiple root if and only if it is at the same time a root of the derivative $a'(\lambda)$ of the polynomial $a(\lambda)$.

PROBLEM 1.71. Prove that a polynomial $a(\lambda) \in \mathbb{F}[\lambda]$ of degree 2 or 3 is irreducible in the ring $\mathbb{F}[\lambda]$ if and only if it has no roots in the field $\mathbb{F}$.

Now we define vector spaces. To that end, we will generalize the notion of operation: multiplication of a vector by an element of the ground field (scalar) is not a binary operation that is defined on pairs of elements from the same set; this operation assigns a vector to any pair (scalar, vector).

DEFINITION 1.72. A *linear* or *vector* space over a field $\mathbb{F}$ is a set V with two operations: addition (a binary operation on V) and multiplication by elements of $\mathbb{F}$ (scalars) having the following properties:

1) the set V is an abelian group with respect to addition;
2) $a(\boldsymbol{v}_1 + \boldsymbol{v}_2) = a\boldsymbol{v}_1 + a\boldsymbol{v}_2$ for any $a \in \mathbb{F}$, $\boldsymbol{v}_1, \boldsymbol{v}_2 \in V$;
3) $(a + b)\boldsymbol{v} = a\boldsymbol{v} + b\boldsymbol{v}$ for any $a, b \in \mathbb{F}$, $\boldsymbol{v} \in V$;
4) $ab\boldsymbol{v} = a(b\boldsymbol{v})$ for any $a, b \in \mathbb{F}$, $\boldsymbol{v} \in V$;
5) $1\boldsymbol{v} = \boldsymbol{v}$ for any $\boldsymbol{v} \in V$.

Elements of a vector space are called *vectors*; elements of the field $\mathbb{F}$ are called scalars.

EXAMPLE 1.73. The set $\mathbb{F}^n$ of columns of length n of elements in a field $\mathbb{F}$ is a vector space over the field $\mathbb{F}$ with respect to the operations given by the formulas

$$\begin{aligned}(u^{(1)}, u^{(2)}, \dots, u^{(n)})^\top &+ (v^{(1)}, v^{(2)}, \dots, v^{(n)})^\top \\ &= (u^{(1)} + v^{(1)}, u^{(2)} + v^{(2)}, \dots, u^{(n)} + v^{(n)})^\top, \\ a(u^{(1)}, u^{(2)}, \dots, u^{(n)})^\top &= (au^{(1)}, au^{(2)}, \dots, au^{(n)})^\top.\end{aligned}$$

The sign $\top$ denotes transposition of matrices, i.e., $(u^{(1)}, \dots, u^{(n)})^\top$ is the column vector $\begin{pmatrix} u^{(1)} \\ \cdots \\ u^{(n)} \end{pmatrix}$.

EXAMPLE 1.74. The set $\mathcal{F}(M, \mathbb{F})$ of all functions on the set M with values in the field $\mathbb{F}$ is a vector space over $\mathbb{F}$ with respect to the operations defined by the formulas

$$(f_1 + f_2)(m) = f_1(m) + f_2(m), \quad (af)(m) = af(m),$$

where $f_1, f_2, f \in \mathcal{F}(M, \mathbb{F})$, $m \in M$, $a \in \mathbb{F}$.

EXAMPLE 1.75. Let $\mathbb{K}$ be a subfield of the field $\mathbb{F}$. Then $\mathbb{F}$ becomes a vector space over $\mathbb{K}$ if we define multiplication of elements from $\mathbb{F}$ by elements from $\mathbb{K}$ as multiplication in $\mathbb{F}$. The dimension of the vector space (see Definition 1.82) $\mathbb{F}$ over the field $\mathbb{K}$ is called the *extension degree* of $\mathbb{F}$ over $\mathbb{K}$. If this dimension is finite, then we say that $\mathbb{F}$ is an extension of finite degree of the field $\mathbb{K}$. The extension degree is denoted by $[\mathbb{F} : \mathbb{K}]$.

We will present some corollaries of axioms for vector space (which involve multiplication by scalars and not only the abelian group structure). All the proofs are similar to those of analogous properties of rings.

1. $a\mathbf{0} = \mathbf{0}$ for any $a \in \mathbb{F}$ (here $\mathbf{0}$ is the zero vector of the space V).
2. $a(-\boldsymbol{v}) = -a\boldsymbol{v}$ for any $a \in \mathbb{F}$, $\boldsymbol{v} \in V$.
3. $a(\boldsymbol{v}_1 - \boldsymbol{v}_2) = a\boldsymbol{v}_1 - a\boldsymbol{v}_2$ for any $a \in \mathbb{F}$, $\boldsymbol{v}_1, \boldsymbol{v}_2 \in V$.
4. $0 \cdot \boldsymbol{v} = \mathbf{0}$ for any $\boldsymbol{v} \in V$.
5. $(-1) \cdot \boldsymbol{v} = -\boldsymbol{v}$ for any $\boldsymbol{v} \in V$.
6. $(a - b)\boldsymbol{v} = a\boldsymbol{v} - b\boldsymbol{v}$ for any $a, b \in \mathbb{F}$, $\boldsymbol{v} \in V$.

DEFINITION 1.76. Vector spaces V and U over the field $\mathbb{F}$ are called *isomorphic* if there exists a one-to-one mapping

$$\psi \colon V \to U$$

such that

1) $\psi(\boldsymbol{v}_1 + \boldsymbol{v}_2) = \psi(\boldsymbol{v}_1) + \psi(\boldsymbol{v}_2)$ for any $\boldsymbol{v}_1, \boldsymbol{v}_2 \in V$;
2) $\psi(a\boldsymbol{v}) = a\psi(\boldsymbol{v})$ for any $a \in \mathbb{F}$, $\boldsymbol{v} \in V$.

The mapping ψ is called an isomorphism of the vector spaces V and U.

Any expression of the form

$$a_1\boldsymbol{v}_1 + \cdots + a_n\boldsymbol{v}_n,$$

where $a_1, \ldots, a_n \in \mathbb{F}$, is called a *linear combination* of the vectors $\boldsymbol{v}_1, \boldsymbol{v}_2, \ldots, \boldsymbol{v}_n \in V$. We say that a vector $\boldsymbol{u} \in V$ *is linearly expressed in terms of vectors* $\boldsymbol{v}_1, \boldsymbol{v}_2, \ldots, \boldsymbol{v}_n \in V$ if $\boldsymbol{u}$ is their linear combination.

DEFINITION 1.77. A system of vectors $\{\boldsymbol{v}_1, \boldsymbol{v}_2, \ldots, \boldsymbol{v}_n\} \subset V$ is called a *basis* of V if each vector $\boldsymbol{u} \in V$ can be uniquely linearly expressed in terms of $\boldsymbol{v}_1, \boldsymbol{v}_2, \ldots, \boldsymbol{v}_n$. The coefficients of this expression are called the *coordinates of the vector* $\boldsymbol{u}$ in the basis $\{\boldsymbol{v}_1, \boldsymbol{v}_2, \ldots, \boldsymbol{v}_n\}$.

EXAMPLE 1.78. The column vectors

$$\boldsymbol{e}_1 = (1, 0, \ldots, 0)^\top, \quad \boldsymbol{e}_2 = (0, 1, \ldots, 0)^\top, \quad \ldots, \quad \boldsymbol{e}_n = (0, 0, \ldots, 1)^\top,$$

form *the canonical basis* of the space $\mathbb{F}^n$. The coordinates of a vector $\boldsymbol{u} = (u^{(1)}, u^{(2)}, \ldots, u^{(n)})^\top$ in this basis are the elements $u^{(1)}, u^{(2)}, \ldots, u^{(n)}$ of the field. It is clear that there are also other bases in the space $\mathbb{F}^n$.

The description of vector spaces up to isomorphism is rather simple.

LEMMA 1.79. *If a vector space V over a field $\mathbb{F}$ has a basis of n vectors, then V is isomorphic to the space $\mathbb{F}^n$.*

PROOF. Let $\{\boldsymbol{v}_1, \boldsymbol{v}_2, \dots, \boldsymbol{v}_n\}$ be a basis of the space V. We consider the mapping

$$\psi \colon V \to \mathbb{F}^n$$

which maps any vector to the column of its coordinates in the basis $\{\boldsymbol{v}_1, \boldsymbol{v}_2, \dots, \boldsymbol{v}_n\}$. Obviously, this is a one-to-one mapping. It is easy to check that ψ satisfies the definition of isomorphism of vector spaces. □

PROBLEM 1.80. Prove that the set of all linear combinations of vectors $\boldsymbol{v}_1, \boldsymbol{v}_2, \dots, \boldsymbol{v}_k$ from an arbitrary vector space V is a subspace in V.

PROBLEM 1.81. Prove that two sets of linearly independent vectors generating the same subspace have the same cardinality.

DEFINITION 1.82. The number of linearly independent vectors in a set that generate a finite vector space V is called the *dimension* of the space V and is denoted by $\dim V$.

PROBLEM 1.83. Prove that every set of linearly independent vectors in a finite vector space can always be extended to a basis.

DEFINITION 1.84. A vector space V is a *direct sum* of its subspaces $W_1, \dots, W_t$ (notation: $V = W_1 \dotplus W_2 \dotplus \cdots \dotplus W_t$) if $W_i \cap W_j = \{0\}$ for $i \neq j$ and any vector from V can be uniquely represented as a sum of vectors from W_i, $i = 1, \dots, t$.

EXAMPLE 1.85. Let $\{\boldsymbol{v}_1, \dots, \boldsymbol{v}_n\}$ be a basis of V and suppose that the set $\{1, \dots, n\}$ is partitioned into t disjoint subsets,

$$\{\boldsymbol{v}_1, \dots, \boldsymbol{v}_n\} = \{\boldsymbol{v}_{1,1}, \dots, \boldsymbol{v}_{1,k_1}\} \cup \{\boldsymbol{v}_{2,1}, \dots, \boldsymbol{v}_{2,k_2}\} \cup \cdots \cup \{\boldsymbol{v}_{t,1}, \dots, \boldsymbol{v}_{t,k_t}\}.$$

Then $V = W_1 \dotplus W_2 \dotplus \cdots \dotplus W_t$, where W_i is the subspace generated by the vectors $\boldsymbol{v}_{i,1}, \dots, \boldsymbol{v}_{i,k_i}$, $i = 1, \dots, t$.

Now we proceed to the last algebraic system that we consider in this section. Combining the notions of a vector space and a ring, we obtain the notion of an algebra.

DEFINITION 1.86. A set A with addition, (inner) multiplication, and multiplication by elements of a field $\mathbb{F}$ is called an *algebra* over the field $\mathbb{F}$ if

1) the set A is a vector space with respect to addition and multiplication by elements of $\mathbb{F}$;
2) the set A is a ring with respect to addition and inner multiplication;
3) $(av_1)v_2 = v_1(av_2) = a(v_1v_2)$ for any $a \in \mathbb{F}$, $v_1, v_2 \in A$.

A subset of an algebra is called a *subalgebra* if it is both a subspace and a subring. A mapping of algebras is called an isomorphism if it is an isomorphism both of vector spaces and rings.

EXAMPLE 1.87. The set M_n of $(n \times n)$-matrices with entries from the field $\mathbb{F}$ is an algebra of dimension n^2 over the field $\mathbb{F}$.

EXAMPLE 1.88. The ring of polynomials $\mathbb{F}[\lambda]$ over a field $\mathbb{F}$ is an infinite-dimensional algebra over the field $\mathbb{F}$. Its basis is the infinite set of monomials λ^n, $n = 0, 1, \dots$.

EXAMPLE 1.89. The quotient ring $\mathbb{F}[\lambda]/(a(\lambda))$, where $a(\lambda)$ is an irreducible polynomial over the field $\mathbb{F}$, is an algebra of dimension $n = \deg a(\lambda)$ over the field $\mathbb{F}$. The residue classes $1 + (a(\lambda))$, $\lambda + (a(\lambda))$, $\dots$, $\lambda^{n-1} + (a(\lambda))$ form its basis.

For further information on properties of the algebraic systems introduced here, we refer the reader to standard university textbooks, e.g., [**132**], [**250**], [**317**].

1.2. Construction of finite fields

The theory of finite fields is a good illustration of the general theory of fields because in this case many questions can be given an explicit answer. Moreover, the theory of finite fields is closely related to number theory, coding theory, and cryptography.

In §1.1 we already discussed an important class of finite fields. Namely (cf. Example 1.47 and Theorem 1.50), we proved that for any prime number p the quotient ring $\mathbb{Z}/(p)$ is a finite field of p elements that can be identified with the Galois field $\mathbb{F}_p$.

Let us find out what is the possible number of elements of a finite field $\mathbb{F}$.

THEOREM 1.90. *Let $\mathbb{F}$ be a finite field. Then $\mathbb{F}$ contains p^n elements, where the prime number p is the characteristic of the field $\mathbb{F}$ and the natural number $n = [\mathbb{F} : \mathbb{F}_p]$ is the degree of the extension of the field $\mathbb{F}$ over its prime subfield $\mathbb{F}_p$.*

PROOF. It is clear that the characteristic of the field $\mathbb{F}$ is a prime number p: otherwise the number of elements of the prime subfield contained in $\mathbb{F}$ would be infinite. Therefore $\mathbb{F}$ contains a prime subfield $\mathbb{F}_p$, which is isomorphic to $\mathbb{Z}/(p)$. Besides, the extension degree $[\mathbb{F} : \mathbb{F}_p]$ of the field $\mathbb{F}$ over $\mathbb{F}_p$ must be finite since the total number of elements in $\mathbb{F}$ is finite. Put $[\mathbb{F} : \mathbb{F}_p] = n$. Considering $\mathbb{F}$ as a linear space over $\mathbb{F}_p$, choose an arbitrary basis $u_1, \dots, u_n$ in $\mathbb{F}$ (with respect to $\mathbb{F}_p$). Then each element $a \in \mathbb{F}$ is uniquely represented as a linear combination $a = a^{(1)}u_1 + \cdots + a^{(n)}u_n$, $a^{(i)} \in \mathbb{F}_p$, $i = 1, 2, \dots, n$, and every such linear combination is an element of $\mathbb{F}$. Therefore the total number of elements of the field is equal to the number of all possible row vectors $(a^{(1)}, \dots, a^{(n)})$ in which $a^{(i)}$, $i = 1, 2, \dots, n$ can be arbitrary elements of the field $\mathbb{F}_p$. As each $a^{(i)}$ may have exactly p different values, the total number of elements of the field $\mathbb{F}$ (i.e., row vectors $(a^{(1)}, \dots, a^{(n)})$) is equal to p^n. □

LEMMA 1.91. *Let $\mathbb{F}'$ be a finite field containing a subfield $\mathbb{F}$ of q elements. Then $\mathbb{F}'$ contains q^m elements, where $m = [\mathbb{F}' : \mathbb{F}]$.*

PROOF. Similar to the proof of Theorem 1.90. □

LEMMA 1.92. *If $\mathbb{F} = \mathbb{F}_q$ is a finite field of q elements, then any element $a \in \mathbb{F}_q$ satisfies the equation $a^q = a$.*

PROOF. For $a = 0$ the identity $a^q = a$ is trivial. Let a be an arbitrary nonzero element of the field $\mathbb{F}_q$. Consider all nonzero elements

$$a_1, a_2, \dots, a_{q-1} \tag{1.7}$$

of $\mathbb{F}_q$. Multiplying all elements from (1.7) by a, we get the elements

$$aa_1, aa_2, \dots, aa_{q-1}, \tag{1.8}$$

which belong to the field $\mathbb{F}_q$, are nonzero and distinct (indeed, if $aa_i = aa_j$ for $i \neq j$, then dividing both sides of the equation by a, we will get $a_i = a_j$, which contradicts the choice (1.7)). Therefore, in (1.8) all elements of (1.7) are contained only once. Thus the products of elements from (1.7) and (1.8) are equal, i.e.,

$$a^{q-1}\prod_{i=1}^{q-1} a_i = \prod_{i=1}^{q-1} a_i.$$

Canceling the nonzero factor $\prod_{i=1}^{q-1} a_i$ in the last equality, we get $a^{q-1} = 1$, and multiplying it by a, we get the desired equation. □

LEMMA 1.93. *If $\mathbb{F}' = \mathbb{F}_q$ is a finite field of q elements and $\mathbb{F}$ is a subfield of the field $\mathbb{F}_q$, then the polynomial $\lambda^q - \lambda \in \mathbb{F}[\lambda]$ factors in $\mathbb{F}_q[\lambda]$ as follows:*

$$\lambda^q - \lambda = \prod_{a\in\mathbb{F}_q} (\lambda - a).$$

(The field $\mathbb{F}_q$ is called the field of decomposition *of the polynomial $\lambda^q - \lambda$ over $\mathbb{F}$.)*

PROOF. A polynomial $\lambda^q - \lambda$ of degree q has no more than q different roots in the field $\mathbb{F}_q$. According to Lemma 1.92, there *are* q different roots: they are all elements of the field $\mathbb{F}_q$. Thus, the polynomial can be factorized in the field $\mathbb{F}_q$ as in the lemma and cannot be completely factorized in any smaller field. □

We will now prove the existence theorem.

THEOREM 1.94. *For any prime number p and any natural n, there exists a finite field of p^n elements.*

Any finite field of $q = p^n$ elements is isomorphic to a quotient ring $\mathbb{F}_p[\lambda]/(a(\lambda))$, where $\mathbb{F}_p$ is a prime field of p elements and $a(\lambda)$ is an irreducible (over $\mathbb{F}_p$) polynomial of degree n (for a proof of the existence of an irreducible polynomial over $\mathbb{F}_p$ of a given degree n, see Section 1.3 below, Lemma 1.143).

PROOF. Let $\mathbb{F}_p$ be a prime field of p elements and $a(\lambda) \in \mathbb{F}_p[\lambda]$ be an irreducible polynomial over $\mathbb{F}_p$ of $\deg a(\lambda) = n$. Without loss of generality, we will assume that the polynomial $a(\lambda)$ is monic. According to Lemma 1.64, the quotient ring $\mathbb{F}_p[\lambda]/(a(\lambda))$ is a field. Elements of this field are residue classes $r(\lambda) + (a(\lambda))$, where $r(\lambda)$ is a polynomial in $\mathbb{F}_p[\lambda]$ of degree not greater than $n - 1$, i.e., $r(\lambda) = r_0 + r_1\lambda + \cdots + r_{n-1}\lambda^{n-1}$. Obviously, there are exactly q^n such classes.

Let $\mathbb{F}_q$ be an arbitrary field of $q = p^n$ elements. As $\operatorname{char}\mathbb{F}_q = \operatorname{char}\mathbb{F}_p[\lambda]/(a(\lambda)) = p$, the prime fields contained in the fields $\mathbb{F}_q$ and $\mathbb{F}_p[\lambda]/(a(\lambda))$ are isomorphic. Hence, the fields $\mathbb{F}_q$ and $\mathbb{F}_p[\lambda]/(a(\lambda))$ can be considered as extensions of the same field $\mathbb{F}_p$. As follows from Lemma 1.93, the elements of the fields $\mathbb{F}_q$ and $\mathbb{F}_p[\lambda]/(a(\lambda))$ are roots of the same polynomial $\lambda^q - \lambda \in \mathbb{F}_p[\lambda]$.

Let $\alpha = \lambda + (a(\lambda))$ be an element of the field $\mathbb{F}_p[\lambda]/(a(\lambda))$. Obviously, $a(\alpha) = 0$, i.e., the element α is a root of the polynomial $a(\lambda)$. As $\gcd(a(\lambda), \lambda^q - \lambda) \neq 1$ (see Problem 1.66) and $a(\lambda)$ is an irreducible polynomial, the polynomial $a(\lambda)$ divides $\lambda^q - \lambda$. Since all elements of the field $\mathbb{F}_q$ are roots of the polynomial $\lambda^q - \lambda$ and $a(\lambda)$ divides $\lambda^q - \lambda$, there exists an element $\beta \in \mathbb{F}_q$ such that $a(\beta) = a_0 + a_1\beta + \cdots + a_{n-1}\beta^{n-1} + \beta^n = 0$. It is clear that β cannot be a root of any polynomial from $\mathbb{F}_p[\lambda]$ of degree less than n. Hence, all the p^n linear combinations $b_0 + b_1\beta + \cdots + b_{n-1}\beta^{n-1}$, $b_i \in \mathbb{F}_p$, $i = 0, 1, \ldots, n-1$, are different and represent all the elements of the field $\mathbb{F}_q$.

We define a one-to-one correspondence $\psi\colon \mathbb{F}_p[\lambda]/(a(\lambda)) \to \mathbb{F}_q$ by setting $\psi(b_0 + b_1\alpha + \cdots + b_{n-1}\alpha^{n-1}) = b_0 + b_1\beta + \cdots + b_{n-1}\beta^{n-1}$, $b_i \in \mathbb{F}_p$, $i = 0, 1, \ldots, n-1$. A simple check shows that ψ is an isomorphism of these fields. $\square$

Using arguments from the proof of Theorem 1.94 and Lemma 1.64, we can construct the field $\mathbb{F}_{q^n}$ for any finite field $\mathbb{F}_q$, with the help of a monic and irreducible (over $\mathbb{F}_q$) polynomial of degree n.

Theorem 1.94 allows us to talk about a uniquely determined field of q elements. We will denote it by $\mathbb{F}_q$, where q is a power of a prime number p, which is the characteristic of this field.

LEMMA 1.95. *Let $\mathbb{F}$ be an arbitrary field of characteristic p. Then for any $a, b \in \mathbb{F}$ and for any natural number l, one has*

$$(a+b)^{p^l} = a^{p^l} + b^{p^l}.$$

PROOF. Since the binomial coefficient

$$\binom{p}{k} = \frac{p(p-1)\cdots(p-k+1)}{1\cdot 2\cdots k},$$

$1 \le k \le p-1$, is an integer divisible by p, one has $\binom{p}{k} \equiv 0 \pmod p$. Therefore, by the binomial formula, which holds in any field, we have

$$(a+b)^p = a^p + \binom{p}{1}a^{p-1}b + \cdots + \binom{p}{p-1}ab^{p-1} + b^p = a^p + b^p.$$

We finish by induction on l. To do this, we use the identity

$$(a+b)^{p^l} = ((a+b)^{p^{l-1}})^p,$$

from which the induction step follows. $\square$

THEOREM 1.96. *Let $\mathbb{F}_q$ be a finite field of $q = p^n$ elements. Then any subfield of the field $\mathbb{F}_q$ has order p^m, where m is a natural number divisible by n.*

Conversely, if m is a natural number dividing n, then the field $\mathbb{F}_q$ contains exactly one subfield of order p^m.

PROOF. Let $\mathbb{F}$ be a subfield of the field $\mathbb{F}_q$. Since $\operatorname{char}\mathbb{F} = p$, the field $\mathbb{F}$ must have order p^m, where m is a natural number not greater than n. It follows from Lemma 1.91 that the number $q = p^n$ must be a power of p^m; thus, m must divide n.

Conversely, let $\mathbb{F}_q$ be a field of order $q = p^n$ and suppose that $n = m\cdot k$. Then the number $p^m - 1$ divides the number $p^n - 1$, whence the polynomial $\lambda^{p^m} - \lambda$ divides $\lambda^{p^n} - \lambda$; the roots of the latter are all elements of the field $\mathbb{F}_q$. Since $\lambda^{p^m} - \lambda$ divides $\lambda^{p^n} - \lambda$, the p^m elements of $\mathbb{F}_q$ are roots of the polynomial $\lambda^{p^m} - \lambda$. Denote this set of roots by $M \subset \mathbb{F}_q$, and suppose that $a, b \in M$. Then $(a+b)^{p^m} = a^{p^m} + b^{p^m} = a + b$ according to Lemma 1.95, i.e. $a + b \in M$, and $(a\cdot b^{-1})^{p^m} = a^{p^m}\cdot(b^{-1})^{p^m} = a^{p^m}\cdot(b^{p^m})^{-1} = a\cdot b^{-1} \in M$. It is obvious that $0, 1 \in M$ as well. Thus, the roots of the polynomial $\lambda^{p^m} - \lambda$ in the field $\mathbb{F}_q$ form a subfield $M = \mathbb{F}_{p^m}$ of order p^m. The uniqueness of this subfield follows from the fact that in any subfield $\mathbb{F}'$ of order p^m the identity $a^{p^m} = a$ holds and thus $\mathbb{F}' = M$. $\square$

EXAMPLE 1.97. A subfield of an arbitrary finite field $\mathbb{F}_{p^n}$ can be found by making a list of all natural numbers that are divisors of n. Figure 1.1 illustrates the structures of subfields of the fields $\mathbb{F}_{p^{12}}$ and $\mathbb{F}_{p^{30}}$ (inclusion relations are given in the diagrams).

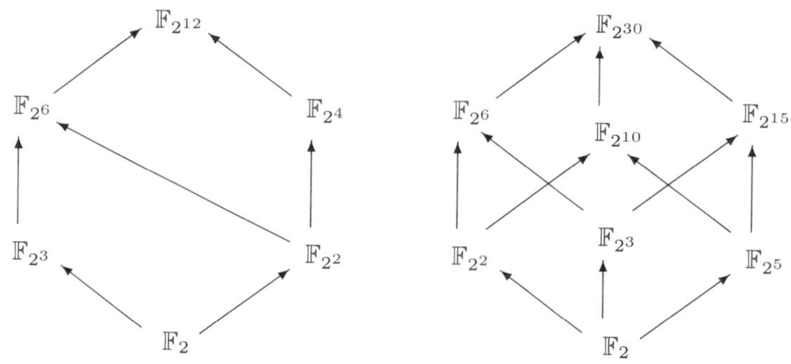

FIGURE 1.1

THEOREM 1.98. *The multiplicative group $\mathbb{F}_q^*$ of nonzero elements of an arbitrary finite field $\mathbb{F}_q$ is cyclic.*

PROOF. If the number $q-1$ is prime, then the statement of the theorem is trivial because the orders of all elements of the group $\mathbb{F}_q^*$ (except 1) are equal to $q-1$, and so any element different from 1 generates $\mathbb{F}_q^*$.

Let $q-1$ be a composite number; suppose its prime factorization is

$$q-1=\prod_{i=1}^{s} p_i^{\nu_i},$$

where p_i, $i=1,2,\ldots,s$, are different primes and ν_i, $i=1,2,\ldots,s$, are natural numbers. The polynomial $\lambda^{(q-1)/p_i}-1$ divides the polynomial $\lambda^{q-1}-1$, and among the $q-1$ elements of the group $\mathbb{F}_q^*$ there must be at least one element that is not a root of the polynomial $\lambda^{(q-1)/p_i}-1$ as this polynomial can have no more than $(q-1)/p_i$ roots. For any $i=1,2,\ldots,s$ choose an element a_i with the following property: $a_i^{(q-1)/p_i}\neq 1$. Let $b_i=a_i^{(q-1)/p_i^{\nu_i}}$ and $b=\prod_{i=1}^{s} b_i$. We will show that the order of the element b is equal to $q-1$ and that the group $\mathbb{F}_q^*$ is cyclic. It is clear that $b_i^{p_i^{\nu_i}}=1$, $i=1,2,\ldots,s$, so the order of b_i divides $p_i^{\nu_i}$. Let this order be equal to $p_i^{n_i}$. If n_i is less than ν_i, then $b_i^{p_i^{\nu_i-1}}=1$. However, $b_i^{p_i^{\nu_i-1}}=a_i^{(q-1)/p_i}\neq 1$. Therefore, the order of the element b_i is equal to $p_i^{\nu_i}$. Suppose that $b^n=1$ for some natural n. We will show that this implies the congruence $n\equiv 0 \pmod{p_i^{\nu_i}}$ for all $i=1,2,\ldots,s$. Indeed, for any i the following equality holds:

$$b^{n\prod_{j\neq i} p_j^{\nu_j}}=1.$$

Substituting $\prod_{i=1}^{s} b_i$ for b and using the equality $b_j^{p_j^{\nu_j}}=1$, we get

$$b_i^{n\prod_{j\neq i} p_j^{\nu_j}}=1.$$

Comparing the degrees, we get

$$n\cdot\prod_{j\neq i} p_j^{\nu_j}\equiv 0 \pmod{p_i^{\nu_i}}.$$

Since $\gcd\Bigl(p_i^{\nu_i},\prod_{j\neq i} p_j^{\nu_j}\Bigr)=1$, the latter congruence is equivalent to

$$n\equiv 0 \pmod{p_i^{\nu_i}}.$$

In other words, n is divisible by $p_i^{\nu_i}$ for any $i = 1, 2, \ldots, s$, i.e., it is divisible by $q-1 = p_1^{\nu_1} \ldots p_s^{\nu_s}$. Thus, the order of the element b is equal to $q-1$ and $\mathbb{F}_q^* = \langle b \rangle$. $\square$

DEFINITION 1.99. A generator of the cyclic group $\mathbb{F}_q^*$ is called a *primitive* element of the field $\mathbb{F}_q$.

It follows from Theorem 1.15 that the field $\mathbb{F}_q$ contains $\varphi(q-1)$ primitive elements, where φ is Euler's function.

Let $\mathbb{F}_{q^m}$ be an extension of the field $\mathbb{F}_q$. We introduce a function from $\mathbb{F}_{q^m}$ to $\mathbb{F}_q$ which is important for many applications.

DEFINITION 1.100. Let $a \in \mathbb{F}_{q^m}$. The *trace* $\mathrm{Tr}_{q^m/q}(a)$ of an element a over $\mathbb{F}_q$ is the element defined by the equality

$$\mathrm{Tr}_{q^m/q}(a) = a + a^q + a^{q^2} + \cdots + a^{q^{m-1}}.$$

If $q = p$ is a prime number (i.e., $\mathbb{F}_q$ is a prime subfield of the field $\mathbb{F}_{q^m}$), then $\mathrm{Tr}_{q^m/q}(a)$ is called the *absolute trace* of the element a; the absolute trace is denoted by $\mathrm{Tr}_m(a)$.

PROBLEM 1.101. Prove that the trace $\mathrm{Tr}_{q^m/q}$ has the following properties:

1) $\mathrm{Tr}_{q^m/q}(a+b) = \mathrm{Tr}_{q^m/q}(a) + \mathrm{Tr}_{q^m/q}(b)$ for any $a, b \in \mathbb{F}_{q^m}$;

2) $\mathrm{Tr}_{q^m/q}(ca) = c\,\mathrm{Tr}_{q^m/q}(a)$ for any $c \in \mathbb{F}_q$ and $a \in \mathbb{F}_{q^m}$;

3) $\mathrm{Tr}_{q^m/q}$ is a linear mapping from $\mathbb{F}_{q^m}$ onto $\mathbb{F}_q$, where $\mathbb{F}_{q^m}$ and $\mathbb{F}_q$ are considered as linear spaces over $\mathbb{F}_q$;

4) $\mathrm{Tr}_{q^m/q}(a) = ma$ for any $a \in \mathbb{F}_q$;

5) $\mathrm{Tr}_{q^m/q}(a^q) = \mathrm{Tr}_{q^m/q}(a)$ for any $a \in \mathbb{F}_{q^m}$.

THEOREM 1.102. *The equality* $\mathrm{Tr}_{q^m/q}(a) = 0$ *holds if and only if there exists an element* $b \in \mathbb{F}_{q^m}$ *such that* $a = b^q - b$.

PROOF. The sufficiency of this condition immediately follows from the result of Problem 1.101(5).

We will prove the converse. Suppose that an element $a \in \mathbb{F}_{q^m}$ is such that $\mathrm{Tr}_{q^m/q}(a) = 0$. Let b be a root of the polynomial $\lambda^q - \lambda - a$ from an extension of the field $\mathbb{F}_{q^m}$. Then $b^q - b = a$ and

$$\begin{aligned}
0 = \mathrm{Tr}_{q^m/q}(a) &= a + a^q + a^{q^2} + \cdots + a^{q^{m-1}} \\
&= (b^q - b) + (b^q - b)^q + (b^q - b)^{q^2} + \cdots + (b^q - b)^{q^{m-1}} \\
&= (b^q - b) + (b^{q^2} - b^q) + (b^{q^3} - b^{q^2}) + \cdots + (b^{q^m} - b^{q^{m-1}}) \\
&= b^{q^m} - b,
\end{aligned}$$

and so $b \in \mathbb{F}_{q^m}$. $\square$

PROBLEM 1.103. Suppose that $\mathbb{F}_q \subset \mathbb{F}_{q'} \subset \mathbb{F}_{q''}$. Show that for all $a \in \mathbb{F}_{q''}$ the following equality holds:

$$\mathrm{Tr}_{q''/q}(a) = \mathrm{Tr}_{q'/q}(\mathrm{Tr}_{q''/q'}(a)).$$

Another interesting mapping from a finite field to its subfield is defined as a product.

DEFINITION 1.104. Suppose that $a \in \mathbb{F}_{q^m}$. The element

$$\mathrm{N}_{q^m/q}(a) = a \cdot a^q \cdot a^{q^2} \cdots a^{q^{m-1}}.$$

is called the *norm* $\mathrm{N}_{q^m/q}(a)$ over $\mathbb{F}_q$ of the element a.

If $q = p$ is a prime number (i.e., $\mathbb{F}_q$ is a prime subfield of the field $\mathbb{F}_{q^m}$), then $\mathrm{N}_{q^m/q}(a)$ is called the *absolute norm* of the element a and is denoted by $\mathrm{N}_m(a)$.

PROBLEM 1.105. Prove that the norm $\mathrm{N}_{q^m/q}$ has the following properties:
1. $\mathrm{N}_{q^m/q}(a \cdot b) = \mathrm{N}_{q^m/q}(a) \cdot \mathrm{N}_{q^m/q}(b)$ for any $a, b \in \mathbb{F}_{q^m}$.
2. The norm $\mathrm{N}_{q^m/q}$ is a mapping from $\mathbb{F}_{q^m}$ to $\mathbb{F}_q$ which maps $\mathbb{F}^*_{q^m}$ to $\mathbb{F}^*_q$.
3. $\mathrm{N}_{q^m/q}(a) = a^m$ for any $a \in \mathbb{F}_q$.
4. $\mathrm{N}_{q^m/q}(a^q) = \mathrm{N}_{q^m/q}(a)$ for any $a \in \mathbb{F}_{q^m}$.

PROBLEM 1.106. Suppose that $\mathbb{F}_q \subset \mathbb{F}_{q'} \subset \mathbb{F}_{q''}$. Show that for all $a \in \mathbb{F}_{q''}$ the following equality holds:

$$\mathrm{N}_{q''/q}(a) = \mathrm{N}_{q'/q}(\mathrm{N}_{q''/q'}(a)).$$

All possible linear mappings from $\mathbb{F}_{q^m}$ to $\mathbb{F}_q$ can be described with the help of the trace $\mathrm{Tr}_{q^m/q}$.

THEOREM 1.107. *Let $\mathbb{F}_{q^m}$ be an extension of the finite field $\mathbb{F}_q$ (both fields are considered as linear spaces over $\mathbb{F}_q$). Then any linear mapping from $\mathbb{F}_{q^m}$ to $\mathbb{F}_q$ is of the form L_b, $b \in \mathbb{F}_{q^m}$, where $L_b(a) = \mathrm{Tr}_{q^m/q}(ba)$; if $b \neq c$, then $L_b \neq L_c$.*

PROOF. According to Problem 1.101(3), any mapping L_b is a linear mapping from $\mathbb{F}_{q^m}$ to $\mathbb{F}_q$. If $b, c \in \mathbb{F}_{q^m}$ and $b \neq c$, then

$$L_b(a) - L_c(a) = \mathrm{Tr}_{q^m/q}(ba) - \mathrm{Tr}_{q^m/q}(ca) = \mathrm{Tr}_{q^m/q}((b-c)a) \neq 0$$

for a suitable element $a \in \mathbb{F}_{q^m}$ since $\mathrm{Tr}_{q^m/q}$ maps $\mathbb{F}_{q^m}$ onto $\mathbb{F}_q$. Thus, we have q^m different linear mappings L_b from $\mathbb{F}_{q^m}$ to $\mathbb{F}_q$.

Choose a basis $\{u_1, \ldots, u_m\}$ of the vector space $\mathbb{F}_{q^m}$ over $\mathbb{F}_q$. Any linear mapping L from $\mathbb{F}_{q^m}$ to $\mathbb{F}_q$ is uniquely determined by the images of the elements of the basis $\{L(u_1), \ldots, L(u_m)\}$. The number of all such n-tuples is q^m. Thus, all linear mappings from $\mathbb{F}_{q^m}$ to $\mathbb{F}_q$ are the mappings L_b, $b \in \mathbb{F}_{q^m}$. □

DEFINITION 1.108. Suppose that $\mathbb{F}_q \subset \mathbb{F}_{q^m}$. Two bases $\{u_1, \ldots, u_m\}$ and $\{v_1, \ldots, v_m\}$ of the vector space $\mathbb{F}_{q^m}$ over $\mathbb{F}_q$ are called *dual* if for all $1 \leq i, j \leq m$ one has

$$\mathrm{Tr}_{q^m/q}(u_i \cdot v_j) = \begin{cases} 0 & \text{for } i \neq j; \\ 1 & \text{for } i = j. \end{cases}$$

Let $\{u_1, \ldots, u_m\}$ be a basis of $\mathbb{F}_{q^m}$ over $\mathbb{F}_q$. An arbitrary element $a \in \mathbb{F}_{q^m}$ can be uniquely represented as

$$a = c_1(a)u_1 + \cdots + c_m(a)u_m,$$

where $c_i(a) \in \mathbb{F}_q$, $i = 1, 2, \ldots, m$. The mappings $c_i(a)$ are obviously linear mappings from $\mathbb{F}_{q^m}$ to $\mathbb{F}_q$. Therefore, by virtue of Theorem 1.107, there exist elements $v_i \in \mathbb{F}_{q^m}$ such that $c_i(a) = \mathrm{Tr}_{q^m/q}(v_i a)$ for all $a \in \mathbb{F}_{q^m}$. If we put $a = u_i$, $i = 1, \ldots, m$, we see that $\mathrm{Tr}_{q^m/q}(v_j u_i)$ is equal to 0 for $i \neq j$ and to 1 for $i = j$. The elements $\{v_1, \ldots, v_m\}$ are a basis of the vector space $\mathbb{F}_{q^m}$ over $\mathbb{F}_q$, because if

$$d_1 v_1 + \cdots + d_m v_m = 0, \quad d_i \in \mathbb{F}_q, \quad i = 1, \ldots, m,$$

then, multiplying this relation by u_i, $i = 1, \ldots, m$, and applying the trace $\mathrm{Tr}_{q^m/q}$, we get $d_i = 0$, $i = 1, \ldots, m$. This shows that for any basis of the field $\mathbb{F}_{q^m}$ over $\mathbb{F}_q$ there exists a dual basis.

There exist many different bases of $\mathbb{F}_{q^m}$ over $\mathbb{F}_q$. We will focus on two most important types of bases.

DEFINITION 1.109. Let $\mathbb{F}_q \subset \mathbb{F}_{q^m}$.

1. A *polynomial* basis is a basis of the vector space $\mathbb{F}_{q^m}$ over $\mathbb{F}_q$ of the form $\{1, a, a^2, \ldots, a^{m-1}\}$, where $a \in \mathbb{F}_{q^m}$. In this construction a will often be a primitive element of the field $\mathbb{F}_{q^m}$.

2. A *normal* basis is a basis of $\mathbb{F}_{q^m}$ over $\mathbb{F}_q$ of the form $\{a, a^q, a^{q^2}, \ldots, a^{q^{m-1}}\}$ for a suitably chosen element $a \in \mathbb{F}_{q^m}$.

The elements of a finite field can be represented in different ways (see, for example, [**225**]). We will give only one construction based on Theorem 1.94. According to this theorem, an arbitrary finite field $\mathbb{F}_q$, $q = p^m$ (p is the characteristic of the field), is isomorphic to the quotient ring $\mathbb{F}_q[x]$ modulo a monic irreducible (over $\mathbb{F}_p$) polynomial $a(x) = a_0 + a_1 x + \cdots + a_{m-1}x^{m-1} + x^m$, $\deg a(x) = m$. The element $\alpha = x + (a(x))$ of the field $\mathbb{F}_q[x]/(a(x))$ is a root of the polynomial $a(x)$, i.e., $a(\alpha) = 0$. Thus, the elements of order q of the field in question can be represented canonically, with the help of the polynomial basis $\{1, \alpha, \alpha^2, \ldots, \alpha^{m-1}\}$, as linear combinations $c_0 + c_1\alpha + \cdots + c_{m-1}\alpha^{m-1}$, $c_i \in \mathbb{F}_p$, $i = 1, 2, \ldots, m-1$. Addition of two elements of the field $\mathbb{F}_q$ is addition of the corresponding coefficients in its canonical representation. The operation of multiplication can be performed in two steps. First, the ordinary multiplication of elements represented in a polynomial basis is performed, and then the resulting expression is reduced to the canonical form using the relation $\alpha^m = -(a_0 + a_1\alpha + \cdots + a_{m-1}\alpha^{m-1})$.

There are two abelian groups related to the finite field $\mathbb{F}_q$: the additive group and the multiplicative group. Thus, one should not confuse the characters (see Definition 1.19) of these groups.

A character of the additive group of the field $\mathbb{F}_q$, $q = p^m$ (p is the characteristic of the field $\mathbb{F}_q$), is called an *additive character* of the field $\mathbb{F}_q$. We will identify the prime field contained in $\mathbb{F}_q$ with $\mathbb{F}_p = \mathbb{Z}/(p)$. Let $\mathrm{Tr}_m\colon \mathbb{F}_q \to \mathbb{F}_p$ be the absolute trace from $\mathbb{F}_q$ to $\mathbb{F}_p$ (see Definition 1.100). Consider the function

$$\chi_1(c) = e^{2\pi \boldsymbol{i}\, \mathrm{Tr}_m(c)/p}$$

($c \in \mathbb{F}_q$). It is easy to check that this function is an additive character of $\mathbb{F}_q$. This character is called the *canonical additive* character of $\mathbb{F}_q$. Any additive character of $\mathbb{F}_q$ can be expressed in terms of χ_1.

LEMMA 1.110. *Let $\mathbb{F}_q$, $q = p^m$, be a finite field, $b \in \mathbb{F}_q$. Then the function χ_b defined by the equation $\chi_b(c) = \chi_1(bc)$ for all $c \in \mathbb{F}_q$, is an additive character of the field $\mathbb{F}_q$. Any additive character of the field $\mathbb{F}_q$ equals χ_b for some $b \in \mathbb{F}_q$.*

PROOF. The first statement of the lemma is proved by a direct check. According to Problem 1.101(3), the function Tr_m maps the field $\mathbb{F}_q$ onto $\mathbb{F}_p$ and χ_1 is a nontrivial character. Thus, for $a, b \in \mathbb{F}_q$, $a \neq b$, there exists an element $c \in \mathbb{F}_q$ such that

$$\frac{\chi_a(c)}{\chi_b(c)} = \frac{\chi_1(ac)}{\chi_1(bc)} = \chi_1((a-b)c) \neq 1.$$

Therefore, χ_a and χ_b are different characters. When the element b runs over the field $\mathbb{F}_q$, we get q different additive characters. On the other hand, according to Theorem 1.24 there exist exactly q different additive characters of the field $\mathbb{F}_q$. □

If we put $b = 0$ in Lemma 1.110, we obtain the trivial character χ_0, having the property $\chi_0(c) = 1$ for all $c \in \mathbb{F}_q$.

Any character of the multiplicative group $\mathbb{F}_q^*$ of the field $\mathbb{F}_q$ is called a *multiplicative character* of the field $\mathbb{F}_q$. Since the group $\mathbb{F}_q^*$ is cyclic of order $q-1$ (see Theorem 1.98), all the characters can be described in the following way.

LEMMA 1.111. *Let b be a primitive element of the field $\mathbb{F}_q$. Then for any j, $0 \le j \le q-2$, the function ψ_j defined by*

$$\psi_j(b^k) = e^{2\pi \boldsymbol{i} jk/(q-1)}$$

for all k, $0 \le k \le q-2$, is a multiplicative character of the field $\mathbb{F}_q$. Any multiplicative character of the field $\mathbb{F}_q$ equals ψ_j for some j, $0 \le j \le q-2$.

PROOF. The lemma immediately follows from Problem 1.21. □

For any choice of the primitive element b in Lemma 1.111, the character ψ_0 is always the trivial multiplicative character, i.e., $\psi_0(c) = 1$ for all $c \in \mathbb{F}_q^*$.

PROBLEM 1.112. Prove that the group of multiplicative characters of a finite field $\mathbb{F}_q$ is cyclic of order $q-1$ with identity element ψ_0.

EXAMPLE 1.113. Suppose that $\operatorname{char} \mathbb{F}_q = 2$, i.e., $q = 2^n$. Identify the field $\mathbb{F}_q$ with $\mathbb{F}_2[\lambda]/(a(\lambda))$, the ring of residues modulo the irreducible (over $\mathbb{F}_2$) polynomial $a(\lambda) = a_0 \oplus a_1\lambda \oplus \cdots \oplus a_{n-1}\lambda^{n-1} \oplus \lambda^n$, $\deg a(\lambda) = n$, where $\oplus$ denotes addition modulo 2. Then the additive group of the field $\mathbb{F}_{2^n}$ is an elementary abelian 2-group (see [**155**]) and the additive characters have the form

$$\chi_b(c) = e^{2\pi \boldsymbol{i} \operatorname{Tr}_n(bc)/2} = (-1)^{\operatorname{Tr}_n(bc)},$$

where $c \in \mathbb{F}_{2^n}$, $c = c_0 \oplus c_1\alpha \oplus \cdots \oplus c_{n-1}\alpha^{n-1}$, $\alpha = \lambda \oplus (a(\lambda)) \in \mathbb{F}_{2^n}$, $c_i \in \mathbb{F}_2$, $i = 1, \dots, n-1$. According to Problem 1.101(3), the function Tr_n is a linear mapping from $\mathbb{F}_{2^n}$ to $\mathbb{F}_2$; it can be written in the form

$$\begin{aligned}\operatorname{Tr}_n(bc) &= \operatorname{Tr}_n(bc_0 \oplus bc_1\alpha \oplus \cdots \oplus bc_{n-1}\alpha^{n-1}) \\ &= c_0 \operatorname{Tr}_n(b) \oplus c_1 \operatorname{Tr}_n(b\alpha) \oplus \cdots \oplus c_{n-1} \operatorname{Tr}_n(b\alpha^{n-1}) \\ &= c_0b'_0 \oplus c_1b'_1 \oplus \cdots \oplus c_{n-1}b'_{n-1} = \langle c, b' \rangle,\end{aligned}$$

where $\langle \cdot\,, \cdot \rangle$ is the *scalar product*, and $b \mapsto b'$ is a 1-1 mapping of the field $\mathbb{F}_{2^n}$ into itself. Hence, $\chi_b(c) = (-1)^{\langle c, b' \rangle}$.

The group $\mathbb{F}_{2^n}^*$ is cyclic of order $2^n - 1$. There are exactly $\varphi(2^n - 1)$ primitive elements of the field $\mathbb{F}_{2^n}$, and each of them generates this group. If $2^n - 1 = p$ is prime, any element of the field $\mathbb{F}_{2^n}$ except 0 and 1 generates the group $\mathbb{F}_{2^n}^*$. Let b be a primitive element of the field $\mathbb{F}_{2^n}$; one has

$$\psi_j(b^k) = e^{2\pi \boldsymbol{i} jk/(2^n-1)}$$

for all k, $0 \le k \le 2^n - 2$.

Consider now the structure of the ring $\mathbb{F}_q[\lambda]/(a(\lambda))$ of remainders modulo some polynomial. This ring plays an important role when describing some classes of codes. In view of these applications, we will give several new definitions and prove some properties of finite rings.

DEFINITION 1.114. A finite additive ring R is called *reducible* if there exist proper ideals J_i, $i = 1, 2, \ldots, t$, $t \geq 2$, such that

$$R = J_1 \dotplus J_2 \dotplus \cdots \dotplus J_t$$

($\dotplus$ denotes direct sum of groups, i.e., $J_i \cap J_j = \{0\}$, $i \neq j$, and any element of the ring R is uniquely representable as a sum of elements of the ideals J_i, $i = 1, 2, \ldots, t$).

PROBLEM 1.115. Let R be a finite commutative reducible ring with identity e and $R = J_1 \dotplus J_2 \dotplus \cdots \dotplus J_t$ is the corresponding expansion. Prove that:

1) if $r = a_1 + \cdots + a_t$ and $r' = a_1' + \cdots + a_t'$, where a_i, $a_i' \in J_i$, then

$$rr' = a_1 a_1' + \cdots + a_t a_t';$$

2) any ideal J_i, $i = 1, 2, \ldots, t$, if considered as a subring of the ring R, contains its own identity e_i. Besides, $e = e_1 + \cdots + e_t$, $e_i e_j = 0$ for $i \neq j$.

DEFINITION 1.116. A finite commutative ring R, which cannot be represented as a direct sum of its ideals is called *irreducible*.

DEFINITION 1.117. Let R be a finite commutative ring with identity e. An element $a \in R$ is called a *proper idempotent* if $a^2 = a$, $a \neq 0, e$.

PROBLEM 1.118. Prove that any element from a finite field R is an idempotent of certain degree.

LEMMA 1.119. *A finite commutative ring R with identity is reducible if and only if it contains a proper idempotent.*

PROOF. Let R be reducible and let $R = J_1 \dotplus J_2$ be the corresponding decomposition (without loss of generality, we can assume that $t = 2$); denote by e the identity of the ring R. According to Problem 1.115(2), $e = e_1 + e_2$, where e_i, $i = 1, 2$, are the identities of the ideals J_1 and J_2 as subrings of R; thus, they are proper idempotents.

Conversely, suppose that $a \in R$, $a \neq 0, e$ and $a^2 = a$. We consider the ideals $J_1 = aR$ and $J_2 = (e - a)R$ of the ring R. It is easy to check that J_1 and J_2 are proper ideals of R and $R = J_1 + J_2$. If $b \in J_1 \cap J_2$, then $b = ar_1 = (e - a)r_2$, $r_1, r_2 \in R$. Multiplying the latter equation by a, we get $ar_1 = 0$, i.e., $b = 0$. Therefore, $J_1 \cap J_2 = \{0\}$, i.e., $R = J_1 \dotplus J_2$ is a direct sum of subgroups and R is a reducible ring. □

EXAMPLE 1.120. Let $a(\lambda) \in \mathbb{F}_q[\lambda]$ be a normalized polynomial, $\deg a(\lambda) = n$, with factorization

$$a(\lambda) = b_1^{k_1}(\lambda) b_2^{k_2}(\lambda) \cdots b_t^{k_t}(\lambda),$$

where $b_i(\lambda)$, $i = 1, 2, \ldots, t$, are different monic irreducible polynomials over $\mathbb{F}_q$, $k_i \geq 1$, $i = 1, 2, \ldots, t$, $\deg b_i(\lambda) = n_i$. Obviously, the quotient ring $R = \mathbb{F}_q[\lambda]/(a(\lambda))$ is a finite commutative ring with identity $1 \in \mathbb{F}_q$ and $\#R = q^n$. Besides, the quotient ring R consists of the equivalence classes of the type $\{c(\lambda)\} = c(\lambda) + (a(\lambda))$, where $c(\lambda) \in \mathbb{F}_q[\lambda]$ and $\deg c(\lambda) \leq n - 1$. Consider the polynomials

$$m_i(\lambda) = \frac{a(\lambda)}{b_i^{k_i}(\lambda)} = b_1^{k_1}(\lambda) \cdots b_{i-1}^{k_{i-1}}(\lambda) \cdot b_{i+1}^{k_{i+1}}(\lambda) \cdots b_t^{k_t}(\lambda),$$

$i = 1, 2, \ldots, t$. It is easy to see that $\gcd(m_1(\lambda), \ldots, m_t(\lambda)) = 1$. Hence, by Problem 1.58, in the ring $\mathbb{F}_q[\lambda]$, there are polynomials $u_1(\lambda), u_2(\lambda), \ldots, u_t(\lambda) \in \mathbb{F}_q[\lambda]$ such that

$$m_1(\lambda)u_1(\lambda) + \cdots + m_t(\lambda)u_t(\lambda) = 1.$$

We denote $m_i(\lambda)u_i(\lambda)$ by $e_i(\lambda)$, $i = 1, 2, \ldots, t$. It is easy to check that

$$[e_i(\lambda)]^2 \equiv e_i(\lambda) \pmod{a(\lambda)}, \qquad i = 1, 2, \ldots, t,$$
$$e_i(\lambda) \cdot e_j(\lambda) \equiv 0 \pmod{a(\lambda)}, \qquad i \neq j,$$

i.e., the elements $e_i(\lambda)$ are idempotents in the ring R and $R = J_1 \dotplus J_2 \dotplus \cdots \dotplus J_t$, where $J_i = (e_i(\lambda))$ is the ideal generated by the idempotent $e_i(\lambda)$, $i = 1, 2, \ldots, t$.

Let us investigate the structure of the ideal J_i. An arbitrary element $\{c(\lambda)\}$ of this ideal can be written as

$$\{c(\lambda)\} = q(\lambda)e_i(\lambda) + (a(\lambda)) = q(\lambda)m_i(\lambda)u_i(\lambda) + (a(\lambda)),$$

where $q(\lambda) \in \mathbb{F}_q[\lambda]$, $\deg q(\lambda) \leq n_i - 1$ and $\#J_i = q^{n_i}$. Define a mapping θ from the ideal J_i to the ring $\mathbb{F}_q[\lambda]/(b_i(\lambda))$ by the formula

$$\theta(\{c(\lambda)\}) = \theta(q(\lambda)e_i(\lambda) + (a(\lambda))) = q(\lambda) + (b_i(\lambda)).$$

It is easy to check that θ is a homomorphism from J_i to the ring $\mathbb{F}_q/(b_i(\lambda))$. Since $\#J_i = \#(\mathbb{F}_q/(b_i(\lambda))) = q^{n_i}$ and $\mathbb{F}_q/(b_i(\lambda))$ is a field (see Lemma 1.64), the ideal J_i is isomorphic to this field. Hence, all ideals are minimal (an ideal of a commutative ring is called minimal if it does not contain any proper ideals of the ring different from itself and $\{0\}$).

1.3. Polynomials over finite fields

Let $\mathbb{F}_q$ be a finite field, let $\mathbb{F}_{q'}$ ($q' = q^m$, m is a natural number) be its extension, and suppose that $b \in \mathbb{F}_{q'}$.

DEFINITION 1.121. The monic polynomial of minimum degree $m_b(\lambda) \in \mathbb{F}_q[\lambda]$ for which $m_b(b) = 0$ is called a *minimal polynomial of the element* b over the field $\mathbb{F}_q$.

It is clear that there exist monic polynomials from $\mathbb{F}_q[\lambda]$ for which b is a root. For example, $\lambda^{q'} - \lambda \in \mathbb{F}_q[\lambda]$ is such a polynomial, for which b is a root according to Lemma 1.93. Therefore, there exist polynomials of minimum degree with this property. Suppose that there exist two minimal polynomials for the element b, i.e., $m_b(\lambda)$ and $m_b'(\lambda)$; we have $\deg m_b(\lambda) = \deg m_b'(\lambda)$, $m_b(\lambda) \neq m_b'(\lambda)$. Then the polynomial $\widetilde{m}_b(\lambda) = m_b(\lambda) - m'(\lambda)$ obviously satisfies the condition $\widetilde{m}_b(b) = 0$ and $\deg \widetilde{m}_b(\lambda) < \deg m_b(\lambda)$, which contradicts the minimality of the degree of the polynomial $m_b(\lambda)$. Thus, the minimal polynomial $m_b(\lambda)$ of the element b over the field $\mathbb{F}_q$ is uniquely determined.

Suppose now that the minimal polynomial $m_b(\lambda)$ is reducible, i.e., $m_b(\lambda) = m_b'(\lambda) \cdot m_b''(\lambda)$, where $m_b'(\lambda)$, $m_b''(\lambda)$ are monic polynomials such that $\deg m_b'(\lambda) < \deg m_b(\lambda)$ and $\deg m_b''(\lambda) < \deg m_b(\lambda)$. Then in the field $\mathbb{F}_{q'}$ the following relation holds: $m_b(b) = m_b'(b)m_b''(b) = 0$. Since the field does not contain zerodivisors, one has either $m_b'(b) = 0$ or $m_b''(b) = 0$. In both cases we get a contradiction with the minimality of the polynomial $m_b(\lambda)$. Thus, the polynomial $m_b(\lambda)$ is irreducible.

PROBLEM 1.122. Let $a(\lambda) \in \mathbb{F}_q[\lambda]$ be an irreducible polynomial over the field $\mathbb{F}_q$, $\mathbb{F}_q \subset \mathbb{F}_{q'}$, and let $b \in \mathbb{F}_{q'}$ be a root of the polynomial $a(\lambda)$. Prove that b is a root of a polynomial $c(\lambda) \in \mathbb{F}_q[\lambda]$ if and only if $a(\lambda)$ divides $c(\lambda)$.

THEOREM 1.123. *For any field $\mathbb{F}_q$ (q is power of a prime number) the polynomial $x^{q^n} - x$ is equal to the product of all normalized irreducible polynomials over $\mathbb{F}_q$ whose degrees divide n.*

PROOF. Let $a(\lambda)$ be a normalized irreducible polynomial over $\mathbb{F}_q$ of $\deg a(\lambda) = d$ and $d \mid n$. The case $a(\lambda) = \lambda$ is trivial. Suppose that $a(\lambda) \neq \lambda$. According to Lemma 1.64, we can generate the field $\mathbb{F}_{q^d}$ using the polynomial $a(\lambda)$. In this field there exists an element $b \neq 0$ for which $a(\lambda)$ is the minimal polynomial, i.e., $a(\lambda) = m_b(\lambda)$. Since b is also a root of the polynomial $\lambda^{q^d-1} - 1$, we infer that $a(\lambda)$ divides $\lambda^{q^d-1} - 1$. To complete the proof we need the following lemmas.

LEMMA 1.124. *If t, r, and s are natural numbers, $t \geq 2$, $r \geq 1$, $s \geq 1$, then $t^s - 1$ divides $t^r - 1$ if and only if s divides r.*

PROOF. One has $r = su + v$, where $0 \leq v < s$. Then

$$\frac{t^r - 1}{t^s - 1} = t^v \cdot \frac{t^{us} - 1}{t^s - 1} + \frac{t^v - 1}{t^s - 1}.$$

Suppose that $t^s - 1$ divides $t^r - 1$. Since $t^s - 1$ divides $t^{us} - 1$ and the last summand is less than one, the equality is possible only if $v = 0$. Therefore, s divides r.

The converse statement is obvious. □

LEMMA 1.125. *Over an arbitrary field $\mathbb{F}_q$, the polynomial $\lambda^s - 1$ divides the polynomial $\lambda^r - 1$ if and only if s divides r.*

PROOF. One has $r = su + v$, $0 \leq v < s$. Hence,

$$\lambda^{su+v} - 1 = \lambda^v(\lambda^{us} - 1) + (\lambda^v - 1).$$

Suppose that $\lambda^s - 1$ divides $\lambda^r - 1$. Since $\lambda^s - 1$ divides $\lambda^{us} - 1$ and $\deg \lambda^v - 1 < s$, the last equality is possible only if $v = 0$. Therefore, s divides r.

The converse is obvious. □

According to Lemmas 1.124 and 1.125, $q^d - 1$ divides $q^n - 1$ and $\lambda^{q^d-1} - 1$ divides $\lambda^{q^n-1} - 1$. Therefore, $a(\lambda)$ divides $\lambda^{q^n} - \lambda$.

Conversely, let $a(\lambda)$ divide $\lambda^{q^n} - \lambda$, $\deg a(\lambda) = d$. Suppose again that $a(\lambda) \neq \lambda$. We generate the field $\mathbb{F}_{q^d}$ using the polynomial $a(\lambda)$. Let $\alpha \in \mathbb{F}_{q^d}$ be a root of $a(\lambda)$ and β a primitive element of the field $\mathbb{F}_{q^d}$, so that

$$\beta = b_0 + b_1\alpha + \cdots + b_{d-1}\alpha^{d-1}, \quad b_i \in \mathbb{F}_q, \quad i = 0, 1, \ldots, d-1. \tag{1.9}$$

Since $a(\alpha) = 0$, one has $\alpha^{q^n} = \alpha$. According to Lemma 1.95, it follows from (1.9) that $\beta^{q^n} = \beta$. Therefore, $\beta^{q^n-1} = 1$ and the order of the element β, which is equal to $q^d - 1$, must divide $q^n - 1$. Thus, d divides n by Lemma 1.124. □

PROBLEM 1.126. Let $\mathbb{F}_q \subset \mathbb{F}_{q^n}$, let β be a primitive element of the field $\mathbb{F}_{q^n}$, and let $m_\beta(\lambda)$ be its minimal polynomial over $\mathbb{F}_q$. Prove that $\deg m_\beta(\lambda) = n$.

DEFINITION 1.127. Let $\mathbb{F}_{q^n}$ be an extension of the field $\mathbb{F}_q$, and suppose that $\alpha \in \mathbb{F}_{q^n}$. Then the elements

$$\alpha, \alpha^q, \ldots, \alpha^{q^{n-1}}$$

are called the *conjugates* of the element α with respect to the field $\mathbb{F}_q$.

The conjugate elements of an element $\alpha \in \mathbb{F}_{q^n}$ over the field $\mathbb{F}_q$ are all distinct if and only if the minimal polynomial of the element α over $\mathbb{F}_q$ has degree n. If this is not true, then the degree d of the minimal polynomial of the element α is a proper divisor of the number n, and then among the conjugate elements of α with respect to $\mathbb{F}_q$, only the elements $\alpha, \alpha^q, \ldots, \alpha^{q^{d-1}}$ will be distinct; each of them is repeated in the sequence of conjugate elements n/d times.

DEFINITION 1.128. An automorphism of the field $\mathbb{F}_{q^n}$ which fixes all the elements of the field $\mathbb{F}_q$ is called an *automorphism of the field* $\mathbb{F}_{q^n}$ *over the field* $\mathbb{F}_q$.

THEOREM 1.129. *The automorphisms of the field $\mathbb{F}_{q^n}$ over the field $\mathbb{F}_q$ form a cyclic group with respect to composition. This group is called the* Galois group *of order n. The elements of this group are automorphisms $\sigma_0, \sigma_1, \ldots, \sigma_{n-1}$ (Frobenius automorphisms) defined by the formula $\sigma_j(a) = a^{q^j}$ for $a \in \mathbb{F}_{q^n}$, $0 \le j \le n-1$.*

PROOF. It is easy to check that for any j, $0 \le j \le n-1$, the σ_j are automorphisms of $\mathbb{F}_{q^n}$ over $\mathbb{F}_q$. Now let σ be an arbitrary automorphism of $\mathbb{F}_{q^n}$ over $\mathbb{F}_q$ and let b be a primitive element of the field $\mathbb{F}_{q^n}$; denote by $m_b(\lambda) = \lambda^n + m_{n-1}\lambda^{n-1} + \cdots + m_0 \in \mathbb{F}_q[\lambda]$ its minimal polynomial over $\mathbb{F}_q$. Then

$$0 = \sigma(b^n + m_{n-1}b^{n-1} + \cdots + m_0) = \sigma(b)^n + m_{n-1}\sigma(b)^{n-1} + \cdots + m_0,$$

so the element $\sigma(b) \in \mathbb{F}_{q^n}$ is also a root of the polynomial $m_b(\lambda)$. Since b is a root of $m_b(\lambda)$, it is easy to check using Lemma 1.95 that $b^q, b^{q^2}, \ldots, b^{q^{n-1}}$ are also roots of the polynomial $m_b(\lambda)$. We will show that they are all distinct. Indeed, suppose that $b^{q^j} = b^{q^k}$ for some j, k, $0 \le j < k \le n-1$. Raising this equality to the power of q^{n-k}, we get

$$b^{q^{n-k+j}} = b^{q^n} = b.$$

In view of Problem 1.122, the polynomial $m_b(\lambda)$ divides the polynomial $\lambda^{q^{n-k+j}} - \lambda$. By Theorem 1.123, this is possible only in the case where n divides $n-k+j$. Since $0 < n-k+j < n$, we arrive at a contradiction. Therefore, $b^q, b^{q^2}, \ldots, b^{q^{n-1}}$ are all roots of the polynomial $m_b(\lambda)$, and $\sigma(b) = b^{q^j}$ for some j, $0 \le j < k \le n-1$. Since σ is an automorphism, for any $a \in \mathbb{F}_{q^n}$ one has $\sigma(a) = a^{q^j}$ and $\sigma = \sigma_j$. It is clear that the automorphisms $\{\sigma_j \mid j = 0, 1, \ldots, n-1\}$ form a cyclic group with respect to composition, generated by σ_1. □

Let $m_\alpha(\lambda)$ be the minimal polynomial of the element $\alpha \in \mathbb{F}_{q^n}$ over the field $\mathbb{F}_q$. Its degree $d = \deg m_\alpha(\lambda)$ is a divisor of the number n.

DEFINITION 1.130. The polynomial $a(\lambda) = m_\alpha(\lambda)^{\frac{n}{d}}$ is called the *characteristic polynomial of the element* α over $\mathbb{F}_{q^n}$.

The observation made right after Definition 1.127 and Theorem 1.129 show that the roots of the polynomial $m_\alpha(\lambda)$ in $\mathbb{F}_{q^n}$ are the elements $\alpha, \alpha^q, \ldots, \alpha^{q^{d-1}}$, and the roots of the characteristic polynomial $a(\lambda)$ in the field $\mathbb{F}_{q^n}$ are exactly the conjugates of α over $\mathbb{F}_q$. Therefore,

$$a(\lambda) = \lambda^n + a_{n-1}\lambda^{n-1} + \cdots + a_0 = (\lambda - \alpha)(\lambda - \alpha^q)\cdots(\lambda - \alpha^{q^{n-1}}).$$

Comparing the coefficients in the last equality, one gets the following new expressions for the trace and the norm:

$$\mathrm{Tr}_{q^n/q}(\alpha) = -a_{n-1}; \quad \mathrm{N}_{q^n/q}(\alpha) = (-1)^n a_0.$$

Now we introduce another parameter which characterizes a polynomial.

LEMMA 1.131. *If $a(\lambda) \in \mathbb{F}_q[\lambda]$ is a polynomial of degree $n \geq 1$ and $a(0) \neq 0$, then there exists a natural number $e \leq q^n - 1$, for which the polynomial $\lambda^e - 1$ is divisible by $a(\lambda)$.*

PROOF. There are $q^n - 1$ nonzero elements in the quotient ring $\mathbb{F}_q[\lambda]/(a(\lambda))$ (i.e., residue classes modulo the ideal $(a(\lambda))$). Since each of the q^n residue classes $\lambda^j + (a(\lambda))$, $j = 0, 1, \ldots, q^n - 1$, is a nonzero element of this quotient ring, there exist integers r and s, $0 \leq r < s \leq q^n - 1$ such that $\lambda^s \equiv \lambda^r \pmod{a(\lambda)}$. Since $\gcd(\lambda, a(\lambda)) = 1$, one has $\lambda^{s-r} \equiv 1 \pmod{a(\lambda)}$. Thus, the polynomial $\lambda^{s-r} - 1$ is divisible by $a(\lambda)$, where $0 < s - r \leq q^n - 1$. □

Since the polynomial $\lambda - 1$ is divisible by any nonzero constant polynomial, we can include constant polynomials in the following definition.

DEFINITION 1.132. Let $a(\lambda) \in \mathbb{F}_q[\lambda]$ be a nonzero polynomial. If $a(0) \neq 0$, then the smallest natural number e for which polynomial $a(\lambda)$ divides $\lambda^e - 1$, is called the *period* (or the *order*) of the polynomial $a(\lambda)$ and is denoted by $\operatorname{ord}(a(\lambda))$. If $a(0) = 0$, then the polynomial $a(\lambda)$ is uniquely representable as $a(\lambda) = \lambda^k b(\lambda)$, where k is natural number, $b(\lambda) \in \mathbb{F}_q[\lambda]$, $b(0) \neq 0$. In this case order $\operatorname{ord}(a(\lambda))$ of the polynomial $a(\lambda)$ is equal to the order $\operatorname{ord}(b(\lambda))$ of the polynomial $b(\lambda)$.

LEMMA 1.133. *Let $a(\lambda) \in \mathbb{F}_q[\lambda]$ be an irreducible polynomial of degree n for which $a(0) \neq 0$. The order of this polynomial coincides with the order of any root of this polynomial in the multiplicative group $\mathbb{F}_{q^n}^*$ of the field $\mathbb{F}_{q^n}$.*

PROOF. According to Theorem 1.123, the field $\mathbb{F}_{q^n}$ contains all the roots of the polynomial $a(\lambda)$. If α is a root of $a(\lambda)$, then Theorem 1.129 shows that the elements $\alpha^q, \ldots, \alpha^{q^{n-1}}$ are also roots of the polynomial $a(\lambda)$. Since $\gcd(q, q^n - 1) = 1$, the orders of the elements $\alpha^q, \ldots, \alpha^{q^{n-1}}$ in the group $\mathbb{F}_{q^n}^*$ coincide.

Let e be the order of the element α in the group $\mathbb{F}_{q^n}^*$. According to Problem 1.122, the equality $a^e = 1$ is possible if and only if $a(\lambda)$ divides $\lambda^e - 1$. The result now follows from the definitions of order of an element and of a polynomial. □

PROBLEM 1.134. Prove that the period of an irreducible polynomial $a(\lambda) \in \mathbb{F}_q[\lambda]$ of degree n over $\mathbb{F}_q$ divides the number $q^n - 1$.

There is a relation between the orders of two polynomials that are "dual" to each other (see the following definition).

DEFINITION 1.135. For a polynomial $a(\lambda) = a_0 + a_1\lambda + \cdots + a_{n-1}\lambda^{n-1} + a_n\lambda^n \in \mathbb{F}_q[\lambda]$, $\deg a(\lambda) = n$, the polynomial

$$a^*(\lambda) = \lambda^n a\Big(\frac{1}{\lambda}\Big) = a_0\lambda^n + a_1\lambda^{n-1} + \cdots + a_{n-1}\lambda + a_n$$

is called *dual* to the polynomial $a(\lambda)$.

LEMMA 1.136. *Let $a(\lambda)$ be a nonzero polynomial from $\mathbb{F}_q[\lambda]$; denote the dual polynomial by $a^*(\lambda)$. Then* $\operatorname{ord}(a(\lambda)) = \operatorname{ord}(a^*(\lambda))$.

PROOF. If $a(0) \neq 0$, then the statement follows from the fact that the polynomial $a(\lambda)$ divides $\lambda^e - 1$ if and only if the polynomial $a^*(\lambda)$ divides $\lambda^e - 1$.

If $a(0) = 0$, then $a(\lambda) = \lambda^k b(\lambda)$, where k is a natural number, $b(\lambda) \in \mathbb{F}_q[\lambda]$, and $b(0) \neq 0$. From what we have proved above and the definition of period it follows that $\operatorname{ord}(a(\lambda)) = \operatorname{ord}(b(\lambda)) = \operatorname{ord}(b^*(\lambda)) = \operatorname{ord}(a^*(\lambda))$, where the last identity follows from the identity $a^*(\lambda) = b^*(\lambda)$. □

As the period of any polynomial of degree n from $\mathbb{F}_q[\lambda]$ is at most $q^n - 1$, it is interesting to look closely at the polynomials for which this bound is attained.

DEFINITION 1.137. A polynomial $a(\lambda) \in \mathbb{F}_q[\lambda]$ of degree $n \geq 1$ is called a *primitive polynomial* over the field $\mathbb{F}_q$ if it is a minimal polynomial over $\mathbb{F}_q$ of a primitive element of an extension $\mathbb{F}_{q^n}$ of the field $\mathbb{F}_q$.

Thus, a primitive polynomial of degree n over $\mathbb{F}_q$ is a monic polynomial which is irreducible over $\mathbb{F}_q$ and has a root $\alpha \in \mathbb{F}_{q^n}$ that generates the multiplicative group $\mathbb{F}^*_{q^n}$ of the field $\mathbb{F}_{q^n}$.

THEOREM 1.138. *A polynomial $a(\lambda) \in \mathbb{F}_q[\lambda]$ of degree n is primitive over $\mathbb{F}_q$ if and only if it is a normalized polynomial such that $a(0) \neq 0$ and $\operatorname{ord}(a(\lambda)) = q^n - 1$.*

PROOF. Let $a(\lambda)$ be a primitive polynomial over $\mathbb{F}_q$, i.e., a monic polynomial satisfying the condition $a(0) \neq 0$. Since $a(\lambda)$ is irreducible over $\mathbb{F}_q$, it follows from Lemma 1.133 and from the fact that its root is a primitive element of the extension $\mathbb{F}_{q^n}$ of the field $\mathbb{F}_q$, that $\operatorname{ord}(a(\lambda)) = q^n - 1$.

Conversely, let $a(\lambda) \in \mathbb{F}_q[\lambda]$ be a normalized polynomial such that $a(0) \neq 0$ and $\operatorname{ord}(a(\lambda)) = q^n - 1$. Therefore, $n \geq 1$. We will show that the polynomial $a(\lambda)$ is irreducible over $\mathbb{F}_q$. Assume the converse. Then either $a(\lambda)$ is a power of an irreducible polynomial or it is a product of two coprime polynomials of a positive degree. In the first case $a(\lambda) = b^k(\lambda)$, where $k \geq 2$, the polynomial $b(\lambda) \in \mathbb{F}_q[\lambda]$ is irreducible over $\mathbb{F}_q$, $b(0) \neq 0$. Since $a(\lambda)$ divides $\lambda^{q^n-1} - 1$ and since $\lambda^{q^n-1} - 1$ is the product of all normalized irreducible polynomials from $\mathbb{F}_q[\lambda]$ whose degree divides n, except the polynomial λ, it follows that the polynomial $a(\lambda)$ cannot be a power of an irreducible polynomial. In the second case $a(\lambda) = b_1(\lambda) \cdot b_2(\lambda)$, where $b_1(\lambda)$, $b_2(\lambda)$, $b_1(0) \neq 0$, $b_2(0) \neq 0$, are coprime polynomials from $\mathbb{F}_q[\lambda]$ of positive degrees n_1 and n_2 respectively. If $\operatorname{ord}(b_i(\lambda)) = e_i$, $i = 1, 2$, then $b_i(\lambda)$ divides $\lambda^{e_i} - 1$, $i = 1, 2$. By virtue of Lemma 1.125 we may assume that $\lambda^{e_i} - 1$ divides $\lambda^{e_i e_2} - 1$, $i = 1, 2$, but since $\gcd(b_1(\lambda), b_2(\lambda)) = 1$, we conclude that $b_1(\lambda)b_2(\lambda)$ divides $\lambda^{e_1 e_2} - 1$, whence $\operatorname{ord}(a(\lambda)) \leq \operatorname{ord}(b_1(\lambda))\operatorname{ord}(b_2(\lambda))$. According to Lemma 1.131, the inequalities $e_1 \leq q^{n_1} - 1$, $e_2 \leq q^{n_2} - 1$ hold, which gives $\operatorname{ord}(a(\lambda)) \leq (q^{n_1} - 1)(q^{n_2} - 1) < q^{n_1+n_2} - 1 = q^n - 1$. We arrive at a contradiction. Therefore, the polynomial $a(\lambda)$ is irreducible. By Lemma 1.133, the orders of all its roots are equal to $q^n - 1$, i.e., it is a minimal polynomial of a primitive element over $\mathbb{F}_q$, whence a primitive polynomial. □

Let us denote by $\mathcal{N}_q(d)$ the number of normalized irreducible polynomials in $\mathbb{F}_q[\lambda]$ of degree d. Then the identity $q^n = \sum_{d|n} d\mathcal{N}_q(d)$ follows from Theorem 1.123, where the summation is over all positive divisors d of the number n. To find the number $\mathcal{N}_q(d)$ we will use the Möbius function.

DEFINITION 1.139. The *Möbius function* is the function μ defined on the set of natural numbers $\mathbb{N}$ as follows:

$$\mu(n) = \begin{cases} 1 & \text{if } n = 1; \\ (-1)^k & \text{if } n \text{ is a product of } k \text{ different prime numbers;} \\ 0 & \text{if } n \text{ is divisible by a square of a prime.} \end{cases}$$

LEMMA 1.140. *If $n \in \mathbb{N}$, then*

$$\sum_{d|n} \mu(d) = \begin{cases} 1 & \text{if } n = 1; \\ 0 & \text{if } n > 1. \end{cases}$$

PROOF. The case $n = 1$ is trivial. Suppose $n > 1$, and let $n = p_1 e^{t_1} \cdot p_2^{t_2} \cdots p_s^{t_s}$ be the canonical factorization of n into prime factors, where the p_i, $i = 1, 2, \ldots, s$, are different primes. It is clear that the divisors d of the number n for which $\mu(d) \neq 0$ are products of different prime numbers occurring in this factorization. Hence,

$$\begin{aligned} \sum_{d|n} \mu(d) &= \mu(1) + \sum_{i=1}^{s} \mu(p_i) + \sum_{1 \le i_1 < i_2 \le s} \mu(p_{i_1} p_{i_2}) + \cdots + \mu(p_1 p_2 \cdots p_s) \\ &= 1 + \binom{s}{1}(-1) + \binom{s}{2}(-1)^2 + \cdots + \binom{s}{s}(-1)^s \\ &= [1 + (-1)]^s = 0. \end{aligned}$$

□

LEMMA 1.141. *Suppose that h and H are two functions defined over $\mathbb{N}$ with values in an abelian group $\mathfrak{G}$ with additive operation. Then the equality*

$$H(n) = \sum_{d|n} h(d) \quad \textit{for all } n \in \mathbb{N}$$

holds if and only if

$$h(n) = \sum_{d|n} \mu\left(\frac{n}{d}\right) H(d) = \sum_{d|n} \mu(d) H\left(\frac{n}{d}\right) \qquad \textit{for all } n \in \mathbb{N}.$$

PROOF. Using the expressions for the function $H(n)$ and Lemma 1.140, we get

$$\begin{aligned} \sum_{d|n} \mu\left(\frac{n}{d}\right) H(d) &= \sum_{d|n} \mu(d) H\left(\frac{n}{d}\right) = \sum_{d|n} \mu(d) \sum_{c|\frac{n}{d}} h(c) \\ &= \sum_{c|n} \sum_{d|\frac{n}{c}} \mu(d) h(c) = \sum_{c|n} h(c) \sum_{d|\frac{n}{c}} \mu(d) = h(n) \end{aligned}$$

for all $n \in \mathbb{N}$.

The converse statement can be proven similarly. □

LEMMA 1.142. *The number $\mathcal{N}_q(n)$ of monic polynomials of degree n in the ring $\mathbb{F}_q[\lambda]$ is given by the following formula:*

$$\mathcal{N}_q(n) = \frac{1}{n} \sum_{d|n} \mu\left(\frac{n}{d}\right) q^d = \frac{1}{n} \sum_{d|n} \mu(d) q^{n/d}.$$

PROOF. Applying Lemma 1.141 for the case $\mathfrak{G} = \mathbb{Z}$ (the group of integers), $h(n) = n\mathcal{N}_q(n)$, and $H(n) = q^n$ for all $n \in \mathbb{N}$, we get the lemma. □

Earlier, in Theorem 1.94, when proving the existence of the field of p^n elements for an arbitrary prime p and natural number n, we used without proof the fact that there exists an irreducible polynomial of degree n over the field $\mathbb{F}_p$. Now, applying the above results, we will prove a more general statement.

LEMMA 1.143. *For any finite field $\mathbb{F}_q$ and any natural number n there exists an irreducible polynomial of degree n in the ring $\mathbb{F}_q[\lambda]$.*

PROOF. Taking into account that $\mu(1) = 1$ and $\mu(d) \geq -1$ for all $d \in \mathbb{N}$ and using Lemma 1.142, we get

$$\mathcal{N}_q(n) \geq \frac{1}{n}(q^n - q^{n-1} - q^{n-2} - \cdots - q) = \frac{1}{n}\left(q^n - \frac{q^n - q}{q-1}\right) > 0. \qquad \square$$

We conclude this section with some remarks on divisors of the polynomial $\lambda^n - 1$ over $\mathbb{F}_q$ and the structure of the ring $R_{q,n} = \mathbb{F}_q[\lambda]/(\lambda^n - 1)$. They will be helpful when we will study the construction of cyclic codes.

Suppose that $\mathbb{F}_q$ is a finite field, n is a natural number and $\gcd(q, n) = 1$. Then the smallest natural number l for which

$$q^l \equiv 1 \pmod{n},$$

is called the *index* of q modulo n.

According to Lemma 1.125, the polynomial $\lambda^n - 1$ divides the polynomial $\lambda^{q^l-1} - 1$ and does not divide $\lambda^{q^s-1} - 1$ for any s, $0 < s < l$. Since the roots of the polynomial $\lambda^{q^l} - \lambda = \lambda(\lambda^{q^l-1} - 1)$ are all elements of the field $\mathbb{F}_{q^l}$ (see Lemma 1.93), the roots of the polynomial $\lambda^n - 1$, called the *roots of unity of degree* n, lie in the extension $\mathbb{F}_{q^l}$ of the field $\mathbb{F}_q$ and do not lie in any smaller field. The derivative of the polynomial $\lambda^n - 1$ is equal to $n\lambda^{n-1}$, and it is coprime with $\lambda^n - 1$, since $\gcd(q, n) = 1$. Thus, the polynomial $\lambda^n - 1$ has (see Problem 1.70) n different roots (including the unity).

LEMMA 1.144. *Let $a \in \mathbb{F}_{q^l}$ be an arbitrary root of the polynomial $\lambda^n - 1$. Then*

$$\sum_{i=0}^{n-1} a^i = \begin{cases} 0 & \text{if } a \neq 1; \\ n & \text{if } a = 1. \end{cases}$$

PROOF. If $a = 1$, the sum is equal to n. We assume that $a \neq 1$. Then

$$\sum_{i=0}^{n-1} a^i = \frac{1 - a^n}{1 - a} = 0. \qquad \square$$

PROBLEM 1.145. Prove that the roots of the polynomial $\lambda^n - 1$ have the following properties:

1. They form a cyclic subgroup of the group $\mathbb{F}_{q^l}^*$; i.e., in $\mathbb{F}_{q^l}$ there exists an element α (called a *primitive nth root of unity*) such that

$$\lambda^n - 1 = \prod_{i=0}^{n-1}(\lambda - \alpha^i).$$

2. They form the whole multiplicative group of the field if and only if $n = q^l - 1$.

From now on in this section, α will denote a primitive nth root of unity.

We proceed now with the investigation of the factorization of the polynomial $\lambda^n - 1$. First we introduce the notion of cyclotomic class.

DEFINITION 1.146. Let n be a natural number and q a power of a prime number. A *cyclotomic class* $\mathcal{C}_{n,s}$ modulo n over $\mathbb{F}_q$, where $s \in \{0, 1, \ldots, n-1\}$, is the set

$$\mathcal{C}_{n,s} = \{s, sq, sq^2, \ldots, sq^{l_s - 1}\}$$

($sq^{l_s} \equiv s \pmod{n}$).

We will index cyclotomic classes by their smallest elements (this choice of representatives is convenient but not unique). The set of all integers modulo n is the union of disjoint cyclotomic classes

$$\{0, 1, 2, \ldots, n-1\} = \bigcup_s \mathcal{C}_{n,s},$$

where s runs over the set of representatives of cyclotomic classes modulo n.

PROBLEM 1.147. Prove that if $\gcd(s, n) = 1$, then $l_s = l$, where l is the exponent to which q belongs modulo n.

PROBLEM 1.148. Let $a(\lambda) = \prod_{i \in K}(\lambda - \alpha^i)$, where $K \subset \{0, 1, 2, \ldots, n-1\}$. Prove that the coefficients of the polynomial $a(\lambda)$ lie in the field $\mathbb{F}_q$ if and only if for any $i \in K$ the inclusion $iq \pmod{n} \in K$ holds.

Let α^s be a root of $\lambda^n - 1$ and let its minimal polynomial be $m_{\alpha^s}(\lambda) = m_s(\lambda)$ (since the element α is fixed, we will use the brief notation for the minimal polynomial). As was shown in Theorem 1.129, the elements $\{\alpha^{sq^i} \mid i = 0, 1, \ldots, l_s - 1\}$ are roots of the polynomial $m_s(\lambda)$, so

$$m_s(\lambda) = \prod_{i \in \mathcal{C}_{n,s}} (\lambda - \alpha^i).$$

Hence,

$$\lambda^n - 1 = \prod_s m_s(\lambda),$$

where s runs over the whole set of representatives of the cyclotomic classes modulo n.

We will now describe the structure of the ring $R_{q,n} = \mathbb{F}_q[\lambda]/(\lambda^n - 1)$. According to Example 1.120, for any polynomial $\theta_s(\lambda) = (\lambda^n - 1)/m_s(\lambda)$ there exists a polynomial u_s such that $e_s(\lambda) = u_s(\lambda)\theta_s(\lambda)$ is an idempotent in the ring. Besides, $\sum_s e_s(\lambda) = 1$ and $e_{s_1}(\lambda)e_{s_2}(\lambda) = 0$ for $s_1 \neq s_2$ in the ring $R_{q,n}$. Finally, the ring $R_{q,n}$ itself is a direct sum of ideals $J_s(e_s(\lambda))$, where s runs over the whole set of representatives of the cyclotomic classes modulo n.

Comments to Chapter 1

Trying as we may to make the exposition self-contained, the reader may encounter mathematical questions that are not answered in this chapter. To help such a reader, we list some references here.

Answers to various algebraic questions which the reader may have can be found in the remarkable books [**219, 406**]. A comprehensive reference on finite fields (with applications) is [**225**]. A shorter exposition may be found in [**9, 264**]. The general theory of fields (not only for the case of finite fields) can be found in [**423**]. All the necessary definitions and facts of group theory are contained in [**155**]. Information about classical groups, which play an important role in the classification of Boolean functions, in the studies of Reed–Muller codes, and cryptology, can be found in [**107**]; for permutation groups see [**376, 410**].

To become familiar with algorithmic aspects of realization of the operations over different algebraic structures, see, for example, [**400, 284**].

Rudiments of probability theory, statistics, and information theory are used only in two chapters. The necessary material can be found, for example, in [**353, 94, 123, 223**]. Information about coding theory problems will be presented in Chapters 4, 5.

CHAPTER 2

Boolean Functions

2.1. Basic concepts and definitions

Let $\mathbb{F}_q$ ($q = p^n$, p prime) be a finite field of q elements. In this book we will work with finite fields of characteristic 2.

For an arbitrary natural number $n \in \mathbb{N}$ put $V_n = \mathbb{F}_2 \times \cdots \times \mathbb{F}_2$ (n times); this is the vector space of vectors of length n with entries from the field $\mathbb{F}_2$ (the binary field). We will denote the operations in the the field $\mathbb{F}_2$ by $\cdot$ for multiplication modulo 2 and $\oplus$ for addition modulo 2. We will also need the operation $\vee$ (see Table 2.1).

TABLE 2.1

$\cdot$	0	1
0	0	0
1	0	1

$\oplus$	0	1
0	0	1
1	1	0

$\vee$	0	1
0	0	1
1	1	1

For convenience, we will often omit the sign $\cdot$ in algebraic expressions, implying that $a \cdot b = ab$. For addition of vectors from V_n modulo 2, we will also use the notation $\oplus$. It will become clear from the context in which space the operation is performed. Let us fix the notation for two vectors from V_n:

$$0 = (0, \dots, 0)^\top, \qquad 1 = (1, \dots, 1)^\top.$$

We will call the elements $\boldsymbol{x}$ of the vector space V_n "vectors". We will always have in mind column vectors.

Choose a basis of the space V_n as follows:

$$\boldsymbol{e}_1 = (1, 0, \dots, 0)^\top, \quad \boldsymbol{e}_2 = (0, 1, \dots, 0)^\top, \quad \dots, \quad \boldsymbol{e}_n = (0, 0, \dots, 1)^\top. \tag{2.1}$$

Then the vector $\boldsymbol{x} \in V_n$ can be represented as

$$\boldsymbol{x} = x^{(1)} \cdot \boldsymbol{e}_1 \oplus \cdots \oplus x^{(n)} \cdot \boldsymbol{e}_n, \tag{2.2}$$

where the coefficients $x^{(i)}$ ($i = 1, \dots, n$) from the field $\mathbb{F}_2$ are called the coordinates of the vector $\boldsymbol{x}$ with respect to the given basis. It is clear that the i-component of the vector $\boldsymbol{x}$ and its ith coordinate in the basis (2.1) coincide. We will not distinguish between them from now on; however, this may not be true if a different basis is used.

DEFINITION 2.1. A *Boolean function* of n variables is a mapping from V_n to $\mathbb{F}_2$. We will denote the set of all Boolean functions of n variables by $\mathcal{F}_n$.

We will use the following notation for the value of a Boolean function f at a vector $\boldsymbol{x} \in V_n$:

$$f(\boldsymbol{x}) = f(x^{(1)} \cdot \boldsymbol{e}_1 \oplus \cdots \oplus x^{(n)} \cdot \boldsymbol{e}_n) = f(x^{(1)}, \dots, x^{(n)}).$$

On the right-hand side of this equation, we write not the transposed vector $\boldsymbol{x}$, but the row vector formed by the coordinates of $\boldsymbol{x}$ in the basis (2.1).

We will denote by $\boldsymbol{x}^\top$ the transposed vector $\boldsymbol{x}$ (the same notation will be used for arbitrary matrices). In particular, we will write

$$\boldsymbol{x} = (x^{(1)}, \ldots, x^{(n)})^\top.$$

To define a function $f(\boldsymbol{x}) = f(x^{(1)}, \ldots, x^{(n)})$ it suffices to know its values at each vector $\boldsymbol{x}$, i.e., to fill Table 2.2.

TABLE 2.2

$x^{(1)}$	$x^{(2)}$	$\ldots$	$x^{(n-1)}$	$x^{(n)}$	$f(x^{(1)}, x^{(2)}, \ldots, x^{(n-1)}, x^{(n)})$
0	0	$\ldots$	0	0	$f(0, 0, \ldots, 0, 0)$
1	0	$\ldots$	0	0	$f(1, 0, \ldots, 0, 0)$
0	1	$\ldots$	0	0	$f(0, 1, \ldots, 0, 0)$
$\ldots$	$\ldots$	$\ldots$	$\ldots$	$\ldots$	$\ldots$
1	1	$\ldots$	1	1	$f(1, 1, \ldots, 1, 1)$

We will call the coordinates $x^{(1)}, x^{(2)}, \ldots, x^{(n)}$ the variables or the arguments of the function f. For convenience, we will use the lexicographic order to place the values of the variables in this table, also called a *truth table*. This method of description of a Boolean function is called the tabular method. It is clear that there is a 1-1 correspondence between the sets $\mathcal{F}_n$ and V_{2^n}. That is why the following proposition holds.

PROPOSITION 2.2. *The cardinality $\#\mathcal{F}_n$ of the set of all Boolean functions of n variables $x^{(1)}, x^{(2)}, \ldots, x^{(n)}$ equals 2^{2^n}.*

As we increase the number of variables, the table defining the function becomes larger and more complicated. The number of functions of n variables grows fast:

$$\#\mathcal{F}_1 = 4, \qquad \#\mathcal{F}_2 = 16, \qquad \#\mathcal{F}_3 = 256, \qquad \#\mathcal{F}_4 = 65536, \qquad \ldots$$

The set $\mathcal{F}_n$ contains functions that actually depend on a smaller number of variables. For convenience, we adopt the following definition.

DEFINITION 2.3. A function $f(x^{(1)}, x^{(2)}, \ldots, x^{(i-1)}, x^{(i)}, x^{(i+1)}, \ldots, x^{(n)}) \in \mathcal{F}_n$ *essentially depends* on the variable $x^{(i)}$ if there exist values $\alpha^{(1)}, \alpha^{(2)}, \ldots, \alpha^{(i-1)}$, $\alpha^{(i+1)}, \ldots, \alpha^{(n)}$ of the variables $x^{(1)}$, $x^{(2)}, \ldots, x^{(i-1)}$, $x^{(i+1)}, \ldots, x^{(n)}$ such that

$$f(\alpha^{(1)}, \alpha^{(2)}, \ldots, \alpha^{(i-1)}, 0, \alpha^{(i+1)}, \ldots, \alpha^{(n)}) \neq f(\alpha^{(1)}, \alpha^{(2)}, \ldots, \alpha^{(i-1)}, 1, \alpha^{(i+1)}, \ldots, \alpha^{(n)}).$$

In this case, the variable $x^{(i)}$ is called *essential*. If $x^{(i)}$ is not an essential variable, it is called *nonessential*, or *fictitious*.

Let the variable $x^{(i)}$ be nonessential for the function $f(x^{(1)}, \ldots, x^{(n)})$. From the table of the function $f(x^{(1)}, \ldots, x^{(n)})$ we generate a new one by deleting all rows starting with $\alpha^{(1)}, \alpha^{(2)}, \ldots, \alpha^{(i-1)}, 1, \alpha^{(i+1)}, \ldots, \alpha^{(n)}$ and the column for the variable $x^{(i)}$. The new table will correspond to a function $g(x^{(1)}, x^{(2)}, \dots, x^{(i-1)}, x^{(i+1)}, \dots, x^{(n)})$ of $n-1$ variables. We will say that the function $g(x^{(1)}, x^{(2)}, \dots, x^{(i-1)}, x^{(i+1)}, \dots, x^{(n)})$ is obtained from the function $f(x^{(1)}, \dots, x^{(n)})$ by deleting the fictitious variable $x^{(i)}$ and that the function $f(x^{(1)}, \dots, x^{(n)})$ is obtained from $g(x^{(1)}, x^{(2)}, \dots, x^{(i-1)}, x^{(i+1)}, \dots, x^{(n)})$ by adding the fictitious variable $x^{(i)}$.

DEFINITION 2.4. A function f' obtained by replacing some of the variables in f with constants 0 or 1, is called a *subfunction* of the Boolean function f.

If we replace the variables $x^{(i_1)}, \dots, x^{(i_s)}$ with constants $\sigma^{(i_1)}, \dots, \sigma^{(i_s)}$ ($s \leq n$) in a function f, then the function obtained is denoted by $f^{\sigma^{(i_1)}, \dots, \sigma^{(i_s)}}_{x^{(i_1)}, \dots, x^{(i_s)}}$. The truth table of the subfunction $f^{\sigma^{(i_1)}, \dots, \sigma^{(i_s)}}_{x^{(i_1)}, \dots, x^{(i_s)}}$ is obtained from the truth table of the function f by choosing the rows that start with

$$(\alpha^{(1)}, \dots, \alpha^{(i_1-1)}, \sigma^{(i_1)}, \alpha^{(i_1+1)}, \dots, \alpha^{(i_s-1)}, \sigma^{(i_s)}, \alpha^{(i_s+1)}, \dots, \alpha^{(n)})$$

and deleting the columns corresponding to the variables $x^{(i_1)}, \dots, x^{(i_s)}$.

THEOREM 2.5. *The set $\bar{\mathcal{F}}_n$ of functions of n variables essentially depending on all variables has cardinality*

$$\#\bar{\mathcal{F}}_n = \sum_{i=0}^{n} (-1)^i \binom{n}{i} 2^{2^{n-i}},$$

where $\binom{n}{i} = \frac{n!}{i!(n-i)!}$ are binomial coefficients.

PROOF. In the proof of the theorem we will use a well-known combinatorial identity for binomial coefficients:

$$\sum_{i=1}^{m} (-1)^{i+1} \binom{n}{i} \binom{n-i}{m-i} = \binom{n}{m}. \tag{2.3}$$

We will prove the theorem by induction on n.

Induction base. Obviously, when $n = 1$, two functions essentially depend on their variable and two other depend on it nonessentially.

Induction step. The number of functions of n variables essentially depending on all variables is equal to the difference between the number of all functions of n variables and the number of functions essentially depending on a smaller number of variables. Thus we get

$$\#\bar{\mathcal{F}}_n = 2^{2^n} - \binom{n}{1} \#\bar{\mathcal{F}}_{n-1} - \binom{n}{2} \#\bar{\mathcal{F}}_{n-2} - \dots - \binom{n}{n} \#\bar{\mathcal{F}}_0,$$

where $\binom{n}{k} \#\bar{\mathcal{F}}_{n-k}$ is the number of functions of n variables essentially depending on $n-k$ variables. By the induction hypothesis for $\#\bar{\mathcal{F}}_{n-k}$, $k = 1, 2, \dots, n$, one has

$$\#\bar{\mathcal{F}}_n = 2^{2^n} - \binom{n}{1} \sum_{i=0}^{n-1} (-1)^i \binom{n-1}{i} 2^{2^{n-1-i}} - \dots$$
$$- \binom{n}{k} \sum_{i=0}^{n-k} (-1)^i \binom{n-k}{i} 2^{2^{n-k-i}} - \dots - \binom{n}{n} \sum_{i=0}^{0} (-1)^i \binom{0}{i} 2^{2^{0-i}}.$$

Collecting the terms with equal powers of 2, we get

$$\begin{aligned}\#\bar{\mathcal{F}}_n &= 2^{2^n} - \left[\binom{n}{1}\binom{n-1}{0}\right]2^{2^{n-1}} - \cdots - \left[\sum_{i=1}^{k}(-1)^{i+k}\binom{n}{i}\binom{n-i}{k-i}\right]2^{2^{n-k}}\\ &\quad - \cdots - \left[\sum_{i=1}^{n}(-1)^{i+n}\binom{n}{i}\binom{n-i}{n-i}\right]2^{2^{n-n}}\\ &= 2^{2^n} + (-1)^1\left[\binom{n}{1}\binom{n-1}{0}\right]2^{2^{n-1}} + \cdots\\ &\quad + (-1)^k\left[\sum_{i=1}^{k}(-1)^{i+1}\binom{n}{i}\binom{n-i}{k-i}\right]2^{2^{n-k}} + \cdots\\ &\quad + (-1)^n\left[\sum_{i=1}^{n}(-1)^{i+1}\binom{n}{i}\binom{n-i}{n-i}\right]2^{2^{n-n}}.\end{aligned}$$

Using (2.3) for $m = 1, 2, \ldots, n$, we get

$$\begin{aligned}\#\bar{\mathcal{F}}_n &= \binom{n}{0}2^{2^n} + (-1)^1\binom{n}{1}2^{2^{n-1}}\\ &\quad + \cdots + (-1)^k\binom{n}{k}2^{2^{n-k}} + \cdots + (-1)^n\binom{n}{n}2^{2^{n-n}}\\ &= \sum_{i=0}^{n}(-1)^i\binom{n}{i}2^{2^{n-i}}.\end{aligned}$$

□

Consider polynomials from $\mathbb{F}_2[x^{(1)}, \ldots, x^{(n)}]$ in which the exponent of each variable is at most 1:

$$P(x^{(1)}, \ldots, x^{(n)}) = a_0 \oplus \left(\bigoplus_{i=1}^{n} a_i x^{(i)}\right) \oplus \left(\bigoplus_{1\le i<j\le n} a_{ij}x^{(i)}x^{(j)}\right) \oplus \cdots \oplus a_{12\ldots n}x^{(1)}\cdots x^{(n)}, \tag{2.4}$$

where $a_0, a_i, a_{ij}, \ldots, a_{12\ldots n}$ are elements of the field $\mathbb{F}_2$, or equivalently

$$P(x^{(1)}, \ldots, x^{(n)}) = \bigoplus_{(\alpha^{(1)},\ldots,\alpha^{(n)})^\top \in V_n} b(\alpha^{(1)}, \ldots, \alpha^{(n)})(x^{(1)})^{\alpha^{(1)}} \cdots (x^{(n)})^{\alpha^{(n)}}, \tag{2.5}$$

where $b(\alpha^{(1)}, \ldots, \alpha^{(n)}) \in \mathbb{F}_2$ for all $(\alpha^{(1)}, \ldots, \alpha^{(n)})^\top \in V_n$.

It is obvious that in (2.5), b is a Boolean function.

We will also use a brief notation

$$\boldsymbol{x}^{\boldsymbol{\alpha}} = (x^{(1)})^{\alpha^{(1)}} \cdots (x^{(n)})^{\alpha^{(n)}}, \tag{2.6}$$

where $\boldsymbol{x} = (x^{(1)}, \ldots, x^{(n)})^\top$, $\boldsymbol{\alpha} = (\alpha^{(1)}, \ldots, \alpha^{(n)})^\top$.

THEOREM 2.6. *Every function $f \in \mathcal{F}_n$ can be uniquely represented as a polynomial from $\mathbb{F}_2[x^{(1)}, \ldots, x^{(n)}]$ with powers of all variables less than or equal to 1:*

$$f(x^{(1)}, \ldots, x^{(n)}) = c_0 \oplus \left(\bigoplus_{i=1}^{n} c_i x^{(i)}\right) \oplus \left(\bigoplus_{1\le i<j\le n} c_{ij}x^{(i)}x^{(j)}\right) \oplus \cdots \oplus c_{12\ldots n}x^{(1)}\cdots x^{(n)}, \tag{2.7}$$

(Zhegalkin polynomial) or, in an equivalent form,

$$f(x^{(1)},\dots,x^{(n)}) = \bigoplus_{(\alpha^{(1)},\dots,\alpha^{(n)})^\top \in V_n} g_f(\alpha^{(1)},\dots,\alpha^{(n)})(x^{(1)})^{\alpha^{(1)}}\cdots(x^{(n)})^{\alpha^{(n)}} \tag{2.8}$$

(algebraic normal form, ANF).

PROOF. It is clear that every polynomial from the set described in the theorem defines a mapping from V_n to $\mathbb{F}_2$, i.e., a Boolean function. On the other hand, according to (2.7), (2.8), there are exactly 2^{2^n} polynomials, and different polynomials induce different mappings. □

With the notation (2.6) we can present formula (2.8) in the form

$$f(\boldsymbol{x}) = \bigoplus_{\boldsymbol{\alpha}\in V_n} g_f(\boldsymbol{\alpha})\boldsymbol{x}^{\boldsymbol{\alpha}}.$$

This 1-1 correspondence $f(\boldsymbol{x}) \leftrightarrow g_f(\boldsymbol{x})$ between Boolean functions $f(\boldsymbol{x})$ and $g_f(\boldsymbol{x})$ in (2.8) is sometimes called the *Möbius transform.* If a function $g_f(\boldsymbol{x})$ is given, then the function $f(\boldsymbol{x})$ can be obtained with the aid of (2.8). Let us find a formula for the inverse mapping.

Introduce a partial order on vectors from V_n. We will say that a vector $\boldsymbol{x}$ precedes a vector $\boldsymbol{y}$ (notation: $\boldsymbol{x} \preccurlyeq \boldsymbol{y}$) if $x^{(i)} \le y^{(i)}$ for all i, $1 \le i \le n$, in the usual arithmetical sense ($0 \le 0$, $0 \le 1$, $1 \le 1$). If vectors $\boldsymbol{x}$ and $\boldsymbol{y}$ are different and $\boldsymbol{x} \preccurlyeq \boldsymbol{y}$, we will say that $\boldsymbol{x}$ strictly precedes $\boldsymbol{y}$ and denote this by $\boldsymbol{x} \prec \boldsymbol{y}$.

PROBLEM 2.7. Prove that in (2.6) the following equality holds:

$$\boldsymbol{a}^{\boldsymbol{\varepsilon}} = \begin{cases} 1 & \text{if } \boldsymbol{\varepsilon} \preccurlyeq \boldsymbol{a};\\ 0 & \text{otherwise}\end{cases}$$

for all $\boldsymbol{\varepsilon}, \boldsymbol{a} \in V_n$.

PROBLEM 2.8. Prove that

$$\boldsymbol{x} \preccurlyeq \boldsymbol{y} \iff \boldsymbol{x}\oplus\boldsymbol{1} \succcurlyeq \boldsymbol{y}\oplus\boldsymbol{1}.$$

We will also define weights of vectors.

DEFINITION 2.9. The *(Hamming) weight* of a vector $\boldsymbol{x} = (x^{(1)},\dots,x^{(n)})^\top$ is the number of nonzero coordinates $x^{(i)}$; it is denoted by $\operatorname{wt}(\boldsymbol{x})$.

THEOREM 2.10. *If a function $f(\boldsymbol{x}) \in \mathcal{F}_n$ is given, then the function $g_f(\boldsymbol{u}) \in \mathcal{F}_n$ is obtained by the following formula:*

$$g_f(\boldsymbol{u}) = \bigoplus_{\boldsymbol{x}\in V_n,\ \boldsymbol{x}\preccurlyeq\boldsymbol{u}} f(\boldsymbol{x}).$$

PROOF. We will proceed by induction on the weight of the vector u. If $\boldsymbol{u} = \boldsymbol{0}$, then clearly $g_f(\boldsymbol{0}) = f(\boldsymbol{0})$. Let the theorem hold for all vectors of weight less than $(\operatorname{wt}\boldsymbol{u})$. According to the definition (2.8) of the function $g_f(\boldsymbol{x})$ and the induction hypothesis, we have

$$f(\boldsymbol{u}) = \bigoplus_{\boldsymbol{x}\in V_n,\,\boldsymbol{x}\preccurlyeq\boldsymbol{u}} g_f(\boldsymbol{x}) = \bigoplus_{\boldsymbol{x}\in V_n,\,\boldsymbol{x}\prec\boldsymbol{u}}\ \bigoplus_{\boldsymbol{y}\in V_n,\,\boldsymbol{y}\preccurlyeq\boldsymbol{x}} f(\boldsymbol{y}) \oplus g_f(\boldsymbol{u}).$$

It is easy to see that in the sum $\bigoplus_{\boldsymbol{x}\in V_n,\, \boldsymbol{x}\prec\boldsymbol{u}} \bigoplus_{\boldsymbol{y}\in V_n,\, \boldsymbol{y}\preccurlyeq\boldsymbol{x}} f(\boldsymbol{y})$ a term $f(\boldsymbol{y})$ is present only if $\boldsymbol{y} \prec \boldsymbol{u}$, and that if it is, this occurs $2^{\mathrm{wt}(\boldsymbol{u})-\mathrm{wt}(\boldsymbol{y})} - 1$ times, which is an odd number. Thus,

$$\bigoplus_{\boldsymbol{x}\in V_n,\boldsymbol{x}\prec\boldsymbol{u}} \bigoplus_{\boldsymbol{y}\in V_n,\boldsymbol{y}\preccurlyeq\boldsymbol{x}} f(\boldsymbol{y}) = \bigoplus_{\boldsymbol{y}\in V_n,\ \boldsymbol{y}\prec\boldsymbol{u}} f(\boldsymbol{y}),$$

whence

$$g_f(\boldsymbol{u}) = \bigoplus_{\boldsymbol{x}\in V_n,\boldsymbol{x}\preccurlyeq\boldsymbol{u}} f(\boldsymbol{x}). \qquad \square$$

REMARK 2.11. It follows from the formulas $f(\boldsymbol{u}) = \bigoplus_{\boldsymbol{x}\in V_n,\boldsymbol{x}\preccurlyeq\boldsymbol{u}} g_f(\boldsymbol{x})$ and $g_f(\boldsymbol{u}) = \bigoplus_{\boldsymbol{x}\in V_n,\boldsymbol{x}\preccurlyeq\boldsymbol{u}} f(\boldsymbol{x})$ that the Möbius transform is an involution of $\mathcal{F}_n$, i.e., it is equal to its own inverse.

DEFINITION 2.12. An expression $x^{(i_1)} \cdots x^{(i_k)}$ (when $c_{i_1 i_2 \dots i_k} = 1$) in (2.7) or an expression $(x^{(1)})^{\alpha^{(1)}} \cdots (x^{(n)})^{\alpha^{(n)}}$ (when $g_f(\alpha^{(1)}, \dots, \alpha^{(n)}) = 1$) in (2.8) is called a *summand in the Zhegalkin polynomial (or ANF)* of the function f. The weight $\mathrm{wt}(\boldsymbol{\alpha})$ of the vector $\boldsymbol{\alpha} = (\alpha^{(1)}, \dots, \alpha^{(n)})^\top$ is called the *length of the summand* $(x^{(1)})^{\alpha^{(1)}} \cdots (x^{(n)})^{\alpha^{(n)}}$.

DEFINITION 2.13. The *algebraic degree of a function* f (denoted by $\deg f$) is the number of variables in the longest summand of the Zhegalkin polynomial (or ANF) of the function f.

DEFINITION 2.14. The *algebraic degree of the function* f *in a variable* $x^{(i)}$ (denoted by $\deg(f, x^{(i)})$) is the number of variables in the longest summand containing $x^{(i)}$ in the Zhegalkin polynomial of the function f.

It is clear that if $\deg(f, x^{(i)}) = 0$, then the variable $x^{(i)}$ is nonessential for the function f. If $\deg(f, x^{(i)}) = 1$, we will say that the function f *depends on the variable* $x^{(i)}$ linearly. If $\deg(f, x^{(i)}) > 1$, we will say that the function f depends on the variable $x^{(i)}$ nonlinearly. We will say that the summands of length 1 in a Zhegalkin polynomial are *linear summands*.

Now we introduce a class of Boolean functions.

DEFINITION 2.15. A Boolean function $f(x^{(1)}, \dots, x^{(n)})$ of n variables is called *functionally separable* if there exist functions $h(y^{(1)}, \dots, y^{(s)})$ of s variables and functions $g(z^{(1)}, \dots, z^{(n-s+1)})$ of $n-s+1$ variables such that for $1 < s < n-1$ the following equality holds:

$$f(x^{(1)}, \dots, x^{(n)}) = g(h(x^{(1)}, \dots, x^{(s)}), x^{(s+1)}, \dots, x^{(n)}). \tag{2.9}$$

We will show that most Boolean functions are not functionally separable.

THEOREM 2.16. *The quotient*

$$\frac{\text{the number of functionally separable Boolean functions of } n \text{ variables}}{\#\mathcal{F}_n}$$

tends to zero as $n \to \infty$.

PROOF. There are $\binom{n}{s}$ ways to choose the s variables in Definition 2.15. Since the function h (resp. g) depends on s (resp. $n-s+1$) variables, there exist 2^{2^s}

choices for h and $2^{2^{n-s+1}}$ choices for g. Hence, the total number of functionally separable functions is at most

$$\sum_{s=2}^{n-2} \binom{n}{s} 2^{2^s} \cdot 2^{2^{n-s+1}} \leq (n-3)\binom{n}{2} 2^{2^2} \cdot 2^{2^{n-1}} \leq 8n(n-1)(n-3)2^{2^{n-1}},$$

and the ratio of this number to 2^{2^n} tends to zero as $n \to \infty$: this follows from the inequality

$$\binom{n}{s} 2^{2^s} \cdot 2^{2^{n-s+1}} \leq \binom{n}{2} 2^{2^2} \cdot 2^{2^{n-1}},$$

which holds for all s: $1 < s < n-1$. $\square$

An important special case of functional separability, which we will use later, is a representation of the function $f \in \mathcal{F}_{n+m}$, after a nondegenerate linear substitution of the variables, as a sum of two functions $h \in \mathcal{F}_n$ and $g \in \mathcal{F}_m$ of disjoint sets of variables:

(2.10) $$f(Az) = h(\boldsymbol{x}) \oplus g(\boldsymbol{y}), \quad \boldsymbol{x} \in V_n, \quad \boldsymbol{y} \in V_m, \quad \boldsymbol{z} \in V_{n+m}.$$

PROBLEM 2.17 (Open problem). Find effective necessary and sufficient conditions that will show when a given Boolean function can be represented in the form (2.10).

We conclude this section with various examples of Boolean functions.

Table 2.3 shows some functions which are often used in cybernetics, coding theory, and cryptography. They play the role of elementary functions.

TABLE 2.3

x	0 (constant 0)	1 (constant 1)	x (identity)	$\bar{x}$ (negation)
0	0	1	0	1
1	0	1	1	0

$x^{(1)}$	$x^{(2)}$	$x^{(1)} \cdot x^{(2)}$ (conjunction, multiplication modulo 2)	$x^{(1)} \vee x^{(2)}$ (disjunction)	$x^{(1)} \to x^{(2)}$ (implication)	$x^{(1)} \oplus x^{(2)}$ (addition modulo 2)	$x^{(1)}/x^{(2)}$ (Sheffer stroke)
0	0	0	0	1	0	1
0	1	0	1	1	1	1
1	0	0	1	0	1	1
1	1	1	1	1	0	0

Denote by $\mathcal{A}_n$ the set of Boolean functions from $\mathcal{F}_n$ for which $\deg f \leq 1$:

$$\mathcal{A}_n = \{f \in \mathcal{F}_n \mid \deg f \leq 1\}.$$

We will call such functions *affine functions*; $\mathcal{A}_n$ is the set (class) of affine functions. Affine functions for which $f(\mathbf{0}) = 0$ are called *linear functions*:

$$\mathcal{L}_n = \{f \in \mathcal{A}_n \mid f(\mathbf{0}) = 0\}.$$

The Zhegalkin polynomial of an arbitrary affine function f has the form

(2.11) $$f(x^{(1)}, \ldots, x^{(n)}) = a^{(1)}x^{(1)} \oplus \cdots \oplus a^{(n)}x^{(n)} \oplus b,$$

where $a^{(i)}$ $(i = 1, \dots, n)$, $b \in \mathbb{F}_2$. In the case of linear functions, $b = 0$.

THEOREM 2.18. $\#\mathcal{A}_n = 2^{n+1}, \;\; \#\mathcal{L}_n = 2^n$.

PROOF. The assertion immediately follows from (2.11). □

Consider the class of functions $f \in \mathcal{F}_n$ whose value does not depend on any permutation of its variables. We will denote this class of functions by $\mathcal{S}_n$ and will call it the class of *symmetric functions*.

THEOREM 2.19. $\#\mathcal{S}_n = 2^{n+1}$.

The proof of the theorem follows from the fact that the value of a symmetric function depends only on the weight of the vector of variables.

THEOREM 2.20. *A Boolean function* $f(x^{(1)}, \dots, x^{(n)})$ *is symmetric if and only if*

$$\begin{aligned} f(x^{(1)}, x^{(2)}, \dots, x^{(n)}) &= f(x^{(2)}, x^{(1)}, x^{(3)}, \dots, x^{(n)}) \\ &= f(x^{(2)}, x^{(3)}, \dots, x^{(n)}, x^{(1)}) = \cdots \end{aligned} \tag{2.12}$$

for any $x^{(1)}, x^{(2)}, \dots, x^{(n)} \in \mathbb{F}_2$.

The theorem follows from the fact that the transpositions from (2.12) and a complete n-cycle generate the symmetric group in n symbols (see [**155, 410**]).
Later on, other classes of Boolean functions will be introduced.

DEFINITION 2.21. A Boolean function $f \in \mathcal{F}_n$ is called (c_0, c_1)-*regular* (or simply *regular*) if
1) for any $\boldsymbol{x} \in V_n$ such that $f(\boldsymbol{x}) = 0$ the following equality holds:

$$\#\{\boldsymbol{y} \in V_n \mid \operatorname{dist}(\boldsymbol{x}, \boldsymbol{y}) = 1, \;\; f(\boldsymbol{y}) = 1\} = c_0;$$

2) for any $\boldsymbol{x} \in V_n$ such that $f(\boldsymbol{x}) = 1$ the following equality holds:

$$\#\{\boldsymbol{y} \in V_n \mid \operatorname{dist}(\boldsymbol{x}, \boldsymbol{y}) = 1, \;\; f(\boldsymbol{y}) = 0\} = c_1.$$

If a Boolean function is (c, c)-regular, we will say that it is c-*regular*.

2.2. Numerical and metric characteristics

Many properties of Boolean functions from $\mathcal{F}_n$ are naturally studied in terms of the geometry of vector spaces over the field $\mathbb{F}_2$.

PROPOSITION 2.22. *The set* $\mathcal{F}_n$ *is a vector space over the field* $\mathbb{F}_2$*;* $\dim \mathcal{F}_n = 2^n$.

Proof of this proposition follows directly from Theorem 2.6.
In Theorem 2.6, one of the possible bases of the space $\mathcal{F}_n$ is given. This basis consists of functions

$$\{1;\; x^{(i)}\; (i = 1, \dots, n);\; x^{(i)}x^{(j)}\; (1 \le i < j \le n); \dots; x^{(1)} \cdots x^{(n)}\}. \tag{2.13}$$

Later, when we study different representations of the elements from $\mathcal{F}_n$, we will work with other bases, too.

DEFINITION 2.23. The *Hamming distance* (or simply *distance*) between the vectors $\boldsymbol{x}$ and $\boldsymbol{y}$ is equal to the number of positions at which they differ. Notation: $\operatorname{dist}(\boldsymbol{x}, \boldsymbol{y})$.

It is clear that

$$\text{dist}(\boldsymbol{x}, \boldsymbol{y}) = \text{wt}(\boldsymbol{x} \oplus \boldsymbol{y}). \tag{2.14}$$

Here is a list of simple properties of the Hamming distance:
1) $\text{dist}(\boldsymbol{x}, \boldsymbol{y}) = \text{dist}(\boldsymbol{y}, \boldsymbol{x})$ for any $\boldsymbol{x}, \boldsymbol{y} \in V_n$ (symmetry of the distance);
2) $\text{dist}(\boldsymbol{x}, \boldsymbol{y}) > 0$ for $\boldsymbol{x} \neq \boldsymbol{y}$; $\text{dist}(\boldsymbol{x}, \boldsymbol{x}) = 0$ for any $\boldsymbol{x} \in V_n$;
3) $\text{dist}(\boldsymbol{x}, \boldsymbol{z}) \leq \text{dist}(\boldsymbol{x}, \boldsymbol{y}) + \text{dist}(\boldsymbol{y}, \boldsymbol{z})$ for any $\boldsymbol{x}, \boldsymbol{y}, \boldsymbol{z} \in V_n$ (*triangle inequality*).

PROBLEM 2.24. Prove the triangle inequality for the Hamming distance. Prove that the inequality is strict if and only if for any i, $1 \leq i \leq n$, one has either $x^{(i)} = y^{(i)}$ or $z^{(i)} = y^{(i)}$ (or both).

Since the Boolean functions form a vector space V_{2^n}, one can give similar definitions for them, too.

DEFINITION 2.25. The *weight* $\text{wt}(f)$ *of a Boolean function* f is defined as follows:

$$\text{wt}(f) = \#\{\boldsymbol{x} \in V_n \mid f(\boldsymbol{x}) = 1\}.$$

DEFINITION 2.26. A function $f \in \mathcal{F}_n$ for which $\text{wt}(f) = 2^{n-1}$ is called *balanced*.

PROBLEM 2.27. Put $g(x^{(1)}, \ldots, x^{(n)}, y) = f(x^{(1)}, \ldots, x^{(n)})$ (i.e., y is a dummy variable for the function $g(x^{(1)}, \ldots, x^{(n)}, y)$). Prove that

$$\text{wt}(g(x^{(1)}, \ldots, x^{(n)}, y)) = 2\,\text{wt}(f(x^{(1)}, \ldots, x^{(n)})).$$

PROBLEM 2.28. Suppose that $f(x^{(1)}, \ldots, x^{(n)}) \in \mathcal{F}_n$; put $g(x^{(1)}, \ldots, x^{(n-1)}, z^{(1)}, z^{(2)}) = f(x^{(1)}, \ldots, x^{(n-1)}, z^{(1)} \oplus z^{(2)})$. Prove that $\text{wt}(g) = 2\,\text{wt}(f)$.

PROBLEM 2.29. Suppose that $f(x^{(1)}, \ldots, x^{(n)}) \in \mathcal{F}_n$. Prove that the function $g(x^{(1)}, \ldots, x^{(n)}, y) = f(x^{(1)}, \ldots, x^{(n)}) \oplus y$ is balanced.

PROBLEM 2.30. Let $f(x^{(1)}, \ldots, x^{(n)}) \in \mathcal{F}_n$. Prove that $\text{wt}(f) \equiv 1 \pmod 2$ if and only if $\deg f = n$.

The result of Problem 2.30 could be stated as follows: a monomial $x^{(1)} \cdots x^{(n)}$ makes the weight of the Boolean function $f(x^{(1)}, \ldots, x^{(n)})$ even. This is related to the following interesting question.

PROBLEM 2.31 (Open problem). Is it possible to isolate monomials of a Boolean function that are "responsible" for divisibility of the weight of the function by 2^k, $k = 2, 3, 4, \ldots$?

DEFINITION 2.32. The *distance between Boolean functions* f and g, $\text{dist}(f, g)$, is defined as

$$\text{dist}(f, g) = \#\{\boldsymbol{x} \in V_n \mid f(\boldsymbol{x}) \neq g(\boldsymbol{x})\}.$$

It is clear that

$$\text{dist}(f, g) = \text{wt}(f \oplus g).$$

DEFINITION 2.33. The *scalar product* (inner product) of two vectors $\boldsymbol{x}, \boldsymbol{u} \in V_n$ is defined as follows:

$$\langle \boldsymbol{x}, \boldsymbol{u} \rangle = \bigoplus_{i=1}^{n} x^{(i)} u^{(i)}.$$

For an arbitrary set $M \subseteq V_n$, we will denote by $M^\perp$ the set

$$M^\perp = \{\boldsymbol{y} \in V_n \mid \langle \boldsymbol{y}, \boldsymbol{x}\rangle = 0 \text{ for all } \boldsymbol{x} \in M\}.$$

DEFINITION 2.34. The *Fourier transform* of a Boolean function f is the integer-valued function $\bar{W}_f$ on V_n defined as follows:

$$\bar{W}_f(\boldsymbol{u}) = \sum_{\boldsymbol{x}\in V_n} f(\boldsymbol{x})(-1)^{\langle \boldsymbol{x},\boldsymbol{u}\rangle}$$

(addition on the right-hand side is the usual addition of integers). For any $\boldsymbol{u} \in V_n$, the value $\bar{W}_f(\boldsymbol{u})$ is called a *Fourier coefficient*.

For an arbitrary Boolean function f, we will denote by $\exp f$ the integer-valued function

$$\exp f(\boldsymbol{x}) = (-1)^{f(\boldsymbol{x})}.$$

The function $\exp f$ is also known under the name of *sign function.*

DEFINITION 2.35. The *Walsh–Hadamard transform* of a Boolean function f is the integer-valued function W_f on V_n defined by the formula

$$W_f(\boldsymbol{u}) = \sum_{\boldsymbol{x}\in V_n} \exp f(\boldsymbol{x})(-1)^{\langle \boldsymbol{x},\boldsymbol{u}\rangle} = \sum_{\boldsymbol{x}\in V_n} (-1)^{f(\boldsymbol{x})\oplus\langle \boldsymbol{x},\boldsymbol{u}\rangle}$$

(the addition is again the usual addition of integers). For any $\boldsymbol{u} \in V_n$, the value $W_f(\boldsymbol{u})$ is called a *Walsh–Hadamard coefficient*.

Fourier or Walsh–Hadamard coefficients are often called *spectral coefficients*.

THEOREM 2.36. *Fourier and Walsh–Hadamard coefficients satisfy the relation*

$$\bar{W}_f(\boldsymbol{u}) = 2^n\delta(\boldsymbol{u}) - 2\bar{W}_f(\boldsymbol{u}),$$

where $\delta(\boldsymbol{u})$ is Dirac's δ-function:

$$\delta(\boldsymbol{u}) = \begin{cases} 1 & \text{if } \boldsymbol{u} = \boldsymbol{0}; \\ 0 & \text{if } \boldsymbol{u} \neq \boldsymbol{0}. \end{cases}$$

PROOF.

$$\begin{aligned} W_f(\boldsymbol{u}) &= \sum_{\boldsymbol{x}\in V_n} (-1)^{f(\boldsymbol{x})\oplus\langle \boldsymbol{x},\boldsymbol{u}\rangle} = \sum_{\substack{\boldsymbol{x}\in V_n\\ f(\boldsymbol{x})=0}} (-1)^{\langle \boldsymbol{x},\boldsymbol{u}\rangle} - \sum_{\substack{\boldsymbol{x}\in V_n\\ f(\boldsymbol{x})=1}} (-1)^{\langle \boldsymbol{x},\boldsymbol{u}\rangle} \\ &= \sum_{\boldsymbol{x}\in V_n} (-1)^{\langle \boldsymbol{x},\boldsymbol{u}\rangle} - 2\sum_{\substack{\boldsymbol{x}\in V_n\\ f(\boldsymbol{x})=1}} (-1)^{\langle \boldsymbol{x},\boldsymbol{u}\rangle}. \end{aligned}$$

The sum $\sum_{\boldsymbol{x}\in V_n}(-1)^{\langle \boldsymbol{x},\boldsymbol{u}\rangle}$ is the sum of values of a character of an elementary abelian 2-group. Therefore, this sum is equal to zero if the vector $\boldsymbol{u}$ is not a zero-vector (i.e., if the character is nontrivial), and is equal to 2^n if the vector $\boldsymbol{u}$ is zero (i.e., if the character is trivial). Since the sum $\sum_{\boldsymbol{x}\in V_n,\ f(\boldsymbol{x})=1}(-1)^{\langle \boldsymbol{x},\boldsymbol{u}\rangle}$ coincides with $\bar{W}_f(\boldsymbol{u})$, the theorem follows. □

Both Walsh–Hadamard and Fourier coefficients have their advantages and disadvantages. We will mainly use Walsh–Hadamard coefficients.

THEOREM 2.37. *For the Walsh–Hadamard transform the following inversion formula holds:*

$$\exp f(\boldsymbol{x}) = 2^{-n} \sum_{\boldsymbol{u} \in V_n} W_f(\boldsymbol{u})(-1)^{\langle \boldsymbol{x}, \boldsymbol{u} \rangle}.$$

PROOF. Rewrite the right-hand side of the formula as follows:

$$\begin{aligned} 2^{-n} \sum_{\boldsymbol{u} \in V_n} W_f(\boldsymbol{u})(-1)^{\langle \boldsymbol{x}, \boldsymbol{u} \rangle} &= 2^{-n} \sum_{\boldsymbol{u} \in V_n} \sum_{\boldsymbol{y} \in V_n} (-1)^{f(\boldsymbol{y}) \oplus \langle \boldsymbol{y}, \boldsymbol{u} \rangle} (-1)^{\langle \boldsymbol{x}, \boldsymbol{u} \rangle} \\ &= 2^{-n} \sum_{\boldsymbol{y} \in V_n} (-1)^{f(\boldsymbol{y})} \sum_{\boldsymbol{u} \in V_n} (-1)^{\langle \boldsymbol{x} \oplus \boldsymbol{y}, \boldsymbol{u} \rangle}. \end{aligned}$$

The sum $\sum_{\boldsymbol{u} \in V_n} (-1)^{\langle \boldsymbol{x} \oplus \boldsymbol{y}, \boldsymbol{u} \rangle}$ is equal to zero if $\boldsymbol{x} \neq \boldsymbol{y}$, and is equal to 2^n if $\boldsymbol{x} = \boldsymbol{y}$. Taking this into account, we get

$$2^{-n} \sum_{\boldsymbol{u} \in V_n} W_f(\boldsymbol{u})(-1)^{\langle \boldsymbol{x}, \boldsymbol{u} \rangle} = 2^{-n}(-1)^{f(\boldsymbol{x})} 2^n = \exp f(\boldsymbol{x}). \qquad \square$$

Thus, Walsh–Hadamard coefficients uniquely determine a Boolean function. On the other hand, not every set of 2^n numbers is the set of Walsh–Hadamard coefficients for a Boolean function. Moreover, answering the question whether a given set of numbers is a set of Walsh–Hadamard coefficients is not easy.

PROBLEM 2.38. Let $W \colon V_n \to \mathbb{Z}$ be an integer-valued function on V_n. Prove that a Boolean function $f \in \mathcal{F}_n$ such that $W(\boldsymbol{u}) = W_f(\boldsymbol{u})$ for any $\boldsymbol{u} \in V_n$, exists if and only if the following *orthogonality equations* are satisfied:

$$\sum_{\boldsymbol{u} \in V_n} W(\boldsymbol{u}) W(\boldsymbol{u} \oplus \boldsymbol{v}) = \begin{cases} 0 & \text{if } \boldsymbol{v} \neq \boldsymbol{0}; \\ 2^{2n} & \text{if } \boldsymbol{v} = \boldsymbol{0}. \end{cases}$$

In relation to Theorem 2.36, we observe that inversion formulas exist for Fourier coefficients as well (cf. Theorem 2.37).

PROBLEM 2.39. Let $L \subseteq V_n$ be an arbitrary subspace. Prove that the following equation holds:

$$\sum_{\boldsymbol{x} \in L} (-1)^{\langle \boldsymbol{x}, \boldsymbol{y} \rangle} = \begin{cases} 2^{\dim L} & \text{if } \boldsymbol{y} \in L^{\perp}; \\ 0 & \text{otherwise.} \end{cases}$$

The following lemma is often useful.

LEMMA 2.40. *Suppose that $f \in \mathcal{F}_n$, $L \subseteq V_n$ is an arbitrary subspace, and $\boldsymbol{a}, \boldsymbol{b} \in V_n$ are arbitrary vectors. Then*

$$\sum_{\boldsymbol{x} \in \boldsymbol{a} \oplus L} (-1)^{f(\boldsymbol{x}) \oplus \langle \boldsymbol{b}, \boldsymbol{x} \rangle} = 2^{\dim L - n} (-1)^{\langle \boldsymbol{a}, \boldsymbol{b} \rangle} \sum_{\boldsymbol{y} \in \boldsymbol{b} \oplus L^{\perp}} W_f(\boldsymbol{y})(-1)^{\langle \boldsymbol{a}, \boldsymbol{y} \rangle}. \tag{2.15}$$

PROOF. To prove the lemma, it suffices to invoke Theorem 2.37, to change the order of summation, and use the result of Problem 2.39:

$$\begin{aligned} \sum_{\boldsymbol{x} \in \boldsymbol{a} \oplus L} (-1)^{f(\boldsymbol{x}) \oplus \langle \boldsymbol{b}, \boldsymbol{x} \rangle} &= \sum_{\boldsymbol{x} \in \boldsymbol{a} \oplus L} \Big(2^{-n} \sum_{\boldsymbol{y} \in V_n} W_f(\boldsymbol{y})(-1)^{\langle \boldsymbol{x}, \boldsymbol{y} \rangle} \Big) (-1)^{\langle \boldsymbol{b}, \boldsymbol{x} \rangle} \\ &= 2^{-n} \sum_{\boldsymbol{y} \in V_n} W_f(\boldsymbol{y}) \sum_{\boldsymbol{x} \in \boldsymbol{a} \oplus L} (-1)^{\langle \boldsymbol{x}, \boldsymbol{y} \oplus \boldsymbol{b} \rangle} \\ &= 2^{\dim L - n} (-1)^{\langle \boldsymbol{a}, \boldsymbol{b} \rangle} \sum_{\boldsymbol{y} \in \boldsymbol{b} \oplus L^{\perp}} W_f(\boldsymbol{y})(-1)^{\langle \boldsymbol{a}, \boldsymbol{y} \rangle}. \qquad \square \end{aligned}$$

Relation (2.15) is most useful in the case $\boldsymbol{a} = \boldsymbol{b} = \boldsymbol{0}$:

$$\sum_{\boldsymbol{x}\in L} (-1)^{f(\boldsymbol{x})} = 2^{\dim L - n} \sum_{\boldsymbol{y}\in L^{\perp}} W_f(\boldsymbol{y}). \tag{2.16}$$

Since $(-1)^{f(\boldsymbol{x})} = 1 - 2f(\boldsymbol{x})$, we can easily get one more useful formula,

$$\sum_{\boldsymbol{x}\in L^{\perp}} W_f(\boldsymbol{y}) = 2^n - 2^{n-\dim L+1}\,\mathrm{wt}(f|_L), \tag{2.17}$$

where $\mathrm{wt}(f|_L) = \#\{\boldsymbol{x} \in L \mid f(\boldsymbol{x}) = 1\}$.

PROBLEM 2.41. Suppose that $f \in \mathcal{F}_n$. Prove that for a vector $\boldsymbol{u} \in V_n$ with $\mathrm{wt}(\boldsymbol{u}) = k$ and coordinates u^i (of the vector u) equal to 1 in positions $i_1, \ldots, i_k$, the following equality holds:

$$\sum_{\boldsymbol{x}\in V_n,\ \boldsymbol{x}\preccurlyeq\boldsymbol{u}} W_f(\boldsymbol{x}) = 2^n - 2^{k+1}\,\mathrm{wt}\big(f^{0,\ldots,0}_{i_1,\ldots,i_k}\big).$$

In particular,

$$W_f(\boldsymbol{0}) = 2^n - 2\,\mathrm{wt}(f).$$

PROBLEM 2.42. Suppose that $f \in \mathcal{F}_n$ and that, for a fixed subspace $L_0 \subseteq V_n$, there exists an integer-valued function $W(\boldsymbol{y})$, $\boldsymbol{y} \in V_n$, such that for arbitrary vectors $\boldsymbol{a}, \boldsymbol{b} \in V_n$ the following equality holds:

$$\sum_{\boldsymbol{x}\in\boldsymbol{a}\oplus L_0} (-1)^{f(\boldsymbol{x})\oplus\langle\boldsymbol{b},\boldsymbol{x}\rangle} = 2^{\dim L_0 - n}(-1)^{\langle\boldsymbol{a},\boldsymbol{b}\rangle} \sum_{\boldsymbol{y}\in\boldsymbol{b}\oplus L_0^{\perp}} W(\boldsymbol{y})(-1)^{\langle\boldsymbol{a},\boldsymbol{y}\rangle}.$$

Prove that $W_f(\boldsymbol{b}) = W(\boldsymbol{b})$ for all $\boldsymbol{b} \in V_n$.

THEOREM 2.43 (Parseval's equation). *Walsh–Hadamard coefficients satisfy the following relation:*

$$\sum_{\boldsymbol{u}\in V_n} W_f^2(\boldsymbol{u}) = 2^{2n}.$$

PROOF.

$$\begin{aligned}
\sum_{\boldsymbol{u}\in V_n} W_f^2(\boldsymbol{u}) &= \sum_{\boldsymbol{u}\in V_n}\Big(\sum_{\boldsymbol{x}\in V_n}(-1)^{f(\boldsymbol{x})\oplus\langle\boldsymbol{x},\boldsymbol{u}\rangle}\Big)^2\\
&= \sum_{\boldsymbol{u}\in V_n}\Big(\sum_{\boldsymbol{x},\boldsymbol{y}\in V_n}(-1)^{f(\boldsymbol{x})\oplus\langle\boldsymbol{x},\boldsymbol{u}\rangle\oplus f(\boldsymbol{y})\oplus\langle\boldsymbol{y},\boldsymbol{u}\rangle}\Big)\\
&= \sum_{\boldsymbol{x},\boldsymbol{y}\in V_n}(-1)^{f(\boldsymbol{x})\oplus f(\boldsymbol{y})}\sum_{\boldsymbol{u}\in V_n}(-1)^{\langle\boldsymbol{x}\oplus\boldsymbol{y},\boldsymbol{u}\rangle} = 2^{2n}.
\end{aligned}$$ □

PROBLEM 2.44. Prove that for any $\boldsymbol{u} \neq \boldsymbol{0}$, the Walsh–Hadamard coefficients satisfy the equality

$$\sum_{\boldsymbol{x}\in V_n} W_f(\boldsymbol{x}) W_f(\boldsymbol{x}\oplus\boldsymbol{u}) = 0.$$

Let l be an arbitrary affine function:

$$l(\boldsymbol{x}) = a^{(1)}x^{(1)} \oplus \cdots \oplus a^{(n)}x^{(n)} \oplus b = \langle\boldsymbol{a},\boldsymbol{x}\rangle \oplus b,$$

where $\boldsymbol{a} \in V_n$, $b \in \mathbb{F}_2$. We will write the affine function l in the form $l = l_{\boldsymbol{a},b}$. It is easy to prove the following statement.

PROPOSITION 2.45.

$$W_{l_{\boldsymbol{a},b}}(\boldsymbol{u}) = \begin{cases} (-1)^b 2^n & \text{if } \boldsymbol{u} = \boldsymbol{a}; \\ 0 & \text{if } \boldsymbol{u} \neq \boldsymbol{a}. \end{cases}$$

PROBLEM 2.46. Prove that the coefficients of the Zhegalkin polynomial for the function $f \in \mathcal{F}_n$ can be expressed in terms of Walsh–Hadamard coefficients as follows:

$$g_f(\boldsymbol{u}) = 2^{\mathrm{wt}(\boldsymbol{u})-1} - 2^{\mathrm{wt}(\boldsymbol{u})-n-1} \sum_{\boldsymbol{\alpha} \preccurlyeq \boldsymbol{u} \oplus \mathbf{1}} W_f(\boldsymbol{\alpha}) \pmod 2.$$

PROBLEM 2.47. Prove that if $W_f(\boldsymbol{\alpha}) \equiv 0 \pmod{2^k}$ for all $\boldsymbol{\alpha} \in V_n$, then $\deg f \leq n - k + 1$.

When calculating Walsh–Hadamard coefficients, the following lemma is often used.

LEMMA 2.48. *Suppose that a Boolean function $h \in \mathcal{F}_{n+m}$ of $n + m$ variables is expressed as the sum of two functions $f \in \mathcal{F}_n$ and $g \in \mathcal{F}_m$ of n and m nonintersecting variables respectively:*

$$h(\boldsymbol{x}, \boldsymbol{y}) = f(\boldsymbol{x}) \oplus g(\boldsymbol{y}),$$

where $\boldsymbol{x} \in V_n$, $\boldsymbol{y} \in V_m$. Then

$$W_h(\boldsymbol{\alpha}, \boldsymbol{\beta}) = W_f(\boldsymbol{\alpha}) W_g(\boldsymbol{\beta})$$

for all $\boldsymbol{\alpha} \in V_n$, $\boldsymbol{\beta} \in V_m$.

PROOF.

$$\begin{aligned} W_h(\boldsymbol{\alpha}, \boldsymbol{\beta}) &= \sum_{\boldsymbol{x} \in V_n,\ \boldsymbol{y} \in V_m} (-1)^{h(\boldsymbol{x},\boldsymbol{y})} (-1)^{\langle (\boldsymbol{x},\boldsymbol{y}), (\boldsymbol{\alpha},\boldsymbol{\beta}) \rangle} \\ &= \sum_{\boldsymbol{x} \in V_n,\ \boldsymbol{y} \in V_m} (-1)^{f(\boldsymbol{x}) \oplus g(\boldsymbol{y})} (-1)^{\langle \boldsymbol{x},\boldsymbol{\alpha} \rangle \oplus \langle \boldsymbol{y},\boldsymbol{\beta} \rangle} \\ &= \sum_{\boldsymbol{x} \in V_n} (-1)^{f(\boldsymbol{x})} (-1)^{\langle \boldsymbol{x},\boldsymbol{\alpha} \rangle} \sum_{\boldsymbol{y} \in V_m} (-1)^{g(\boldsymbol{y})} (-1)^{\langle \boldsymbol{y},\boldsymbol{\beta} \rangle} \\ &= W_f(\boldsymbol{\alpha}) W_g(\boldsymbol{\beta}). \end{aligned}$$

□

Let $M \subset \mathcal{F}_n$ be an arbitrary subset of $\mathcal{F}_n$.

DEFINITION 2.49. The *distance from a Boolean function $f \in \mathcal{F}_n$ to a set M* is defined as

$$\mathrm{dist}(f, M) = \min_{g \in M} \mathrm{dist}(f, g).$$

From the definition of Walsh–Hadamard coefficients one can see that

$$W_f(\boldsymbol{u}) = W_{f \oplus l_{\boldsymbol{u},0}}(\mathbf{0}).$$

Using the result of Problem 2.41, it is easy to see that

$$W_f(\boldsymbol{u}) = 2^n - 2\,\mathrm{wt}(f \oplus l_{\boldsymbol{u},0}) = 2^n - 2\,\mathrm{dist}(f, l_{\boldsymbol{u},0}).$$

Hence,

$$\mathrm{dist}(f, l_{\boldsymbol{u},0}) = 2^{n-1} - \frac{1}{2} W_f(\boldsymbol{u}), \quad \mathrm{dist}(f, l_{\boldsymbol{u},1}) = 2^{n-1} + \frac{1}{2} W_f(\boldsymbol{u}).$$

DEFINITION 2.50. The distance between a Boolean function $f \in \mathcal{F}_n$ and the set of affine functions $\mathcal{A}_n$ is called the *nonlinearity of the function* f and is denoted by N_f.

PROPOSITION 2.51. *We have*

$$N_f = 2^{n-1} - \frac{1}{2} \max_{\boldsymbol{u} \in V_n} |W_f(\boldsymbol{u})| .$$

This equality immediately follows from the argument above.

PROBLEM 2.52. Suppose that $f_1 \in \mathcal{F}_{n_1}$, $f_2 \in \mathcal{F}_{n_2}$ and set $f(\boldsymbol{x}, \boldsymbol{y}) = f_1(\boldsymbol{x}) \oplus f_2(\boldsymbol{y})$, $\boldsymbol{x} \in V_{n_1}$, $\boldsymbol{y} \in V_{n_2}$. Prove that

$$N_f \geq 2^{n_2} N_{f_1} + 2^{n_1} N_{f_2} - 2 N_{f_1} N_{f_2}.$$

Now we discuss yet another representation of a Boolean function. To wit, let $\delta_{\boldsymbol{u}}(\boldsymbol{x})$ be the shift of Dirac's δ-function by the vector $\boldsymbol{u} \in V_n$:

$$\delta_{\boldsymbol{u}}(\boldsymbol{x}) = \delta(\boldsymbol{u} \oplus \boldsymbol{x}).$$

It is clear that

$$\delta_{\boldsymbol{u}}(\boldsymbol{x}) = \prod_{i=1}^{n} z_i(\boldsymbol{u}, \boldsymbol{x}), \quad \text{where} \quad z_i(\boldsymbol{u}, \boldsymbol{x}) = \begin{cases} x^{(i)} & \text{if } u^{(i)} = 1; \\ 1 - x^{(i)} & \text{if } u^{(i)} = 0. \end{cases}$$

For any Boolean function $f(\boldsymbol{x}) \in \mathcal{F}_n$ the following equality holds:

$$f(\boldsymbol{x}) = \sum_{\boldsymbol{u} \in V_n} f(\boldsymbol{u}) \delta_{\boldsymbol{u}}(\boldsymbol{x}). \tag{2.18}$$

Using the representation of the function $\delta_{\boldsymbol{u}}(\boldsymbol{x})$ ($\boldsymbol{u} \in V_n$), the expression (2.18) can be reduced to a polynomial with integer coefficients:

$$f(\boldsymbol{x}) = f(x^{(1)}, \ldots, x^{(n)}) = \sum_{\boldsymbol{u} \in V_n} \lambda_f(\boldsymbol{u}) (x^{(1)})^{u^{(1)}} \cdots (x^{(n)})^{u^{(n)}} \tag{2.19}$$

in which the exponent of each variable is at most 1. It is clear that different Boolean functions correspond to different integer polynomials of the form (2.19). However, note that an arbitrary integer polynomial of the form (2.19) need not correspond to a Boolean function over V_n, as its values over V_n could be different from 0 and 1.

DEFINITION 2.53. The *numerical normal form of a Boolean function* (NNF) $f \in \mathcal{F}_n$ is a polynomial with integer coefficients defined as follows:

$$f(\boldsymbol{x}) = \sum_{\boldsymbol{u} \in V_n} \lambda_f(\boldsymbol{u}) (x^{(1)})^{u^{(1)}} \cdots (x^{(n)})^{u^{(n)}}.$$

From Problem 2.7 we see that

$$f(\boldsymbol{x}) = \sum_{\boldsymbol{u} \in V_n, \ \boldsymbol{u} \preccurlyeq \boldsymbol{x}} \lambda_f(\boldsymbol{u}).$$

PROBLEM 2.54. Suppose that $f \in \mathcal{F}_n$. Prove that for any $\boldsymbol{u} \in V_n$ the following equality holds:

$$\lambda_f(\boldsymbol{u}) = (-1)^{\mathrm{wt}(\boldsymbol{u})} \sum_{\boldsymbol{x} \in V_n, \ \boldsymbol{u} \preccurlyeq \boldsymbol{x}} (-1)^{\mathrm{wt}(\boldsymbol{x})} f(\boldsymbol{x}). \tag{2.20}$$

(*Hint*: First calculate the coefficients of the numerical normal form of the function $\delta_{\boldsymbol{u}}(\boldsymbol{x})$.)

If the numerical normal form of a Boolean function f is known, we can easily obtain the Zhegalkin polynomial of this function by reducing the numerical normal form modulo 2. Conversely, from the Zhegalkin polynomial of a function f one can obtain its numerical normal form.

THEOREM 2.55. *Suppose that* $f \in \mathcal{F}_n$ *and*

$$f(x^{(1)}, \dots, x^{(n)}) = \bigoplus_{\boldsymbol{u} \in V_n} g_f(\boldsymbol{u})(x^{(1)})^{u^{(1)}} \cdots (x^{(n)})^{u^{(n)}}$$

is Zhegalkin polynomial of f. *Then for any* $\boldsymbol{u} \in V_n$,

$$\lambda_f(\boldsymbol{u}) = \sum_{k=1}^{n} (-2)^{k-1} \sum_{\substack{\boldsymbol{v}_1, \dots, \boldsymbol{v}_k \in V_n, \\ \boldsymbol{v}_1 \vee \cdots \vee \boldsymbol{v}_k = \boldsymbol{u}}} g_f(\boldsymbol{v}_1) \cdots g_f(\boldsymbol{v}_k). \tag{2.21}$$

PROOF. Using the relation $a \oplus b = a + b - 2ab$ repeatedly, simplify the algebraic normal form; after all cancelations, the result will follow. □

THEOREM 2.56. *Suppose that* $f \in \mathcal{F}_n$, $\deg f = d \neq 0$. *Then for any* $\boldsymbol{u} \in V_n$,

$$\lambda_f(\boldsymbol{u}) = \sum_{k=\left\lfloor \frac{\mathrm{wt}(\boldsymbol{u})-1}{d} \right\rfloor + 1}^{n} (-2)^{k-1} \sum_{\substack{\boldsymbol{v}_1, \dots, \boldsymbol{v}_k \in V_n, \\ \boldsymbol{v}_1 \vee \cdots \vee \boldsymbol{v}_k = \boldsymbol{u}}} g_f(\boldsymbol{v}_1) \cdots g_f(\boldsymbol{v}_k). \tag{2.22}$$

PROOF. Let $\{\boldsymbol{v}_1, \dots, \boldsymbol{v}_k\}$ be a set of k different vectors from the space V_n. If $\mathrm{wt}(\boldsymbol{v}_1 \vee \cdots \vee \boldsymbol{v}_k) > kd$, then there exists an index j, $1 \le j \le k$ such that $\mathrm{wt}(\boldsymbol{v}_j) > d$. Then for an arbitrary natural $k < (\mathrm{wt}(\boldsymbol{u}))/d$, i.e., for $k \le \lfloor (\mathrm{wt}(\boldsymbol{u}) - 1)/d \rfloor$, each term in the inner sum (2.21) has coefficient $g_f(\boldsymbol{v}_j)$ with $\mathrm{wt}(\boldsymbol{v}_j) > d$, which equals 0 if $\deg f = d$. Hence, the terms in (2.21) corresponding to $k = 1, 2, \dots, \lfloor (\mathrm{wt}(\boldsymbol{u}) - 1)/d \rfloor$ are all zero. □

COROLLARY 2.57. *In Theorem* 2.56 *any coefficient* $\lambda_f(\boldsymbol{u})$, $\boldsymbol{u} \in V_n$, *is divisible by* $2^{\lfloor (\mathrm{wt}(\boldsymbol{u})-1)/d \rfloor}$.

There exists a natural relation between the coefficients of a numerical normal form and the Fourier coefficients.

Let $f(\boldsymbol{x}) = \sum_{\boldsymbol{u} \in V_n} \lambda_f(\boldsymbol{u})(x^{(1)})^{u^{(1)}} \cdots (x^{(n)})^{u^{(n)}}$ be the numerical normal form of a Boolean function f. Then

$$\begin{aligned}
\bar{W}_f(\boldsymbol{a}) &= \sum_{\boldsymbol{x} \in V_n} f(\boldsymbol{x})(-1)^{\langle \boldsymbol{a}, \boldsymbol{x} \rangle} \\
&= \sum_{\boldsymbol{x} \in V_n} \sum_{\boldsymbol{u} \in V_n} \lambda_f(\boldsymbol{u})(x^{(1)})^{u^{(1)}} \cdots (x^{(n)})^{u^{(n)}} (-1)^{\langle \boldsymbol{a}, \boldsymbol{x} \rangle} \\
&= \sum_{\boldsymbol{u} \in V_n} \lambda_f(\boldsymbol{u}) \sum_{\boldsymbol{x} \in V_n, \boldsymbol{u} \preccurlyeq \boldsymbol{x}} (-1)^{\langle \boldsymbol{a}, \boldsymbol{x} \rangle}.
\end{aligned}$$

Replacing (in the first line of the above formula) summation over $\boldsymbol{x}$ by summation over $\boldsymbol{x} \oplus \mathbf{1}$, we obtain

$$\begin{aligned}\bar{W}_f(\boldsymbol{a}) &= \sum_{\boldsymbol{u}\in V_n} \lambda_f(\boldsymbol{u}) \sum_{\boldsymbol{x}\in V_n,\ \boldsymbol{x}\preccurlyeq \boldsymbol{u}\oplus\mathbf{1}} (-1)^{\langle \boldsymbol{a},\boldsymbol{x}\oplus\mathbf{1}\rangle} \\ &= \sum_{\boldsymbol{u}\in V_n} \lambda_f(\boldsymbol{u}) \sum_{\boldsymbol{x}\in V_n,\boldsymbol{x}\preccurlyeq \boldsymbol{u}\oplus\mathbf{1}} (-1)^{\langle \boldsymbol{a},\boldsymbol{x}\rangle\oplus \mathrm{wt}(\boldsymbol{a})} \\ &= (-1)^{\mathrm{wt}(\boldsymbol{a})} \sum_{\boldsymbol{u}\in V_n,\ \boldsymbol{a}\preccurlyeq \boldsymbol{u}} 2^{n-\mathrm{wt}(\boldsymbol{u})}\lambda_f(\boldsymbol{u}).\end{aligned}$$

Thus, we proved the following proposition.

PROPOSITION 2.58.

$$\bar{W}_f(\boldsymbol{a}) = (-1)^{\mathrm{wt}(\boldsymbol{a})} \sum_{\boldsymbol{u}\in V_n,\ \boldsymbol{a}\preccurlyeq \boldsymbol{u}} 2^{n-\mathrm{wt}(\boldsymbol{u})}\lambda_f(\boldsymbol{u}),$$

and in particular

$$\mathrm{wt}(f) = \bar{W}_f(\mathbf{0}) = \sum_{\boldsymbol{u}\in V_n} 2^{n-\mathrm{wt}(\boldsymbol{u})}\lambda_f(\boldsymbol{u}).$$

PROPOSITION 2.59.

$$\lambda_f(\boldsymbol{u}) = 2^{-n}(-2)^{\mathrm{wt}(\boldsymbol{u})} \sum_{\boldsymbol{x}\in V_n,\ \boldsymbol{x}\preccurlyeq \boldsymbol{u}} \bar{W}_f(\boldsymbol{x}).$$

PROOF. Using Problem 2.54, we obtain

$$\lambda_f(\boldsymbol{u}) = (-1)^{\mathrm{wt}(\boldsymbol{u})} \sum_{\boldsymbol{y}\in V_n,\ \boldsymbol{y}\preccurlyeq \boldsymbol{u}} (-1)^{\mathrm{wt}(\boldsymbol{y})} f(\boldsymbol{y}).$$

Applying the inversion formula to the Fourier coefficients, we see that

$$\begin{aligned}\lambda_f(\boldsymbol{u}) &= 2^{-n}(-1)^{\mathrm{wt}(\boldsymbol{u})} \sum_{\boldsymbol{y}\in V_n,\boldsymbol{y}\preccurlyeq \boldsymbol{u}} \left(\sum_{\boldsymbol{x}\in V_n} (-1)^{\mathrm{wt}(\boldsymbol{y})\oplus\langle \boldsymbol{x},\boldsymbol{y}\rangle} \bar{W}_f(\boldsymbol{x})\right) \\ &= 2^{-n}(-1)^{\mathrm{wt}(\boldsymbol{u})} \sum_{\boldsymbol{x}\in V_n} \bar{W}_f(\boldsymbol{x}) \left(\sum_{\boldsymbol{y}\in V_n,\boldsymbol{y}\preccurlyeq \boldsymbol{u}} (-1)^{\langle \boldsymbol{y},\boldsymbol{x}\oplus\mathbf{1}\rangle}\right).\end{aligned}$$

Using Problem 2.39, we finally obtain:

$$\lambda_f(\boldsymbol{u}) = 2^{-n}(-2)^{\mathrm{wt}(\boldsymbol{u})} \sum_{\boldsymbol{x}\in V_n,\boldsymbol{x}\preccurlyeq \boldsymbol{u}} \bar{W}_f(\boldsymbol{x}).$$ □

PROBLEM 2.60. Find expressions relating the coefficients of a numerical normal form and the Walsh–Hadamard coefficients.

Now we will give some examples of computing coefficients of a numerical normal form for a Boolean function.

Suppose that $l_{\boldsymbol{a},0} \in \mathcal{L}_n$, $\boldsymbol{a} \in V_n$. Using Theorem 2.55, we obtain:

$$\lambda_{l_{\boldsymbol{a},0}}(\boldsymbol{u}) = \begin{cases} (-2)^{\mathrm{wt}(\boldsymbol{u})-1} & \text{if } \boldsymbol{u} \preccurlyeq \boldsymbol{a} \text{ and } \boldsymbol{u} \neq \mathbf{0}; \\ 0 & \text{otherwise.}\end{cases}$$

Let $l_{\boldsymbol{a},1} \in \mathcal{A}_n$, $\boldsymbol{a} \in V_n$. Then

$$\lambda_{l_{\boldsymbol{a},1}}(\boldsymbol{u}) = \begin{cases} -(-2)^{\mathrm{wt}(\boldsymbol{u})-1} & \text{if } \boldsymbol{u} \preccurlyeq \boldsymbol{a} \text{ and } \boldsymbol{u} \neq \boldsymbol{0}; \\ 1 & \text{if } \boldsymbol{u} = \boldsymbol{0}; \\ 0 & \text{otherwise.} \end{cases}$$

Suppose that $\deg f = 2$. It is easy to show that in this case the coefficients $\lambda_f(\boldsymbol{u})$ are either 0 or

$$\begin{cases} \pm 1 & \text{if } \boldsymbol{u} = \boldsymbol{0}; \\ \pm 2^i, \ \frac{\mathrm{wt}(\boldsymbol{u})-2}{2} \leq i \leq \mathrm{wt}(\boldsymbol{u}) - 1, & \text{otherwise.} \end{cases}$$

Suppose that $r \in \{0, 1, \dots, n\}$ and let f be a Boolean function assuming value 1 only at the vectors of weight r. For this function,

$$\lambda_f(\boldsymbol{u}) = (-1)^{\mathrm{wt}(\boldsymbol{u})-r} \binom{\mathrm{wt}(\boldsymbol{u})}{r}.$$

Every symmetric Boolean function is a sum of functions of this type.

Now we introduce another representation of Boolean functions that is often used in applications to coding and cryptography. Put $q = 2^n$. Denote by $\mathcal{K} \subset \mathbb{F}_q[\boldsymbol{x}]$ the set of polynomials of degree $\leq q-2$ with zero constant term. Every polynomial from $\mathcal{K}$ defines a self-mapping of $\mathbb{F}_q$. In the field $\mathbb{F}_q$, choose a basis $\boldsymbol{\theta}_1, \boldsymbol{\theta}_2, \dots, \boldsymbol{\theta}_n$ over $\mathbb{F}_2$. Any $\boldsymbol{x} \in \mathbb{F}_q$ can be looked at as a vector whose components are coordinates in this basis $\boldsymbol{\theta}_1, \boldsymbol{\theta}_2, \dots, \boldsymbol{\theta}_n$:

$$\boldsymbol{x} = \bigoplus_{i=1}^{n} x^{(i)} \boldsymbol{\theta}_i = (\boldsymbol{\theta}_1, \ \boldsymbol{\theta}_2, \dots, \boldsymbol{\theta}_n) \begin{pmatrix} x^{(1)} \\ x^{(2)} \\ \dots \\ x^{(n)} \end{pmatrix},$$

where $x^{(i)} \in \mathbb{F}_2$, $i = 1, 2, \dots, n$.

For a polynomial $P \in \mathcal{K}$ and trace Tr_n, consider $\mathrm{Tr}_n(P(\boldsymbol{x}))$ (recall that $\mathrm{Tr}_n(\boldsymbol{\alpha}) = \boldsymbol{\alpha} + \boldsymbol{\alpha}^2 + \dots + \boldsymbol{\alpha}^{2^{n-1}}$). Since the values of Tr_n lie in $\mathbb{F}_2$, $\mathrm{Tr}_n(P(\boldsymbol{x}))$ is a function on V_n with values in $\mathbb{F}_2$, i.e., a Boolean function of the form

$$\begin{aligned} \mathrm{Tr}_n(P(\boldsymbol{x})) &= \mathrm{Tr}_n(P(x^{(1)}\boldsymbol{\theta}_1 \oplus x^{(2)}\boldsymbol{\theta}_2 + \dots + x^{(n)}\boldsymbol{\theta}_n)) \\ &= f_P(x^{(1)}, x^{(2)}, \dots, x^{(n)}) = f_P(\boldsymbol{x}). \end{aligned}$$

Recall, as noted above, that on the right-hand side of the latter equation $\boldsymbol{x}$ is considered as a vector from V_n.

Since $\#\mathcal{K} > \#\mathcal{F}_n$ for $n > 1$, it is clear that there exist different polynomials $P, Q \in \mathcal{K}$ such that $f_P(\boldsymbol{x}) = f_Q(\boldsymbol{x})$ for all x, i.e., that P and Q induce the same Boolean function. Therefore, it is natural to select in $\mathcal{K}$ a canonical set of polynomials corresponding to different Boolean functions.

DEFINITION 2.61. Polynomials $P, Q \in \mathbb{F}_q[\boldsymbol{x}]$ are called *trace equivalent* if

$$\mathrm{Tr}_n(P(\boldsymbol{x})) = \mathrm{Tr}_n(Q(\boldsymbol{x}))$$

for every $\boldsymbol{x} \in \mathbb{F}_q$.

PROBLEM 2.62. Prove that $\mathrm{Tr}_n(P(\boldsymbol{x})) = 0$ for every $\boldsymbol{x} \in \mathbb{F}_q$ if and only if $P(\boldsymbol{x}) = Q(\boldsymbol{x}) + Q^2(\boldsymbol{x})$ for some polynomial $Q \in \mathbb{F}_q[\boldsymbol{x}]$.

Recall the concept of cyclotomic class (see §1.3). The set of residues modulo $2^n - 1$ splits into subsets called *cyclotomic classes* modulo $2^n - 1$. The cyclotomic class $\mathcal{C}_{n,s}$ in which s is the smallest number consists of numbers

$$\mathcal{C}_{n,s} = \{s, s \cdot 2, s \cdot 2^2, \dots, s \cdot 2^{n_s - 1}\},$$

where n_s is the smallest integer such that $s \cdot 2^{n_s} \equiv s \pmod{2^n - 1}$. We will call the number s the *representative* of this cyclotomic class. We will denote the set of all nonzero representatives of cyclotomic classes by Π_n.

The cardinality n_s of the cyclotomic class $\mathcal{C}_{n,s}$ divides n (this property means that in the field $\mathbb{F}_{2^n}$ there exists a unique subfield $\mathbb{F}_{2^{n_s}}$ of order n_s). The relative trace $\mathrm{Tr}_{2^n/2^{n_s}}$ from the field $\mathbb{F}_{2^n}$ to the field $\mathbb{F}_{2^{n_s}}$ is an $\mathbb{F}_{2^{n_s}}$-linear map with kernel consisting of elements of the form $b \oplus b^{2^{n_s}}$, $b \in \mathbb{F}_{2^n}$. Each coset of the additive group of the field $\mathbb{F}_{2^n}$ contains only one element $c_s \in \mathbb{F}_{2^{n_s}}$.

DEFINITION 2.63. We will say that a polynomial $P(\boldsymbol{x}) \in \mathcal{K}$ is *cyclotomically reduced* (or is a cr-polynomial) if it has the form

$$P(\boldsymbol{x}) = \bigoplus_{s \in \Pi_n} c_s \boldsymbol{x}^s, \quad c_s \in \mathbb{F}_{2^{n_s}}.$$

THEOREM 2.64. 1. *Every polynomial from $\mathcal{K}$ is trace equivalent to a cyclotomically reduced polynomial.*

2. *Different cyclotomically reduced polynomials are not trace equivalent.*

PROOF. Put $P(\boldsymbol{x}) = \bigoplus_{i=1}^{2^n-2} a_i \boldsymbol{x}^i \in \mathcal{K}$. We will represent this polynomial in the form

$$P(\boldsymbol{x}) = \bigoplus_{s \in \Pi_n} \bigoplus_{j=0}^{n_s - 1} a_{s \cdot 2^j} \boldsymbol{x}^{s \cdot 2^j} = \bigoplus_{s \in \Pi_n} P_s(\boldsymbol{x}).$$

Fix an element $s \in \Pi_n$ and consider the polynomial $P_s(\boldsymbol{x})$, which is called *cyclotomically homogeneous*:

$$P_s(\boldsymbol{x}) = \bigoplus_{j=0}^{n_s - 1} a_{s \cdot 2^j} \boldsymbol{x}^{s \cdot 2^j}.$$

Since trace is invariant under the Frobenius automorphism, for every integer t the following equality holds:

$$\mathrm{Tr}_n(a_{s \cdot 2^j} \boldsymbol{x}^{s \cdot 2^j}) = \mathrm{Tr}_n(a_{s \cdot 2^j}^{2^t} \boldsymbol{x}^{s \cdot 2^{j+t}})$$

for all $s \in \Pi_n$, $j = 0, \dots, n_s - 1$, $a_{s \cdot 2^j} \in \mathbb{F}_{2^n}$. In particular, for $t = n_s - j$, one has (see the definition of n_s):

$$\mathrm{Tr}_n(a_{s \cdot 2^j} \boldsymbol{x}^{s \cdot 2^j}) = \mathrm{Tr}_n(a_{s \cdot 2^j}^{2^{n_s - j}} \boldsymbol{x}^s).$$

Linearity of the trace implies the following chain of equalities:

$$\mathrm{Tr}_n(P_s(\boldsymbol{x})) = \bigoplus_{j=0}^{n_s - 1} \mathrm{Tr}_n(a_{s \cdot 2^j} \boldsymbol{x}^{s \cdot 2^j}) = \bigoplus_{j=0}^{n_s - 1} \mathrm{Tr}_n(a_{s \cdot 2^j}^{2^{n_s - j}} \boldsymbol{x}^s) = \mathrm{Tr}_n\left(\boldsymbol{x}^s \cdot \bigoplus_{j=0}^{n_s - 1} a_{s \cdot 2^j}^{2^{n_s - j}}\right).$$

Thus, any cyclotomically homogeneous $P_s(\boldsymbol{x})$ is trace equivalent to the following monomial:

$$\left(\bigoplus_{j=0}^{n_s - 1} a_{s \cdot 2^j}^{2^{n_s - j}}\right) \boldsymbol{x}^s.$$

With similar calculations for every s from Π_n we obtain the first part of the theorem.

PROBLEM 2.65. Show that distinct cyclotomically reduced polynomials are not trace equivalent.

Using the result of Problem 2.65 we complete the proof of the theorem. □

If an arbitrary cr-polynomial P satisfies the relation $P(\mathbf{0}) = 0$, then $f_P(\mathbf{0}) = 0$. Moreover, $\deg f_P < n$. Therefore, cr-polynomials are not enough for representation of all Boolean functions on $\mathcal{F}_n$. Denote by $\mathcal{O}_n$ the set of polynomials $P' \in \mathbb{F}_{2^n}[\boldsymbol{x}]$ having the form

$$P'(\boldsymbol{x}) = c_0\boldsymbol{\alpha} \oplus P(\boldsymbol{x}) \oplus c_{2^n-1}\boldsymbol{\alpha}\boldsymbol{x}^{2^n-1},$$

where $P(\boldsymbol{x})$ is an arbitrary cr-polynomial, $\boldsymbol{\alpha} \in \mathbb{F}_{2^n}$ is such that $\mathrm{Tr}_n(\boldsymbol{\alpha}) = 1$, and c_0 and c_{2^n-1} are elements of the field $\mathbb{F}_2$.

PROBLEM 2.66. 1. Prove that distinct polynomials in the set $\mathcal{O}_n$ are not trace equivalent.

2. Prove that every polynomial in $\mathbb{F}_{2^n}[\boldsymbol{x}]$ is trace equivalent to a polynomial in $\mathcal{O}_n$.

LEMMA 2.67. *For every Boolean function $f \in \mathcal{F}_n$ there exists a unique polynomial $P \in \mathcal{O}_n$ such that*

$$f_P = f.$$

PROOF. Put $\Pi'_n = \Pi_n \cup \{0, 2^n - 1\}$. From the definition of the set $\mathcal{O}_n$ we have

$$\#\mathcal{O}_n = \prod_{s\in\Pi'_n} 2^{n_s} = 2^{\sum_{s\in\Pi'_n} n_s},$$

where $n_0 = n_{2^n-1} = 1$. Let $M_{2,n}(t) = \#\{s \in \Pi'_n \mid n_s = t\}$. Then, using the inclusion-exclusion principle and Möbius inversion formula for cyclic sequences (see [**156**]), we get

$$\#\mathcal{O}_n = 2^{\sum_{s\in\Pi'_n} n_s} = 2^{\sum_{t|n} t\cdot M_{2,n}(t)} = 2^{2^n} = \#\mathcal{F}_n.$$

The relation obtained for the cardinality of $\mathcal{O}_n$ and the result of Problem 2.66 prove the lemma. □

The polynomials in $\mathcal{O}_n$ will be called polynomials representing Boolean functions in $\mathcal{F}_n$.

We will consider two more important notions often used in the study of Boolean functions.

DEFINITION 2.68. The *derivative of a Boolean function* f in the direction of $\boldsymbol{u}$ ($\boldsymbol{u} \neq \mathbf{0}$, $\boldsymbol{u} \in V_n$) is the following Boolean function:

$$(D_{\boldsymbol{u}}f)(\boldsymbol{x}) = f(\boldsymbol{x} \oplus \boldsymbol{u}) \oplus f(\boldsymbol{x}), \quad \boldsymbol{x} \in V_n.$$

DEFINITION 2.69. The derivative of a Boolean function f in the direction of a subset $L \subset V_n$ is the following Boolean function:

$$(D_L f)(\boldsymbol{x}) = \bigoplus_{\boldsymbol{u}\in L} f(\boldsymbol{x} \oplus \boldsymbol{u}).$$

It is easy to see that if $\boldsymbol{u}_1, \ldots, \boldsymbol{u}_k$ $(k \leq n)$ is a basis of the space L, then

$$D_L f = D_{\boldsymbol{u}_k} D_{\boldsymbol{u}_{k-1}} \cdots D_{\boldsymbol{u}_1} f.$$

PROBLEM 2.70. Prove the following properties of the derivatives:

1. For an arbitrary subspace $L \subset V_n$ and any $f \in \mathcal{F}_n$, $\boldsymbol{u} \in L$, $\boldsymbol{x} \in V_n$, one has
$$D_L f(\boldsymbol{x}) = D_L f(\boldsymbol{x} \oplus \boldsymbol{u}).$$
2. For any $f, g \in \mathcal{F}_n$ and any subspace $L \subset V_n$,
$$D_L(f \oplus g) = D_L(f) \oplus D_L(g).$$
3. For any $f \in \mathcal{F}_n$ and $\boldsymbol{u}, \boldsymbol{v}, \boldsymbol{x} \in V_n$,
$$D_{\boldsymbol{u}\oplus\boldsymbol{v}} f(\boldsymbol{x}) = D_{\boldsymbol{u}} f(\boldsymbol{x}) \oplus D_{\boldsymbol{v}} f(\boldsymbol{x} \oplus \boldsymbol{u}).$$
4. The equation
$$D_{\boldsymbol{u}} f \equiv 0$$
holds for all $\boldsymbol{u} \in V_n$ if and only if $f \equiv \text{const}$.
5. The equality
$$D_{\boldsymbol{u}} f \equiv \text{const}$$
holds for all $\boldsymbol{u} \in V_n$ if and only if f is an affine function.

PROBLEM 2.71. Let $\varphi_{\boldsymbol{u}}(\boldsymbol{x}) \in \mathcal{F}_n$, $\boldsymbol{u} \in V_n$, be a set of Boolean functions. Prove that the existence of a Boolean function $f \in \mathcal{F}_n$ such that
$$D_{\boldsymbol{u}} f(\boldsymbol{x}) = \varphi_{\boldsymbol{u}}(\boldsymbol{x})$$
for any $\boldsymbol{u}, \boldsymbol{x} \in V_n$ is equivalent to the condition that
$$\varphi_{\boldsymbol{u}\oplus\boldsymbol{v}}(\boldsymbol{x}) = \varphi_{\boldsymbol{u}}(\boldsymbol{x}) \oplus \varphi_{\boldsymbol{v}}(\boldsymbol{x} \oplus \boldsymbol{u})$$
for all $\boldsymbol{u}, \boldsymbol{v}, \boldsymbol{x} \in V_n$.

(This condition can be thought of as a condition for differentiability of the system of Boolean functions $\varphi_{\boldsymbol{u}}(\boldsymbol{x}) \in \mathcal{F}_n$, $\boldsymbol{u} \in V_n$. For the general case of functions on a finite abelian group this condition can be found in [**234**].)

To conclude this section we note that we do not claim completeness. There exist various other representations of Boolean functions which we do not mention here but which play an important role in various branches of discrete mathematics. (For example, such is the disjunctive normal form.) In this book we are interested in the representations that are used in cryptology and coding theory.

2.3. Autocorrelation and crosscorrelation

The crosscorrelation and autocorrelation of Boolean functions are among the most important tools for the study of cryptographic properties of Boolean functions. These tools allow us to characterize various properties of Boolean functions as well as to find useful relations between those properties.

DEFINITION 2.72. Let $f, g \in \mathcal{F}_n$. The integer-valued function
$$\Delta_{f,g}(\boldsymbol{u}) = \sum_{x \in V_n} (-1)^{f(x) \oplus g(x \oplus u)},$$
defined on the space V_n, is called the *crosscorrelation function* of f and g.

The following proposition is immediate.

PROPOSITION 2.73. *For any $\boldsymbol{u} \in V_n$ and any $f, g \in \mathcal{F}_n$,*
$$\Delta_{f,g}(\boldsymbol{u}) = \Delta_{g,f}(\boldsymbol{u}).$$

DEFINITION 2.74. Let $f, g \in \mathcal{F}_n$. The integer-valued function

$$\Delta_f(\boldsymbol{u}) = \sum_{x \in V_n} (-1)^{f(x) \oplus f(x \oplus u)},$$

defined on V_n is called the *autocorrelation function* of f.

It is easy to see that the autocorrelation function of f in $\boldsymbol{u} \in V_n$ coincides with the zero Walsh–Hadamard coefficient of the derivative of the function f in the direction $\boldsymbol{u}$:

$$\Delta_f(\boldsymbol{u}) = W_{D_u f}(0).$$

The following proposition is immediate.

PROPOSITION 2.75. *For any* $\boldsymbol{u} \in V_n$ *and any* $f \in \mathcal{F}_n$,

$$\Delta_{f,f}(\boldsymbol{u}) = \Delta_f(\boldsymbol{u}).$$

Not every function on V_n (i.e., not every 2^n-tuple of numbers) can be the autocorrelation function of a Boolean function. Moreover, it is not easy to answer the question whether a given 2^n-tuple of numbers is a value of an autocorrelation function.

PROBLEM 2.76 (open problem). Let $\Delta : V_n \to \mathbb{Z}$ be an integer-valued function on V_n such that $\Delta(\mathbf{0}) = 2^n$. Find efficient necessary and sufficient conditions for the existence of a Boolean function $f \in \mathcal{F}_n$ such that $\Delta_f = \Delta$.

Definition 2.35 of the Walsh–Hadamard transform can be rewritten in terms of matrices. To this end we need the following definition.

DEFINITION 2.77. A *Sylvester-type Hadamard matrix* is a $(2^n \times 2^n)$ matrix H_n (where n is a natural number) such that

$$H_1 = \begin{pmatrix} 1 & 1 \\ 1 & -1 \end{pmatrix}, \qquad H_n = H_1 \otimes H_{n-1} \qquad \text{for} \quad n > 1,$$

where $\otimes$ is the Kronecker (tensor) product of matrices.

The most important properties of these matrices follow from the relation

$$H_n = H_n^\top = 2^n H_n^{-1}.$$

In particular, the last relation implies that

$$H_n H_n^\top = H_n H_n = 2^n E_{2^n},$$

where E_{2^n} is the $(2^n \times 2^n)$ identity matrix.

The matrix $H_n = (h_{\boldsymbol{u},\boldsymbol{v}}), \boldsymbol{u}, \boldsymbol{v} \in V_n$, with rows and columns labeled by vectors of the space V_n, is essentially the character table of the elementary abelian 2-group, i.e., $h_{\boldsymbol{u},\boldsymbol{v}} = (-1)^{\langle \boldsymbol{u},\boldsymbol{v} \rangle}$. In order to enumerate the rows and columns of this matrix conveniently, the one-to-one mapping between V_n and $\mathbb{Z}_{2^n}$ is used; e.g., the vector $\boldsymbol{u}$ corresponds to the integer $k(\boldsymbol{u}) = u^{(1)} + u^{(2)} \cdot 2 + \cdots + u^{(n)} \cdot 2^{n-1}$. Then, using Definitions 2.35 and 2.77, the Walsh–Hadamard coefficients can be represented as follows:

$$(W_f(0), \ldots, W_f(2^n - 1)) = ((-1)^{f(0)}, \ldots, (-1)^{f(2^n-1)}) \cdot H_n.$$

DEFINITION 2.78. Two Boolean functions $f, g \in \mathcal{F}_n$ are called *perfectly uncorrelated* if for any $\boldsymbol{u} \in V_n$ the following equality holds:

$$\Delta_{f,g}(\boldsymbol{u}) = 0.$$

DEFINITION 2.79. Two Boolean functions $f, g \in \mathcal{F}_n$ are called *uncorrelated of degree* k, $0 \leq k \leq n$, if for any $\boldsymbol{u} \in V_n$ such that $0 \leq \mathrm{wt}(u) \leq k$ the following equality holds:

$$\Delta_{f,g}(\boldsymbol{u}) = 0.$$

Now we state the so-called crosscorrelation theorem. This theorem is important because it can be used as a tool to obtain new cryptographic properties of Boolean functions as well as results that have already been derived using other arguments.

THEOREM 2.80 (Crosscorrelation theorem). *Suppose that* $f, g \in \mathcal{F}_n$. *Then*

$$(\Delta_{f,g}(0), \ldots, \Delta_{f,g}(2^{n-1}))H_n = (W_f(0)W_g(0), \ldots, W_f(2^n-1)W_g(2^n-1)).$$

PROOF. We will perform the following simple transformations:

$$\begin{aligned}\sum_{\boldsymbol{u}\in V_n} \Delta_{f,g}(\boldsymbol{u})(-1)^{\langle \boldsymbol{u},\boldsymbol{v}\rangle} &= \sum_{\boldsymbol{u}\in V_n}\sum_{\boldsymbol{x}\in V_n}(-1)^{f(\boldsymbol{x})\oplus g(\boldsymbol{x}\oplus\boldsymbol{u})}(-1)^{\langle \boldsymbol{u},\boldsymbol{v}\rangle}\\ &= \sum_{\boldsymbol{x}\in V_n}(-1)^{f(\boldsymbol{x})}\sum_{\boldsymbol{u}\in V_n}(-1)^{g(\boldsymbol{x}\oplus\boldsymbol{u})\oplus\langle \boldsymbol{u},\boldsymbol{v}\rangle}.\end{aligned}$$

Changing the variables $\boldsymbol{z} = \boldsymbol{x} \oplus \boldsymbol{u}$, we get:

$$\begin{aligned}\sum_{\boldsymbol{x}\in V_n}(-1)^{f(\boldsymbol{x})}\sum_{\boldsymbol{u}\in V_n}(-1)^{g(\boldsymbol{x}\oplus\boldsymbol{u})\oplus\langle \boldsymbol{u},\boldsymbol{v}\rangle} &= \sum_{\boldsymbol{x}\in V_n}(-1)^{f(\boldsymbol{x})}\sum_{\boldsymbol{z}\in V_n}(-1)^{g(\boldsymbol{z})\oplus\langle \boldsymbol{z}\oplus\boldsymbol{x},\boldsymbol{v}\rangle}\\ &= \sum_{\boldsymbol{x}\in V_n}(-1)^{f(\boldsymbol{x})\oplus\langle \boldsymbol{x},\boldsymbol{v}\rangle}\sum_{\boldsymbol{z}\in V_n}(-1)^{g(\boldsymbol{z})\oplus\langle \boldsymbol{z},\boldsymbol{v}\rangle}\\ &= W_f(\boldsymbol{v})W_g(\boldsymbol{v}).\end{aligned}$$

□

The following statement is actually a Boolean analog of the well-known Wiener–Khinchin theorem for continuous functions.

COROLLARY 2.81. *Suppose that* $f \in \mathcal{F}_n$. *Then*

$$(\Delta_f(0), \ldots, \Delta_f(2^n-1))H_n = (W_f^2(0), \ldots, W_f^2(2^n-1)).$$

PROOF. Apply Theorem 2.80 with $f = g$. □

PROBLEM 2.82. Let a Boolean function $h \in \mathcal{F}_{n+m}$ of $n+m$ variables be represented as a sum of two functions $f \in \mathcal{F}_n$ and $g \in \mathcal{F}_m$ of n and m variables, respectively,

$$h(\boldsymbol{x}, \boldsymbol{y}) = f(\boldsymbol{x}) \oplus g(\boldsymbol{y}),$$

where $\boldsymbol{x} \in V_n$, $\boldsymbol{y} \in V_m$. Prove that

$$\Delta_h(\alpha, \beta) = \Delta_f(\alpha)\Delta_g(\beta),$$

for all $\alpha \in V_n$, $\beta \in V_m$.

COROLLARY 2.83. *Suppose that* $f, g \in \mathcal{F}_n$. *Then*

$$2^n(\Delta_{f,g}(0), \ldots, \Delta_{f,g}(2^n-1)) = (W_f(0)W_g(0), \ldots, W_f(2^n - 1W_g(2^n-1))H_n.$$

The last corollary implies Parseval's equality

$$\sum_{\boldsymbol{u}\in V_n} W_f^2(\boldsymbol{u}) = 2^n\Delta_f(0) = 2^{2n}.$$

COROLLARY 2.84 (Convolution theorem). *If* $h(\boldsymbol{x}) = f(\boldsymbol{x}) \oplus g(\boldsymbol{x})$, *where* $f, g \in \mathcal{F}_n$, *then*

$$W_h(\boldsymbol{u}) = \frac{1}{2^n} \sum_{\boldsymbol{x} \in V_n} W_f(\boldsymbol{x}) W_g(\boldsymbol{x} \oplus \boldsymbol{u})$$

for any $\boldsymbol{u} \in V_n$.

PROOF. Put $g_1(\boldsymbol{x}) = g(\boldsymbol{x}) \oplus \langle \boldsymbol{u}, \boldsymbol{x} \rangle$. Then $W_{g_1}(\boldsymbol{v}) = W_g(\boldsymbol{v} \oplus \boldsymbol{u})$ for all $\boldsymbol{v} \in V_n$ and $W_h(\boldsymbol{u}) = \Delta_{f,g_1}(0)$. Using Corollary 2.83, we find that

$$W_h(\boldsymbol{u}) = \frac{1}{2^n} \sum_{\boldsymbol{x} \in V_n} W_f(\boldsymbol{x}) W_{g_1}(\boldsymbol{x}) = \frac{1}{2^n} \sum_{\boldsymbol{x} \in V_n} W_f(\boldsymbol{x}) W_g(\boldsymbol{x} \oplus \boldsymbol{u}). \qquad \square$$

For any Boolean function $f \in \mathcal{F}_n$ consider the set of its nonzero Walsh–Hadamard coefficients

$$V_n \setminus CI_f = \{\boldsymbol{u} \in V_n \mid W_f(\boldsymbol{u}) \neq 0\}.$$

This set is very important for cryptographic applications: later on, many important cryptographic properties will be defined in terms of nonzero Walsh–Hadamard coefficients. The following set will be studied in detail:

$$CI_f = \{\boldsymbol{v} \in V_n \mid W_f(\boldsymbol{v}) = 0\}.$$

DEFINITION 2.85. One says that the Boolean functions $f, g \in \mathcal{F}_n$ have *disjoint spectra* if

$$V_n \setminus CI_f \cap V_n \setminus CI_g = \emptyset.$$

PROBLEM 2.86. Prove that Boolean functions $f, g \in \mathcal{F}_n$ have disjoint spectra if and only if $W_f(\boldsymbol{u}) W_g(\boldsymbol{u}) = 0$ for any $\boldsymbol{u} \in V_n$.

LEMMA 2.87. *Boolean functions* $f, g \in \mathcal{F}_n$ *have disjoint spectra if and only if they are perfectly uncorrelated.*

PROOF. Suppose the functions f and g have disjoint spectra, i.e., $W_f(\boldsymbol{u}) W_g(\boldsymbol{u}) = 0$ for all $u \in V_n$ (see Problem 2.86). It follows from Corollary 2.83 that

$$\Delta_{f,g}(\boldsymbol{u}) = \frac{1}{2^n} \sum_{x \in V_n} W_f(x) W_g(x) (-1)^{\langle x, u \rangle} = 0.$$

The converse statement follows from Theorem 2.80. $\square$

LEMMA 2.88. *Suppose that* $f, g \in \mathcal{F}_n$. *Then*

$$\sum_{u \in V_n} \Delta_{f,g}^2(u) \leq 2^{3n}.$$

PROOF. Put

$$\begin{aligned} \boldsymbol{W} &= (W_f(0)W_g(0), \ldots, W_f(2^n - 1)W_g(2^n - 1)), \\ \Delta &= (\Delta_{f,g}(0), \ldots, \Delta_{f,g}(2^n - 1)). \end{aligned}$$

By Theorem 2.80,

$$\langle \boldsymbol{W}, \boldsymbol{W} \rangle = \langle \Delta H_n, \Delta H_n \rangle = \langle \Delta H_n H_n^T, \Delta \rangle = 2^n \langle \Delta, \Delta \rangle.$$

Here the inner (scalar) product $\langle \cdot, \cdot \rangle$ is taken over the field of real numbers.

One has

$$\begin{aligned}2^n \sum_{\boldsymbol{u}\in V_n} \Delta_{f,g}^2(\boldsymbol{u}) &= \sum_{\boldsymbol{u}\in V_n} W_f^2(\boldsymbol{u})W_g^2(\boldsymbol{u}) \\ &\le \left(\sum_{\boldsymbol{u}\in V_n} W_f^2(\boldsymbol{u})\right)\left(\sum_{\boldsymbol{u}\in V_n} W_g^2(\boldsymbol{u})\right) \\ &= 2^{2n}\cdot 2^{2n} = 2^{4n}.\end{aligned}$$

The last equality implies the statement of the lemma. □

THEOREM 2.89. *Suppose that $f, g \in \mathcal{F}_n$ and L is a subspace of V_n. Then*

$$\sum_{x\in L} W_f(x)W_g(x) = 2^{\dim L} \sum_{u\in L^\perp} \Delta_{f,g}(\boldsymbol{u}).$$

PROOF. Using Theorem 2.80 we have

$$\sum_{\boldsymbol{x}\in L} W_f(\boldsymbol{x})W_g(\boldsymbol{x}) = \sum_{\boldsymbol{x}\in L}\sum_{\boldsymbol{u}\in V_n} \Delta_{f,g}(\boldsymbol{u})(-1)^{\langle \boldsymbol{u},\boldsymbol{x}\rangle} = \sum_{\boldsymbol{u}\in V_n} \Delta_{f,g}(\boldsymbol{u}) \sum_{\boldsymbol{x}\in L}(-1)^{\langle \boldsymbol{u},\boldsymbol{x}\rangle}.$$

To complete the proof of the lemma it suffices to apply Problem 2.39. □

THEOREM 2.90. *For any $f, g \in \mathcal{F}_n$ the following inequality holds:*

$$N\Delta_{f,g}\cdot NW_{f,g} \ge \max_{\boldsymbol{u}\in V_n} |\,\Delta_{f,g}(\boldsymbol{u})|,$$

where $N\Delta_{f,g}(\boldsymbol{u}) = |\{\boldsymbol{u}\in V_n \mid \Delta_{f,g}(\boldsymbol{u}) \ne 0\}|$ and $NW_{f,g} = |\{\boldsymbol{u}\in V_n \mid W_f(\boldsymbol{u})W_g(\boldsymbol{u}) \ne 0\}|$.

PROOF. The case $\Delta_{f,g}(\boldsymbol{u}) = 0$ for any $\boldsymbol{u}\in V_n$ is easy. Suppose now that there exists a vector $\boldsymbol{u}$ such that $\Delta_{f,g}(\boldsymbol{u}) \ne 0$. Then for a vector $\boldsymbol{u}_0 \in V_n$ we have

$$\Delta_{f,g}(\boldsymbol{u}_0) = \max_{\boldsymbol{u}\in V_n} |\Delta_{f,g}(\boldsymbol{u})| > 0.$$

Consider the crosscorrelation between the functions $f(\boldsymbol{x})$ and $g_1(\boldsymbol{x}) = g(\boldsymbol{x}\oplus\boldsymbol{u}_0)$. We have that $\Delta_{f,g_1}(\boldsymbol{0}) = \Delta_{f,g}(\boldsymbol{u}_0)$, i.e., the maximum of the absolute value of the crosscorrelation function is attained at $\boldsymbol{0}$. If $\Delta_{f,g}(\boldsymbol{u}) < 0$, consider the correlation between the functions $f(\boldsymbol{x})$ and $g(\boldsymbol{x}\oplus\boldsymbol{u}_0)\oplus 1$. It is easy to see that no operations performed above change the values of $N\Delta_{f,g}$ and $NW_{f,g}$. Thus, without loss of generality we can assume that $\Delta_{f,g}(\boldsymbol{0}) > 0$ and $\Delta_{f,g}(\boldsymbol{0}) = \max_{\boldsymbol{u}\in V_n} |\Delta_{f,g}(\boldsymbol{u})|$. Using Corollary 2.83, we see that

$$2^n\Delta_{f,g}(\boldsymbol{0}) = \sum_{\boldsymbol{u}\in V_n} W_f(\boldsymbol{u})W_g(\boldsymbol{u}).$$

Since $\Delta_{f,g}(\boldsymbol{0}) > 0$, not all $W_f(\boldsymbol{u})W_g(\boldsymbol{u})$ are equal to zero, and so there exist a vector $\boldsymbol{z}_0$ and $\varepsilon\in\mathbb{F}_2$ such that

$$W_f(\boldsymbol{z}_0)W_g(\boldsymbol{z}_0) = (-1)^\varepsilon \max_{\boldsymbol{u}\in V_n} |W_f(\boldsymbol{u})W_g(\boldsymbol{u})|.$$

Since the value $\Delta_{f,g}(\boldsymbol{u})$ is less than or equal to 2^n, we have

$$2^n N\Delta_{f,g} \ge \sum_{\boldsymbol{u}\in V_n} \Delta_{f,g}(\boldsymbol{u}) = W_f(\boldsymbol{0})W_g(\boldsymbol{0}).$$

Put $h_1(\boldsymbol{x}) = f(\boldsymbol{x})\oplus\langle \boldsymbol{z}_0,\boldsymbol{x}\rangle$ and $h_2(\boldsymbol{x}) = g(\boldsymbol{x})\oplus\langle \boldsymbol{z}_0,\boldsymbol{x}\rangle\oplus\varepsilon$. Then

$$W_{h_1}(\boldsymbol{0})W_{h_2}(\boldsymbol{0}) = (-1)^\varepsilon W_f(\boldsymbol{z}_0)W_g(\boldsymbol{z}_0) = \max_{\boldsymbol{u}\in V_n} |W_f(\boldsymbol{u})W_g(\boldsymbol{u})|.$$

It is also clear that

$$N\Delta_{f,g} = |\{\boldsymbol{u} \in V_n \mid \Delta_{f,g}(\boldsymbol{u}) \neq 0\}| = |\{\boldsymbol{u} \in V_n \mid \Delta_{h_1,h_2} \neq 0\}| = N\Delta_{h_1,h_2}.$$

Hence,

$$\begin{aligned} 2^n N\Delta_{f,g} &= 2^n N\Delta_{h_1,h_2} \geq \sum_{\boldsymbol{u}\in V_n} \Delta_{h_1,h_2}(\boldsymbol{u}) = W_{h_1}(\boldsymbol{0})W_{h_2}(\boldsymbol{0}) \\ &= W_f(\boldsymbol{z}_0)W_g(\boldsymbol{z}_0) = \max_{\boldsymbol{u}\in V_n} |W_f(\boldsymbol{u})W_g(\boldsymbol{u})|. \end{aligned} \tag{2.23}$$

We also have that

$$\begin{aligned} (2^n - NW_{f,g}) &\geq \frac{\sum_{\boldsymbol{u}\in V_n} W_f(\boldsymbol{u})W_g(\boldsymbol{u})}{\max_{\boldsymbol{u}\in V_n} |W_f(\boldsymbol{u})W_g(\boldsymbol{u})|} = \frac{2^n \Delta_{f,g}(0)}{\max_{\boldsymbol{u}\in V_n} |W_f(\boldsymbol{u})W_g(\boldsymbol{u})|} \\ &= \frac{2^n \max_{\boldsymbol{u}\in V_n} \Delta_{f,g}(0)}{\max_{\boldsymbol{u}\in V_n} |W_f(\boldsymbol{u})W_g(\boldsymbol{u})|} \end{aligned} \tag{2.24}$$

Multiplying (2.23) and (2.24) we obtain the theorem. □

If we substitute $f = g$ in Theorem 2.90, we obtain the following corollary.

COROLLARY 2.91. *For any $f \in \mathcal{F}_n$ the following inequality holds:*

$$N\Delta_f \cdot NW_f \geq 2^n, \tag{2.25}$$

where

$$N\Delta_f = \#|\{\boldsymbol{u} \in V_n \mid \Delta_f(\boldsymbol{u}) \neq 0\}|$$

and

$$NW_f = \#|\{\boldsymbol{u} \in V_n \mid W_f(\boldsymbol{u}) \neq 0\}| = \#|V_n \setminus CI_f|.$$

REMARK 2.92. The inequality (2.25) could be in a sense regarded as an analog of the well-known Heisenberg's uncertainty principle in quantum physics. As we will see in the chapters to follow, in order to ensure certain cryptographic properties of a Boolean function it is necessary for the function to have a small number of zero Walsh–Hadamard coefficients, but to ensure other properties it has to have a small number of zeros of the autocorrelation function. The inequality (2.25) shows that it is impossible to have both, i.e., if a Boolean function is "good" in one aspect, it will be "bad" in another.

2.4. Group algebra of Boolean functions

The set of Boolean functions $\mathcal{F}_n$ can be made a commutative ring by defining the convolution operation $*$ as follows:

$$(f * g)(\boldsymbol{x}) = \bigoplus_{\boldsymbol{y}\in V_n} f(\boldsymbol{x} \oplus \boldsymbol{y})g(\boldsymbol{y}) = \bigoplus_{\boldsymbol{z}\in V_n} f(\boldsymbol{z})g(\boldsymbol{z} \oplus \boldsymbol{x})$$

for any $\boldsymbol{x} \in V_n$. This ring is called a *group algebra* and is denoted by $\mathbb{F}_2 V_n$. In the general case the group algebra is built from two objects: a group and a field. When working with Boolean functions it is sometimes more convenient to use the properties of group algebras.

As an example we will show how differentiation of Boolean functions can be interpreted in terms of convolution. By Definition 2.68 the derivative of a Boolean function $f \in \mathcal{F}_n$ in the direction of a subspace $L \subset V_n$ is equal to

$$(D_L f)(\boldsymbol{x}) = \bigoplus_{\boldsymbol{u}\in L} f(\boldsymbol{x} \oplus \boldsymbol{u}).$$

It follows from the definition of convolution that

$$D_L f = f * I_L,$$

where I_L is the indicator function for the subspace L:

$$I_L(\boldsymbol{x}) = \begin{cases} 1 & \text{if } \boldsymbol{x} \in L; \\ 0 & \text{if } \boldsymbol{x} \notin L. \end{cases}$$

Thus, convolution of Boolean functions can be viewed as a generalization of differentiation, and one can expect that many properties proved for Boolean derivatives have analogues for convolution.

Any Boolean function $f \in \mathcal{F}_n$ can be related to the convolution operator S_f in $\mathbb{F}_2 V_n$ as follows:

$$S_f g = f * g.$$

The set of operators S_f for all f is the so-called *product algebra* of $\mathbb{F}_2 V_n$, which is isomorphic to the algebra $\mathbb{F}_2 V_n$ itself. The operator S_f is linear and its properties determine the properties of the function f.

The following property of the group algebra $\mathbb{F}_2 V_n$ is essentially a theorem from functional analysis.

THEOREM 2.93. *The convolution operators are the only linear operators that commute with the shift operators, i.e., they are the solutions of the operational equation*

$$St_{\boldsymbol{u}} = t_{\boldsymbol{u}} S \tag{2.26}$$

for all $\boldsymbol{u} \in V_n$. *Here,* $S\colon \mathbb{F}_2 V_n \to \mathbb{F}_2 V_n$ *is a linear operator and* $t_{\boldsymbol{u}}\colon \mathcal{F}_n \to \mathcal{F}_n$, $\boldsymbol{u} \in V_n$, *is the shift operator defined by the formula*

$$(t_{\boldsymbol{u}} f)(\boldsymbol{x}) = f(\boldsymbol{x} \oplus \boldsymbol{u})$$

for all $f \in \mathcal{F}_n$, $\boldsymbol{x} \in V_n$.

PROOF. Let $S = S_f$ be the convolution operator corresponding to the Boolean function $f \in \mathcal{F}_n$. For any $g \in \mathcal{F}_n$, $\boldsymbol{x} \in V_n$ one has

$$(S_f t_{\boldsymbol{u}} g)(\boldsymbol{x}) = \sum_{\boldsymbol{y} \in V_n} f(\boldsymbol{x} \oplus \boldsymbol{y}) g(\boldsymbol{y} \oplus \boldsymbol{u}) = \sum_{\boldsymbol{z} \in V_n} f(\boldsymbol{x} \oplus \boldsymbol{u} \oplus \boldsymbol{z}) g(\boldsymbol{z}) = (t_{\boldsymbol{u}} S_f g)(\boldsymbol{x}).$$

Let the linear operator $S\colon \mathbb{F}_2 V_n \to \mathbb{F}_2 V_n$ be a solution of (2.26). Consider the functional $l\colon \mathcal{F}_n \to \mathbb{F}_2$

$$l(g) = (Sg)(\mathbf{0}).$$

Since this functional is linear, there exists a Boolean function f such that

$$l(g) = \sum_{\boldsymbol{x} \in V_n} f(\boldsymbol{x}) g(\boldsymbol{x})$$

for all $g \in \mathcal{F}_n$. Using this equality and the operational equation for an arbitrary $g \in \mathcal{F}_n$ we obtain:

$$\sum_{\boldsymbol{x} \in V_n} f(\boldsymbol{x}) g(\boldsymbol{x} \oplus \boldsymbol{u}) = (St_{\boldsymbol{u}} g)(\mathbf{0}) = (t_{\boldsymbol{u}} Sg)(\mathbf{0}) = (Sg)(\boldsymbol{u}),$$

or, in other words, $f * g = Sg$, i.e., $S = S_f$. ☐

Note that the shift operator can also be considered as a convolution operator:

$$t_{\boldsymbol{u}} = S_{\delta_{\boldsymbol{u}}},$$

where $\delta_{\boldsymbol{u}}$ is defined by $\delta_{\boldsymbol{u}} = t_{\boldsymbol{u}}\delta$ and δ is the Dirac function.

Any convolution operator is a linear combination of shift operators; to wit, the operator S_f is the sum of the operators $t_{\boldsymbol{u}}$ over all vectors $\boldsymbol{u}$ such that $f(\boldsymbol{u}) = 1$.

Let the functions $f(\boldsymbol{x})$ and $g(\boldsymbol{x})$ be given via their Zhegalkin polynomials. We will find an expression for $f * g$. Consider $f(\boldsymbol{x} \oplus \boldsymbol{y})g(\boldsymbol{y})$ as a function of $\boldsymbol{y}$ with coefficients that are Boolean functions of $\boldsymbol{x}$. After addition modulo 2 over all $\boldsymbol{y} \in V_n$ all monomials of $\boldsymbol{y}$ will cancel out except for the leading term $y^{(1)}y^{(2)} \cdots y^{(n)}$. Thus, we proved the following lemma.

LEMMA 2.94. *The Boolean function $(f * g)(\boldsymbol{x})$ is equal to the coefficient of the leading term $y^{(1)}y^{(2)} \cdots y^{(n)}$ of $f(\boldsymbol{x} \oplus \boldsymbol{y})g(\boldsymbol{y})$.*

COROLLARY 2.95. *If $\deg f + \deg g < n$, then $f * g \equiv 0$. If $\deg f + \deg g \geq n$, then $\deg f * g \leq \deg f + \deg g - n$.*

We now consider some examples.

EXAMPLE 2.96. Denote the monomial $x^{(i_1)}x^{(i_2)} \cdots x^{(i_k)}$ by $\boldsymbol{x}^{\boldsymbol{a}}$, where $\boldsymbol{a} \in V_n$ is defined as follows:

$$a^{(j)} = 1 \iff j \in \{i_1, i_2, \ldots, i_k\}.$$

Then

$$\boldsymbol{x}^{\boldsymbol{a}} * \boldsymbol{x}^{\boldsymbol{b}} = \begin{cases} (\boldsymbol{x} \oplus 1)^{\boldsymbol{a}\cdot\boldsymbol{b}} & \text{if } \boldsymbol{a} \vee \boldsymbol{b} = 1; \\ 0 & \text{otherwise.} \end{cases}$$

Here $\boldsymbol{a}\cdot\boldsymbol{b} = (a^{(1)}b^{(1)}, \ldots, a^{(n)}b^{(n)})$. Note that (the expansion of) $(\boldsymbol{x}\oplus\boldsymbol{y})^{\boldsymbol{a}}\boldsymbol{y}^{\boldsymbol{b}}$ contains the monomial $y^{(1)}y^{(2)} \cdots y^{(n)}$ if and only if $\boldsymbol{a} \vee \boldsymbol{b} = 1$, or equivalently if and only if

$$\boldsymbol{a} = (\boldsymbol{b} \oplus 1) \oplus \boldsymbol{ab}.$$

In this case,

$$(\boldsymbol{x} \oplus \boldsymbol{y})^{\boldsymbol{a}}\boldsymbol{y}^{\boldsymbol{b}} = (\boldsymbol{x} \oplus \boldsymbol{y})^{\boldsymbol{b}\oplus 1}\boldsymbol{y}^{\boldsymbol{b}}(\boldsymbol{x} \oplus \boldsymbol{y})^{\boldsymbol{a}\cdot\boldsymbol{b}}$$

and the coefficient at $y^{(1)}y^{(2)} \cdots y^{(n)}$ is equal to $(\boldsymbol{x} \oplus 1)^{\boldsymbol{a}\cdot\boldsymbol{b}}$.

EXAMPLE 2.97. Let $f(\boldsymbol{x}) = \bigoplus_{\boldsymbol{a}\in V_n} g_f(\boldsymbol{a})\boldsymbol{x}^{\boldsymbol{a}}$ be an arbitrary function. Then

$$f * \boldsymbol{x}^{\boldsymbol{b}} = \bigoplus_{\boldsymbol{a} \succcurlyeq \boldsymbol{b}\oplus 1} g_f(\boldsymbol{a})(\boldsymbol{x} \oplus 1)^{\boldsymbol{a}\cdot\boldsymbol{b}}.$$

This expression is obtained from Example 2.96 by using the linearity of the convolution and the equivalence of the equation $\boldsymbol{a} = (\boldsymbol{b} \oplus 1) \oplus \boldsymbol{ab}$ to $\boldsymbol{a} \succcurlyeq \boldsymbol{b} \oplus 1$.

EXAMPLE 2.98. Let $f(\boldsymbol{x}) = \bigoplus_{\substack{\boldsymbol{a}\in V_n \\ \mathrm{wt}(\boldsymbol{a})=r}} g_f(\boldsymbol{a})\boldsymbol{x}^{\boldsymbol{a}}$ be a homogeneous monomial of degree r. Then

$$f * \boldsymbol{x}^{\boldsymbol{b}} = \begin{cases} g_f(\boldsymbol{b} \oplus 1) & \text{if } \mathrm{wt}(\boldsymbol{b}) = n - r; \\ \bigoplus g_f(\boldsymbol{a}_i)(x^{(i)} \oplus 1) & \text{if } \mathrm{wt}(\boldsymbol{b}) = n - r + 1, \end{cases}$$

where the sum is taken over all the $n - r + 1$ values of i for which $b^{(i)} = 1$; for all these values i the vector $\boldsymbol{a}_i$ is given by the equality

$$\boldsymbol{a}_i = (\boldsymbol{b} \oplus 1) \oplus \boldsymbol{e}_i, \qquad e_i^{(j)} = 1 \iff i = j.$$

Note that the linearity of convolution implies the following equality:

$$(f * g)(\boldsymbol{x}) = (f * gh)(\boldsymbol{x}) \oplus (f * g(h \oplus 1))(\boldsymbol{x}) \qquad \forall f, g, h \in \mathcal{F}_n.$$

This equation enables us to compute the convolution of a fixed function f with a function of algebraic degree r if we are given the convolution of f with a function of large algebraic degree. In particular, if $f * g \equiv 0$ for all g with $\deg g = r + 1$, then $f * g \equiv 0$ for all g with $\deg g \leq r$. From this relation as well as from Corollary 2.95 and Example 2.98 the following lemma can be derived.

LEMMA 2.99. *The algebraic degree of a Boolean function $f \in \mathcal{F}_n$ is equal to r if and only if the following two conditions are satisfied:*
1) *$f * g \equiv 0$ for all g with $\deg g = n - r - 1$;*
2) *$f * g \not\equiv 0$ for at least one g with $\deg g = n - r$.*

Note that since the convolution is linear, it suffices to check the first condition of Lemma 2.99 for the elements of a basis of the vector space of functions g with algebraic degree $n - r - 1$.

The Reed–Muller code $RM(r, n)$ (see Chapter 5) is defined as the set of all binary vectors with length 2^n corresponding to Boolean functions with algebraic degree at most r. Thus, $RM(r, n)$ is a linear subspace of $\mathcal{F}_n$, so any Boolean function can be reduced modulo $RM(r, n)$.

We will write

$$f \equiv g \pmod{RM(r, n)}$$

if $\deg(f \oplus g) \leq r$. Using the linearity of convolution and Lemma 2.99 we obtain the following corollary.

COROLLARY 2.100. *For an arbitrary Boolean function $f, g \in \mathcal{F}_n$, the relation*

$$f \equiv g \pmod{RM(r, n)}$$

is satisfied if and only if

$$f * h \equiv g * h \pmod{RM(r, n)}$$

for all h with $\deg h = n - r - 1$.

It suffices to check the last condition for all g in some basis (of the vector space of functions g with algebraic degree $n - r - 1$).

Note that the condition $\deg g = n - r - 1$ in Lemma 2.99 and the condition $\deg h = n - r - 1$ in Corollary 2.100 can be replaced with the condition $\deg g = n - r$ if instead of the equation $f * g \equiv 0$ we demand that the equation $f * g \equiv C$ be satisfied (the constant C depends on g). So, we obtain the following lemma.

LEMMA 2.101. *The algebraic degree of a Boolean function $f \in \mathcal{F}_n$ is equal to r if and only if the following two conditions are satisfied:*
1) *$f * g \equiv C$ for all g with $\deg g = n - r$;*
2) *$f * g \not\equiv C$ for at least one g with $\deg g = n - r + 1$.*

COROLLARY 2.102. *For arbitrary Boolean functions $f, g \in \mathcal{F}_n$ the relation*

$$f \equiv g \pmod{RM(r, n)}$$

is satisfied if and only if

$$f * h \equiv g * h \pmod{RM(r, n)}$$

for all h with $\deg h = n - r$.

2.5. Cryptographic properties of Boolean functions and mappings

The goal of this section is to define the main subject of the book, namely cryptographic properties of Boolean functions and mappings. In keeping with the mathematical style of the exposition, we will try to use as little as possible specific cryptographic and technical concepts and notions related to problems of creating concrete cryptographic systems for information protection. Such concepts are, for example, classification of ciphers (for cryptographic systems), key management, and many others. A complete exposition of the problems related to these concepts can be found in [**10, 150**]. We will just remark that some of the properties of Boolean functions considered in the book, which at first sight look abstract, were developed when working with applied problems of design and analysis of ciphers. On the other hand, examples are known when results of abstract study of Boolean functions proved to be useless in cryptography.

We will introduce two of the most widely used symmetric ciphers.

A *stream cipher* is an encryption scheme which produces a pseudo-random sequence (also known as key stream) based on a key and possibly an initialization vector. The key stream is then XOR-ed (added modulo 2) with the ciphered information (which may be a plain or ciphered text).

A *block cipher* is an encryption scheme that, based on a key and on (repeated) permutations of high complexity, transforms n-bit blocks of information (which may be a plain or ciphered text). The main difference is that the ciphered information is an input for the block cipher and hence the process mixes both the key and the information.

REMARK 2.103. Our description of these two classes of ciphers was informal; its aim was just to give the reader the general idea. More detailed information about these ciphers, which are part of the family of symmetric ciphers, will be given in Chapter 9. We will only be interested in various transformations of information that are performed during the encryption process.

DEFINITION 2.104. Systems used for transformation of information in a particular cipher will be called *cryptographic (discrete)* Boolean functions.

In mathematical arguments we will still use the terms "Boolean function" and "mapping".

In 1949, C. Shannon published his seminal work [**355**]. In this paper Shannon, who was both a cryptographer and a great scientist, stated the basic principles of constructing cryptographic functions.

The first principle is diffusion. It means that " ... statistical structure of messages which leads to its redundancy is 'dissipated' into long range statistics, i.e., into statistical structure involving long combinations of letters in the cryptogram."

The second principle is confusion: it means "... to make the relation between the simple statistics of the cryptogram and the simple description of the key a very complex and involved one."

After giving examples for these principles (without formalizing them), C. Shannon makes the conclusion that "... for studying a secret system it is necessary to apply one of the above-considered principles or both of them."

C. Shannon also introduced the mixing principle in cryptography, which is well known in ergodic theory. Although a mixing transformation can occur only in a space with infinitely many points, C. Shannon gives the following cryptographic interpretation: "Speaking loosely, however, we can think of a mixing transformation as one which distributes any reasonably cohesive region in the space fairly uniformly over the entire space. If the first region could be described in simple terms, the second would require very complex ones." Commenting on this principle, C. Shannon shows how to create a good mixing: "Good mixing transformations are often formed by repeated products of two simple noncommuting operations."

We would like to emphasize two points. First, the principles for building cryptographic functions, as stated by C. Shannon, are empirical; they have no rigorous mathematical statement. This is clear from the explanations given by Shannon in his paper. Second, these principles apply not only to a particular type of ciphers but also to any cryptographic function. For example, the family of block ciphers on which contemporary authors demonstrate Shannon's principles, appeared later than the principles themselves; moreover, block ciphers are constructed according to these principles.

Attempts of some authors to formalize Shannon's principles as a fixed complete set of requirements to a cryptographic function are unlikely to succeed. The mathematical content of Shannon's principles is still being developed.

Now we continue the introduction to some important features of cryptographic functions. More detailed study of these features will be done in later chapters.

Let $\Phi\colon V_n \to V_m$ be a mapping from V_n to V_m. We will denote by $\mathcal{F}_{n,m}$ the set of all mappings from V_n to V_m. Obviously, the mapping Φ could be written as

$$\begin{aligned} y^{(1)} &= f_1(x^{(1)}, \ldots, x^{(n)}), \\ &\cdots\cdots\cdots\cdots\cdots \\ y^{(m)} &= f_m(x^{(1)}, \ldots, x^{(n)}), \end{aligned}$$

where f_i $(i = 1, 2, \ldots, m)$ are the coordinate Boolean functions defining the mapping Φ. We will write this mapping as $\Phi = (f_1, \ldots, f_m)$.

Balanced mapping.

DEFINITION 2.105. A mapping $\Phi \in \mathcal{F}_{n,m}$ $(m \le n)$ is called *balanced* if for any $\boldsymbol{v} \in V_m$ the following equality holds:

$$\#\{\boldsymbol{u} \in V_n \mid \Phi(\boldsymbol{u}) = \boldsymbol{v}\} = 2^{n-m}.$$

It is obvious that a balanced mapping must be surjective. This means that there does not exist a Boolean function g of m variables different from the constant function 0, such that

$$g(f_1(\boldsymbol{u}), \ldots, f_m(\boldsymbol{u})) \equiv 0.$$

Such functions $f_1, \ldots, f_m$ are called *algebraically independent*. If $n = m$, that is, surjection and one-to-one mapping of Φ are the same, algebraic independence of functions $f_1, \ldots, f_m$ is equivalent to the fact that the mapping $\Phi = (f_1, \ldots, f_n)$ is one-to-one.

PROBLEM 2.106. Prove that a mapping $\Phi = (f_1, \ldots, f_m)$ is surjective if and only if the functions $f_1, \ldots, f_m$ are algebraically independent.

If $m = 1$, the mapping $\Phi = f$ is a Boolean function; it is called balanced if

$$\mathrm{wt}(f) = 2^{n-1}.$$

Nonlinearity of cryptographic mappings.

DEFINITION 2.107. The *nonlinearity of the mapping* Φ is the integer N defined by the formula

$$N_\Phi = \min_{\boldsymbol{c}\in V_m, \boldsymbol{c}\neq \boldsymbol{0}} \{N_{c^{(1)}f_1\oplus\cdots\oplus c^{(m)}f_m}\},$$

where $N_{c^{(1)}f_1\oplus\cdots\oplus c^{(m)}f_m}$ is the nonlinearity of the Boolean function $c^{(1)}f_1 \oplus \cdots \oplus c^{(m)}f_m$ (see Definition 2.50 in §2.2).

Correlation-immune and resilient functions.

DEFINITION 2.108. A Boolean function $f \in \mathcal{F}_n$ is called *correlation-immune of order* m, $1 \leq m \leq n$ if for every $1 \leq i_1 < \cdots < i_m \leq n$, $a^{(1)}, \ldots, a^{(m)} \in \mathbb{F}_2$, one has

$$\mathrm{wt}\big(f_{i_1,\ldots,i_m}^{a^{(1)},\ldots,a^{(m)}}\big) = \frac{\mathrm{wt}(f)}{2^m},$$

where $f_{i_1,\ldots,i_m}^{a^{(1)},\ldots,a^{(m)}}$ is considered as a function of $n-m$ variables.

A function which is balanced and correlation-immune of mth order is also called m-resilient.

Resilient mappings. The notion of resiliency for a mapping $\Phi \in \mathcal{F}_{n,m}$ is a generalization of resiliency for functions; informally, it is the measure of the ability of mappings to keep the "balanced" property when some of the coordinates of the argument are fixed.

DEFINITION 2.109. A mapping $\Phi = (f_1, \ldots, f_m) \in \mathcal{F}_{n,m}$ is called (n, m, t)-*resilient* if for every $1 \leq i_1 < \cdots < i_t \leq n$, $a^{(1)}, \ldots, a^{(t)} \in \mathbb{F}_2$, the mapping $\Phi' = \big((f_1)_{i_1,\ldots,i_t}^{a^{(1)},\ldots,a^{(t)}}, \ldots, (f_m)_{i_1,\ldots,i_t}^{a^{(1)},\ldots,a^{(t)}}\big) \in \mathcal{F}_{n-t,m}$ is resilient.

Strict avalanche criteria (SAC). Propagation criteria (PC).

DEFINITION 2.110. One says that a function $f \in \mathcal{F}_n$ satisfies:

1) SAC *criterion* if $f(\boldsymbol{x} \oplus \boldsymbol{u}) \oplus f(\boldsymbol{x})$ is a balanced function for any $\boldsymbol{u} \in V_n$ such that $\mathrm{wt}(\boldsymbol{u}) = 1$;

2) SAC(k) *criterion*, where $1 \leq k < n$, if the subfunction $f_{i_1,\ldots,i_k}^{a^{(1)},\ldots,a^{(k)}}$ satisfies SAC for any $1 \leq i_1 < \cdots < i_k \leq n$, $a^{(1)}, \ldots, a^{(k)} \in \mathbb{F}_2$;

3) PC(l) *criterion*, or PC of degree l ($1 < l \leq n$), if $f(\boldsymbol{x} \oplus \boldsymbol{u}) \oplus f(\boldsymbol{x})$ is a balanced function for any $\boldsymbol{u} \in V_n$ such that $1 \leq \mathrm{wt}(\boldsymbol{u}) \leq l$;

4. PC(l) *criterion* of order k, $1 < l \leq n$, $1 \leq k < n$ (PC of degree l and order k or PC(l) of order k), if the subfunction $f_{i_1,\ldots,i_k}^{a^{(1)},\ldots,a^{(k)}}$ satisfies PC(l) for any $1 \leq i_1 < \cdots < i_k \leq n$, $a^{(1)}, \ldots, a^{(k)} \in \mathbb{F}_2$.

It is easy to see that there are relations between the criteria described above. We will discuss them in more detail in the following sections.

Linear structures of mappings.

DEFINITION 2.111. The mapping $\Phi \in \mathcal{F}_{n,l}$ *has a linear structure* if there exists a vector $\boldsymbol{a} \neq \boldsymbol{0}$ in the space V_n such that

$$D_{\boldsymbol{a}}\Phi(\boldsymbol{x}) = \Phi(\boldsymbol{x}) \oplus \Phi(\boldsymbol{x} \oplus \boldsymbol{a}) = \boldsymbol{b}_{\boldsymbol{a}} = \text{const}, \quad \boldsymbol{b}_{\boldsymbol{a}} \in V_l,$$

for any $\boldsymbol{x} \in V_n$. Such a vector $\boldsymbol{a}$ is called a *linear translator* of the mapping Φ.

PROBLEM 2.112. Prove that $\boldsymbol{a} \in V_n$ is a linear translator for a Boolean function $f \in \mathcal{F}_n$, i.e., $D_{\boldsymbol{a}} f = \varepsilon$, if and only if for all $\boldsymbol{z}$ such that $\langle \boldsymbol{z}, \boldsymbol{a} \rangle = \varepsilon \oplus 1$, the equality $W_f(\boldsymbol{z}) = 0$ holds.

PROBLEM 2.113. Prove that $\boldsymbol{a} \in V_n$, $\boldsymbol{a} \neq \boldsymbol{0}$, is a linear translator for a Boolean function $f \in \mathcal{F}_n$ if and only if $|\Delta_f(\boldsymbol{a})| = 2^n$.

DEFINITION 2.114. The subspace $L_\Phi = \{\boldsymbol{a} \in V_n \mid D_{\boldsymbol{a}}\Phi = \text{const}\}$ is called the *linearity subspace of the mapping* Φ, or the *space of linear translators*.

Clearly, for any affine mapping Φ its linearity subspace coincides with the whole space V_m.

PROBLEM 2.115. Let the mapping $\Phi \in \mathcal{F}_{n_1+n_2,l}$ be represented as a sum of two mappings $\Phi_1 \in \mathcal{F}_{n_1,l}$ and $\Phi_2 \in \mathcal{F}_{n_2,l}$ depending the disjoint sets of n_1 and n_2 variables:

$$\Phi(\boldsymbol{x}, \boldsymbol{y}) = \Phi_1(\boldsymbol{x}) \oplus \Phi_2(\boldsymbol{y}).$$

Prove that

$$L_\Phi = L_{\Phi_1} \dot{+} L_{\Phi_2}.$$

The presence of nontrivial linear structures in a mapping Φ shows that this mapping is somewhat "weak". This fact is reflected in the following lemma.

LEMMA 2.116. *A mapping* $\Phi \in \mathcal{F}_{n,l}$ *has a linear structure with* $\dim L_\Phi = r \geq 1$ *if and only if there exists a* $(n \times n)$ *matrix* A *such that*

$$\Psi(\boldsymbol{x}) = \Phi(A\boldsymbol{x}) = x^{(1)}\boldsymbol{b}_1 \oplus \cdots \oplus x^{(r)}\boldsymbol{b}_r \oplus \Phi^*(x^{(r+1)}, \dots, x^{(n)}), \tag{2.27}$$

where $\boldsymbol{b}_i \in V_n$, $1 \leq i \leq r$, *and* $\Phi^* \in \mathcal{F}_{n-r,l}$ *is a mapping which does not have a linear structure.*

PROOF. Suppose that the mapping Φ has a linear structure L_Φ with $\dim L_\Phi = r \leq 1$. Let $\boldsymbol{a}_1, \dots, \boldsymbol{a}_r$ be a basis of the linear space L_Φ. The vectors $\boldsymbol{a}_1, \dots, \boldsymbol{a}_r$ can be included in a family of linearly independent vectors $\boldsymbol{a}_{r+1}, \dots, \boldsymbol{a}_n$ to form a basis of the whole V_n. Let A be the matrix of the linear transformation of the space V_n which transforms the canonical basis $\{\boldsymbol{e}_1, \dots, \boldsymbol{e}_n\}$ to a basis $\{\boldsymbol{a}_1, \dots, \boldsymbol{a}_n\}$ such that $A\boldsymbol{e}_i = \boldsymbol{a}_i$, $i = 1, 2, \dots, n$. Then the following equalities hold:

$$\begin{aligned}\Psi(x^{(1)}, x^{(2)}, \dots, x^{(n)}) &= \Phi(A(x^{(1)}, x^{(2)}, \dots, x^{(n)})) \\ &= \Phi(A(x^{(1)}\boldsymbol{e}_1 \oplus x^{(2)}\boldsymbol{e}_2 \oplus \cdots \oplus x^{(n)}\boldsymbol{e}_n)) \\ &= \Phi(x^{(1)}\boldsymbol{a}_1 \oplus x^{(2)}\boldsymbol{a}_2 \oplus \cdots \oplus x^{(n)}\boldsymbol{a}_n).\end{aligned}$$

Thus, the vectors $x^{(1)}\boldsymbol{a}_1, x^{(2)}\boldsymbol{a}_2, \dots, x^{(r)}\boldsymbol{a}_r$ are linear translators of the mapping Φ because

$$\Phi(\boldsymbol{u}) \oplus \Phi(\boldsymbol{u} \oplus x^{(i)}\boldsymbol{a}_i) = x^{(i)}\boldsymbol{b}_i, \quad \boldsymbol{b}_i \in V_l,$$

for any $\boldsymbol{u} \in V_n$, $i = 1, 2, \dots, r$. Therefore,

$$\begin{aligned}\Phi(x^{(1)}\boldsymbol{a}_1 \oplus \cdots \oplus x^{(n)}\boldsymbol{a}_n) &= x^{(1)}\boldsymbol{b}_1 \oplus \Phi(x^{(2)}\boldsymbol{a}_2 \oplus \cdots \oplus x^{(n)}\boldsymbol{a}_n) \\ &= x^{(1)}\boldsymbol{b}_1 \oplus x^{(2)}\boldsymbol{b}_2 \oplus \Phi(x^{(3)}\boldsymbol{a}_3 \oplus \cdots \oplus x^{(n)}\boldsymbol{a}_n) = \dots \\ &= x^{(1)}\boldsymbol{b}_1 \oplus \cdots \oplus x^{(r)}\boldsymbol{b}_r \oplus \Phi(x^{(r+1)}\boldsymbol{a}_{r+1} \oplus \cdots \oplus x^{(n)}\boldsymbol{a}_n),\end{aligned}$$

i.e., the equality (2.27) holds. It is clear that the mapping

$$\Phi^*(x^{(r+1)}, \dots, x^{(n)}) = \Phi(x^{(r+1)}\boldsymbol{a}_{r+1} \oplus \cdots \oplus x^{(n)}\boldsymbol{a}_n)$$

does not have linear structure; otherwise we would get a contradiction with the fact that $\dim L_\Phi = r$.

The converse statement is obvious. □

The above lemma says that, by a linear transformation of the variables, the "linear" part in the mapping Φ can be isolated.

PROBLEM 2.117. Prove that the following three definitions of the set $\mathcal{F}_n^0$ consisting of Boolean functions with a nontrivial linear structure, are equivalent:

1. $f \in \mathcal{F}_n^0$ if f has a nontrivial linear structure.

2. $f \in \mathcal{F}_n^0$ if f is linearly equivalent to a Boolean function independent of at least one variable.

3. $f \in \mathcal{F}_n^0$ if there exist variables $x^{(i_1)}, \dots, x^{(i_k)}$ on which the function f depends *quasi-linearly* in the following sense: $f(\boldsymbol{x}) \neq f(\boldsymbol{y})$ for any two vectors $\boldsymbol{x}, \boldsymbol{y} \in V_n$ such that

$$x^{(i_1)} \oplus \dots \oplus x^{(i_k)} \neq y^{(i_1)} \oplus \dots \oplus y^{(i_k)},$$

$x^{(j)} = y^{(j)}$ for all $j \notin \{i_1, \dots, i_k\}$.

PROBLEM 2.118. Prove that any function $f \in \mathcal{F}_n^0$ is affine equivalent to a function of the form

$$\varepsilon \cdot x^{(1)} \oplus g(x^{(2)}, \dots, x^{(k)}),$$

where $\varepsilon \in \mathbb{F}_2$, $k \leq n$, and $g \in \mathcal{F}_{k-1}$ does not have a nontrivial linear structure.

PROBLEM 2.119. Prove that any function $f \in \mathcal{F}_n^0$ is balanced.

PROBLEM 2.120. Suppose that $f \in \mathcal{F}_n^0$, $\dim L_f^0 = r$, and that in an appropriate basis the function f can be represented as

$$f(\boldsymbol{y}, \boldsymbol{z}) = l(\boldsymbol{y}) \oplus g(\boldsymbol{z}),$$

where $\boldsymbol{y} \in V_r$, $\boldsymbol{z} \in V_{n-r}$, $l(\boldsymbol{y}) \in \mathcal{F}_r$ is an affine function, and $g(\boldsymbol{z}) \in \mathcal{F}_{n-r}$ is a function without linear structure. Prove that

$$N_f = 2^r N_g.$$

The notion of space of linear structures can be generalized as follows.

DEFINITION 2.121. The *space of r-nonlinearity* L_f^r, $r < \deg f$, of a function $f \in \mathcal{F}_n$ is the set of vectors

$$L_f^r = \{\boldsymbol{a} \in V_n \mid \deg D_{\boldsymbol{a}} f(\boldsymbol{x}) \leq r\}.$$

PROBLEM 2.122. Prove that for any $r < \deg f$ the set L_f^r is a linear subspace in V_n.

It is obvious that L_f^0 coincides with the space of linear translators of the function f, and L_f^{-1} coincides with the stabilizer of the group of shifts of function f.

PROBLEM 2.123. Prove that if $L_f^0 \neq L_f^{-1}$, then $\dim L_f^0 = \dim L_f^{-1} + 1$.

PROBLEM 2.124. Prove that if $L_f^r \neq L_f^{r-1}$, then $\dim L_f^r \geq \dim L_f^{r-1} + r + 1$.

It is easy to see that the subspaces of r-nonlinearity form a *flag of subsets*, i.e., they are nested as follows:

$$L_f^{-1} \subseteq L_f^0 \subseteq \dots \subseteq L_f^{\deg f - 1} = V_n.$$

DEFINITION 2.125. For a function $f \in \mathcal{F}_n$, denote by $q_f \colon V_n \times V_n \to \mathbb{F}_2$ the following mapping, called the mapping *associated* with the function f:

$$q_f(\boldsymbol{u}, \boldsymbol{v}) = D_{\boldsymbol{u}} D_{\boldsymbol{v}} f(\boldsymbol{0}) = D_{\boldsymbol{u}} f(\boldsymbol{v}) \oplus D_{\boldsymbol{u}} f(\boldsymbol{0}) = f(\boldsymbol{0}) \oplus f(\boldsymbol{u}) \oplus f(\boldsymbol{v}) \oplus f(\boldsymbol{u} \oplus \boldsymbol{v}).$$

The linear subspace

$$\ker q_f = \{\boldsymbol{u} \in V_n \mid q_f(\boldsymbol{u}, \boldsymbol{v}) = 0 \text{ for any } \boldsymbol{v} \in V_n\} \subset V_n$$

will be called the *kernel* of the associated mapping.

PROBLEM 2.126. Prove that $L_f = \ker q_f$ for any $f \in \mathcal{F}_n$.

Linearity index of a mapping. When constructing Boolean mappings for cryptographic applications one uses special methods. One of these methods is to construct nonlinear mappings from linear (affine) mappings with the help of the so-called time clock motion.

DEFINITION 2.127. Let V_m, V_l, V_k be vector spaces over the field $\mathbb{F}_2$, let $h = (h_1, \dots, h_k)$ be a linear mapping from V_m to V_k, and let $(h_i,\ i = 1, \dots, k)$ be coordinate functions of this linear mapping. Suppose $A = \{A_{\boldsymbol{u}} \in \mathcal{A}_{m,l} \mid \boldsymbol{u} \in V_k\}$ is a combination of affine mappings $A_{\boldsymbol{u}}$ belonging to the set of affine mappings $\mathcal{A}_{m,l}$ from V_m to V_l. We define the mapping $\Phi \colon V_m \to V_l$ by the formula

$$\Phi(\boldsymbol{x}) = A_{h(\boldsymbol{x})}(\boldsymbol{x}) \tag{2.28}$$

for any $\boldsymbol{x} \in V_m$. We will say that Φ is a *linear branching* with the *branching space* V_k and with the *branching mapping* h. The mappings $A_{\boldsymbol{u}}$, $\boldsymbol{u} \in V_k$, are called *branched mappings*.

It is easy to see that any mapping Ψ from V_m to V_l can be represented as a linear branching (2.28). Indeed, it is enough to put $m = k$, $f(\boldsymbol{x}) = \boldsymbol{x}$, and $A_{\boldsymbol{x}}(\boldsymbol{x}) = \Psi(\boldsymbol{x})$ for every $\boldsymbol{x} \in V_n$. Obviously, when linear branchings are studied, it is enough to consider the case where the mapping $h \colon V_m \to V_k$ is surjective and $m \geq k$.

One should note that the representation of an arbitrary mapping as a linear branching is not always unambiguous.

DEFINITION 2.128. The *index of linearity of a mapping* Φ, denoted by $\mathrm{ill}(\Phi)$, is the minimum dimension of the branching space in a representation of the mapping Φ in the form (2.28).

This parameter in a sense characterizes the deviation of the mapping Φ from the set of linear mappings. In this sense, the mappings that are maximally distant from the linear mappings are the mappings that maximize the quantity $\mathrm{ill}(\Phi)$.

If we want to calculate $\mathrm{ill}(\Phi)$ directly from the definition, we will have to inspect all possible representations of the mapping Φ in the form (2.28), which is, of course, quite time consuming. That is why the following problem is of interest.

PROBLEM 2.129 (open problem). Develop an efficient algorithm for calculating (or estimating) $\mathrm{ill}(\Phi)$ for classes of mappings Φ that are important for applications.

EXAMPLE 2.130. We will consider the well-known (see, for example, [**10**]) stream cipher A5, applied in the European mobile communications system GSM (Group Special Mobile). More precisely, we will be interested in the nonlinear mapping applied in this cipher. The transformation Φ in question is a one-to-one self-mapping of the space V_{64}. The vector space V_{64} is split into the components

V_{19}, V_{22}, and V_{23}, i.e., $V_{64} = V_{19} \times V_{22} \times V_{23}$. Accordingly, any vector $\boldsymbol{x} \in V_{64}$ will be presented as

$$\boldsymbol{x}^\top = (\boldsymbol{x}_1^\top, \boldsymbol{x}_2^\top, \boldsymbol{x}_3^\top) = \big((x_1^{(1)}, \dots, x_1^{(19)}), (x_2^{(1)}, \dots, x_2^{(22)}), (x_3^{(1)}, \dots, x_3^{(23)})\big),$$

where $\boldsymbol{x}_1 \in V_{19}$, $\boldsymbol{x}_2 \in V_{22}$, $\boldsymbol{x}_3 \in V_{23}$. Let B_1 be a linear transformation of the space V_{19}, realized by a linear polynomial feedback shift register $f_1(\lambda) = \lambda^{19} \oplus \lambda^5 \oplus \lambda^2 \oplus \lambda \oplus 1$; let B_2 be a linear transformation of the subspace V_{22}, realized by a linear polynomial feedback shift register $f_2(\lambda) = \lambda^{22} \oplus \lambda \oplus 1$; and let B_3 be a linear transformation of the subspace V_{23}, realized by a linear polynomial feedback shift register $f_3(\lambda) = \lambda^{23} \oplus \lambda^{15} \oplus \lambda^2 \oplus \lambda \oplus 1$. The branching space has dimension $k = 3$ and the branching mapping l is

$$l(\boldsymbol{x}) = (l_1(\boldsymbol{x}_1), l_2(\boldsymbol{x}_2), l_3(\boldsymbol{x}_3))^\top = (x_1^{10}, x_2^{11}, x_3^{12})^\top.$$

The branched mappings $A_{\boldsymbol{u}}$, $\boldsymbol{u} \in V_3$, have the form

$$A_{000} = A_{111} = \begin{pmatrix} B_1 & 0 & 0 \\ 0 & B_2 & 0 \\ 0 & 0 & B_3 \end{pmatrix}, \qquad A_{001} = A_{110} = \begin{pmatrix} B_1 & 0 & 0 \\ 0 & B_2 & 0 \\ 0 & 0 & E_{23} \end{pmatrix},$$

$$A_{010} = A_{101} = \begin{pmatrix} B_1 & 0 & 0 \\ 0 & E_{22} & 0 \\ 0 & 0 & B_3 \end{pmatrix}, \qquad A_{100} = A_{011} = \begin{pmatrix} E_{19} & 0 & 0 \\ 0 & B_2 & 0 \\ 0 & 0 & B_3 \end{pmatrix},$$

where E_s is the identity mapping of the space V_s to itself.

PROBLEM 2.131. The linearity index of the space Φ applied in the cipher A5 is at most 3: $\mathrm{ill}(\Phi) \leq 3$. Compute the exact value of the linearity index $\mathrm{ill}(\Phi)$.

PROBLEM 2.132. Suppose that $l = 1$, so that the mapping is a Boolean function from $\mathcal{F}_m$. Prove that $\mathrm{ill}(f) \leq t$ if and only if, after an affine change of variables, the function f can be represented as

$$\begin{aligned} &f(x^{(1)}, \dots, x^{(n-t)}, y^{(1)}, \dots, y^{(t)}) \\ &\quad = g_1(y^{(1)}, \dots, y^{(t)})x^{(1)} \oplus \dots \oplus g_{n-t}(y^{(1)}, \dots, y^{(t)})x^{(n-t)} \oplus f_1(y^{(1)}, \dots, y^{(t)}). \end{aligned}$$

PROBLEM 2.133. Prove that for any Boolean function f the inequality $\deg f \leq 1 + \mathrm{ill}(f)$ holds.

DEFINITION 2.134. We will say that a Boolean function $f \in \mathcal{F}_n$ *is given as a linear branching* if there exist numbers t, s, $n = t + s$ and mappings $\Phi \in \mathcal{F}_{t,s}$, $h \in \mathcal{F}_s$ such that

$$f(\boldsymbol{z}) = f(\boldsymbol{x}, \boldsymbol{y}) = \langle \boldsymbol{x}, \Phi(\boldsymbol{y}) \rangle \oplus h(\boldsymbol{y})$$

for any $\boldsymbol{z} = (\boldsymbol{x}, \boldsymbol{y}) \in V_n$, $\boldsymbol{x} \in V_t$, $\boldsymbol{y} \in V_s$.

PROBLEM 2.135. Let $f \in \mathcal{F}_n$ be given as a linear branching. Prove that for any $\boldsymbol{u} = (\boldsymbol{v}, \boldsymbol{w}) \in V_n$, $\boldsymbol{v} \in V_t$, $\boldsymbol{w} \in V_s$ the following equality holds:

$$W_f(\boldsymbol{u}) = 2^t \sum_{\boldsymbol{y} \in \Phi^{-1}(\boldsymbol{v})} (-1)^{h(\boldsymbol{y}) \oplus \langle \boldsymbol{y}, \boldsymbol{w} \rangle}.$$

PROBLEM 2.136. Let $f \in \mathcal{F}_n$ be given as a linear branching. Prove that for any $\boldsymbol{v} \in V_t$ the following inequality holds:

$$2^t \sqrt{\#\Phi^{-1}(\boldsymbol{v})} \leq \max_{\boldsymbol{w} \in V_s} |W_f(\boldsymbol{v}, \boldsymbol{w})| \leq 2^t \#\Phi^{-1}(\boldsymbol{v}).$$

Boolean functions without prohibitions.

DEFINITION 2.137. A Boolean function $f \in \mathcal{F}_n$ is called a *function without prohibitions* if, for each natural number N and every set $y^{(1)}, \dots, y^{(N)} \in \mathbb{F}_2$, the system of equations

$$\begin{aligned} y^{(1)} &= f(x^{(1)}, x^{(2)}, \dots, x^{(n)}), \\ y^{(2)} &= f(x^{(2)}, x^{(3)}, \dots, x^{(n+1)}), \\ &\dots\dots\dots\dots\dots\dots\dots\dots \\ y^{(N)} &= f(x^{(N)}, x^{(N+1)}, \dots, x^{(N+n-1)}) \end{aligned}$$

has a solution with respect to $x^{(1)}, x^{(2)}, \dots, x^{(N+n-1)}$.

Reducibility of properties of Boolean mappings. Let $\Phi = (f_1, \dots, f_m)$ be an arbitrary mapping from $\mathcal{F}_{n,m}$. We will denote by $\mathrm{pr}_i \colon \mathcal{F}_{n,m} \to \mathcal{F}_{n,1} = \mathcal{F}_n$ $(i = 1, 2, \dots, m)$ the projection operator which maps any mapping Φ to its coordinate function $\mathrm{pr}_i \Phi = f_i$. In some cases, properties of the mappings from $\mathcal{F}_{n,m}$ can be expressed via properties of its coordinate functions.

DEFINITION 2.138. A property $\varkappa$ is called *reducible* if the mapping $\Phi \in \mathcal{F}_{n,m}$ has the property $\varkappa$ if and only if each nonzero linear combination of its coordinates has the property $\varkappa$.

The simplest example of a reducible property for $\Phi \in \mathcal{F}_{n,m}$, $n \geq m$, is balancedness.

THEOREM 2.139. *A mapping $\Phi = (f_1, f_2, \dots, f_m) \in \mathcal{F}_{n,m}$, $n \geq m \geq 1$, f_i, $i = 1, 2, \dots, m$ is balanced if and only if for any nonzero vector $\boldsymbol{b} = (b^{(1)}, b^{(2)}, \dots, b^{(m)})^\top \in V_m$ the Boolean function $f_{\boldsymbol{b}} = \bigoplus_{i=1}^m b^{(i)} f_i$ is balanced.*

PROOF. For a vector $\boldsymbol{a} = (a^{(1)}, \dots, a^{(m)})^\top \in V_m$, put $\Phi^{-1}(\boldsymbol{a}) = \{\boldsymbol{x} \in V_n \mid \Phi(\boldsymbol{x}) = \boldsymbol{a}\}$. If the mapping Φ is balanced, i.e., $\#\Phi^{-1}(\boldsymbol{a}) = \#\Phi^{-1}(\boldsymbol{a}')$ for any $\boldsymbol{a}, \boldsymbol{a}' \in V_m$, then for any nonzero vector $\boldsymbol{b} = (b^{(1)}, \dots, b^{(m)})^\top \in V_m$ the following equality holds:

$$\sum_{\boldsymbol{c} \in V_n} (-1)^{b^{(1)} f_1(\boldsymbol{c}) \oplus \dots \oplus b^{(m)} f_m(\boldsymbol{c})} = \sum_{\boldsymbol{a} \in V_m} \#\Phi^{-1}(\boldsymbol{a}) (-1)^{b^{(1)} a^{(1)} \oplus \dots \oplus b^{(m)} a^{(m)}} = 0.$$

Therefore, the function $b^{(1)} f_1 \oplus \dots \oplus b^{(m)} f_m$ is balanced.

Conversely, if $f_{\boldsymbol{b}}$ is balanced, then for any vector $\boldsymbol{a} = (a^{(1)}, \dots, a^{(m)})^\top \in V_m$ we get

$$\begin{aligned} &\#\Phi^{-1}(\boldsymbol{a}) \\ &\quad= \frac{1}{2^m} \sum_{\boldsymbol{c} \in V_n} \Bigg(\sum_{b^{(1)} \in \mathbb{F}_2} (-1)^{b^{(1)}(f_1(\boldsymbol{c}) \oplus a^{(1)})} \Bigg) \cdots \Bigg(\sum_{b^{(m)} \in \mathbb{F}_2} (-1)^{b^{(m)}(f_m(\boldsymbol{c}) \oplus a^{(m)})} \Bigg) \\ &\quad= \frac{1}{2^m} \sum_{\boldsymbol{b} \in V_m} (-1)^{b^{(1)} a^{(1)} \oplus \dots \oplus b^{(m)} a^{(m)}} \sum_{\boldsymbol{c} \in V_n} (-1)^{b^{(1)} f_1(\boldsymbol{c}) \oplus \dots \oplus b^{(m)} f_m(\boldsymbol{c})} \\ &\quad= \frac{2^n}{2^m} = 2^{n-m}. \end{aligned}$$

□

Now we will describe a typical way to introduce properties of discrete mappings. Let $\varkappa$ be a property and $\Xi = \{\xi\}$ a set of operators:

$$\xi \colon \mathcal{F}_{n,m} \to \mathcal{F}_{n,m}.$$

DEFINITION 2.140. A mapping $\Phi \in \mathcal{F}_{n,m}$ has a *secondary property* $(\varkappa, \Xi)$ if each mapping $\xi(\Phi)$, $\xi \in \Xi$, has the property $\varkappa$.

To generate secondary properties, the following operators are often used: operators fixing some variables and operators taking a derivative in the direction of a linear subspace L.

An operator fixing some variables $\mathrm{con}_J^{\boldsymbol{a}}$ is defined by a set of numbers $J = \{j_1, \dots, j_l\}$, $1 \le j_1 < \dots < j_l \le n$ and binary vectors $\boldsymbol{a} = (a^{(1)}, \dots, a^{(l)})^\top \in V_l$; it is given by the formula

$$\mathrm{con}_J^{\boldsymbol{a}}(\Phi) = \big((f_1)_{x^{(j_1)},\dots,x^{(j_l)}}^{a^{(1)},\dots,a^{(l)}}, \dots, (f_m)_{x^{(j_1)},\dots,x^{(j_l)}}^{a^{(1)},\dots,a^{(l)}}\big).$$

Operators taking a Boolean derivative in the direction of the subspace $L \subset V_n$ act as follows:

$$D_L\Phi(\boldsymbol{x}) = \bigoplus_{\boldsymbol{u} \in L} \Phi(\boldsymbol{x} \oplus \boldsymbol{u}).$$

The operators $\mathrm{con}_J^{\boldsymbol{a}}$ and D_L are linear, and they commute with projection operators:

$$\mathrm{con}_J^{\boldsymbol{a}} \mathrm{pr}_i = \mathrm{pr}_i \mathrm{con}_J^{\boldsymbol{a}}, \quad D_L \mathrm{pr}_i = \mathrm{pr}_i D_L.$$

We will give examples of secondary properties generated by the property of balancedness and by operators of fixing some of the variables and of taking Boolean derivatives.

EXAMPLE 2.141. Suppose that all mappings obtained from a mapping Φ by fixing some of the variables are balanced. The corresponding secondary property is called correlation immunity (see Definition 2.108).

EXAMPLE 2.142. Suppose that all derivatives of a mapping Φ in the direction of the subspace $\{\boldsymbol{0}, \boldsymbol{u}\}$, $\boldsymbol{u} \in \mathrm{PC}$ are balanced. In this case, we say that the mapping Φ satisfies a PC criterion (see Definition 2.110).

In these examples, the reducibility of the secondary properties was proved in [**419**]. We will prove a more general theorem, which includes this result as a special case.

THEOREM 2.143. *Suppose that $\varkappa$ is a reducible property of mappings and Ξ is a set of linear operators commuting with projection operators. Then the secondary property $(\varkappa, \Xi)$ is reducible.*

PROOF. Suppose that a mapping $\Phi \in \mathcal{F}_{n,m}$ has the property $(\varkappa, \Xi)$. This means that any mapping $\xi(\Phi)$, $\xi \in \Xi$, has the property $\varkappa$. Since $\varkappa$ is reducible, any nonzero linear combination of coordinate functions $\xi(\Phi)$ has the property $\varkappa$; i.e., for any collection of $b_i \in \mathbb{F}_2$ not all of which are zero, the function $\bigoplus_{i=1}^n b_i \mathrm{pr}_i \xi(\Phi)$ has the property $\varkappa$. Since the operator ξ is linear and commutes with projection operators, we have

$$\bigoplus_{i=1}^n b_i \mathrm{pr}_i \xi(\Phi) = \bigoplus_{i=1}^n b_i \xi \mathrm{pr}_i \Phi = \xi\Big(\bigoplus_{i=1}^n b_i \mathrm{pr}_i \Phi\Big).$$

Thus, for any nonzero linear combination $\bigoplus_{i=1}^{n} b_i \operatorname{pr}_i \Phi$ and any operator $\xi \in \Xi$, the function $\xi\left(\bigoplus_{i=1}^{n} b_i \operatorname{pr}_i \Phi\right)$ has property $\varkappa$. This means that any nonzero linear combination $\bigoplus_{i=1}^{n} b_i \operatorname{pr}_i \Phi$ has the secondary property $(\varkappa, \Xi)$.

It is easy to check that the above argument can be reversed. □

To conclude this section we will make an important note. A large variety of characteristics of cryptographic functions important in practice conflict with each other: their optimal values cannot be attained simultaneously. If a cryptographic function is optimal relative to a property, it is quite possible that relative to some other properties it will be poor. We will discuss various examples of such conflicts later.

2.6. Covering sequences of Boolean functions

Derivatives of Boolean functions (see Definition 2.68) are widely used in their study. They will be used in the study of functions with certain cryptographic properties.

DEFINITION 2.144. Suppose that n is a natural number, $\boldsymbol{\theta} = (\theta_{\boldsymbol{u}})$, $\boldsymbol{u} \in V_n$, $\theta_{\boldsymbol{u}} \in \mathbb{Z}$. We will say that $\boldsymbol{\theta}$ is a *covering sequence, of the Boolean function* $f \in \mathcal{F}_n$ if

$$\sum_{\boldsymbol{u} \in V_n} \theta_{\boldsymbol{u}} D_{\boldsymbol{u}} f(\boldsymbol{x}) = \rho = \text{const} \tag{2.29}$$

for each $\boldsymbol{x} \in V_n$. In this case, we will also say that the Boolean function f allows for the covering sequence $\boldsymbol{\theta}$. The number ρ is called the *level of this covering sequence*. If $\rho \neq 0$, we will say that the covering sequence $\boldsymbol{\theta}$ is *nontrivial*.

Although $\boldsymbol{\theta}$ is actually an integer-valued function on V_n ($\boldsymbol{\theta}\colon V_n \to \mathbb{Z}$), it is better to use the term "covering sequence" to avoid confusion with Boolean functions.

Note that the set of integer sequences $\boldsymbol{\theta} = (\theta_{\boldsymbol{u}})$, $\boldsymbol{u} \in V_n$, $\theta_{\boldsymbol{u}} \in \mathbb{Z}$ of length 2^n is a module over the ring of integers; it is clear that the following relations are satisfied:

$$(a+b)\boldsymbol{\theta} = a\boldsymbol{\theta} + b\boldsymbol{\theta}, \quad a(\boldsymbol{\theta} + \boldsymbol{\mu}) = a\boldsymbol{\theta} + a\boldsymbol{\mu}$$

for any $a, b \in \mathbb{Z}$, $\boldsymbol{\theta} = (\theta_{\boldsymbol{u}})$, $\boldsymbol{\mu} = (\mu_{\boldsymbol{u}})$, $\boldsymbol{u} \in V_n$, and $\theta_{\boldsymbol{u}}, \mu_{\boldsymbol{u}} \in \mathbb{Z}$.

PROBLEM 2.145. Prove that the set of covering sequences (of all levels) for a given Boolean function $f \in \mathcal{F}_n$ is a submodule of the module of all integer sequences of length 2^n.

PROBLEM 2.146. Prove that for any Boolean function there exists a trivial covering sequence.

Some properties of Boolean functions can easily be expressed in the language of covering sequences.

EXAMPLE 2.147. Suppose that a Boolean function $f \in \mathcal{F}_n$ have a linear structure $\boldsymbol{u}_0 \in V_n$, i.e., $D_{\boldsymbol{u}_0} f(\boldsymbol{x}) = f(\boldsymbol{x}) \oplus f(\boldsymbol{x} \oplus \boldsymbol{u}_0) = c_{\boldsymbol{u}_0}$. Then it is easy to see that the function f allows for a covering sequence $\boldsymbol{\theta}$ of the form: $\theta_{\boldsymbol{u}_0} = 1$; $\theta_{\boldsymbol{u}} = 0$, $\boldsymbol{u} \neq \boldsymbol{u}_0$. The level ρ of this covering sequence is equal to $c_{\boldsymbol{u}_0}$.

PROBLEM 2.148. Show that different Boolean functions can have the same covering sequence.

Let us consider the question of how covering sequences could be characterized in terms of Walsh–Hadamard coefficients. For a sequence $\boldsymbol{\theta} = (\theta_{\boldsymbol{u}})$, $\boldsymbol{u} \in V_n$, we will denote the Fourier transform by $\overline{W}_{\boldsymbol{\theta}}(\boldsymbol{b}) = \sum_{\boldsymbol{u}\in V_n} \theta_{\boldsymbol{u}}(-1)^{\langle \boldsymbol{b},\boldsymbol{u}\rangle}$, $\boldsymbol{b} \in V_n$. For the Boolean function $f \in \mathcal{F}_n$ as in §2.3, we put $V_n \setminus CI_f = \{\boldsymbol{v} \in V_n \mid W_f(\boldsymbol{v}) \neq 0\}$.

THEOREM 2.149. *Suppose that $f \in \mathcal{F}_n$ and $\boldsymbol{\theta} = (\theta_{\boldsymbol{u}})$, $\boldsymbol{u} \in V_n$ is an integer sequence. Then the function f allows for $\boldsymbol{\theta}$ as a covering sequence of level ρ if and only if*

$$V_n \setminus CI_f \subseteq \{\boldsymbol{b} \in V_n \mid \overline{W}_{\boldsymbol{\theta}}(\boldsymbol{b}) = \overline{W}_{\boldsymbol{\theta}}(\boldsymbol{0}) - 2\rho\},$$

or equivalently

$$\{\boldsymbol{b} \in V_n \mid \overline{W}_{\boldsymbol{\theta}}(\boldsymbol{b}) \neq \overline{W}_{\boldsymbol{\theta}}(\boldsymbol{0}) - 2\rho\} \subseteq \{\boldsymbol{b} \in V_n \mid W_f(\boldsymbol{b}) = 0\}.$$

PROOF. Since $(-1)^{D_{\boldsymbol{u}}f(\boldsymbol{x})} = 1 - 2D_{\boldsymbol{u}}f(\boldsymbol{x})$, it is easy to infer from Definition 2.144 that the function f allows for a covering sequence $\boldsymbol{\theta}$ of level ρ if and only if for each $\boldsymbol{x} \in V_n$ the following equality holds:

$$\sum_{\boldsymbol{u}\in V_n} \theta_{\boldsymbol{u}}(-1)^{f(\boldsymbol{x}\oplus\boldsymbol{u})} = \Big(\sum_{\boldsymbol{u}\in V_n} \theta_{\boldsymbol{u}} - 2\rho\Big)(-1)^{f(\boldsymbol{x})}.$$

Integer-valued functions are equal if and only if their Fourier transforms are equal, i.e., if for any $\boldsymbol{b} \in V_n$ one has

$$\sum_{\boldsymbol{x}\in V_n}\sum_{\boldsymbol{u}\in V_n} \theta_{\boldsymbol{u}}(-1)^{f(\boldsymbol{x}\oplus\boldsymbol{u})\oplus\langle \boldsymbol{b},\boldsymbol{x}\rangle} = \Big(\sum_{\boldsymbol{u}\in V_n} \theta_{\boldsymbol{u}} - 2\rho\Big)\sum_{\boldsymbol{x}\in V_n}(-1)^{f(\boldsymbol{x})\oplus\langle \boldsymbol{b},\boldsymbol{x}\rangle},$$

which is equivalent to

$$\Big(\sum_{\boldsymbol{u}\in V_n} \theta_{\boldsymbol{u}}(-1)^{\langle \boldsymbol{b},\boldsymbol{u}\rangle}\Big)W_f(\boldsymbol{b}) = \Big(\sum_{\boldsymbol{u}\in V_n} \theta_{\boldsymbol{u}} - 2\rho\Big)W_f(\boldsymbol{b}).$$

Thus, the function f allows a covering sequence $\boldsymbol{\theta}$ of level ρ if and only if for any $\boldsymbol{b} \in V_n$ one has

$$[\overline{W}_{\boldsymbol{\theta}}(\boldsymbol{b}) - (\overline{W}_{\boldsymbol{\theta}}(\boldsymbol{0}) - 2\rho)]W_f(\boldsymbol{b}) = 0,$$

i.e.,

$$\{\boldsymbol{b} \in V_n \mid \overline{W}_{\boldsymbol{\theta}}(\boldsymbol{b}) \neq \overline{W}_{\boldsymbol{\theta}}(\boldsymbol{0}) - 2\rho\} \cap \{\boldsymbol{b} \in V_n \mid W_f(\boldsymbol{b}) \neq 0\} = \varnothing.$$

Now the statement of the theorem is obvious. □

REMARK 2.150. We can also note that

$$\{\boldsymbol{b} \in V_n \mid \overline{W}_{\boldsymbol{\theta}}(\boldsymbol{b}) \neq \overline{W}_{\boldsymbol{\theta}}(\boldsymbol{0}) - 2\rho\} = \Big\{\boldsymbol{b} \in V_n \;\Big|\; \sum_{\boldsymbol{u}\in V_n,\ \langle \boldsymbol{u},\boldsymbol{b}\rangle=1} \theta_{\boldsymbol{u}} \neq \rho\Big\}.$$

Sometimes it is convenient to use Theorem 2.149 in the following form.

COROLLARY 2.151. *A function f allows for $\boldsymbol{\theta}$ as a covering sequence of level ρ if and only if for all $\boldsymbol{b} \in V_n$ such that $\sum_{\boldsymbol{u}\in V_n,\ \langle \boldsymbol{u},\boldsymbol{b}\rangle=1} \theta_{\boldsymbol{u}} \neq \rho$, the equality $W_f(\boldsymbol{b}) = 0$ holds, i.e., if and only if $f \oplus l_{\boldsymbol{b},0}$ is a balanced function.*

The following corollary is obvious.

COROLLARY 2.152. *A Boolean function is balanced if and only if it allows a nontrivial covering sequence.*

Theorem 2.149 enables one to easily construct a class of covering sequences.

THEOREM 2.153. *Let $M \subset V_n$ be an arbitrary set. An integer sequence $\boldsymbol{\theta}^M$ with elements $\theta_{\boldsymbol{u}}^M = \sum_{\boldsymbol{z}\in M}(-1)^{\langle \boldsymbol{z},\boldsymbol{u}\rangle}$ is a covering sequence for any Boolean function $f \in \mathcal{F}_n$ such that $V_n \setminus CI_f \subseteq M$. The level of the covering sequence $\boldsymbol{\theta}^M$ is equal to*

$$\rho = \begin{cases} -2^{n-1} & \textit{if } f \textit{ is a balanced function;} \\ 0 & \textit{otherwise.} \end{cases}$$

PROOF. Calculate the Fourier transform of the sequence $\boldsymbol{\theta}^M$:

$$\overline{W}_{\boldsymbol{\theta}^M}(\boldsymbol{b}) = \sum_{\boldsymbol{u}\in V_n} \theta_{\boldsymbol{u}}^M (-1)^{\langle \boldsymbol{b},\boldsymbol{u}\rangle} = \sum_{\boldsymbol{u}\in V_n}\sum_{\boldsymbol{z}\in M} (-1)^{\langle \boldsymbol{z}\oplus\boldsymbol{b},\boldsymbol{u}\rangle}$$

$$= \sum_{\boldsymbol{z}\in M}\sum_{\boldsymbol{u}\in V_n} (-1)^{\langle \boldsymbol{z}\oplus\boldsymbol{b},\boldsymbol{u}\rangle} = \begin{cases} 2^n, & \text{if } \boldsymbol{b} \in M; \\ 0 & \text{otherwise.} \end{cases}$$

Thus, the Fourier transform $\overline{W}_{\boldsymbol{\theta}^M}(\boldsymbol{b})$ is constant on M, and if $V_n \setminus CI_f \subseteq M$, then according to Theorem 2.149 the function f allows $\boldsymbol{\theta}^M$ as a covering sequence of level

$$\rho = \frac{1}{2}(\overline{W}_{\boldsymbol{\theta}^M}(\mathbf{0}) - \overline{W}_{\boldsymbol{\theta}^M}(\boldsymbol{b})) = \begin{cases} -2^{n-1} & \text{if } f \text{ is a balanced function;} \\ 0 & \text{otherwise.} \end{cases} \qquad \square$$

We note that the level of the covering sequence $\boldsymbol{\theta}^M$ constructed in Theorem 2.153 is very high. When studying the properties of Boolean functions, we will use covering sequences of low levels.

Comments to Chapter 2

In this book we hardly ever discuss functional concepts in the algebra of logic (see [**411, 412**]), the theory of disjunctive normal forms (see [**177, 401**]), or complexity theory of Boolean functions (see [**285**]). These theories also have various applications in cryptology, but they must be treated elsewhere.

Returning to the different ways for representation of Boolean functions considered in this book, it is necessary to add one more, the hypergraph representation. We refer the reader, for example, to [**433**].

Derivatives of Boolean functions are widely used throughout the book. A significant part of cryptographic and code-theoretic properties of Boolean functions can be expressed in terms of this notion. An attempt to develop differential and integral calculus for Boolean functions can be found, for example, in [**38**].

The probabilistic and statistical suppositions about distributions on domains of Boolean functions and Boolean mappings that are used in the book correspond to uniform distributions. It is interesting to find out how various characteristics of Boolean functions change if the distribution changes. Some related problems are considered in [**276**].

CHAPTER 3

Classifications of Boolean Functions

A group-theoretic classification of Boolean functions has various technical and cryptographic applications. A well-known example, which plays an important practical role, is the Harvard's catalogue. The classification of Boolean functions of five variables is also known. The existence of such catalogues facilitates analysis and synthesis of discrete devices and enables choosing functions with the best characteristics for applications in various particular cases.

3.1. Group equivalence of mappings. Polya's theorem

A large part of combinatorial analysis deals with enumeration: counting the number of possibilities, the number of solutions, the number of sets of a given type, etc. Since Polya's theorem plays an important role in enumeration and, in particular, in classification of Boolean functions, we will present it for the case where the objects are mappings of arbitrary finite sets.

Let A and B be finite sets of cardinalities $\#A = l$ and $\#B = m$. We will denote the set of all mappings (we will call them functions) from A to B by B^A. The number of such functions is equal to m^l.

Let $\mathfrak{S}_A$ be the symmetric group of permutations of elements of the set A, and let $\mathfrak{G} \subset \mathfrak{S}_A$ be a subgroup. The order of the group $\mathfrak{G}$ satisfies the following inequalities:

$$1 \leq \#\mathfrak{G} \leq l!\,.$$

It is well known that for a given permutation $\mathfrak{g}$, the set A can be uniquely partitioned into orbits of the subgroup generated by the permutation $\mathfrak{g}$. If t is the length of an orbit and if a is an element of this orbit, then the orbit consists of the elements

$$a, \quad \mathfrak{g}a, \quad \mathfrak{g}^2 a, \quad \dots, \quad \mathfrak{g}^{t-1}a,$$

where $\mathfrak{g}^k = \mathfrak{g}\cdots\mathfrak{g}$ (k times). If there are $c_1(\mathfrak{g})$ orbits of length 1, $c_2(\mathfrak{g})$ orbits of length 2, etc., we say that the permutation $\mathfrak{g}$ has the type $\{c_1(\mathfrak{g}), c_2(\mathfrak{g}), \dots, c_l(\mathfrak{g})\}$. It is clear that

$$c_1(\mathfrak{g}) + 2c_2(\mathfrak{g}) + \cdots + lc_l(\mathfrak{g}) = l$$

for any $\mathfrak{g} \in \mathfrak{G}$, as the sum of the lengths of all orbits is equal to the number of elements of A.

For any element $\mathfrak{g} \in \mathfrak{G}$, consider the monomial $\lambda_1^{c_1(\mathfrak{g})}\lambda_2^{c_2(\mathfrak{g})}\cdots\lambda_l^{c_l(\mathfrak{g})}$ in l variables $\lambda_1, \lambda_2, \dots, \lambda_l$.

DEFINITION 3.1. The polynomial

$$P_{\mathfrak{G}}(\lambda_1, \lambda_2, \dots, \lambda_l) = \frac{1}{\#\mathfrak{G}} \sum_{\mathfrak{g}\in\mathfrak{G}} \lambda_1^{c_1(\mathfrak{g})}\lambda_2^{c_2(\mathfrak{g})}\cdots\lambda_l^{c_l(\mathfrak{g})}$$

is called the *orbit index* (or orbit index polynomial) of a group of permutations $\mathfrak{G}$.

EXAMPLE 3.2. If $\mathfrak{G}$ contains only the identical permutation, we have

$$P_{\mathfrak{G}}(\lambda_1, \lambda_2, \dots, \lambda_l) = \lambda_1^l,$$

as the identical permutation has type $\{l, 0, 0, \dots, 0\}$.

EXAMPLE 3.3. Here is a very instructive example from [**19**]. Let A be the set of vertices of a cube, so $l = 8$, and let $\mathfrak{G}$ be the set of those permutations of A which are induced by rotations of the cube. There are 24 such permutations. They can be classified into five categories:

a) identity;

b) three rotations through the angle π around the lines connecting centers of the opposite faces;

c) six rotations through the angle $\frac{\pi}{2}$ around the lines connecting the centers of the opposite faces;

d) six rotations through the angle π around the lines connecting the midpoints of the opposite edges;

e) eight rotations through the angle $\frac{2}{3}\pi$ around the lines connecting opposite vertices.

In case a) there are eight orbits of length 1; in case b) there are four orbits of length 2; in case c) there are two orbits of length 4; in case d) there are four orbits of length 2; in case e) there are two orbits of length 1 and two orbits of length 3. Thus, the orbit index is

$$P_{\mathfrak{G}}(\lambda_1, \lambda_2, \dots, \lambda_8) = \frac{1}{24}(\lambda_1^8 + 9\lambda_2^4 + 6\lambda_4^2 + 8\lambda_1^2\lambda_3^2).$$

Now let A be the set of edges of the same cube and $\mathfrak{G}$ be the group of all 24 permutations of the elements of A induced by rotations of the cube. This time $l = 12$. The rotations are the same as before. Let us look at the permutation types. The identity rotation is a permutation of type $\{12, 0, 0, \dots, 0\}$. Permutations of case b) are of type $\{0, 6, 0, \dots, 0\}$; rotations of case c) induce permutations of type $\{0, 0, 0, 3, 0, \dots, 0\}$; for case d), the type is $\{2, 5, 0, \dots, 0\}$, and for case e), it is $\{0, 0, 4, 0, \dots, 0\}$. Thus, the orbit index is

$$P_{\mathfrak{G}}(\lambda_1, \lambda_2, \dots, \lambda_{12}) = \frac{1}{24}(\lambda_1^{12} + 3\lambda_2^6 + 6\lambda_4^3 + 6\lambda_1^2\lambda_2^5 + 8\lambda_3^4).$$

The mapping

$$B^A \times \mathfrak{G} \to B^A, \quad (f, \mathfrak{g}) \mapsto f^{\mathfrak{g}}, \quad f \in B^A, \quad \mathfrak{g} \in \mathfrak{G},$$

where $f^{\mathfrak{g}}(a) = f(\mathfrak{g}a)$ ($\mathfrak{g}a$ is the image of a under the action of $\mathfrak{g}$) for each $a \in A$, is the right action of the group $\mathfrak{G}$ on the set B^A: indeed,

$$f^{\mathfrak{g}_1 \cdot \mathfrak{g}_2} = (f^{\mathfrak{g}_1})^{\mathfrak{g}_2}, \quad f^{\mathfrak{e}} = f$$

for all $\mathfrak{g}_1, \mathfrak{g}_2, \mathfrak{e} \in \mathfrak{G}$ ($\mathfrak{e}$ is the identity element of the group $\mathfrak{G}$), $f \in B^A$.

We introduce the following binary relation on the set B^A: $f_1 \overset{\mathfrak{G}}{\sim} f_2$ iff there exists an element $\mathfrak{g} \in \mathfrak{G}$ such that $f_1^{\mathfrak{g}} = f_2$.

It is easy to check that this relation $\overset{\mathfrak{G}}{\sim}$ is an equivalence relation, i.e., that it has the following properties:

a) reflexivity: $f \overset{\mathfrak{G}}{\sim} f$;

b) symmetry: $f_1 \overset{\mathfrak{G}}{\sim} f_2 \Rightarrow f_2 \overset{\mathfrak{G}}{\sim} f_1$;

c) transitivity: $f_1 \overset{\mathfrak{G}}{\sim} f_2,\ f_2 \overset{\mathfrak{G}}{\sim} f_3 \Rightarrow f_1 \overset{\mathfrak{G}}{\sim} f_3$.

Thus, the set of all functions B^A is partitioned into the equivalence classes modulo the relation $\overset{\mathfrak{G}}{\sim}$.

Denote the equivalence class containing the function f by $\{f\}_{\mathfrak{G}}$. It is clear that the cardinality of this class satisfies the following inequalities:

$$1 \leq \#\{f\}_{\mathfrak{G}} \leq \#\mathfrak{G}.$$

The equivalence class $\{f\}_{\mathfrak{G}}$ is sometimes called the $\mathfrak{G}$-*type*. Functions belonging to the same class are called $\mathfrak{G}$-*equivalent*. We will omit the prefix '$\mathfrak{G}$-' if it is clear from the context which group is considered.

If $\#\{f\}_{\mathfrak{G}} = 1$, we will say that the function f is $\mathfrak{G}$-*invariant.*

DEFINITION 3.4. The set

$$\mathcal{J}_{\mathfrak{G}}(f) = \{\mathfrak{g} \in \mathfrak{G} \mid f^{\mathfrak{g}} = f\}$$

is called the *stabilizer of the function* f.

PROBLEM 3.5. Prove that the set $\mathcal{J}_{\mathfrak{G}}(f)$ is a group.

LEMMA 3.6. *The stabilizers of* $\mathfrak{G}$*-equivalent functions* f_1 *and* f_2 *are conjugate in the group* $\mathfrak{G}$, *i.e., there exists an element* $\mathfrak{g} \in \mathfrak{G}$ *such that*

$$\mathfrak{g}\mathcal{J}_{\mathfrak{G}}(f_2)\mathfrak{g}^{-1} = \mathcal{J}_{\mathfrak{G}}(f_1).$$

PROOF. Since f_1 and f_2 are $\mathfrak{G}$-equivalent, there exists an element $\mathfrak{g}$ such that $f_2 = f_1^{\mathfrak{g}}$. Suppose that $\mathfrak{h} \in \mathcal{J}_{\mathfrak{G}}(f_2)$. Then

$$f_1^{\mathfrak{g}\mathfrak{h}\mathfrak{g}^{-1}} = (f_1^{\mathfrak{g}})^{\mathfrak{h}\mathfrak{g}^{-1}} = (f_2)^{\mathfrak{h}\mathfrak{g}^{-1}} = (f_2^{\mathfrak{h}})^{\mathfrak{g}^{-1}} = f_2^{\mathfrak{g}^{-1}} = f_1.$$

Hence, $\mathfrak{g}\mathcal{J}_{\mathfrak{G}}(f_2)\mathfrak{g}^{-1} \subseteq \mathcal{J}_{\mathfrak{G}}(f_1)$. A similar argument with $\mathfrak{g}$ replaced by $\mathfrak{g}^{-1}$ shows that

$$\mathfrak{g}^{-1}\mathcal{J}_{\mathfrak{G}}(f_1)\mathfrak{g} \subseteq \mathcal{J}_{\mathfrak{G}}(f_2),$$

which is equivalent to

$$\mathcal{J}_{\mathfrak{G}}(f_1) \subseteq \mathfrak{g}\mathcal{J}_{\mathfrak{G}}(f_2)\mathfrak{g}^{-1}.$$

This proves the lemma. □

There exists a well-known relation between the cardinality of the class $\{f\}_{\mathfrak{G}}$ and the order of the stabilizer $\mathcal{J}_{\mathfrak{G}}(f)$.

THEOREM 3.7 (Orbit-stabilizer theorem). *For an arbitrary function* $f \in B^A$ *one has*

$$\#\{f\}_{\mathfrak{G}} = \frac{\#\mathfrak{G}}{\#\mathcal{J}_{\mathfrak{G}}(f)}.$$

PROOF. The equivalence class $\{f\}_{\mathfrak{G}}$ consists of functions

$$\{f\}_{\mathfrak{G}} = \{h \mid h = f^{\mathfrak{g}}, \mathfrak{g} \in \mathfrak{G}\}.$$

Let $\mathfrak{g}$ and $\mathfrak{g}'$ be elements of $\mathfrak{G}$ for which $f^{\mathfrak{g}} = f^{\mathfrak{g}'}$. Let the permutation $\mathfrak{g}^{-1}$ act on the functions in both sides of this equality:

$$(f^{\mathfrak{g}})^{\mathfrak{g}^{-1}} = f^{\mathfrak{g}\cdot\mathfrak{g}^{-1}} = f = (f^{\mathfrak{g}'})^{\mathfrak{g}^{-1}} = f^{\mathfrak{g}'\cdot\mathfrak{g}^{-1}}.$$

Therefore, the element $\mathfrak{g}' \cdot \mathfrak{g}^{-1}$ lies in the stabilizer $\mathcal{J}_{\mathfrak{G}}(f)$; in other words, elements $\mathfrak{g}$ and $\mathfrak{g}'$ belong to the same left coset of the subgroup $\mathcal{J}_{\mathfrak{G}}(f)$ in the group $\mathfrak{G}$. On the other hand, it is easy to show that elements from the same left coset of the subgroup $\mathcal{J}_{\mathfrak{G}}(f)$ acting on f give rise to the same function. Thus, the number of different functions in $\{f\}_{\mathfrak{G}}$ coincides with the number of left cosets of the group $\mathcal{J}_{\mathfrak{G}}(f)$, i.e.,

$$\#\{f\}_{\mathfrak{G}} = [\mathfrak{G} : \mathcal{J}_{\mathfrak{G}}(f)] = \frac{\#\mathfrak{G}}{\#\mathcal{J}_{\mathfrak{G}}(f)}. \qquad \square$$

LEMMA 3.8. *Suppose that $\mathfrak{S}_A \supset \mathfrak{G} \supset \mathfrak{H}$. Then for any function f,*

$$\mathcal{J}_{\mathfrak{G}}(f) \supset \mathcal{J}_{\mathfrak{H}}(f).$$

The proof is easy.

A more complicated relation exists between the equivalence classes $\{\dots\}_{\mathfrak{H}}$ and $\{\dots\}_{\mathfrak{G}}$ if $\mathfrak{S}_A \supset \mathfrak{G} \supset \mathfrak{H}$. Put $[\mathfrak{G} : \mathfrak{H}] = s$; let

$$\mathfrak{G} = \mathfrak{g}_1\mathfrak{H} \cup \mathfrak{g}_2\mathfrak{H} \cup \dots \cup \mathfrak{g}_s\mathfrak{H} \tag{3.1}$$

be the decomposition of $\mathfrak{G}$ into the union of right cosets of the subgroup $\mathfrak{H}$ ($\mathfrak{g}_1 = \mathfrak{e}$).

LEMMA 3.9. *If the hypotheses of Lemma 3.8 hold, then*

$$\{f\}_{\mathfrak{G}} = \{f^{\mathfrak{g}_1}\}_{\mathfrak{H}} \cup \{f^{\mathfrak{g}_2}\}_{\mathfrak{H}} \cup \dots \cup \{f^{\mathfrak{g}_s}\}_{\mathfrak{H}}.$$

PROOF. It follows immediately from the decomposition (3.1). $\square$

The decomposition of the set B^A into equivalence classes modulo a group action is called the *group-theoretic classification* of functions. Such a classification is useful if some the relevant properties and characterisitics of functions are invariant under the group action in question.

In this aspect, the so-called Burnside lemma [**19**] is important. It is an essential part of Polya's theorem. This lemma allows one to count the number of equivalence classes with respect to a given group action.

LEMMA 3.10. *Suppose that a group $\mathfrak{G} \subset \mathfrak{S}_A$ acts on B^A and let $\{f_1\}_{\mathfrak{G}}, \dots, \{f_{\mathcal{K}(\mathfrak{G})}\}_{\mathfrak{G}}$ be all the different $\mathfrak{G}$-types. Then the number $\mathcal{K}(\mathfrak{G})$ of $\mathfrak{G}$-types satisfies the equality*

$$\mathcal{K}(\mathfrak{G}) = \frac{1}{\#\mathfrak{G}} \sum_{\mathfrak{g}\in\mathfrak{G}} \mathrm{st}(\mathfrak{g}),$$

where $\mathrm{st}(\mathfrak{g}) = \#\{f \in B^A \mid f^{\mathfrak{g}} = f\}$.

PROOF. Define the following function on $\mathfrak{G} \times B^A$:

$$\omega(\mathfrak{g}, f) = \begin{cases} 1 & \text{if } f^{\mathfrak{g}} = f, \\ 0 & \text{otherwise.} \end{cases}$$

Changing the order of summation, we obtain

$$\begin{aligned} \sum_{\mathfrak{g}\in\mathfrak{G}} \mathrm{st}(\mathfrak{g}) &= \sum_{\mathfrak{g}\in\mathfrak{G}} \#\{f \in B^A \mid f^{\mathfrak{g}} = f\} \\ &= \sum_{\mathfrak{g}\in\mathfrak{G}} \sum_{f\in B^A} \omega(\mathfrak{g}, f) = \sum_{f\in B^A} \sum_{\mathfrak{g}\in\mathfrak{G}} \omega(\mathfrak{g}, f) = \sum_{f\in B^A} \#\mathcal{J}_{\mathfrak{G}}(f). \end{aligned}$$

Since for any function $f' \in \{f\}_{\mathfrak{G}}$ we have $\#\mathcal{J}_{\mathfrak{G}}(f) = \#\mathcal{J}_{\mathfrak{G}}(f')$ (see Lemma 3.6), it follows that

$$\sum_{f\in B^A} \#\mathcal{J}_{\mathfrak{G}}(f) = \sum_{j=1}^{\mathcal{K}(\mathfrak{G})} \sum_{f'\in\{f_j\}_{\mathfrak{G}}} \#\mathcal{J}_{\mathfrak{G}}(f') = \sum_{j=1}^{\mathcal{K}(\mathfrak{G})} \#\mathcal{J}_{\mathfrak{G}}(f_j) \cdot \#\{f_j\}_{\mathfrak{G}}$$
$$= \sum_{j=1}^{\mathcal{K}(\mathfrak{G})} \#\mathfrak{G} = \mathcal{K}(\mathfrak{G}) \cdot \#\mathfrak{G}. \qquad \square$$

Suppose that a weight $w(b)$ is assigned to every element $b \in B$ (this notation is chosen to avoid confusion with the Hamming weight wt). This weight can be a rational number, a variable, or an element of an arbitrary commutative ring. Thus, we can add and multiply weights, multiply a weight by a rational number, and for these operations the associativity, commutativity, and distributivity laws are satisfied.

DEFINITION 3.11. For any function $f \in B^A$ its *weight* $w(f)$ is defined to be the product

$$w(f) = \prod_{a\in A} w(f(a)).$$

Obviously, if functions f and f' belong to the same equivalence class in a group $\mathfrak{G}$, then their weights are equal:

$$w(f) = w(f').$$

As all functions from an equivalence class have the same weight, we can define the weight of a class to be the weight of a function in the class.

DEFINITION 3.12. For any function $f \in B^A$ the *weight of the equivalence class* containing f is defined by the equality

$$w(\{f\}_{\mathfrak{G}}) = w(f).$$

DEFINITION 3.13. The sum

$$\mathrm{en}(B) = \sum_{b\in B} w(b)$$

is called the *enumerator* of the set B.

DEFINITION 3.14. The sum

$$\mathrm{en}(S) = \sum_{f\in S} w(f)$$

is called the *enumerator of an arbitrary set* $S \subseteq B^A$ *of functions*.

PROBLEM 3.15. Prove that

$$\mathrm{en}(B^A) = \left[\sum_{b\in B} w(b)\right]^l = [\mathrm{en}(B)]^l.$$

Suppose that a set A is partitioned into disjoint subsets $A_1, \dots, A_k$, so that

$$\#A = \#A_1 + \cdots + \#A_k.$$

Suppose that S is the set of functions $f \in B^A$ such that f is constant on each of the subsets $A_1, \dots, A_k$.

LEMMA 3.16. *The enumerator of the set $S \subset B^A$ is equal to*

$$\mathrm{en}(S) = \prod_{i=1}^{k} \sum_{b \in B} [w(b)]^{\#A_i}.$$

PROOF. Suppose that $f \in S$. Then the restriction of f to the component A_i ($i = 1, \dots, k$) is the constant

$$f|_{A_i} = b_i, \quad b_i \in B.$$

Hence, the functions f are in a 1-1 correspondence with the sequence $(b_1, \dots, b_k) \in \underbrace{B \times \cdots \times B}_{k} = B^k$. The function corresponding to such a sequence will be denoted by $f = f_{b_1,\dots,b_k}$. Conversely, any sequence $(b_1, \dots, b_k) \in B^k$ corresponds to a unique function $f_{b_1,\dots,b_k} \in S$. Hence,

$$\begin{aligned}
\mathrm{en}(S) &= \sum_{f \in S} w(f) = \sum_{(b_1,\dots,b_k) \in B^k} w(f_{b_1,\dots,b_k}) = \sum_{(b_1,\dots,b_k) \in B^k} \prod_{a \in A} w(f_{b_1,\dots,b_k}(a)) \\
&= \sum_{(b_1,\dots,b_k) \in B^k} \prod_{a_1 \in A_1} w(f_{b_1,\dots,b_k}(a_1)) \cdots \prod_{a_k \in A_k} w(f_{b_1,\dots,b_k}(a_k)) \\
&= \sum_{(b_1,\dots,b_k) \in B^k} [w(b_1)]^{\#A_1} \cdots [w(b_k)]^{\#A_k} \\
&= \sum_{b_1 \in B} [w(b_1)]^{\#A_1} \cdots \sum_{b_k \in B} [w(b_k)]^{\#A_k} = \prod_{i=1}^{k} \sum_{b \in B} [w(b)]^{\#A_i}.
\end{aligned}$$

□

Let us find the enumerator of the set of all $\mathfrak{G}$-types, which we denote by $B^A/\mathfrak{G}$:

$$B^A/\mathfrak{G} = \big\{\{f_1\}_{\mathfrak{G}}, \{f_2\}_{\mathfrak{G}}, \dots, \{f_{\mathcal{K}(\mathfrak{G})}\}_{\mathfrak{G}}\big\}.$$

THEOREM 3.17 (Polya). *We have*

$$\mathrm{en}(B^A/\mathfrak{G}) = \sum_{i=1}^{\mathcal{K}(\mathfrak{G})} w(\{f_i\}_{\mathfrak{G}}) = P_{\mathfrak{G}}\Big(\sum_{b \in B} w(b), \sum_{b \in B} [w(b)]^2, \dots, \sum_{b \in B} [w(b)]^l\Big),$$

where $P_{\mathfrak{G}}(\lambda_1, \dots, \lambda_l)$ is the orbit index of the group $\mathfrak{G}$. In particular, if all weights are equal to 1, then

$$\mathcal{K}(\mathfrak{G}) = P_{\mathfrak{G}}(m, \dots, m).$$

PROOF. Let ω be one of the possible values of the weight of a function, and let M_ω be the set of all functions $f \in B^A$ for which $w(f) = \omega$. It is clear that M_ω is a union of $\mathfrak{G}$-types. Then, similarly to the proof of Lemma 3.10, we can show that the the number of $\mathfrak{G}$-types in M_ω is equal to

$$\frac{1}{\#\mathfrak{G}} \sum_{\mathfrak{g} \in \mathfrak{G}} \mathrm{st}_\omega(\mathfrak{g}), \tag{3.2}$$

where $\mathrm{st}_\omega(\mathfrak{g}) = \#\{f \in B^A \mid w(f) = \omega,\ f^{\mathfrak{g}} = f\}$. All $\mathfrak{G}$-types in M_ω have weight ω. Thus, if we multiply (3.2) by ω and sum over all possible values of ω, we will find the enumerator:

$$\begin{aligned}\mathrm{en}(B^A/\mathfrak{G}) &= \frac{1}{\#\mathfrak{G}}\sum_{\omega}\sum_{\mathfrak{g}\in\mathfrak{G}}\omega(\mathrm{st}_\omega(\mathfrak{g}))\\ &= \frac{1}{\#\mathfrak{G}}\sum_{\mathfrak{g}\in\mathfrak{G}}\sum_{\omega}\omega(\mathrm{st}_\omega(\mathfrak{g})) = \frac{1}{\#\mathfrak{G}}\sum_{\mathfrak{g}\in\mathfrak{G}}\sum_{f\in B^A, f^{\mathfrak{g}}=f} w(f).\end{aligned}$$

In order to find the sum $\sum_{f\in B^A, f^{\mathfrak{g}}=f} w(f)$, we will use the fact that $\mathfrak{g}$ is a permutation of the elements of the set A. Thus, A is the union of orbits whose elements are cyclically presented by the permutation $\mathfrak{g}$. The condition $f^{\mathfrak{g}} = f$ means that

$$f(a) = f(\mathfrak{g}a) = f(\mathfrak{g}^2 a) = \cdots,$$

and f is constant on each orbit. Conversely, each function that is constant on any orbit of the permutation $\mathfrak{g}$ automatically satisfies the condition $f^{\mathfrak{g}} = f$. Let the orbits be $A_1, \ldots, A_k$ and the type of the permutation $\mathfrak{g}$ be $\{c_1(\mathfrak{g}), c_2(\mathfrak{g}), \ldots, c_l(\mathfrak{g})\}$. This means that among the numbers $\#A_1, \ldots, \#A_k$, the number 1 occurs $c_1(\mathfrak{g})$ times, the number 2 occurs $c_2(\mathfrak{g})$ times, and so on.

According to Lemma 3.16, we have

$$\sum_{f\in B^A,\, f^{\mathfrak{g}}=f} w(f) = \prod_{j=1}^{k}\sum_{b\in B}(w(b))^{\#A_j} = \prod_{j=1}^{l}\Big\{\sum_{b\in B}[w(b)]^j\Big\}^{c_j(\mathfrak{g})}.$$

Therefore,

$$\mathrm{en}(B^A/\mathfrak{G}) = \frac{1}{\#\mathfrak{G}}\sum_{\mathfrak{g}\in\mathfrak{G}}\prod_{j=1}^{l}\Big\{\sum_{b\in B}[w(b)]^j\Big\}^{c_j(\mathfrak{g})}. \tag{3.3}$$

It is clear that (3.3) can be obtained by putting

$$\lambda_1 = \sum_{b\in B} w(b), \quad \lambda_2 = \sum_{b\in B}[w(b)]^2, \quad \ldots, \quad \lambda_l = \sum_{b\in B}[w(b)]^l$$

in the expression $P_{\mathfrak{G}}(\lambda_1, \ldots, \lambda_l) = \frac{1}{\#\mathfrak{G}}\sum_{\mathfrak{g}\in\mathfrak{G}}\lambda_1^{c_1(\mathfrak{g})}\lambda_2^{c_2(\mathfrak{g})}\cdots\lambda_l^{c_l(\mathfrak{g})}$ for the orbit index of $\mathfrak{G}$. □

3.2. Classification of Boolean functions of five variables

Now we reutrn from the mappings of finite sets considered in §3.1 to Boolean functions. Denote by A the vector space V_n ($l = 2^n$) (see §3.1) and by B the finite field $\mathbb{F}_2$ ($m = 2$). In this case, the set B^A coincides with $\mathcal{F}_n$. The group is $\mathfrak{S}_A = \mathfrak{S}_{V_n}$, $\#\mathfrak{S}_{V_n} = 2^n!$. The group $\mathfrak{G}$ from the previous section is a subgroup of $\mathfrak{S}_{V_n}$.

To begin with we inroduce five important subgroups of the group $\mathfrak{S}_{V_n}$ that appear in various classifications of Boolean functions. General expressions for cycle indexes of these subgroups look rather complicated, and in order to derive them we will need subtle group-theoretic and combinatorial arguments. We will introduce expressions for cycle indexes and numbers of equivalence classes only for $n = 1, 2, 3, 4, 5, 6$ (see Tables 3.1–3.4). The only exception is the inverted variables group, for which we will get general expressions. We refer the reader interested in details of computation of cycle indexes to [**160**].

Inverted variables group $\mathfrak{J}_n$ (shift group). This group consists of 2^n permutations of the form

$$\mathfrak{g}_{\boldsymbol{a}}\colon (x^{(1)},\ldots,x^{(n)})^\top \mapsto (x^{(1)}\oplus a^{(1)},\ldots,x^{(n)}\oplus a^{(n)})^\top,$$

where $\boldsymbol{a} = (a^{(1)},\ldots,a^{(n)})^\top \in V_n$. Cyclic structure of elements of this group may be either $\{2^n,0,0,\ldots,0\}$ for the identity permutation or $\{0,2^{n-1},0,\ldots,0\}$ for any permutation different from the identity.

Therefore, the cycle index of this group is

$$P_{\mathfrak{J}_n}(\lambda_1,\ldots,\lambda_{2^n}) = \frac{1}{2^n}(\lambda_1^{2^n} + (2^n-1)\lambda_2^{2^{n-1}}),$$

which yields the following expression for the number of equivalence classes modulo $\mathfrak{J}_n$:

$$\mathcal{K}(\mathfrak{J}_n) = P_{\mathfrak{J}_n}(2,2,\ldots,2) = \frac{1}{2^n}(2^{2^n} + (2^n-1)2^{2^{n-1}}).$$

Group of permutations of variables $\mathfrak{S}_n$. This group consists of $n!$ permutations of the form

$$\mathfrak{g}\colon (x^{(1)},\ldots,x^{(n)})^\top \mapsto (x^{(j_1)},\ldots,x^{(j_n)})^\top,$$

where $\begin{pmatrix} 1, & \ldots, & n \\ j_1, & \ldots, & j_n\end{pmatrix}$ is a permutation of the set $\{1,\ldots,n\}$.

TABLE 3.1

n	$\mathcal{K}(\mathfrak{S}_n)$	$P_{\mathfrak{S}_n}(\lambda_1,\ldots,\lambda_{2^n})$
1	4	λ_1^2
2	12	$\frac{1}{2}(\lambda_1^4 + \lambda_1^2\lambda_2)$
3	80	$\frac{1}{6}(\lambda_1^8 + 3\lambda_1^4\lambda_2^2 + 2\lambda_1^2\lambda_3^2)$
4	3 984	$\frac{1}{24}(\lambda_1^{16} + 6\lambda_1^8\lambda_2^4 + 3\lambda_1^4\lambda_2^6 + 8\lambda_1^4\lambda_3^4 + 6\lambda_1^2\lambda_2\lambda_4^3)$
5	37 333 248	$\frac{1}{120}(\lambda_1^{32} + 10\lambda_1^{16}\lambda_2^8 + 15\lambda_1^8\lambda_2^{12} + 20\lambda_1^8\lambda_3^8$ $+ 20\lambda_1^4\lambda_2^2\lambda_3^4\lambda_6^2 + 30\lambda_1^4\lambda_2^2\lambda_4^6 + 24\lambda_1^2\lambda_5^6)$
6	25 626 412 338 274 304	$\frac{1}{720}(\lambda_1^{64} + 15\lambda_1^{32}\lambda_2^{16} + 45\lambda_1^{16}\lambda_2^{24}$ $+ 40\lambda_1^{16}\lambda_3^{16} + 15\lambda_1^8\lambda_2^{28} + 120\lambda_1^8\lambda_2^4\lambda_3^8\lambda_6^4$ $+ 90\lambda_1^8\lambda_2^4\lambda_4^{12} + 40\lambda_1^4\lambda_3^{20} + 90\lambda_1^4\lambda_2^6\lambda_4^{12}$ $+ 144\lambda_1^4\lambda_5^{12} + 120\lambda_1^2\lambda_2\lambda_3^2\lambda_6^9)$

Jevons group $\mathfrak{D}_n$. This group consists of $2^n \cdot n!$ permutations of the form

$$\mathfrak{g}\colon (x^{(1)}, \dots, x^{(n)})^\top \mapsto (x^{(j_1)} \oplus a^{(1)}, \dots, x^{(j_n)} \oplus a^{(n)})^\top,$$

where $\left(\begin{smallmatrix} 1, & \dots, & n \\ j_1, & \dots, & j_n \end{smallmatrix}\right)$ is a permutation of the set $\{1, \dots, n\}$ and $\boldsymbol{a} = (a^{(1)}, \dots, a^{(n)})^\top \in V_n$. The group $\mathfrak{D}_n$ is a semidirect product of the groups $\mathfrak{S}_n$ and $\mathfrak{J}_n$.

TABLE 3.2

n	$\mathcal{K}(\mathfrak{D}_n)$	$P_{\mathfrak{D}_n}(\lambda_1, \dots, \lambda_{2^n})$
1	3	$\frac{1}{2}(\lambda_1^2 + \lambda_2)$
2	6	$\frac{1}{8}(\lambda_1^4 + 3\lambda_2^2 + 2\lambda_1^2\lambda_2 + 2\lambda_4)$
3	22	$\frac{1}{48}(\lambda_1^8 + 13\lambda_2^4 + 8\lambda_1^2\lambda_3^2 + 8\lambda_2\lambda_6 + 6\lambda_1^4\lambda_2^2 + 12\lambda_4^2)$
4	402	$\frac{1}{384}(\lambda_1^{16} + 51\lambda_2^8 + 48\lambda_1^2\lambda_2\lambda_4^3 + 48\lambda_8^2 + 12\lambda_1^8\lambda_2^4$ $+ 84\lambda_4^4 + 12\lambda_1^4\lambda_2^6 + 32\lambda_1^4\lambda_3^4 + 96\lambda_2^2\lambda_6^2)$
5	1 228 158	$\frac{1}{3\,840}(\lambda_1^{32} + 231\lambda_2^{16} + 20\lambda_1^{16}\lambda_2^8 + 520\lambda_4^8 + 80\lambda_1^8\lambda_3^8$ $+ 720\lambda_2^4\lambda_6^4 + 160\lambda_1^4\lambda_2^2\lambda_3^4\lambda_6^2 + 320\lambda_4^2\lambda_{12}^2$ $+ 240\lambda_1^4\lambda_2^2\lambda_4^6 + 480\lambda_8^4 + 240\lambda_2^4\lambda_4^6$ $+ 60\lambda_1^8\lambda_2^{12} + 384\lambda_1^2\lambda_5^6 + 384\lambda_2\lambda_{10}^3)$
6	400 507 806 843 728	$\frac{1}{46\,080}(\lambda_1^{64} + 1053\lambda_2^{32} + 30\lambda_1^{32}\lambda_2^{16} + 4920\lambda_4^{16}$ $+ 180\lambda_1^{16}\lambda_2^{24} + 160\lambda_1^{16}\lambda_3^{16} + 5280\lambda_2^8\lambda_6^8$ $+ 120\lambda_1^8\lambda_2^{28} + 960\lambda_1^8\lambda_2^4\lambda_3^8\lambda_6^4 + 3840\lambda_4^4\lambda_{12}^4$ $+ 720\lambda_1^8\lambda_2^4\lambda_4^{12} + 5760\lambda_8^8 + 2160\lambda_2^8\lambda_4^{12} + 640\lambda_1^4\lambda_3^{20}$ $+ 1920\lambda_2^2\lambda_6^{10} + 1440\lambda_1^4\lambda_2^6\lambda_4^{12} + 2304\lambda_1^4\lambda_2^{12}$ $+ 6912\lambda_2^2\lambda_{10}^6 + 3840\lambda_1^2\lambda_2\lambda_3^2\lambda_6^9 + 3840\lambda_4\lambda_{12}^5)$

Group of linear transformations $\mathfrak{GL}(V_n)$ (general linear group). This group consists of $\prod_{i=0}^{n-1}(2^n - 2^i)$ permutations of the form

$$\mathfrak{g}\colon (x^{(1)}, \dots, x^{(n)})^\top \mapsto \left(\bigoplus_{j=1}^{n} a_{1j}x^{(j)}, \dots, \bigoplus_{j=1}^{n} a_{nj}x^{(j)}\right)^\top,$$

where $A = \big(a_{ij}\big)$ is an invertible $(n \times n)$ matrix, $a_{ij} \in \mathbb{F}_2$, $i, j = 1, 2, \dots, n$. In the matrix form these permutations can be presented as follows:

$$\mathfrak{g}\boldsymbol{x} = A\boldsymbol{x}.$$

TABLE 3.3

n	$\mathcal{K}(\mathfrak{GL}(V_n))$	$P_{\mathfrak{GL}(V_n)}(\lambda_1,\dots,\lambda_{2^n})$
1	4	λ_1^2
2	8	$\frac{1}{6}(\lambda_1^4+3\lambda_1^2\lambda_2+2\lambda_1\lambda_3)$
3	20	$\frac{1}{168}(\lambda_1^8+21\lambda_1^4\lambda_2^2+42\lambda_1^2\lambda_2\lambda_4+56\lambda_1^2\lambda_3^2+48\lambda_1\lambda_7)$
4	92	$\frac{1}{20\,160}(\lambda_1^{16}+105\lambda_1^8\lambda_2^4+210\lambda_1^4\lambda_2^6$ $+1260\lambda_1^4\lambda_2^2\lambda_4^2+2520\lambda_1^2\lambda_2\lambda_4^3+112\lambda_1\lambda_3^5$ $+1680\lambda_1\lambda_3\lambda_6^2+1120\lambda_1^4\lambda_3^4+3360\lambda_1^2\lambda_2\lambda_3^2\lambda_6$ $+5760\lambda_1^2\lambda_7^2+1344\lambda_1\lambda_5^3+2688\lambda_1\lambda_{15})$
5	2 744	$\frac{1}{9\,999\,360}(\lambda_1^{32}+465\lambda_1^{16}\lambda_2^8+26\,040\lambda_1^8\lambda_2^4\lambda_4^4$ $+312\,480\lambda_1^4\lambda_2^2\lambda_4^6+624\,960\lambda_1^2\lambda_2\lambda_4^3\lambda_8^2+6510\lambda_1^8\lambda_2^{12}$ $+78\,120\lambda_1^4\lambda_2^6\lambda_4^4+19\,840\lambda_1^8\lambda_3^8+416\,640\lambda_1^4\lambda_2^2\lambda_3^4\lambda_6^2$ $+833\,280\lambda_1^2\lambda_2\lambda_3^2\lambda_4\lambda_6\lambda_{12}+55\,552\lambda_1^2\lambda_3^{10}$ $+833\,280\lambda_1^2\lambda_3^2\lambda_6^4+476\,160\lambda_1^4\lambda_7^4$ $+1\,428\,480\lambda_1^2\lambda_2\lambda_7^2\lambda_{14}+952\,320\lambda_1\lambda_3\lambda_7\lambda_{21}$ $+666\,624\lambda_1^2\lambda_5^6+1\,333\,248\lambda_1^2\lambda_{15}^2+1\,935\,360\lambda_1\lambda_{31})$

It is known that $\mathcal{K}(\mathfrak{GL}(V_6)) = 950\,988\,216$.

Group of affine transformations $\mathfrak{GA}(V_n)$ (complete affine group). This group consists of $2^n\prod_{i=0}^{n-1}(2^n-2^i)$ permutations of the form

$$\mathfrak{g}\colon (x^{(1)},\dots,x^{(n)})^\top \mapsto \left(\bigoplus_{j=1}^{n} a_{1j}x^{(j)}\oplus a^{(1)},\dots,\bigoplus_{j=1}^{n} a_{nj}x^{(j)}\oplus a^{(n)}\right)^\top,$$

where $A=(a_{ij})$ is an invertible $(n\times n)$ matrix, $a_{ij}\in\mathbb{F}_2$, $i,j=1,2,\dots,n$, and $\boldsymbol{a}=(a^{(1)},\dots,a^{(n)})^\top\in V_n$. In the matrix form these permutations can be presented as follows:

$$\mathfrak{g}\boldsymbol{x}=\boldsymbol{a}\oplus A\boldsymbol{x}.$$

It is known that $\mathcal{K}(\mathfrak{GA}(V_6)) = 15\,768\,919$.

Some generalizations. Let $\mathfrak{G}$ be one of the above groups. Denote by $\mathfrak{GH}_s$, $s\le n$, the following set:

$$\mathfrak{GH}_s=\{(\mathfrak{g},h)\mid \mathfrak{g}\in\mathfrak{G},\ h\in\mathcal{F}_n,\ \deg h\le s\}.$$

TABLE 3.4

n	$\mathcal{K}(\mathfrak{GA}(V_n))$	$P_{\mathfrak{GA}(V_n)}(\lambda_1, \ldots, \lambda_{2^n})$
1	3	$\frac{1}{2}(\lambda_1^2 + \lambda_2)$
2	5	$\frac{1}{24}(\lambda_1^4 + 3\lambda_2^2 + 6\lambda_1^2\lambda_2 + 6\lambda_4 + 8\lambda_1\lambda_3)$
3	10	$\frac{1}{1\,344}(\lambda_1^8 + 49\lambda_2^4 + 42\lambda_1^4\lambda_2^2 + 252\lambda_4^2 + 168\lambda_1^2\lambda_2\lambda_4 + 384\lambda_1\lambda_7 + 224\lambda_1^2\lambda_3^2 + 224\lambda_2\lambda_6)$
4	32	$\frac{1}{322\,560}(\lambda_1^{16} + 645\lambda_2^8 + 210\lambda_1^8\lambda_2^4 + 13\,440\lambda_4^4 + 840\lambda_1^4\lambda_2^6 + 5\,040\lambda_1^4\lambda_2^2\lambda_4^2 + 5\,040\lambda_2^4\lambda_4^2 + 20\,160\lambda_1^2\lambda_2\lambda_4^3 + 20\,160\lambda_8^2 + 1\,792\lambda_1\lambda_3^5 + 26\,880\lambda_1\lambda_3\lambda_6^2 + 26\,880\lambda_1^2\lambda_2\lambda_3^2\lambda_6 + 26\,880\lambda_4\lambda_{12} + 4\,480\lambda_1^4\lambda_3^4 + 13\,440\lambda_2^2\lambda_6^2 + 46\,080\lambda_1^2\lambda_7^2 + 46\,080\lambda_2\lambda_{14} + 43\,008\lambda_1\lambda_{15} + 21\,504\lambda_1\lambda_5^3)$
5	382	$\frac{1}{319\,979\,520}(\lambda_1^{32} + 32\,581\lambda_2^{16} + 930\lambda_1^{16}\lambda_2^8 + 2\,455\,200\lambda_4^8 + 104\,160\lambda_1^8\lambda_2^4\lambda_4^4 + 312\,480\lambda_2^8\lambda_4^4 + 2\,499\,840\lambda_1^4\lambda_2^2\lambda_4^6 + 3\,888\,640\lambda_2^4\lambda_6^4 + 14\,999\,040\lambda_8^4 + 9\,999\,360\lambda_1^2\lambda_2\lambda_4^3\lambda_8^2 + 26\,040\lambda_1^8\lambda_2^{12} + 624\,960\lambda_1^4\lambda_2^6\lambda_4^4 + 79\,360\lambda_1^8\lambda_3^8 + 3\,333\,120\lambda_1^4\lambda_2^2\lambda_3^4\lambda_6^2 + 19\,998\,720\lambda_4^2\lambda_{12}^2 + 13\,332\,480\lambda_1^2\lambda_2\lambda_3^2\lambda_4\lambda_6\lambda_{12} + 888\,832\lambda_1^2\lambda_3^{10} + 14\,221\,312\lambda_2\lambda_6^5 + 13\,332\,480\lambda_1^2\lambda_3^2\lambda_6^4 + 3\,809\,280\lambda_1^4\lambda_7^4 + 11\,427\,840\lambda_2^2\lambda_{14}^2 + 22\,855\,680\lambda_1^2\lambda_2\lambda_7^2\lambda_{14} + 22\,855\,680\lambda_4\lambda_{28} + 30\,474\,240\lambda_1\lambda_3\lambda_7\lambda_{21} + 10\,665\,984\lambda_1^2\lambda_5^6 + 10\,665\,984\lambda_2\lambda_{10}^3 + 21\,331\,968\lambda_1^2\lambda_{15}^2 + 21\,331\,968\lambda_2\lambda_{30} + 61\,931\,520\lambda_1\lambda_{31} + 2\,499\,840\lambda_2^4\lambda_4^6)$

The transformation $(\mathfrak{g}, h)$ acts on the set $\mathcal{F}_n$ as

$$f^{(\mathfrak{g},h)}(\boldsymbol{x}) = f(\mathfrak{g}\boldsymbol{x}) \oplus h(\boldsymbol{x}).$$

It is easy to verify that such a transformation is one-to-one and that composition of two such transformations is a transformation of the same type.

The notion of an invariant of a group $\mathfrak{G}$ is useful when one wants to find a list of representatives of $\mathfrak{G}$-equivalence classes (this problem is called the *constructive enumeration problem*). Let l be a natural number.

DEFINITION 3.18. The mapping $\pi\colon \mathcal{F}_n \to \mathbb{R}^l$ satisfying the equality

$$\pi(f^{\mathfrak{g}}) = \pi(f)$$

for all $f \in \mathcal{F}_n$ and $\mathfrak{g} \in \mathfrak{G}$, is called an *invariant of the group* $\mathfrak{G}$.

It is easy to see that if there exists an invariant π of the group $\mathfrak{G}$ that assumes different values ($\pi(f) \neq \pi(f')$) at the functions f and f', then these functions are not $\mathfrak{G}$-equivalent.

DEFINITION 3.19. An invariant π of the group $\mathfrak{G}$ is called *complete* if from $\pi(f) = \pi(f')$ it follows that functions f and f' are $\mathfrak{G}$-equivalent.

The problem of finding a complete invariant for a given group is very hard, and it is solved only in some special cases. That is why in order to design algorithms for checking equivalence and finding stabilizers, incomplete invariants are used as well. In many cases they help either to obtain the answer or to reduce the number of elements of the group which have to be inspected.

PROBLEM 3.20. Let $\mathfrak{G} \subset \mathfrak{S}_{V_n}$ be one of the above groups. Prove that the weight of the function $\mathrm{wt}(f)$ and the degree of its algebraic normal form $\deg f$ are invariants of the group $\mathfrak{G}$.

The following fact is often useful.

THEOREM 3.21. *As $n \to \infty$, almost all Boolean functions have trivial stabilizer in the group $\mathfrak{GA}(V_n)$.*

PROOF. Suppose that $\mathfrak{g} \in \mathfrak{GA}(V_n)$ and that $\mathfrak{g} = \mathfrak{c}_1 \cdot \mathfrak{c}_2 \cdots \mathfrak{c}_{\nu(\mathfrak{g})}$ is the decomposition of the permutation in disjoint cycles. It is obvious that the number of $\mathfrak{g}$-invariant functions is equal to $2^{\nu(\mathfrak{g})}$. Let M_n be the number of functions of n variables having a nontrivial stabilizer. It is easy to see that

$$M_n \leq (\#\mathfrak{GA}(V_n) - 1)2^{\nu},$$

where $\nu = \max_{\mathfrak{g}\in\mathfrak{GA}(V_n),\mathfrak{g}\neq\mathfrak{e}} \nu(\mathfrak{g})$. We are going to find ν. Let $\mathfrak{g}$ be a nontrivial affine transformation , i.e., $\mathfrak{g} = (A, \boldsymbol{b}) \neq \mathfrak{e}$. Consider the equation $\mathfrak{g}\boldsymbol{x} = A\boldsymbol{x}\oplus\boldsymbol{b} = \boldsymbol{x}$ for the unknown variable $\boldsymbol{x}$. The number of its solutions is the number of cycles of length 1 in the permutation $\mathfrak{g}$. This equation has 2^{n-r} solutions, where $r = \mathrm{rank}(A \oplus E)$. If $r = 0$, then $A = E$ and in this case $\boldsymbol{b} \neq \boldsymbol{0}$. We see that there are no solutions when $r = 0$. When $r \geq 1$, there are at most 2^{n-1} cycles of length 1 in the permutation $\mathfrak{g}$. Thus, the maximum possible number of cycles in the affine permutation will be attained if it has 2^{n-1} cycles of length 1 and all the other cycles have length 2.

PROBLEM 3.22. Prove that there exists an affine permutation $\mathfrak{g}$ from $\mathfrak{GA}(V_n)$ with the type $\{2^{n-1}, 2^{n-2}, 0, \ldots, 0\}$.

Thus, $\nu = 3 \cdot 2^{n-2}$ and

$$\frac{M_n}{2^{2^n}} \leq \frac{\left(2^n \prod_{i=0}^{n-1}(2^n - 2^i) - 1\right) \cdot 2^{3\cdot 2^{n-2}}}{2^{2^n}} \leq 2^{-(2^{n-2}-n^2-n)}.$$

Since $(2^{n-2} - n^2 - n)$ tends to infinity as $n \to \infty$, it is clear that $M_n/2^{2^n} \to 0$ as $n \to \infty$. Therefore, the ratio of functions having nontrivial stabilizer with respect to $\mathfrak{GA}(V_n)$ to all functions tends to 0 as $n \to \infty$. □

COROLLARY 3.23. *Statements similar to Theorem* 3.21 *hold also for the groups* $\mathfrak{T}_n$, $\mathfrak{S}_n$, $\mathfrak{D}_n$, *and* $\mathfrak{GL}(V_n)$.

PROOF. The statement follows from Lemma 3.8. □

A list of representatives of equivalence classes can be created in several steps. First we choose a method to generate a sequence of functions $f_1, f_2, \ldots$ from $\mathcal{F}_n$.

Since we cannot immediately obtain functions from different equivalence classes, we can use various methods for building this sequence. For example, we can search through all Boolean functions in n variables using certain parameters or we can generate this sequence randomly by using, say, the Monte Carlo method, etc.

Then we set $h_1 = f_1$ and calculate the order of the stabilizer $\mathcal{J}_{\mathfrak{G}}(h_1)$. We check whether the function f_2 is equivalent to h_1. If the answer is positive, we continue with f_3; otherwise, we set $h_2 = f_2$ and calculate the order of the stabilizer $\mathcal{J}_{\mathfrak{G}}(h_2)$, and so on. For the function f_i from the initial sequence we check if it is equivalent to one of the functions $h_1, h_2, \ldots, h_s$ already chosen. If it is, we continue with the next function f_{i+1}. If the answer is negative, we set $h_{s+1} = f_i$ and calculate the order of its stabilizer. After that we verify whether the equality

$$\sum_{j=1}^{s+1} \frac{\#\mathfrak{G}}{\#\mathcal{J}_{\mathfrak{G}}(h_j)} = 2^{2^n}$$

holds. If yes, the classification is finished. If the sum in the left-hand side is smaller than 2^{2^n}, we proceed with the function f_{i+1}.

All known classifications of Boolean functions were obtained using this procedure.

It is clear from the expression for the cycle index of the group $\mathfrak{I}_n$ that the use of the groups $\mathfrak{I}_n$ and $\mathfrak{S}_n$ for classification is effective only if $n \leq 3$. The groups $\mathfrak{D}_n$ and $\mathfrak{GL}(V_n)$ are convenient for classification only if $n \leq 4$. Such a classification (the so-called Harvard catalogue) was obtained for the first time in the early 1950s. Tables of all representatives of equivalence classes of functions for all groups for $n \leq 4$ can be found, for example, in [**358**]. For $n = 5$ it turns out that the only group that can be used is $\mathfrak{GA}(V_n)$. A complete classification with respect to this group was obtained in 1975 (see [**372**]); it has representatives of 382 equivalence classes. For classification of Boolean functions of $n \geq 6$ variables this approach proved to be inefficient. Most likely, further classifications could be obtained using generalizations of the group $\mathfrak{GA}(V_n)$, type $\mathfrak{GA}(V_n)\mathfrak{H}_s$ groups.

As an illustration of what was said, we present two classifications of Boolean functions of 6 variables with respect to the groups $\mathfrak{GL}(V_6)\mathfrak{H}_3$ (see Table 3.5) and $\mathfrak{GA}(V_6)\mathfrak{H}_3$ (see Table 3.6), published in [**86**]. For brevity, the index notation for the algebraic normal forms of Boolean functions was used: a monomial is denoted by the sequence of indices of variables occurring in it. For example, by $123 + 456$ we denote the function $x^{(1)}x^{(2)}x^{(3)} \oplus x^{(4)}x^{(5)}x^{(6)}$.

Looking at Table 3.6 we observe that to obtain the classification it suffices to check which of the representatives of equivalence classes in Table 3.5 are equivalent with respect to the group $\mathfrak{GA}(V_6)\mathfrak{H}_3$. This can be done simply by checking affine transformations.

TABLE 3.5. Classification of functions of 6 variables with respect to the group $\mathfrak{GL}(V_6)\mathfrak{H}_3$.

No	Representative of the equivalence class	Cardinality of the class
1	0	1
2	1234	651
3	1234+1256	18 228
4	1234+1256+3456	13 888
5	12345	1 953
6	$\overline{12345}$	63
7	123(45+6)	273 420
8	$\overline{123}(45+6)$	39 060
9	$\overline{1}(2345+236+456)$	874 944
10	1(2345+236+456)	874 944
11	$\overline{123456}$	1
12	$\overline{123456}+\overline{12345}$	63
13	$\overline{123456}+1234$	651
14	$\overline{123456}+12345$	1 953
15	$\overline{123456}+\overline{123}(45+6)$	39 060
16	$\overline{123456}+1234+1256$	18 228
17	$\overline{123456}+123(45+6)$	273 420
18	$\overline{123456}+\overline{1}(2345+236+456)$	874 944
19	$\overline{123456}+1234+1256+3456$	13 888
20	$\overline{123456}+1(2345+236+456)$	874 944
	Total	4 194 304

We present these classifications without proof of their completeness, description of invariants used in its creation, calculations of stabilizers, etc. For these and other details we refer the reader to [**316, 372, 86**]. We also note that we intentionally do not discuss the questions of using computers in these classifications: this could be a subject of a separate monograph.

Classification of Boolean functions is very important in practice since every classification gives rich material for checking various properties, possible values, and distribution of values of parameters for functions from given classes. In addition, a classification allows us to solve problems such as:

- calculation of possible characteristics of functions and exact distributions of these characteristics (such results are interesting if the function in question is a random variable);
- finding the best values of the parameters of functions from a given class in a particular situation;

TABLE 3.6. Classification of functions of 6 variables with respect to the group $\mathfrak{GA}(V_6)\mathfrak{H}_3$.

No	Representatives of equivalence classes	Cardinality of the class
1	0	1
2	1234	651
3	1234+1256	18 228
4	1234+1256+3456	13 888
5	12345	2 016
6	123(45+6)	312 480
7	1(2345+236+456)	1 749 888
8	123456	64
9	123456+12345+12346	41 664
10	123456+1234+1256	1 166 592
11	123456+1234+1256+3456	888 832
	Total	4 194 304

- choosing functions with optimal parameters in a particular situation;
- description of all functions for which given bounds on parameters are satisfied, calculation of the number of such functions, finding their exact description and distribution;
- describing the set of functions with given properties;
- finding a complete system of invariants of functions for a given transformation group.

As of now, the following problems are of great interest for the cryptographic applications.

PROBLEM 3.24 (Open problem). Construct a complete system of affine invariants for the whole set of Boolean functions.

PROBLEM 3.25 (Open problem). Describe all classes of affine equivalence for Boolean functions.

To conclude this section we note that classifications with respect to a group $\mathfrak{G}$ can also be done for arbitrary subsets of the whole set $\mathcal{F}_n$. One such classification will be considered in the next section.

3.3. Classification of quadratic Boolean functions

Properties of quadratic functions over fields of characteristic different from 2 were studied by many authors (see, for example [**250, 76, 352**]). On the other hand, results that pertain to equivalence of quadratic functions over fields of characteristic 2 are not so well presented in the existing literature, and the existing expositions always aim at some particular applications. For example, Dickson's results [**106**] published in 1901 are considered in [**241**] in connection with coding theory applications. Keeping in mind the goal of this book, we will also stick to this approach; our exposition will follow [**241**].

DEFINITION 3.26. A *quadratic function* of n variables is a Boolean function from $\mathcal{F}_n$ whose Zhegalkin polynomial has the form

$$f(\boldsymbol{x}) = f(x^{(1)}, \dots, x^{(n)}) = \bigoplus_{1 \le i < j \le n} a_{ij} x^{(i)} x^{(j)} \oplus \bigoplus_{l=1}^{n} b_l x^{(l)} \oplus \varepsilon,$$

where a_{ij} $(1 \le i < j \le n)$, b_l $(1 \le l \le n)$, ε are elements of the field $\mathbb{F}_2$.

In this definition we intentionally use the term "function" instead of "form": we want to reserve the word "form" for homogeneous Zhegalkin polynomials.

It is clear (see Problem 3.20) that the set of quadratic functions is mapped into itself under the action of an element from the general linear group or the complete affine group. Note that affine functions are a degenerate case of quadratic functions (they are quadratic functions with all coefficients a_{ij} equal to 0).

The polynomial of a quadratic function f can be written in the form

$$f(\boldsymbol{x}) = f(x^{(1)}, \dots, x^{(n)}) = \boldsymbol{x}^\top Q_f \boldsymbol{x} \oplus \boldsymbol{b}_f^\top \boldsymbol{x} \oplus \varepsilon, \tag{3.4}$$

where

$$Q_f = \begin{pmatrix} 0 & a_{1,2} & a_{1,3} & \dots & a_{1,n-1} & a_{1,n} \\ 0 & 0 & a_{2,3} & \dots & a_{2,n-1} & a_{2,n} \\ \dots & \dots & \dots & \dots & \dots & \dots \\ 0 & \dots & \dots & \dots & 0 & a_{n-1,n} \\ 0 & \dots & \dots & \dots & \dots & 0 \end{pmatrix}$$

is an upper triangular $(n \times n)$ matrix with zeros on the main diagonal and $\boldsymbol{b}_f^\top = (b_1, b_2, \dots, b_n)$.

The following relation between a symmetric matrix with zeros on the main diagonal and the matrix Q_f holds:

$$\widetilde{Q}_f = Q_f \oplus Q_f^\top.$$

The matrix $\widetilde{Q}_f$ defines the bilinear form

$$q_f(\boldsymbol{u}, \boldsymbol{v}) = \boldsymbol{u}^\top \widetilde{Q}_f \boldsymbol{v}. \tag{3.5}$$

Using the equality (3.4), it is easy to check that

$$q_f(\boldsymbol{u}, \boldsymbol{v}) = f(\boldsymbol{u} \oplus \boldsymbol{v}) \oplus f(\boldsymbol{u}) \oplus f(\boldsymbol{v}) \oplus \varepsilon. \tag{3.6}$$

DEFINITION 3.27. We say that the bilinear form $q_f(\boldsymbol{u}, \boldsymbol{v})$ is *associated with the quadratic function* f.

It can be seen from (3.5) and(3.6) that the bilinear form $q_f(\boldsymbol{u}, \boldsymbol{v})$ is both *alternating* and symmetric, i.e., for all $\boldsymbol{u}, \boldsymbol{v} \in V_n$ we have

$$q_f(\boldsymbol{u}, \boldsymbol{u}) = 0, \quad q_f(\boldsymbol{u}, \boldsymbol{v}) = q_f(\boldsymbol{v}, \boldsymbol{u}).$$

DEFINITION 3.28. An alternating Boolean bilinear form is called *symplectic*.

DEFINITION 3.29. A symmetric Boolean matrix with zeros on the main diagonal is called a *symplectic matrix*.

Formula (3.5) shows that the symplectic matrices are in one-to-one correspondence with the symplectic bilinear forms. The number of different symplectic $(n \times n)$-matrices (symplectic bilinear forms of n variables) is equal to $2^{\binom{n}{2}}$.

LEMMA 3.30. *The rank of a symplectic matrix is even.*

PROOF. We prove the lemma using induction on n. It is clear that there exist only two symplectic (2×2)-matrices:

$$B_1 = \begin{pmatrix} 0 & 1 \\ 1 & 0 \end{pmatrix}, \quad B_2 = \begin{pmatrix} 0 & 0 \\ 0 & 0 \end{pmatrix}.$$

Moreover, $\operatorname{rank} B_1 = 2$, $\operatorname{rank} B_2 = 0$, i.e., these matrices have even ranks. Assume that the ranks of all symplectic $(m \times m)$-matrices, where $m \leq t$, are even. Let B be a symplectic $((t+1) \times (t+1))$ matrix. The matrix B can be represented as follows:

$$B = \begin{pmatrix} A & \boldsymbol{y} \\ \boldsymbol{y}^\top & 0 \end{pmatrix},$$

where A is a symplectic $(t \times t)$ matrix and $\boldsymbol{y}$ is a column vector from V_t. By the inductive assumption, $\operatorname{rank} A = 2k$. If $\boldsymbol{y}$ linearly depends on the columns of A, then $\operatorname{rank}[A, \boldsymbol{y}] = \operatorname{rank} A = 2k$. Moreover, there exists a column vector $\boldsymbol{x} \in V_t$ such that $A\boldsymbol{x} = \boldsymbol{y}$. Then

$$\boldsymbol{x}^\top [A, \boldsymbol{y}] = \boldsymbol{x}^\top [A, A\boldsymbol{x}] = [\boldsymbol{x}^\top A, \boldsymbol{x}^\top A\boldsymbol{x}].$$

Since A is symplectic, $\boldsymbol{x}^\top A\boldsymbol{x} = 0$ for any $\boldsymbol{x} \in V_t$ and $\boldsymbol{x}^\top A^\top = (A\boldsymbol{x})^\top = \boldsymbol{y}^\top$. Hence,

$$\boldsymbol{x}^\top [A, \boldsymbol{y}] = [\boldsymbol{y}^\top, 0],$$

i.e., the row $[\boldsymbol{y}^\top, 0]$ linearly depends on the first t rows of the matrix $[A, \boldsymbol{y}]$. Thus,

$$\operatorname{rank} \begin{pmatrix} A & \boldsymbol{y} \\ \boldsymbol{y}^\top & 0 \end{pmatrix} = \operatorname{rank}[A, \boldsymbol{y}] = 2k.$$

Suppose now that the column vector $\boldsymbol{y}$ is linearly independent of the rows of A. A similar argument shows that

$$\operatorname{rank} \begin{pmatrix} A & \boldsymbol{y} \\ \boldsymbol{y}^\top & 0 \end{pmatrix} = 2k + 2. \qquad \square$$

Now we will determine the number $\Gamma(n, 2s)$ of all symplectic $(n \times n)$-matrices of rank $2s$. First we will obtain some recurrence relations for $\Gamma(n, r)$, $r = 2s$. Let A be a symplectic $(n \times n)$ matrix of rank r and let

$$B = \begin{pmatrix} A & \boldsymbol{y} \\ \boldsymbol{y}^\top & 0 \end{pmatrix}$$

be an $((n+1) \times (n+1))$ matrix, where $\boldsymbol{y}$ is an arbitrary vector from V_n.

LEMMA 3.31. *Among the 2^n different matrices B, exactly $2^n - 2^r$ matrices have rank $r + 2$ and exactly 2^r matrices have rank r.*

PROOF. If $\boldsymbol{y}$ linearly depends on the columns of A, then, as was shown in the proof of Lemma 3.30, $\operatorname{rank} B = \operatorname{rank} A = r$. If $\boldsymbol{y}$ does not linearly depend on the columns of A, then, as was shown in the proof of Lemma 3.30, $\operatorname{rank} B = r + 2$. $\square$

It follows from Lemma 3.31 that the numbers $\Gamma(n, r)$ satisfy the recurrence relation

$$\Gamma(n+1, r) = 2^r \cdot \Gamma(n, r) + (2^n - 2^{r-2}) \cdot \Gamma(n, r-2). \tag{3.7}$$

THEOREM 3.32. *We have*

$$\Gamma(n, 2s) = \frac{(2^n - 1)(2^{n-1} - 1) \cdots (2^{n-2s+1} - 1)}{(2^{2s} - 1)(2^{2s-2} - 1) \cdots (2^2 - 1)} \cdot 2^{s(s-1)}, \quad s \geq 1, \tag{3.8}$$

and $\Gamma(n, 0) = 1$ for any natural number n.

PROOF. The second part is obvious, and it is easy to verify that $\Gamma(2,2)$ satisfies (3.8).

Suppose that $\Gamma(n,2)$ satisfies (3.8) for all natural numbers $n \leq t$. Then from the recurrence relation (3.7) we have

$$\Gamma(t+1,2) = 2^2 \cdot \Gamma(t,2) + (2^t - 1) \cdot \Gamma(t,0) = 2^2 \cdot \frac{(2^t-1)(2^{t-1}-1)}{(2^2-1)} + (2^t - 1)$$
$$= (2^t - 1) \cdot \frac{2^2 \cdot (2^{t-1} - 1) + 2^2 - 1}{(2^2-1)} = \frac{(2^{t+1}-1)(2^t-1)}{(2^2-1)},$$

i.e., $\Gamma(t+1,2)$ also satisfies (3.8). By induction, $\Gamma(n,2)$ satisfies (3.8) for any natural number n.

Suppose now that $\Gamma(n,2s)$ satisfies (3.8) for any natural $n \leq 2s$. Then, according to (3.7),

$$\Gamma(2(s+1),2(s+1))$$
$$= 2^{2(s+1)} \cdot \Gamma(2s+1,2(s+1)) + (2^{2(s+1)} - 2^{2s}) \cdot \Gamma(2s+1,2s)$$
$$= 2^{2s} \cdot \Gamma(2s+1,2s) = 2^{2s} \cdot \frac{(2^{2s+1}-1)(2^{2s}-1)\cdots(2^2-1)}{(2^{2s}-1)(2^{2s-2}-1)\cdots(2^2-1)} \cdot 2^{s(s-1)}.$$

If we multiply the numerator and denominator by $(2^{2s+2}-1)$ and the numerator by $1 = (2-1)$, we get

$$\Gamma(2(s+1),2(s+1)) = \Gamma(2s+2,2s+2)$$
$$= \frac{(2^{2s+2}-1)(2^{2s+1}-1)\cdots(2^2-1)(2-1)}{(2^{2s+2}-1)(2^{2s}-1)\cdots(2^2-1)} \cdot 2^{s(s+1)},$$

whence $\Gamma(2(s+1),2(s+1))$ satisfies (3.8). Suppose now that $\Gamma(n,2(s+1))$ satisfies (3.8) for $2(s+1) \leq n \leq t$. Then, according to (3.7),

$$\Gamma(t+1,2(s+1)) = 2^{2(s+1)} \cdot \Gamma(t,2(s+1)) + (2^t - 2^{2s}) \cdot \Gamma(t,2s)$$
$$= 2^{2(s+1)} \cdot \frac{(2^t-1)(2^{t-1}-1)\cdots(2^{t-2s-1}-1)}{(2^{2(s+1)}-1)(2^{2s}-1)\cdots(2^2-1)} \cdot 2^{s(s+1)}$$
$$+ \frac{(2^t-1)(2^{t-1}-1)\cdots(2^{t-2s+1}-1)(2^{t-2s}-1)}{(2^{2s}-1)(2^{2s-2}-1)\cdots(2^2-1)} \cdot 2^{s(s+1)}$$
$$= \frac{(2^t-1)(2^{t-1}-1)\cdots(2^{t-2s}-1)}{(2^{2s}-1)(2^{2s-2}-1)\cdots(2^2-1)} \cdot \left(\frac{2^{2(s+1)}(2^{t-2s-1}-1)}{(2^{2(s+1)}-1)} + 1\right) \cdot 2^{s(s+1)}$$
$$= \left(\frac{2^{t+1}-1}{2^{2(s+1)}-1}\right) \cdot \frac{(2^t-1)(2^{t-1}-1)\cdots(2^{t-2s}-1)}{(2^{2s}-1)(2^{2s-2}-1)\cdots(2^2-1)} \cdot 2^{s(s+1)}.$$

Therefore, $\Gamma(t+1,2(s+1))$ and $\Gamma(n,2(s+1))$ satisfy (3.8) for any $n \geq 2(s+1)$. From what was said it follows that if $\Gamma(n,2s)$ satisfies (3.8) for any $n \geq 2s$, then $\Gamma(n,2(s+1))$ satisfies (3.8) for any $n \geq 2(s+1)$. Since the theorem holds for $\Gamma(n,2)$, $n \geq 2$, it also holds for $\Gamma(n,4)$, $n \geq 4$; $\Gamma(n,6)$, $n \geq 6$; etc. □

Now we proceed with the problem of equivalence of quadratic functions with respect to the complete affine group.

THEOREM 3.33. *If B is a symplectic $(n \times n)$ matrix of rank $2s$ $(2s \leq n)$, then there exists a nonsingular $(n \times n)$ matrix R such that*

$$R^\top BR = \begin{pmatrix} A & \mathbf{0} \\ \mathbf{0} & 0 \end{pmatrix},$$

where the $(2s \times 2s)$ matrix A is of the form

$$A = \begin{pmatrix} 0 & 1 & 0 & 0 & \dots & 0 & 0 \\ 1 & 0 & 0 & 0 & \dots & 0 & 0 \\ 0 & 0 & 0 & 1 & \dots & 0 & 0 \\ 0 & 0 & 1 & 0 & \dots & 0 & 0 \\ \dots & \dots & \dots & \dots & \dots & \dots & \dots \\ 0 & 0 & 0 & 0 & \dots & 0 & 1 \\ 0 & 0 & 0 & 0 & \dots & 1 & 0 \end{pmatrix}.$$

PROOF. The theorem is obvious for $n = 1, 2$. Suppose that it holds whenever $n \leq t$ and $s \leq [t/2]$. Then the symplectic $((t+1) \times (t+1))$ matrix B can be written as

$$B = \begin{pmatrix} A & \boldsymbol{y} \\ \boldsymbol{y}^\top & 0 \end{pmatrix},$$

where A is a symplectic $(t \times t)$ matrix, $\boldsymbol{y} \in V_t$. If the rank of A is less than $2[t/2]$, then, by row and column operations, the matrix B can be transformed to

$$\begin{pmatrix} A' & \boldsymbol{y}' & 0 & \dots & 0 \\ (\boldsymbol{y}')^\top & 0 & 0 & \dots & 0 \\ 0 & 0 & 0 & \dots & 0 \\ \dots & \dots & \dots & \dots & \dots \\ 0 & 0 & 0 & \dots & 0 \end{pmatrix},$$

where

$$\begin{pmatrix} A' & \boldsymbol{y}' \\ (\boldsymbol{y}')^\top & 0 \end{pmatrix}$$

is a matrix of size at most $t \times t$ and A' is a symplectic matrix. By the induction hypothesis, such a matrix can be transformed to the form required by the theorem. Thus it is enough to consider only the case of rank $A = 2[t/2]$ (i.e., of the maximum possible rank).

Suppose that $2[t/2] = t$, i.e., $t = 2s$ is an even number. According to the induction hypothesis, the matrix A can be transformed to the given form with the help of a nonsingular $(t \times t)$ matrix R_1:

$$(3.9) \qquad R_1^\top AR_1 = \begin{pmatrix} 0 & 1 & 0 & 0 & \dots & 0 & 0 \\ 1 & 0 & 0 & 0 & \dots & 0 & 0 \\ 0 & 0 & 0 & 1 & \dots & 0 & 0 \\ 0 & 0 & 1 & 0 & \dots & 0 & 0 \\ \dots & \dots & \dots & \dots & \dots & \dots & \dots \\ 0 & 0 & 0 & 0 & \dots & 0 & 1 \\ 0 & 0 & 0 & 0 & \dots & 1 & 0 \end{pmatrix}.$$

Denote by $\widetilde{R}_1$ the nonsingular $((t+1) \times (t+1))$ matrix

$$\widetilde{R}_1 = \begin{pmatrix} R_1 & \mathbf{0} \\ \mathbf{0} & 1 \end{pmatrix}.$$

Then, according to (3.9), the following equalities hold:

$$
\begin{aligned}
B_1 = \widetilde{R}_1^\top B \widetilde{R}_1 &= \begin{pmatrix} R_1^\top & \mathbf{0} \\ \mathbf{0} & 1 \end{pmatrix} \begin{pmatrix} A & \boldsymbol{y} \\ \boldsymbol{y}^\top & 0 \end{pmatrix} \begin{pmatrix} R_1 & \mathbf{0} \\ \mathbf{0} & 1 \end{pmatrix} \\
&= \begin{pmatrix}
0 & 1 & 0 & 0 & \dots & 0 & 0 & a_1 \\
1 & 0 & 0 & 0 & \dots & 0 & 0 & a_2 \\
0 & 0 & 0 & 1 & \dots & 0 & 0 & a_3 \\
0 & 0 & 1 & 0 & \dots & 0 & 0 & a_4 \\
\hdotsfor{8} \\
0 & 0 & 0 & 0 & \dots & 0 & 1 & a_{t-1} \\
0 & 0 & 0 & 0 & \dots & 1 & 0 & a_t \\
a_1 & a_2 & a_3 & a_4 & \dots & a_{t-1} & a_t & 0
\end{pmatrix}.
\end{aligned}
$$

Now it is easy to check that for a nonsingular $((t+1) \times (t+1))$ matrix

$$
R_2 = \begin{pmatrix}
1 & 0 & 0 & 0 & \dots & 0 & 0 & a_2 \\
0 & 1 & 0 & 0 & \dots & 0 & 0 & a_1 \\
0 & 0 & 1 & 0 & \dots & 0 & 0 & a_4 \\
0 & 0 & 0 & 1 & \dots & 0 & 0 & a_3 \\
\hdotsfor{8} \\
0 & 0 & 0 & 0 & \dots & 1 & 0 & a_t \\
0 & 0 & 0 & 0 & \dots & 0 & 1 & a_{t-1} \\
0 & 0 & 0 & 0 & \dots & 0 & 0 & 1
\end{pmatrix}
$$

we have

$$
R_2^\top B_1 R_2 = \begin{pmatrix}
0 & 1 & 0 & 0 & \dots & 0 & 0 & 0 \\
1 & 0 & 0 & 0 & \dots & 0 & 0 & 0 \\
0 & 0 & 0 & 1 & \dots & 0 & 0 & 0 \\
0 & 0 & 1 & 0 & \dots & 0 & 0 & 0 \\
\hdotsfor{8} \\
0 & 0 & 0 & 0 & \dots & 0 & 1 & 0 \\
0 & 0 & 0 & 0 & \dots & 1 & 0 & 0 \\
0 & 0 & 0 & 0 & \dots & 0 & 0 & 0
\end{pmatrix}.
$$

Therefore, for the initial matrix B, the matrix $R = \widetilde{R}_1 R_2$ is the required matrix R that transforms B to the canonical form.

Suppose now that $2[t/2] = t - 1$, i.e., $t = 2s + 1$ is odd. Repeating the above arguments we can show that there exists a nonsingular $((t+1) \times (t+1))$ matrix $\widetilde{R}'_1$ which transforms the matrix B to the form

$$
B'_1 = (\widetilde{R}'_1)^\top B \widetilde{R}'_1 = \begin{pmatrix}
0 & 1 & 0 & 0 & \dots & 0 & 0 & 0 & a_1 \\
1 & 0 & 0 & 0 & \dots & 0 & 0 & 0 & a_2 \\
0 & 0 & 0 & 1 & \dots & 0 & 0 & 0 & a_3 \\
0 & 0 & 1 & 0 & \dots & 0 & 0 & 0 & a_4 \\
\hdotsfor{9} \\
0 & 0 & 0 & 0 & \dots & 0 & 1 & 0 & a_{t-2} \\
0 & 0 & 0 & 0 & \dots & 1 & 0 & 0 & a_{t-1} \\
0 & 0 & 0 & 0 & \dots & 0 & 0 & 0 & \varepsilon \\
a_1 & a_2 & a_3 & a_4 & \dots & a_{t-2} & a_{t-1} & \varepsilon & 0
\end{pmatrix},
$$

where $\varepsilon = 0$ if $\operatorname{rank} B_1' = t-1$, and $\varepsilon = 1$ if $\operatorname{rank} B_1' = t+1$. It is easy check that the $((t+1)\times(t+1))$ matrix

$$R_2' = \begin{pmatrix} 1 & 0 & 0 & 0 & \dots & 0 & 0 & a_2 \\ 0 & 1 & 0 & 0 & \dots & 0 & 0 & a_1 \\ 0 & 0 & 1 & 0 & \dots & 0 & 0 & a_4 \\ 0 & 0 & 0 & 1 & \dots & 0 & 0 & a_3 \\ \dots & & & & & & & \dots \\ 0 & 0 & 0 & 0 & \dots & 0 & 0 & a_{t-1} \\ 0 & 0 & 0 & 0 & \dots & 1 & 0 & a_{t-2} \\ 0 & 0 & 0 & 0 & \dots & 0 & 1 & 0 \\ 0 & 0 & 0 & 0 & \dots & 0 & 0 & 1 \end{pmatrix}$$

transforms B_1' to the canonical form.

Therefore, for the initial matrix B, $R = \widetilde{R}_1' R_2'$ is the matrix which transforms B to the canonical form. □

THEOREM 3.34. *Let f be a quadratic function on $\mathcal{F}_n$ with $\operatorname{rank} \widetilde{Q}_f = 2s$. Then there exists an element $\mathfrak{g}_f \in \mathfrak{GA}(V_n)$ such that $f^{\mathfrak{g}_f}$ has one of the following forms:*

a) *if $\operatorname{rank} \widetilde{Q}_f = 0$, then either $f^{\mathfrak{g}_f}(\boldsymbol{x}) = \varepsilon$, $\varepsilon = 0, 1$, or $f^{\mathfrak{g}_f}(\boldsymbol{x}) = x^{(1)}$;*

b) *if $\operatorname{rank} \widetilde{Q}_f = 2s$, $s \geq 1$, then either $f^{\mathfrak{g}_f}(\boldsymbol{x}) = \sum_{i=1}^{s} x^{(2i-1)}x^{(2i)} \oplus \varepsilon$, $\varepsilon = 0, 1$, or $f^{\mathfrak{g}_f}(\boldsymbol{x}) = \sum_{i=1}^{s} x^{(2i-1)}x^{(2i)} \oplus x^{(2s+1)}$.*

PROOF. Part a) follows from the fact that if $f(\boldsymbol{x}) = \varepsilon = \text{const}$, then $f^{\mathfrak{g}}(\boldsymbol{x}) = f(\boldsymbol{x}) = \varepsilon$ for any element $\mathfrak{g} \in \mathfrak{GA}(V_n)$. If f is not a constant affine function, then $f(\boldsymbol{x}) = \boldsymbol{b}^\top \boldsymbol{x} \oplus \varepsilon$, where $\boldsymbol{b}$ is a nonzero vector. Without loss of generality we will take $b_1 = 1$. Then the element $\mathfrak{g}$ of the group $\mathfrak{GA}(V_n)$ that transforms f to the canonical form will be a nonsingular matrix B such that

$$\mathfrak{g}^{-1} = B^{-1} = \begin{pmatrix} 1 & b_2 & \dots & b_n \\ 0 & 1 & \dots & 0 \\ \dots & & & \dots \\ 0 & 0 & \dots & 1 \end{pmatrix}, \qquad (1, 0, \dots, 0)B^{-1} = (1, b_2, \dots, b_n).$$

Acting by the element $\mathfrak{g}$ on the function f we get

$$f^{\mathfrak{g}}(\boldsymbol{x}) = f(\mathfrak{g}\boldsymbol{x}) = \boldsymbol{b}^\top B\boldsymbol{x} = (1, 0, \dots, 0)\boldsymbol{x} = x^{(1)}.$$

Part b) will be proved in several steps. At each step an element of the group $\mathfrak{GA}(V_n)$ will be chosen so that it brings the initial function closer to the canonical form.

First we will act on the function f by an element of the group $\mathfrak{GA}(V_n)$. This element coincides with the matrix R which transforms the symplectic matrix $\widetilde{Q}_f$ to the canonical form. For $\varepsilon = 0, 1$ we get

$$f^R(\boldsymbol{x}) = f(R\boldsymbol{x}) = \boldsymbol{x}^\top R^\top Q_f R\boldsymbol{x} \oplus \boldsymbol{b}^\top R\boldsymbol{x} \oplus \varepsilon = \sum_{i=1}^{s} x^{(2i-1)}x^{(2i)} \oplus (\boldsymbol{b}')^\top \boldsymbol{x} \oplus \varepsilon.$$

If $\boldsymbol{b}' = \boldsymbol{0}$, then already at the first step we have the desired representation.

At the second step two cases are possible.

Suppose that $\boldsymbol{b}' \neq \boldsymbol{0}$ and all essential variables of the linear function $(\boldsymbol{b}')^\top \boldsymbol{x}$ are independent of the set $\{x^{(1)}, x^{(2)}, \dots, x^{(2s-1)}, x^{(2s)}\}$. Then, as in a), we can find an element of the group $\mathfrak{GA}(V_n)$ that transforms the function f^R into

$$\sum_{i=1}^{s} x^{(2i-1)}x^{(2i)} \oplus x^{(2s+1)} \oplus \varepsilon, \quad \varepsilon = 0, 1.$$

If $\varepsilon = 0$, this is already the canonical form; if $\varepsilon = 1$, it can be transformed to the canonical form by the affine transformation

$$(E, (0, \dots, 0, \overset{2s+1}{1}, 0, \dots, 0)).$$

Suppose now that $\boldsymbol{b}' \neq \boldsymbol{0}$ and some of the essential variables of the linear function $(\boldsymbol{b}')^\top \boldsymbol{x}$ belong to the set $\{x^{(1)}, x^{(2)}, \dots, x^{(2s-1)}, x^{(2s)}\}$. Without loss of generality we may assume that this is $x^{(1)}$ (our argument does not depend on the numbering of the variables). In this case,

$$f^R(\boldsymbol{x}) = x^{(1)}x^{(2)} \oplus x^{(1)} \oplus \cdots .$$

Acting on f^R by the element of the group $\mathfrak{GA}(V_n)$ defined by the nonsingular matrix

$$R_1 = \begin{pmatrix} 1 & 0 & 0 & \dots & 0 \\ 1 & 1 & 0 & \dots & 0 \\ 0 & 0 & 1 & \dots & 0 \\ \dots & \dots & \dots & \dots & \dots \\ 0 & 0 & 0 & \dots & 1 \end{pmatrix},$$

we get

$$f^{RR_1}(\boldsymbol{x}) = x^{(1)}x^{(2)} \oplus \cdots .$$

If $x^{(1)}$ and $x^{(2)}$ belong to the set $\{x^{(1)}, x^{(2)}, \dots, x^{(2s-1)}, x^{(2s)}\}$, then

$$f^R(\boldsymbol{x}) = x^{(1)}x^{(2)} \oplus x^{(1)} \oplus x^{(2)} \oplus \cdots \oplus \varepsilon, \quad \varepsilon = 0, 1.$$

Acting on f^R by the affine transformation $\mathfrak{g} = (E, (1, 1, 0, \dots, 0))$, we get

$$f^{R\mathfrak{g}}(\boldsymbol{x}) = x^{(1)}x^{(2)} \oplus \cdots \oplus \varepsilon \oplus 1, \quad \varepsilon = 0, 1.$$

Using similar arguments for all essential variables of the linear function $(\boldsymbol{b}')^\top \boldsymbol{x}$ from the set $\{x^{(1)}, x^{(2)}, \dots, x^{(2s-1)}, x^{(2s)}\}$, we transform the initial quadratic function to the form

$$\sum_{i=1}^{s} x^{(2i-1)}x^{(2i)} \oplus (\boldsymbol{b}'')^\top \boldsymbol{x} \oplus \varepsilon, \quad \varepsilon = 0, 1,$$

where the essential variables of the linear function $(\boldsymbol{b}'')^\top \boldsymbol{x}$ do not belong to the set $\{x^{(1)}, x^{(2)}, \dots, x^{(2s-1)}, x^{(2s)}\}$. If $\boldsymbol{b}'' = \boldsymbol{0}$, then we get the canonical form. If $\boldsymbol{b}'' \neq \boldsymbol{0}$, then arguing as in the proof of a), we transform the initial function to the form

$$\sum_{i=1}^{s} x^{(2i-1)}x^{(2i)} \oplus x^{(2s+1)} \oplus \varepsilon, \quad \varepsilon = 0, 1.$$

If $\varepsilon = 0$, we already have the canonical form, and if $\varepsilon = 1$, we transform it to the canonical form using the following affine transformation:

$$(E, (0, \dots, 0, \overset{2s+1}{1}, 0, \dots, 0)). \qquad \square$$

PROBLEM 3.35. Prove that the functions from a) and b) of Theorem 3.34 are not $\mathfrak{GA}(V_n)$-equivalent.

The above results show that canonical functions from Theorem 3.34 are representatives of the equivalence classes for quadratic functions under the the action of the group $\mathfrak{GA}(V_n)$. One should note that rank of the bilinear form associated with a quadratic function (or of the corresponding symplectic matrix) is an invariant of the group $\mathfrak{GA}(V_n)$, but not a complete invariant.

PROBLEM 3.36. Prove that each of the following objects forms a complete system of affine invariants for quadratic functions on $\mathcal{F}_n$:

1. The pair consisting of the rank of the associated bilinear form and the dimension of the stabilizer in the complete affine group (sum of these numbers equals n or $n-1$).
2. The pair consisting of the dimension of the moment group in the complete affine group and the number ε_f which equals 1 if f is balanced and 0 otherwise.
3. The dimension of the stabilizer in the complete affine group.

3.4. Classification of homogeneous cubic forms of 8 variables

When we pass from quadratic to cubic functions, the complexity of the classification problem increases sharply. The reason is that there is no adequate mathematical technique; in particular, no algorithms are known for the transformation of third-order tensors to canonical form (for second-order tensors, i.e., matrices, such algorithms were developed long ago). The problem of classification of cubic forms over an arbitrary field is related to difficult algebraic, geometric, and number-theoretical open problems (see [**252**]).

However, there exist classification results for cubic forms in small dimensions. Classifications over the field of real numbers were obtained by Reihel in 1907 for $n = 6$, Schouten in 1933 for $n = 7$, Gurevich in 1932 for $n = 8$, and Vinberg and Elashvili in 1974 for $n = 9$. A similar classification over the field of complex numbers for $n = 7, 8$ was obtained in [**409**].

As in §2.2, we omit the corresponding proofs of these classifications and refer the reader to [**86**]. We will make only a brief general remark. The classification presented in the tables above is for the groups $\mathfrak{GA}(V_n)$, but the set of homogeneous cubic forms is *not* invariant under the action of such a group. That is why homogeneous cubic forms must be considered as elements of the quotient space of the space of cubic Boolean functions modulo the subspace of quadratic Boolean functions. One can also say that the homogeneous cubic forms are classified with respect to the action of the groups $\mathfrak{GA}(V_n)\mathfrak{H}_2$.

In Table 3.7, a classification of homogeneous cubic forms of 6 variables with respect to groups $\mathfrak{GA}(V_n)$ is given.

The invariant used in the classification is the vector $\pi(f) = (\pi_0, \pi_1, \pi_2)$, where $\pi_i = \#M_i$ and M_i is the set of vectors $\boldsymbol{u} \in V_6$ for which the derivative $D_{\boldsymbol{u}}f$ is a quadratic form of rank $2i$, $i = 0, 1, 2$. This invariant is complete.

A classification of homogeneous cubic forms in seven variables under the action of the group $\mathfrak{GA}(V_n)$ is given in Table 3.8. The invariant used here is also the frequency vector $\pi(f) = (\pi_0, \pi_1, \pi_2, \pi_3)$, where $\pi_i = \#M_i$ and M_i is the set of vectors $\boldsymbol{u} \in V_7$ for which the derivative $D_{\boldsymbol{u}}f$ is a quadratic form of rank $2i$, $i = 0, 1, 2, 3$. This invariant is also complete.

A classification of homogeneous cubic forms of eight variables with respect to the group $\mathfrak{GA}(V_8)$ is given in Table 3.9. This is the frequency vector $\pi(f) = (\pi_0, \pi_1, \pi_2, \pi_3)$, where $\pi_i = \#M_i$ and M_i is the set of vectors $\boldsymbol{u} \in V_8$ for which

TABLE 3.7. Classification of homogeneous cubic forms in 6 variables with respect to the group $\mathfrak{GA}(V_6)$.

	Representative of the equivalence class	Cardinality of the class	Order of the moment group	Invariant $\pi(f)$
1	0	1	$2^{15} \cdot 3^4 \cdot 5 \cdot 7^2 \cdot 31$	64, 0, 0
2	123	1 395	$2^{15} \cdot 3^2 \cdot 7^2$	$8, 56, 0$
3	123+145	54 684	$2^{13} \cdot 3^2 \cdot 5$	$2, 30, 32$
4	123+456	357 120	$2^7 \cdot 3^2 \cdot 7^2$	$1, 14, 49$
5	123+135+236	468 720	$2^{11} \cdot 3 \cdot 7$	$1, 7, 56$
6	123+124+135+236+456	166 656	$2^7 \cdot 3^3 \cdot 5 \cdot 7$	$1, 0, 63$
	Total	1 048 576		

TABLE 3.8. Classification of homogeneous cubic forms in 7 variables with respect to the group $\mathfrak{GA}(V_7)$.

	Representative of the equivalence class	Cardinality of the class	Order of the moment group	Invariant $\pi(f)$
1	0	1	$2^{21} \cdot 3^4 \cdot 5 \cdot 7^2 \cdot 31 \cdot 127$	$128, 0, 0, 0$
2	123	11 811	$2^{21} \cdot 3^3 \cdot 5 \cdot 7^2$	$16, 112, 0, 0$
3	123+145	2 314 956	$2^{19} \cdot 3^3 \cdot 5$	$4, 60, 64, 0$
4	123+456	45 354 240	$2^{13} \cdot 3^2 \cdot 7^2$	$2, 28, 98, 0$
5	124+135+236	59 527 440	$2^{17} \cdot 3 \cdot 7$	$2, 14, 112, 0$
6	123+124+135+236+456	21 165 312	$2^{13} \cdot 3^3 \cdot 5 \cdot 7$	$2, 0, 126, 0$
7	123+145+167	1 763 776	$2^{15} \cdot 3^4 \cdot 5 \cdot 7$	$1, 63, 0, 64$
8	123+246+357	2 222 357 760	$2^{13} \cdot 3^2$	$1, 13, 92, 32$
9	123+145+167+246	238 109 760	$2^{15} \cdot 3 \cdot 7$	$1, 7, 56, 64$
10	123+145+246+357	17 778 862 080	$2^{10} \cdot 3^2$	$1, 3, 76, 48$
11	123+124+135+236+456+167	444 471 552	$2^{13} \cdot 3^2 \cdot 5$	$1, 1, 30, 96$
12	123+145+167+246+357	13 545 799 680	$2^6 \cdot 3^3 \cdot 7$	$1, 0, 63, 64$
	Total	34 359 738 368		

the derivative $D_{\boldsymbol{u}}f$ is a quadratic form of rank $2i$, $i = 0, 1, 2, 3$. This invariant is not complete. However, in this case a classification with complete invariants is also known (see [**86**]).

TABLE 3.9. Classification of homogeneous cubic forms in 8 variables with respect to the group $\mathfrak{GA}(V_8)$.

	Representative of the equivalence class	Cardinality of the class	Invariant $\pi(f)$
1	0	1	$256, 0, 0, 0$
2	123	97 155	$32, 224, 0, 0$
3	123+145	84 330 540	$8, 120, 128, 0$
4	123+456	3 855 110 400	$4, 56, 196, 0$
5	124+135+236	5 059 832 400	$4, 28, 224, 0$
6	123+124+135+236+456	1 799 051 520	$4, 0, 252, 0$
7	123+145+167	449 762 880	$2, 126, 0, 128$
8	123+246+357	566 701 228 800	$2, 26, 184, 64$
9	123+145+167+246	60 717 988 800	$2, 14, 112, 128$
10	123+145+246+357	4 533 609 830 400	$2, 6, 152, 96$
11	123+124+135+236+456+167	113 340 245 760	$2, 2, 60, 192$
12	123+145+167+246+357	3 454 178 918 400	$2, 0, 126, 128$
13	123+478+568	2 763 343 134 720	$1, 22, 121, 112$
14	123+145+167+568	2 266 804 915 200	$1, 19, 108, 128$
15	123+246+357+568	18 134 439 321 600	$1, 9, 102, 144$
16	123+246+357+128+138	64 478 006 476 800	$1, 9, 102, 144$
17	123+145+167+357+568	13 600 829 491 200	$1, 7, 56, 192$
18	123+145+478+568	290 151 029 145 600	$1, 6, 89, 160$
19	123+124+135+236+456+167+258	3 626 887 864 320	$1, 5, 10, 240$
20	123+124+135+236+456+178	116 060 411 658 240	$1, 3, 76, 176$
21	123+145+246+357+568	108 806 635 929 600	$1, 3, 76, 176$
22	123+145+246+467+578	1 740 906 174 873 600	$1, 3, 76, 176$
23	123+145+357+478+568	1 740 906 174 873 600	$1, 2, 53, 200$
24	123+246+357+478+568	2 321 208 233 164 800	$1, 1, 58, 196$
25	123+246+357+148+178+258	6 963 624 699 494 400	$1, 1, 58, 196$
26	123+145+167+246+357+568	217 613 271 859 200	$1, 1, 30, 224$
27	123+145+167+246+238+258+348	1 740 906 174 873 600	$1, 1, 30, 224$
28	123+145+167+258+268+378+468	55 266 862 694 400	$1, 0, 63, 192$
29	123+145+246+357+238+678	15 916 856 455 987 200	$1, 0, 49, 206$
30	123+145+246+357+478+568	27 854 498 797 977 600	$1, 0, 35, 220$
31	123+124+135+236+456+167+258+378	12 379 777 243 545 600	$1, 0, 21, 234$
32	123+156+246+256+147+157 +357+348+258+458	497 401 764 249 600	$1, 0, 7, 248$
	Total	72 057 594 037 927 936	

3.5. *RM*-equivalence of Boolean functions

What we have said about the difficulties in classification of cubic Boolean functions under the action of $\mathfrak{GA}(V_n)$ remains valid for functions of a larger degree. These considerations lead to the following definition.

DEFINITION 3.37. We say that two Boolean functions of the same degree d are *RM-equivalent* if they can be obtained from each other by an affine transformation and addition of a function of algebraic degree at most $d-1$.

In other words, *RM*-equivalence is equivalence with respect to the action of the group $\mathfrak{GA}(V_n)\mathfrak{H}_{\deg f-1}$.

RM-equivalence is named after Reed–Muller codes, which will be studied later. We will see that the classification of Boolean functions based on RM-equivalence is really useful. The following definition is an important step in obtaining the classification.

DEFINITION 3.38. We say that a Boolean function f is *nondegenerate* if $L_f^{\deg f-2} = \{\mathbf{0}\}$. In other words, nondegenerate Boolean functions are those whose differentiation in any nonzero direction decreases the algebraic degree by one.

The nondegeneracy property of a Boolean function can be expressed in terms of linear algebra, which helps in the study of these functions. If $\deg f = r$, denote by $[f]_r$ the homogeneous component of degree r of the function f and by $[D_{\boldsymbol{\sigma}} f]_{r-1}$ the homogeneous component of degree $r-1$ of the function $D_{\boldsymbol{\sigma}} f$.

PROBLEM 3.39. Prove that for any $\boldsymbol{x}, \boldsymbol{\sigma}, \boldsymbol{\omega} \in V_n$,

$$[D_{\boldsymbol{\sigma}\oplus\boldsymbol{\omega}} f]_{r-1}(\boldsymbol{x}) = [D_{\boldsymbol{\sigma}} f]_{r-1}(\boldsymbol{x}) \oplus [D_{\boldsymbol{\omega}} f]_{r-1}(\boldsymbol{x}).$$

This equality and Problem 2.70 give

THEOREM 3.40. *The mapping* $D\colon (f, \boldsymbol{\sigma}) \longrightarrow [D_{\boldsymbol{\sigma}} f]_{r-1}$ *is bilinear in* $\boldsymbol{\sigma} \in V_n$ *and* $\deg f = r$.

Nondegeneracy of a Boolean function is connected with the mapping D in the following way: the subspace $L_f^{\deg f-2}$ is exactly the kernel of the linear mapping D when f is fixed:

$$L_f^{\deg f-2} = \{\boldsymbol{\sigma} \in V_n \mid [D_{\boldsymbol{\sigma}} f]_{r-1} = D(f, \boldsymbol{\sigma}) = 0\}.$$

A Boolean function f is nondegenerate if and only if $D(f, \boldsymbol{\sigma}) \neq 0$ for all nonzero $\boldsymbol{\sigma} \in V_n$, i.e., if the linear mapping D (for fixed f) is nondegenerate. The following property of the mapping D is sometimes useful.

LEMMA 3.41. *For any function* φ *with* $\deg \varphi \leq \deg f - 1$ *and for any direction* $\boldsymbol{\sigma} \in V_n$ *the following equality holds:*

$$D(f \oplus \varphi, \boldsymbol{\sigma}) = D(f, \boldsymbol{\sigma}).$$

In particular, Lemma 3.41 implies that if a function f is nondegenerate, then any function $f \oplus \varphi$ with $\deg \varphi \leq \deg f - 1$ is also nondegenerate. This shows that nondegeneracy of a Boolean function is determined by its leading homogenous component. Moreover, it follows from Lemma 3.41 that if $\deg f = r$, then

$$L_f^{r-2} = L_{[f]_r}^{r-2}.$$

It is not hard to verify that

$$D_{\boldsymbol{e}_i}[f]_r(\boldsymbol{x}) = \sum_{\mathrm{wt}(\boldsymbol{\beta})=r-1,\, \boldsymbol{\beta}\in V_n} d_{\boldsymbol{\beta}}^{(i)} \boldsymbol{x}^{\boldsymbol{\beta}},$$

where $\boldsymbol{e}_i = (0, \ldots, 0, \underset{i}{1}, 0, \ldots, 0)^\top$; $\boldsymbol{x}^{\boldsymbol{\beta}} = x^{(j_1)} \cdots x^{(j_{r-1})}$ if the 1's in the vector $\boldsymbol{\beta}$ are in the positions $j_1, \ldots, j_{r-1}$, i.e., if $\boldsymbol{\beta} = (0, \ldots, 0, \underset{j_1}{1}, 0, \ldots, 0, \underset{j_{r-1}}{1}, 0, \ldots, 0)^\top$;

$$d_{\boldsymbol{\beta}}^{(i)} = \begin{cases} 0 & \text{if } \boldsymbol{\beta} \geq \boldsymbol{e}_i \text{ (i.e., the vector } \boldsymbol{\beta} \text{ has 1 in the } i\text{th place)}, \\ c_{\boldsymbol{\beta}\oplus\boldsymbol{e}_i} & \text{otherwise;} \end{cases}$$

and $f(\boldsymbol{x}) = \bigoplus_{\boldsymbol{\alpha}\in V_n} c_{\boldsymbol{\alpha}} \boldsymbol{x}^{\boldsymbol{\alpha}}$.

From the notation we introduced it is obvious that if a function f is fixed, the mapping V_n (with argument D) belongs to the linear span of n homogeneous forms of degree $r-1$ given by the vectors $d_{\boldsymbol{\beta}}^{(1)}, \ldots, d_{\boldsymbol{\beta}}^{(n)}$. The well-known relation between the dimensions of the kernel and of the image of a linear mapping implies

THEOREM 3.42. $\dim L_f^{r-2} = n - \operatorname{rank}(d_{\boldsymbol{\beta}}^{(1)}, \ldots, d_{\boldsymbol{\beta}}^{(n)})$.

The homogeneous forms have the following important property.

LEMMA 3.43. *Suppose that* $\psi(x_1, \ldots, x_n)$ *is a homogeneous form of degree* r. *Then either* x_i *is a nonessential variable or*

$$\deg D_{\boldsymbol{e}_i}\psi = r - 1.$$

For brevity, homogeneous forms of degree r in n variables will be called (n,r)-*forms*. Using Lemma 3.43 it is possible to obtain nondegeneracy of any Boolean function.

THEOREM 3.44. *Suppose that* $\deg f = r$, $\dim L_f^{r-2} = n - k$, $V_n = U \dot{+} W$, *and* $L_f^{r-2} = U$. *Then* $k \geq r$, *and for any choice of the bases of* U *and* W, *the function* $f(\boldsymbol{u}, \boldsymbol{w})$ *can be presented as*

$$f(\boldsymbol{u}, \boldsymbol{w}) = \varphi(\boldsymbol{u}, \boldsymbol{w}) \oplus \psi(\boldsymbol{w}),$$

where $\psi(\boldsymbol{w})$ *is a nondegenerate* (k,r)-*form,* $\deg \varphi \leq r-1$ *and* $L_f^m = L_\varphi^m$ *for all* $m \leq r-2$.

Theorem 3.44 has the following important corollary.

COROLLARY 3.45. *Any Boolean function is RM-equivalent to some nondegenerate form.*

If necessary, one can use an affine change of variables to eliminate nonessential directions for f; therefore, we will assume that $L_f^{-1} = \{\mathbf{0}\}$. For a given function f choose a chain of r-nonlinearity spaces

$$L_f^{r_1} \subset L_f^{r_2} \subset \cdots \subset L_f^{r_s}, \quad L_f^{r_s+1} = V_n$$

such that

$$L_f^{r_i} = L_f^{r_i+1} = \cdots = L_f^{r_{i+1}-1}, \quad L_f^{r_{i+1}-1} \neq L_f^{r_{i+1}}.$$

Put $n - k_i = \dim L_f^{r_i}$. Then $k_1 > k_2 > \cdots > k_s$, $r_1 < r_2 < \cdots < r_s$, $k_s \geq r_s + 2$. It is clear that the set of pairs (k_1, r_1), $(k_2, r_2), \ldots, (k_s, r_s)$ is an affine invariant of a Boolean function. We will call this invariant the *degeneration structure of a Boolean function*.

We are going to find normal forms (with respect to affine transformations of variables) of Boolean function with a given degeneration structure. To begin with, note that we can add to the function f a function of degree at most r_1 without changing the degeneration structure of f. Thus, from now on, without loss of generality, we assume that there are no terms of degree r_1 or less. Let us write down the Zhegalkin polynomials of the function f with respect to all the bases of the space V_n. Denote by f_r^B the sum of monomials of degree greater than r occurring in the Zhegalkin polynomial of the function f in the basis B, and by $k_r^B(f)$ the number of essential variables of the Boolean function f_r^B. It follows from the previous observations that if a pair (k, r) is in the degeneration structure of the function f, then

$$k = \min_B k_r^B(f).$$

Theorem 3.44 reduces the affine classification of Boolean functions to the affine classification of the corresponding nondegenerate forms.

EXAMPLE 3.46. It is easy to verify that the nondegeneracy condition for quadratic forms is equivalent to the classical nondegeneracy condition from linear algebra. It suffices to note that in this case the matrix of the linear map D is the symplectic matrix of the quadratic form. It follows from the known classification of quadratic forms that nondegenerate $(n, 2)$-forms exist only for even n, and that any nondegenerate $(2m, 2)$-form can be transformed by an affine transformation (up to RM-equivalency) to the form

$$x^{(1)}y^{(1)} \oplus x^{(2)}y^{(2)} \oplus \cdots \oplus x^{(m)}y^{(m)}.$$

PROBLEM 3.47. Check that all cubic forms obtained in the classification in §3.4 are nondegenerate.

Comments to Chapter 3

A slightly different proof of Theorem 3.34 can be found in [**22, 225, 264**]. This theorem is indispensable for solving the problem of complete description of weight enumerators of codes $RM(2, m)$ (see Chapter 5).

In 1989 an approach to measuring nonlinearity of Boolean functions was suggested in [**267**]. The authors made an attempt to classify the existing criteria of nonlinearity and to suggest new ones. They use an algebraic idea that any useful nonlinearity measure should be invariant with respect to a given group of transformations of the base space. It is also clear that nonlinearity criteria should be invariant under transformation groups that are as large as possible. According to this approach, a function is considered to be weak if it can be reduced to a cryptographically weak function using a simple (for example, linear or affine) transformation. The most promising group from this point of view is the affine group $\mathfrak{GA}(V_n)$. The criterion of distance from the affine group, the Rothaus criterion, and some other criteria are invariant with respect to this group. Unfortunately, the existence of many equivalence classes modulo $\mathfrak{GA}(V_n)$ for $n > 6$ does not allow an efficient use of these criteria.

Attempts to classify cubic Boolean functions with small number of variables using ideas of RM-equivalence were made in [**86**]. The results in §3.5 concerning RM-equivalence can serve as a theoretical basis for further classification of Boolean functions in the general case.

The problem of classification of Boolean mappings $\Phi \in \mathcal{F}_{m,n}$ is even harder. It is known that Boolean mappings $\Phi, \Psi \in \mathcal{F}_{m,n}$ are equivalent with respect to a pair of groups $\mathfrak{G}$ and $\mathfrak{H}$ acting on V_m and V_n, respectively, if there exist elements $\mathfrak{g} \in \mathfrak{G}$ and $\mathfrak{h} \in \mathfrak{H}$ such that

$$\mathfrak{h} \circ \Phi \circ \mathfrak{g} = \Psi.$$

Contrary to the case of Boolean functions, for $n > 1$ the group $\mathfrak{H}$ can be large, which makes the classification problem much more complicated.

In [**32**] the problem of linear and affine classification of Boolean permutations $\Phi \in \mathcal{F}_{n,n}$ is considered (i.e., the case where $\mathfrak{G} = \mathfrak{H}$ and either $\mathfrak{G} = \mathfrak{GL}(V_n)$ or

$\mathfrak{G} = \mathfrak{GA}(V_n)$). In that work, an algorithm was suggested that, for a given pair of permutations $\Phi, \Psi \in \mathcal{F}_{n,n}$, either finds a pair $\mathfrak{g}_1, \mathfrak{g}_2 \in \mathfrak{G}$ such that

$$\mathfrak{g}_1 \circ \Phi \circ \mathfrak{g}_2 = \Psi$$

or determines that the permutations Φ and Ψ are not equivalent. The asymptotical complexity of this algorithm equals $\mathcal{O}(n^3 2^n)$ for the linear classification problem and $\mathcal{O}(n^3 2^{2n})$ for the affine classification problem (it is easy to see that the complexity of the trivial algorithm of exhaustive search is $\mathcal{O}(2^{n^2})$). In the same work [**32**] the authors generalize this algorithm to mappings $\Phi, \Psi \in \mathcal{F}_{m,n}$ that are not permutations if $|n - m|$ is small. In addition, the authors introduce the notion of almost equivalent mappings, which is important from the cryptographic point of view, and adapt their algorithm to find out whether two mappins are almost linearly (or affinely) equivalent. Here, Boolean mappings $\Phi, \Psi \in \mathcal{F}_{m,n}$ are called *almost equivalent*with respect to the pair of groups $\mathfrak{G}$ and $\mathfrak{H}$ if there exist elements $\mathfrak{g} \in \mathfrak{G}$ and $\mathfrak{h} \in \mathfrak{H}$ such that the equality

$$\mathfrak{h} \circ \Phi \circ \mathfrak{g}(\boldsymbol{x}) = \Psi(\boldsymbol{x})$$

holds for almost all $\boldsymbol{x} \in V_m$.

CHAPTER 4

Linear Codes over the Field $\mathbb{F}_2$

The current chapter is a short digression to the theory of binary linear block codes. Its aim is to expose basic theoretical notions and results for these codes. We will note that in the sections devoted to decoding, the authors intentionally, taking into account the aim of the book, give only algorithmic results and omit technical details pertaining to the realization of coders and decoders. The reader wishing to learn more about these matters is referred to the books [**34, 309, 197**].

4.1. Basic properties of linear block codes

DEFINITION 4.1. A *linear block code* $\boldsymbol{C}$ of length n is a linear subspace of the vector space V_n.

REMARK 4.2. By a code we will always mean a linear block code (when we speak of other codes, it will be mentioned explicitly). Contrary to the standard convention, the elements of a code (code words) will be presented by row vectors. This form of notation will be convenient in the study of Reed–Muller codes.

DEFINITION 4.3. The *dimension* $k_{\boldsymbol{C}}$ *of a code* $\boldsymbol{C}$ is its dimension as a vector space:

$$k_{\boldsymbol{C}} = \dim \boldsymbol{C}.$$

DEFINITION 4.4. The *minimum distance* $d_{\boldsymbol{C}}$ *of a code* (*the code distance*) $\boldsymbol{C}$ is the minimum Hamming distance between different code words:

$$d_{\boldsymbol{C}} = \min \{\operatorname{dist}(\boldsymbol{c}, \boldsymbol{c}') \mid \boldsymbol{c}, \boldsymbol{c}' \in \boldsymbol{C}, \boldsymbol{c} \neq \boldsymbol{c}'\}.$$

As a code $\boldsymbol{C}$ is a vector space, its minimum distance $d_{\boldsymbol{C}}$ satisfies the following equality:

$$d_{\boldsymbol{C}} = \min \{\operatorname{wt}(\boldsymbol{c}) \mid \boldsymbol{c} \in \boldsymbol{C}, \boldsymbol{c} \neq \boldsymbol{0}\}.$$

DEFINITION 4.5. A code of length n, dimension k, and minimum distance d is called an *$[n, k, d]$-code*. If one does not know the minimum distance d or does not care for it, the notation *$[n, k]$-code* will be used.

DEFINITION 4.6. The *code rate* $R_{\boldsymbol{C}}$ *of an* $[n, k]$*-code* $\boldsymbol{C}$ is defined by

$$R_{\boldsymbol{C}} = \frac{k}{n}.$$

Let $\boldsymbol{C} \subset V_n$ be a linear code. A vector $\boldsymbol{x} \in V_n$ is *r-covered by the code* $\boldsymbol{C}$ if there exists a code word $\boldsymbol{c} \in \boldsymbol{C}$ such that $\operatorname{dist}(\boldsymbol{x}, \boldsymbol{c}) \leq r$.

DEFINITION 4.7. The *covering radius* $r_{\boldsymbol{C}}$ *of the code* $\boldsymbol{C}$ is the minimum natural number r such that any vector from V_n is r-covered by the code $\boldsymbol{C}$.

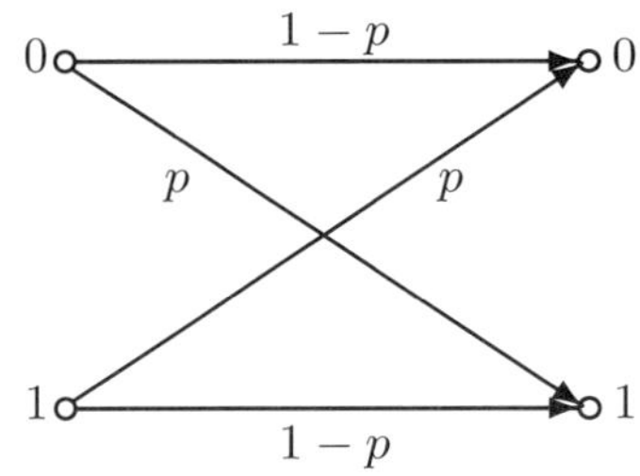

FIGURE 4.1

PROBLEM 4.8. Let $\boldsymbol{C} \subset V_n$ be a linear code and let $d_{\boldsymbol{C}}$ be its minimum distance. Prove that $r_{\boldsymbol{C}} \geq \left\lfloor \frac{d_{\boldsymbol{C}}-1}{2} \right\rfloor$ and that we have equality if and only if the code $\boldsymbol{C}$ is perfect (see Definition 4.44 below).

Elements of any communication system are the sender, the receiver, and the communication channel. Codes are constructed in order to correct errors in communication channels.

We will consider discrete channels on which binary information is transmitted, i.e., the set of input symbols, as well as the set of output symbols, is the field $\mathbb{F}_2$. A symbol $a \in \mathbb{F}_2$ is sent through a channel; it is changed to $a \oplus 1$ with probability p $(0 \leq p \leq 1)$ and remains unchanged with probability $1-p$ (see Figure 4.1). Thus, the probability of change of the sequence sent depends only on the number of changed symbols, but not on the location of the symbols sent. Errors in such channels are called independent.

Such channels are the most studied. They proved to be a good mathematical model which helps in constructing codes for real communication channels.

To find a code correcting t errors it is necessary to find a code $\boldsymbol{C}$ with minimum distance $d_{\boldsymbol{C}}$ satisfying the inequality

$$d_{\boldsymbol{C}} \geq 2t+1.$$

Suppose that at most t errors appeared during the transmission. Then the code word received is at the distance $\leq t$ from the one sent. If the distance to any other code word is greater than t, decoder will correct the error in case we assume that the word received is the closest to the one sent. Sometimes it is possible to correct more than t errors, but it cannot be guaranteed.

Suppose we want to encode messages in order to protect them from errors in the communication channel. A message which is a block of k binary symbols $\boldsymbol{u} = (u^{(1)}, \dots, u^{(k)})$ $(u^{(i)} \in \mathbb{F}_2,\ i = 1, \dots, k)$ will be encoded by a code word $\boldsymbol{c} = (c^{(1)}, \dots, c^{(n)})$ $(c^{(j)} \in \mathbb{F}_2,\ j = 1, \dots, n)$ of an $[n,k]$-code $\boldsymbol{C}$, $n \geq k$.

Let $\boldsymbol{c}_1 = (c_1^{(1)}, \dots, c_1^{(n)}), \dots, \boldsymbol{c}_k = (c_k^{(1)}, \dots, c_k^{(n)})$ be a basis of the vector space $\boldsymbol{C}$. Then to an arbitrary message $\boldsymbol{u} = (u^{(1)}, \dots, u^{(k)})$ there corresponds the code vector

$$\boldsymbol{c} = u^{(1)} \cdot \boldsymbol{c}_1 \oplus \cdots \oplus u^{(k)} \cdot \boldsymbol{c}_k.$$

It is clear that there is a one-to-one correspondence between all possible 2^k messages of length k and the set of code words of an $[n,k]$-code $\boldsymbol{C}$.

Let us continue with the matrix description of a linear block code. For an $[n,k]$-code $\boldsymbol{C}$ consider a $(k \times n)$ matrix G whose rows are basis vectors $\boldsymbol{c}_1, \boldsymbol{c}_2, \dots, \boldsymbol{c}_k$ of the code $\boldsymbol{C}$:

$$G = \begin{pmatrix} \boldsymbol{c}_1 \\ \boldsymbol{c}_2 \\ \cdots \\ \boldsymbol{c}_k \end{pmatrix} = \begin{pmatrix} c_1^{(1)} & c_1^{(2)} & \cdots & c_1^{(n)} \\ c_2^{(1)} & c_2^{(2)} & \cdots & c_2^{(n)} \\ \cdots & \cdots & \cdots & \cdots \\ c_k^{(1)} & c_k^{(2)} & \cdots & c_k^{(n)} \end{pmatrix}.$$

Then the encoding can be expressed by the following algebraic operation:

$$\boldsymbol{c} = \boldsymbol{u} \cdot G.$$

REMARK 4.9. We used a special enumeration of the elements of the matrix G in order to show the role of the basis vectors in the process of encoding. Later we will return to the standard two-index notation (the number of the row and the number of the column).

DEFINITION 4.10. The matrix G is called the *generator matrix of the* $[n, k]$*-code* $\boldsymbol{C}$. The *set of code words of the code* $\boldsymbol{C}$ means the space generated by the rows of the matrix G.

The generator matrix is a concise description of the code. To describe a code $\boldsymbol{C}$, $k \cdot n$ bits defining the elements of the matrix G are necessarily. To describe all the code words of the code $\boldsymbol{C}$ we will need $2^k \cdot n$ bits.

Since the code $\boldsymbol{C}$ is a subspace of V_n, it has an orthogonal complement

$$\boldsymbol{C}^\perp = \{\boldsymbol{x} \in V_n \mid \langle \boldsymbol{x}, \boldsymbol{c} \rangle = 0 \text{ for all } \boldsymbol{c} \in \boldsymbol{C}\}.$$

REMARK 4.11. As opposed to spaces over the field of real numbers, in spaces over finite fields there may exist vectors $\boldsymbol{x} \in V_n$, $\boldsymbol{x} \neq \boldsymbol{0}$ such that $\langle \boldsymbol{x}, \boldsymbol{x} \rangle = 0$. In particular, the space V_n is not necessarily the direct sum of its subspace and the orthogonal complement of the latter.

DEFINITION 4.12. The code $\boldsymbol{C}^\perp$ is called the *dual code* of the code $\boldsymbol{C}$.

PROBLEM 4.13. Prove that the dual code has dimension $n - k$ and that $(\boldsymbol{C}^\perp)^\perp = \boldsymbol{C}$.

Let H be a matrix whose rows are basis vectors of the code $\boldsymbol{C}^\perp$. Then the vector $\boldsymbol{c} \in V_n$ is a code word of the code $\boldsymbol{C}$ if and only if it is orthogonal to any row vector of the matrix H, i.e., if and only if

$$H\boldsymbol{c}^\top = \boldsymbol{0} \in V_{n-k}.$$

DEFINITION 4.14. The $((n-k) \times n)$ matrix H is called the *parity-check matrix of the code* $\boldsymbol{C}$.

THEOREM 4.15. *A code* $\boldsymbol{C}$ *has minimum distance* d *if and only if any* $d-1$ *columns of its parity-check matrix* H *are linearly independent, but there are* d *linearly dependent columns.*

PROOF. A vector $\boldsymbol{c}$ of weight t belongs to the code $\boldsymbol{C}$ if and only if $\boldsymbol{c}H^\top = \boldsymbol{0}$, which is equivalent to linear dependence of some t columns of the matrix H. From this follows the statement of the theorem. □

THEOREM 4.16. *The generator matrix of a code* $\boldsymbol{C}$ *is the parity-check matrix of its dual code* $\boldsymbol{C}^\perp$.

PROBLEM 4.17. Prove Theorem 4.16.

We observe that there are many different ways to choose the generator matrix G and the parity-check matrix H. We will call the following operations *row operations on a binary matrix*:

1. Exchange of two arbitrary rows.

2. Replacement of a row with the sum of this row and another row from the matrix.

It is clear that applying row operations to the generator matrix results in a matrix generating the same code. For a given $[n, k, d]$-code, a new code with the same parameters can be obtained by a permutation of the coordinates of the code words, which is equivalent to the corresponding permutation of columns in the generator matrix.

DEFINITION 4.18. Codes $\boldsymbol{C}$ and $\boldsymbol{C}'$ are *equivalent* if their generator matrices G and G' can be obtained from each other by a permutation of columns and by row operations (such matrices G and G' are called equivalent).

DEFINITION 4.19. The set of permutations of coordinates of code words of the code $\boldsymbol{C}$ (permutations of columns in the generator matrix) that transform every code word to a code word (no necessarily the same) is called the *automorphism group of the code* $\boldsymbol{C}$ and is denoted by $\mathrm{Aut}(\boldsymbol{C})$.

PROBLEM 4.20. Prove that $\mathrm{Aut}(\boldsymbol{C})$ is a group.

PROBLEM 4.21. Prove that for an arbitrary linear block code $\boldsymbol{C}$ one has

$$\mathrm{Aut}(\boldsymbol{C}) = \mathrm{Aut}(\boldsymbol{C}^{\perp}).$$

PROBLEM 4.22. Prove that if a code $\boldsymbol{C}$ is linear and the code $\boldsymbol{C}_1$ is obtained from $\boldsymbol{C}$ by adjoining the all-one code word if it was not initially in the code (i.e., we adjoin the all-one row to the generator matrix of the code $\boldsymbol{C}$), then $\mathrm{Aut}(\boldsymbol{C}_1) \supset \mathrm{Aut}(\boldsymbol{C})$. Furthermore, if the block length of the code $\boldsymbol{C}$ is odd and all code words in $\boldsymbol{C}$ have even weight, then $\mathrm{Aut}(\boldsymbol{C}_1) = \mathrm{Aut}(\boldsymbol{C})$.

LEMMA 4.23. *Let $\boldsymbol{C}$ be a linear $[n, k]$-code with generator matrix G. The permutation of the components of code words defined by an $(n \times n)$ permutation matrix A belongs to the group $\mathrm{Aut}(\boldsymbol{C})$ if and only if $KG = GA$ for an appropriate nonsingular $(k \times k)$ matrix K.*

PROOF. Suppose that $A \in \mathrm{Aut}(\boldsymbol{C})$. Then the matrix $G' = GA$ is the generator matrix of the code $\boldsymbol{C}$, and therefore the matrix G' can be obtained from G by multiplying on the left with some nonsingular $(k \times k)$ matrix K, i.e., $G' = KG = GA$.

Conversely, suppose that $KG = GA$ for a nonsingular $(k \times k)$ matrix K. Then KG is the generator matrix for the code $\boldsymbol{C}$, whence $A \in \mathrm{Aut}(\boldsymbol{C})$. □

PROBLEM 4.24. Prove that the automorphism group of a binary linear $[n, k]$-code is isomorphic to a subgroup of the group $\mathfrak{GL}(V_k)$. (*Hint*: Use Lemma 4.23.)

THEOREM 4.25. *Suppose that $\boldsymbol{C}$ is an $[n, k]$-code. Then the generator matrix G of $\boldsymbol{C}$ is equivalent to a matrix of the form*

$$G' = (E_k \ P), \tag{4.1}$$

where E_k is the $(k \times k)$ identity matrix and P is the $(k \times (n - k))$ matrix.

PROOF. As the rows of the matrix G are linearly independent, row transformations and permutations of columns of G can transform G to the required form. □

DEFINITION 4.26. A generator matrix of the form (4.1) is called a *generator matrix in the systematic form*. The code generated by such a matrix is called *systematic*.

COROLLARY 4.27. *Any linear block code is equivalent to a systematic code.*

PROBLEM 4.28. Let $\boldsymbol{C}$ be a systematic $[n, k]$-code with generator matrix $G = (E_k \quad P)$. Prove that the parity-check matrix H of this code is of the form

$$H = (P^\top \; E_{n-k}).$$

In a systematic code, code words start with *the information symbols* (i.e., the first coordinates are the message). The rest of the symbols are called *check symbols*. Thus, any systematic $[n, k]$-code has k information and $n - k$ check symbols.

We proceed with some inequalities connecting the parameters of a code.

THEOREM 4.29 (Singleton bound). *The minimum distance (minimum weight of a code word) of any $[n, k]$-code satisfies the inequality*

$$d \leq n - k + 1.$$

PROOF. Since an arbitrary $[n, k]$-code is equivalent to a systematic code with the same parameters n, k, and d, we may assume that the code is systematic. In such a code, the code word of minimum weight has weight d. There exist words from the systematic code with one nonzero information symbol and $n - k$ check symbols. The weight of such a code word could not be greater than $n-k+1$. Thus, the minimum distance (minimum weight) cannot be greater than $n - k + 1$. □

The bound given in the theorem is not sharp, and many good codes have minimum distance significantly less than the Singleton bound, but sometimes this estimate is useful. The Singleton bound shows that in order to correct t errors the code must have at least $2t$ check symbols.

DEFINITION 4.30. A code $\boldsymbol{C}$ having minimum distance which satisfies the equation

$$d_{\boldsymbol{C}} = n - k + 1,$$

is called a *code with maximum distance*.

DEFINITION 4.31. Suppose that $\boldsymbol{x} \in V_n$ and $0 \leq r \leq n$. The set of vectors

$$\rho(\boldsymbol{x}, r) = \{\, \boldsymbol{y} \in V_n \mid \operatorname{dist}(\boldsymbol{x}, \boldsymbol{y}) \leq r \,\}$$

is called a *ball* of radius r with center at the the point $\boldsymbol{x}$.

THEOREM 4.32 (Hamming, or sphere-packing, bound). *Suppose that $\boldsymbol{C}$ is an $[n, k]$-code correcting t errors. Then*

$$1 + \binom{n}{1} + \cdots + \binom{n}{t} \leq 2^{n-k}.$$

PROOF. If a code corrects t errors, then balls of radius t centered at code words do not intersect. The number of vectors contained in these balls is

$$2^k \left(1 + \binom{n}{1} + \cdots + \binom{n}{t}\right).$$

This obviously is not greater than the number of vectors in V_n:

$$2^k\left(1+\binom{n}{1}+\cdots+\binom{n}{t}\right)\leq 2^n.$$

The last inequality is equivalent to the statement of the theorem. □

THEOREM 4.33. *If $\boldsymbol{C}$ is an $[n,k]$-code with minimum distance $d_{\boldsymbol{C}}$, then*

$$d_{\boldsymbol{C}}\leq\frac{n\cdot 2^{k-1}}{2^k-1}.$$

PROOF. Suppose that $1\leq i\leq n$ and that $\boldsymbol{C}$ contains code words with zero at the ith place. Let $\boldsymbol{D}$ be a subspace of $\boldsymbol{C}$ consisting of all code words with zero at the ith place. Then the quotient space $\boldsymbol{C}/\boldsymbol{D}$ contains two elements corresponding to the two possible choices for the value of the ith coordinate in a code word. From $\#(\boldsymbol{C}/\boldsymbol{D})=\#\boldsymbol{C}/\#\boldsymbol{D}$ it follows that $\#\boldsymbol{D}=2^{k-1}$, i.e., there are exactly 2^{k-1} code words with zero ith coordinate. Calculating the sum of the weights of the code words from $\boldsymbol{C}$, we see that it is not greater than $n\cdot 2^{k-1}$. As the minimum distance $d_{\boldsymbol{C}}$ of the code $\boldsymbol{C}$ is equal to the minimum weight of a nonzero code word, we infer that the sum of the weights of the words of the code $\boldsymbol{C}$ is not less than $d_{\boldsymbol{C}}\cdot(2^k-1)$, whence the desired inequality. □

THEOREM 4.34 (Gilbert–Varshamov bound). *If*

$$1+\binom{n-1}{1}+\cdots+\binom{n-1}{d-2}<2^{n-k},$$

then there exists an $[n,k]$-code with minimum distance at least d.

PROOF. In view of Theorem 4.15 it suffices to construct an $((n-k)\times n)$ matrix H such that any $d-1$ columns of it are linearly independent. The first column of the matrix H can be any nonzero $(n-k)$-vector. Assume now that we have chosen i columns of the matrix H such that any $d-1$ of them are linearly independent. There are no more than

$$\binom{i}{1}+\cdots+\binom{i}{d-2}$$

different linear combinations of these i columns containing $\leq d-2$ terms. If this number is less than $2^{n-k}-1$, we can add a column different from all of these linear combinations, and any $d-1$ columns of the new $((n-k)\times(i+1))$ matrices will be linearly independent. Now we continue this procedure while

$$1+\binom{i}{1}+\cdots+\binom{i}{d-2}<2^{n-k}.$$ □

THEOREM 4.35 (Elias bound). *The minimum distance $d_{\boldsymbol{C}}$ of any $[n,k]$-block code $\boldsymbol{C}$ satisfies the inequality*

$$d_{\boldsymbol{C}}\leq 2s\left(1-\frac{s}{n}\right)\cdot\frac{N}{N-1},$$

where s is any natural number such that

$$\sum_{j=0}^{s}\binom{n}{j}>2^{n-k}, \tag{4.2}$$

and N is the smallest natural number for which

$$N \geq \frac{1}{2^{n-k}} \sum_{j=0}^{s} \binom{n}{j}. \tag{4.3}$$

PROOF. Let $\boldsymbol{C}$ be an arbitrary $[n, k]$-code. Choose a natural number $s < n/2$ satisfying inequality (4.2). The ball $\rho(\boldsymbol{x}, s) \subset V_n$ with radius s and center at $\boldsymbol{x} \in V_n$ contains $N_{\boldsymbol{x}}$-code vectors. The number of vectors from V_n that are contained in a ball of radius s is equal to $\sum_{j=0}^{s} \binom{n}{j}$; each of the 2^k code words belongs to the same number of such balls. Hence,

$$\sum_{\boldsymbol{x} \in V_n} N_{\boldsymbol{x}} = 2^k \sum_{j=0}^{s} \binom{n}{j}.$$

Denote by N the greatest among the numbers $N_{\boldsymbol{x}}$, $\boldsymbol{x} \in V_n$. One has

$$N \geq \frac{1}{2^{n-k}} \sum_{j=0}^{s} \binom{n}{j} > 1.$$

Observe that there exists a ball of radius s containing more than one code vector (its center $\boldsymbol{x}_0$ is not necessarily a code word). Suppose that this ball $\rho(\boldsymbol{x}_0, s)$ contains the code words $\boldsymbol{c}_1, \boldsymbol{c}_2, \ldots, \boldsymbol{c}_N$. Consider N rows of length n obtained after the following transformation of the code words:

$$\begin{aligned} \boldsymbol{a}_1 &= (a_{11}, a_{12}, \ldots, a_{1n}) = \boldsymbol{c}_1 \oplus \boldsymbol{x}_0; \\ \boldsymbol{a}_2 &= (a_{21}, a_{22}, \ldots, a_{2n}) = \boldsymbol{c}_2 \oplus \boldsymbol{x}_0; \\ &\cdots\cdots\cdots\cdots\cdots\cdots\cdots\cdots \\ \boldsymbol{a}_N &= (a_{N1}, a_{N2}, \ldots, a_{Nn}) = \boldsymbol{c}_N \oplus \boldsymbol{x}_0. \end{aligned} \tag{4.4}$$

It is clear that the Hamming distances between any two of these rows coincide with the distances between the corresponding code words. Put $\mathrm{wt}(\boldsymbol{a}_i) = w_i$, $i = 1, 2, \ldots, N$. Since the radius of the ball is s, the total weight of the rows $\boldsymbol{a}_i$, $i = 1, 2, \ldots, N$, is at most Ns, i.e.,

$$\sum_{i=1}^{N} w_i \leq Ns.$$

Denote by v_j, $j = 1, 2, \ldots, n$, the weight of the column $(a_{1j}, a_{2j}, \ldots, a_{Nj})^{\top}$ (i.e., suppose that this column contains $N - v_j$ zeros and v_j ones). Obviously,

$$\sum_{j=1}^{n} v_j \leq Ns.$$

The total distance d_{total} between the rows (4.4) is equal to the sum of $\binom{N}{2}$ distances between all pairs of them. Besides, the contribution of the jth column to the expression for d_{total} is

$$\sum_{i_1 > i_2} (a_{i_1 j} \oplus a_{i_2 j}) = \binom{N}{2} - \binom{v_j}{2} - \binom{N - v_j}{2} = v_j(N - v_j).$$

Here the sum is a natural number. Summing over j we get

$$d_{\text{total}} = N \sum_{j=1}^{n} v_j - \sum_{j=1}^{n} v_j^2.$$

As $\sum_{j=1}^{n} v_j = Ns - \Delta$, where $\Delta \geq 0$, Jensen's inequality (see [**254**]) for quadratic functions implies that

$$\sum_{j=1}^{n} v_j^2 \geq n\Big(\frac{Ns-\Delta}{n}\Big)^2.$$

Thus, from the expression for d_{total} we obtain that

$$d_{\text{total}} \leq N(Ns-\Delta) - n\Big(\frac{Ns-\Delta}{n}\Big)^2 \leq sN^2\Big(1-\frac{s}{n}\Big) - \Delta N\Big(1-\frac{2s}{n}\Big).$$

Since $s < n/2$, we have

$$d_{\text{total}} \leq sN^2\Big(1-\frac{s}{n}\Big).$$

Of the $\binom{N}{2}$ distances between any two elements of the code contained in the ball $\rho(\boldsymbol{x}_0, s)$, at least one does not exceed the average value of these distances. Denote this average by d_{av}. One has

$$d_{\text{av}} \leq \frac{d_{\text{total}}}{\binom{N}{2}} = 2s\left(1-\frac{s}{n}\right)\frac{N}{N-1}.$$

But $d_{\boldsymbol{C}} \leq d_{\text{av}}$, and this completes the proof. □

Let $\boldsymbol{C}$ be an arbitrary $[n,k]$-code. Denote by A_i, $i = 0, 1, \ldots, n$, the number of its code words of weight i:

$$A_i = \#\{\boldsymbol{c} \in \boldsymbol{C} \mid \text{wt}(\boldsymbol{c}) = i\}.$$

DEFINITION 4.36. The sequence $A_0, A_1, \ldots, A_n$ is called the *weight spectrum of the code* $\boldsymbol{C}$.

DEFINITION 4.37. The polynomial

$$\mathcal{W}_{\boldsymbol{C}}(\lambda, \nu) = \sum_{i=0}^{n} A_i \lambda^{n-i} \nu^i$$

of two variables λ and ν is called the *weight function of the code* $\boldsymbol{C}$.

Observe that the homogeneous polynomial $\mathcal{W}_{\boldsymbol{C}}(\lambda, \nu)$ can be represented in two ways:

$$\mathcal{W}_{\boldsymbol{C}}(\lambda, \nu) = \sum_{i=0}^{n} A_i \lambda^{n-i} \nu^i = \sum_{\boldsymbol{c} \in \boldsymbol{C}} \lambda^{n-\text{wt}(\boldsymbol{c})} \nu^{\text{wt}(\boldsymbol{c})}.$$

If $A'_0, A'_1, \ldots, A'_n$ is the weight spectrum of the dual code $\boldsymbol{C}^{\perp}$, then the weight function of the dual code $\boldsymbol{C}^{\perp}$ is

$$\mathcal{W}_{\boldsymbol{C}^{\perp}}(\lambda, \nu) = \sum_{i=0}^{n} A'_i \lambda^{n-i} \nu^i = \sum_{\boldsymbol{c}' \in \boldsymbol{C}^{\perp}} \lambda^{n-\text{wt}(\boldsymbol{c}')} \nu^{\text{wt}(\boldsymbol{c}')}.$$

Let $f\colon V_n \to U$ be a mapping from the vector space V_n to a vector space U over the field $\mathbb{C}$ of complex numbers. We will define Fourier transform of the mapping f by analogy with Boolean functions (see Definition 2.35):

$$\overline{W}_f(\boldsymbol{u}) = \sum_{\boldsymbol{x} \in V_n} (-1)^{\langle \boldsymbol{u}, \boldsymbol{x} \rangle} f(\boldsymbol{x})$$

for any $\boldsymbol{u} \in V_n$. The following statement can be proved similarly to Lemma 2.40.

LEMMA 4.38. *If $\boldsymbol{C}$ is an $[n,k]$-code, then*

$$\sum_{\boldsymbol{c}'\in\boldsymbol{C}^\perp} f(\boldsymbol{c}') = \frac{1}{\#\boldsymbol{C}}\sum_{\boldsymbol{c}\in\boldsymbol{C}} \overline{W}_f(\boldsymbol{c}).$$

THEOREM 4.39 (MacWilliams identity). *If $\boldsymbol{C}$ is an $[n,k]$-code and $\boldsymbol{C}^\perp$ is its dual $[n,n-k]$-code, then*

$$\mathcal{W}_{\boldsymbol{C}^\perp}(\lambda,\nu) = \frac{1}{\#\boldsymbol{C}}\mathcal{W}_{\boldsymbol{C}^\perp}(\lambda+\nu,\lambda-\nu).$$

PROOF. We will apply Lemma 4.38 to the mapping $f(\boldsymbol{u}) = \lambda^{n-\mathrm{wt}(\boldsymbol{u})}\nu^{\mathrm{wt}(\boldsymbol{u})}$. We have

$$\overline{W}_f(\boldsymbol{u}) = \sum_{\boldsymbol{x}\in V_n} (-1)^{\langle \boldsymbol{u},\boldsymbol{x}\rangle}\lambda^{n-\mathrm{wt}(\boldsymbol{u})}\nu^{\mathrm{wt}(\boldsymbol{u})}.$$

Put $\boldsymbol{u} = (u^{(1)}, u^{(2)}, \dots, u^{(n)})^\top$, $\boldsymbol{x} = (x^{(1)}, x^{(2)}, \dots, x^{(n)})^\top$. Then

$$\begin{aligned}\overline{W}_f(\boldsymbol{u}) &= \sum_{\boldsymbol{x}\in V_n} (-1)^{u^{(1)}x^{(1)}\oplus u^{(2)}x^{(2)}\oplus\cdots\oplus u^{(n)}x^{(n)}}\prod_{i=1}^{n}\lambda^{1-x^{(i)}}\nu^{x^{(i)}}\\ &= \sum_{x^{(1)}=0}^{1}\cdots\sum_{x^{(n)}=0}^{1}\prod_{i=1}^{n}(-1)^{u^{(i)}x^{(i)}}\lambda^{1-x^{(i)}}\nu^{x^{(i)}}.\end{aligned}$$

Since in this case the sum of the products is equal to the product of the sums, the right-hand side of the last equality is equal to

$$\prod_{i=1}^{n}\sum_{w=0}^{1}(-1)^{u^{(i)}w}\lambda^{1-w}\nu^{w}.$$

If $u^{(i)} = 0$, the inner sum is equal to $\lambda+\nu$. If $u^{(i)} = 1$, it is equal to $\lambda-\nu$. Thus,

$$\overline{W}_f(\boldsymbol{u}) = (\lambda+\nu)^{n-\mathrm{wt}(\boldsymbol{u})}(\lambda-\nu)^{\mathrm{wt}(\boldsymbol{u})}.$$

Now from Lemma 4.38 it follows that

$$\sum_{\boldsymbol{u}'\in\boldsymbol{C}^\perp}\lambda^{n-\mathrm{wt}(\boldsymbol{u}')}\nu^{\mathrm{wt}(\boldsymbol{u}')} = \frac{1}{\#\boldsymbol{C}}\sum_{\boldsymbol{u}\in\boldsymbol{C}}(\lambda+\nu)^{n-\mathrm{wt}(\boldsymbol{u})}(\lambda-\nu)^{\mathrm{wt}(\boldsymbol{u})}.$$ □

COROLLARY 4.40. *The following identity (equivalent to Theorem 4.39) holds:*

$$\sum_{j=0}^{n} A_j'\lambda^{n-j}\nu^j = \frac{1}{\#\boldsymbol{C}}\sum_{i=0}^{n} A_i(\lambda+\nu)^{n-i}(\lambda-\nu)^i.$$

The relation between the coefficients A_i and A_i' is of interest. Put

$$(\lambda+\nu)^{n-i}(\lambda-\nu)^i = \sum_{k=0}^{n} P_k(i)\lambda^{n-k}\nu^k.$$

It is easy to see from Corollary 4.40 that

$$A_k' = \frac{1}{\#\boldsymbol{C}}\sum_{i=0}^{n} A_i P_k(i).$$

The expressions $P_k(i)$ are called the *Kravchuk polynomials.*

DEFINITION 4.41. Let n be a natural number. The *Kravchuk polynomials* are given by the formula

$$P_k(z) = \sum_{j=0}^{k} (-1)^j \binom{z}{j} \binom{n-z}{k-j},$$

where z is a free variable, $k = 0, 1, \ldots,$ and

$$\binom{z}{m} = \begin{cases} \frac{z(z-1)\cdots(z-m+1)}{m!} & \text{if } m > 0, \\ 1 & \text{if } m = 0. \end{cases}$$

4.2. The decoding problem

Let $\boldsymbol{C}$ be an $[n, k]$-code. Suppose that a code word $\boldsymbol{c} \in \boldsymbol{C}$ is sent. In the communication channel (see Figure 4.1) this word is corrupted. The receiver obtains the vector $\boldsymbol{y} = \boldsymbol{c} \oplus \boldsymbol{e}$, where $\boldsymbol{e} = (e^{(1)}, e^{(2)}, \ldots, e^{(n)}) \in V_n$ is the error vector.

THEOREM 4.42. *A code $\boldsymbol{C}$ with minimum distance $d_{\boldsymbol{C}}$ can correct any t errors if*

$$d_{\boldsymbol{C}} \geq 2t + 1.$$

PROOF. The decoding rule is to assign to the resulting word $\boldsymbol{y}$ the closest code word from the code $\boldsymbol{C}$. In the space V_n we consider the balls of radius t with centers at every code word of the code $\boldsymbol{C}$. If a pair of balls $\rho(\boldsymbol{c}_1, t)$ and $\rho(\boldsymbol{c}_2, t)$ $(\boldsymbol{c}_1 \neq \boldsymbol{c}_2$, $\boldsymbol{c}_1, \boldsymbol{c}_2 \in \boldsymbol{C})$ has nonempty intersection,

$$\boldsymbol{x} \in \rho(\boldsymbol{c}_1, t) \cap \rho(\boldsymbol{c}_2, t), \quad \boldsymbol{x} \in V_n,$$

then

$$\operatorname{dist}(\boldsymbol{c}_1, \boldsymbol{c}_2) \leq \operatorname{dist}(\boldsymbol{c}_1, \boldsymbol{x}) + \operatorname{dist}(\boldsymbol{x}, \boldsymbol{c}_2) \leq 2t,$$

which contradicts the inequality

$$\operatorname{dist}(\boldsymbol{c}_1, \boldsymbol{c}_2) \geq d_{\boldsymbol{C}} \geq 2t + 1.$$

Therefore, the balls centered at code words are all disjoint. Thus the received word $\boldsymbol{y} = \boldsymbol{c} \oplus \boldsymbol{e}$ will be in the ball $\rho(\boldsymbol{c}, t)$ if $\operatorname{wt}(\boldsymbol{e}) \leq t$, and so the decoding to the nearest code word leads to correct recovery of the code word $\boldsymbol{c}$ sent. □

EXAMPLE 4.43. Suppose that $m \geq 2$ is a natural number and $n = 2^m - 1$. Consider a code $\boldsymbol{C}_m$ of length n with the following parity-check matrix H: it has dimensions $m \times (2^m - 1)$, and its columns are binary representations of the numbers $1, 2, \ldots, 2^m - 1$:

$$H = \begin{pmatrix} 0 & 0 & 0 & \ldots & 1 & 1 \\ 0 & 0 & 0 & \ldots & 1 & 1 \\ \cdots & \cdots & \cdots & \cdots & \cdots & \cdots \\ 0 & 1 & 1 & \ldots & 1 & 1 \\ 1 & 0 & 1 & \ldots & 0 & 1 \end{pmatrix},$$

i.e., the columns of the matrix H are all nonzero vectors of length m. Obviously, $\operatorname{rank} H = m$. Therefore, the dimension $k_{\boldsymbol{C}_m}$ of the code $\boldsymbol{C}_m$ is equal to $2^m - m - 1$. Any two columns of the matrix H are linearly independent. Since H contains the sum of any two columns as a column, Theorem 4.15 shows that $d_{\boldsymbol{C}_m} = 3$. Thus, the code $\boldsymbol{C}_m$ can correct one error. If m is big, the code rate $R_{\boldsymbol{C}_m} = 1 - \frac{m}{2^m - 1}$ is close to 1, but the proportion of the bits which can be corrupted and then corrected is very small. The code $\boldsymbol{C}_m$ is called the *Hamming code*.

DEFINITION 4.44. A code correcting t errors is called *perfect* if it cannot correct any error of weight greater than t.

This definition is equivalent to the following: the balls of radius t centered at code words are disjoint and their union coincides with V_n.

LEMMA 4.45. *The Hamming code* $\boldsymbol{C}_m$ *is a perfect code correcting all errors of weight* 1.

PROOF. Since the Hamming code $\boldsymbol{C}_m$ corrects all errors of weight 1, balls of radius 1 centered at code words do not intersect. There are exactly 2^{2^m-m-1} such balls (the number is equal to the number of the code words). Any ball of radius 1 contains 2^m vectors from V_n, $n = 2^m - 1$. Hence, the total number of vectors contained in the balls is

$$2^m \cdot 2^{2^m-m-1} = 2^{2^m-1} = 2^n. \qquad \square$$

Consider now the decoding algorithm of the Hamming code. Recall that the ith column ($i = 1, 2, \dots, 2^m - 1$) is the binary representation of the number i. Suppose that the jth symbol of the sent code word $\boldsymbol{c} \in \boldsymbol{C}_m$ is corrupted, i.e., the receiver gets the vector $\boldsymbol{y} = \boldsymbol{c} \oplus \boldsymbol{e}$, where $\boldsymbol{e} = (0, \dots, 0, \underset{j}{1}, 0, \dots, 0)$. Put

$$H\boldsymbol{y}^\top = H(\boldsymbol{c} \oplus \boldsymbol{e})^\top = H\boldsymbol{c}^\top \oplus H\boldsymbol{e}^\top.$$

Since $H\boldsymbol{c}^\top = \boldsymbol{0}$,

$$H\boldsymbol{y}^\top = H\boldsymbol{e}^\top.$$

But $H\boldsymbol{e}^\top$ is the jth column of H, and this is the binary representation of the number of the corrupted coordinate of the code word sent. Thus, we have to correct this coordinate.

Consider the dual code $\boldsymbol{C}_m^\perp$. Its generator matrix is the parity-check matrix of the Hamming code. The dimension $k_{\boldsymbol{C}_m^\perp} = \dim \boldsymbol{C}_m^\perp$ is equal to m. It is easy to see that the ith ($i = 1, 2, \dots, m$) column of the parity-check matrix H of the Hamming code represents the values of the linear function $l_{\boldsymbol{e}_i,0}(x^{(1)}, \dots, x^{(m)}) = x^{(i)}$ at nonzero vectors from the space V_n ($\boldsymbol{e}_i$ is the ith vector of the canonical basis of V_m; see §2.1).

It is easy to see that the sum of any m rows of the matrix H is a row of values of a linear function at nonzero vectors of V_m. Therefore, $\mathrm{wt}(\boldsymbol{c}') = 2^{m-1}$ for any $\boldsymbol{c}'$ from $\boldsymbol{C}_m^\perp$. Hence, the code $\boldsymbol{C}_m^\perp$ can correct $2^{m-2} - 1 = \left\lfloor \frac{2^{m-1}-1}{2} \right\rfloor$ errors. Besides, for any $\boldsymbol{c}_1', \boldsymbol{c}_2' \in \boldsymbol{C}_m^\perp$ ($\boldsymbol{c}_1' \neq \boldsymbol{c}_2'$),

$$\mathrm{dist}(\boldsymbol{c}_1', \boldsymbol{c}_2') = 2^{m-1}.$$

DEFINITION 4.46. A code is called *equidistant* if the distance between any two of its code words is the same.

For example, the code $\boldsymbol{C}_m^\perp$ is equidistant. In coding theory this code is called the *binary simplex code* or the maximum length code obtained from feedback shift register. The rate of the code $\boldsymbol{C}_m^\perp$ equals

$$R_{\boldsymbol{C}_m^\perp} = \frac{m}{2^m - 1};$$

if m is big, this is close to 0. The ratio of the bits which may be corrupted to all bits is close to 0.25.

Now we will describe a simple decoding algorithm for linear block codes. To that end, we must introduce a new notation.

DEFINITION 4.47. Let $\boldsymbol{C}$ be an $[n,k]$-code. For any vector $\boldsymbol{a} \in V_n$ the set

$$\boldsymbol{a} \oplus \boldsymbol{C} = \{\boldsymbol{a} \oplus \boldsymbol{c} \mid \boldsymbol{c} \in \boldsymbol{C}\}$$

is called a *coset* of the code $\boldsymbol{C}$.

Since $\boldsymbol{C}$ is a subspace of the vector space V_n, any two cosets of the code $\boldsymbol{C}$ either do not intersect or coincide. The vector space $V_n/\boldsymbol{C}$ consists of cosets. Every coset contains 2^k vectors. The space V_n splits into cosets of the space $\boldsymbol{C}$:

$$V_n = (\boldsymbol{a}_0 \oplus \boldsymbol{C}) \cup (\boldsymbol{a}_1 \oplus \boldsymbol{C}) \cup \cdots \cup (\boldsymbol{a}_s \oplus \boldsymbol{C}),$$

where $\boldsymbol{a}_0 = \boldsymbol{0}$, $s = 2^{n-k} - 1$.

DEFINITION 4.48. Let H be the parity-check matrix of an $[n,k]$-code $\boldsymbol{C}$. Then for an arbitrary vector $\boldsymbol{y} \in V_n$, the vector

$$S(\boldsymbol{y}) = H\boldsymbol{y}^\top$$

of the space V_{n-k} is called the *syndrome vector* of the vector $\boldsymbol{y}$.

If we recall the decoding algorithm for the Hamming code $\boldsymbol{C}_m$ (see Example 4.43), we see that the syndrome of the received vector points to the number of coordinates which have to be corrected. It is clear that vectors $\boldsymbol{y}$ and $\boldsymbol{y}'$ have the same syndrome if and only if they are from the same coset. Also, the syndrome is the zero vector only for the code words of the code $\boldsymbol{C}$.

Let us construct (see Table 4.1) the *standard array* of an $[n,k]$-code $\boldsymbol{C}$ with minimum distance $d_{\boldsymbol{C}} = 2t+1$. In the first row we place all code words of the code $\boldsymbol{C} = \{\boldsymbol{c}_1 = \boldsymbol{0}, \boldsymbol{c}_2, \dots, \boldsymbol{c}_{2^k}\}$. Now look at the ball $\rho(\boldsymbol{0}, t)$. First we choose from this ball a vector $\boldsymbol{v}_1$ at the distance 1 from the zero code word. In the second row we place the vectors of the coset

$$\boldsymbol{v}_1 \oplus \boldsymbol{C} = \{\boldsymbol{v}_1 \oplus \boldsymbol{0}, \boldsymbol{v}_1 \oplus \boldsymbol{c}_2, \dots, \boldsymbol{v}_1 \oplus \boldsymbol{c}_{2^k}\}.$$

The vector $\boldsymbol{v}_1$ is called the *coset leader*. Similarly, we choose from $\rho(\boldsymbol{0}, t)$ a sequence of vectors $\boldsymbol{v}_2, \dots, \boldsymbol{v}_n$ at the distance 1 from the zero code word. We continue filling the rows of the array choosing coset leaders at distances $2, 3, \dots, t$ from the zero coset. It is clear that the cosets obtained in this way do not intersect because if two vectors $\boldsymbol{u}, \boldsymbol{v} \in \rho(\boldsymbol{0}, t)$ are in the same coset, then $\boldsymbol{u} \oplus \boldsymbol{v} \in \boldsymbol{C}$ and $\mathrm{wt}(\boldsymbol{u} \oplus \boldsymbol{v}) = \mathrm{dist}(\boldsymbol{u}, \boldsymbol{v}) \le \mathrm{dist}(\boldsymbol{u}, \boldsymbol{0}) + \mathrm{dist}(\boldsymbol{0}, \boldsymbol{v}) \le 2t$, which contradicts the inequality $\mathrm{wt}(\boldsymbol{u} \oplus \boldsymbol{v}) \ge d_{\boldsymbol{C}} = 2t+1$.

Using all elements of the ball $\rho(\boldsymbol{0}, t)$ as coset leaders, we will fill $j = 1 + \binom{n}{1} + \cdots + \binom{n}{t}$ rows of the standard array.

Let us draw a horizontal line here in the array. If there are no vectors from the space V_n outside of the balls around the code words (i.e., the code $\boldsymbol{C}$ is perfect), then the array is complete. If there are such vectors, we choose one of them, say, $\boldsymbol{v}_{j+1}$, such that it is at the distance $t+1$; we take it as the leader of the coset

$$\boldsymbol{v}_{j+1} \oplus \boldsymbol{C} = \{\boldsymbol{v}_{j+1} \oplus \boldsymbol{0}, \boldsymbol{v}_{j+1} \oplus \boldsymbol{c}_2, \dots, \boldsymbol{v}_{j+1} \oplus \boldsymbol{c}_{2^k}\}$$

and place it in the $(j+1)$th row. Then we check whether the space V_n is exhausted. If not, we choose a leader of the coset $\boldsymbol{v}_{j+2}$ at the distance $t+1$ from the zero code word. It there are no such code words, we choose a leader at the distance $t+2$,

TABLE 4.1. Standard array.

		decoding sphere		decoding area
$\mathbf{0}$	$\boldsymbol{c}_2$	$\boldsymbol{c}_3$	$\cdots$	$\boldsymbol{c}_{2^k}$
$\boldsymbol{v}_1 \oplus \mathbf{0}$	$\boldsymbol{v}_1 \oplus \boldsymbol{c}_2$	$\boldsymbol{v}_1 \oplus \boldsymbol{c}_3$	$\cdots$	$\boldsymbol{v}_1 \oplus \boldsymbol{c}_{2^k}$
$\boldsymbol{v}_2 \oplus \mathbf{0}$	$\boldsymbol{v}_2 \oplus \boldsymbol{c}_2$	$\boldsymbol{v}_2 \oplus \boldsymbol{c}_3$	$\cdots$	$\boldsymbol{v}_2 \oplus \boldsymbol{c}_{2^k}$
$\vdots$	$\vdots$	$\vdots$	$\cdots$	$\vdots$
$\boldsymbol{v}_j \oplus \mathbf{0}$	$\boldsymbol{v}_j \oplus \boldsymbol{c}_2$	$\boldsymbol{v}_j \oplus \boldsymbol{c}_3$	$\cdots$	$\boldsymbol{v}_j \oplus \boldsymbol{c}_{2^k}$
$\boldsymbol{v}_{j+1} \oplus \mathbf{0}$	$\boldsymbol{v}_{j+1} \oplus \boldsymbol{c}_2$	$\boldsymbol{v}_{j+1} \oplus \boldsymbol{c}_3$	$\cdots$	$\boldsymbol{v}_{j+1} \oplus \boldsymbol{c}_{2^k}$
$\vdots$	$\vdots$	$\vdots$	$\cdots$	$\vdots$
$\boldsymbol{v}_{2^{n-k}} \oplus \mathbf{0}$	$\boldsymbol{v}_{2^{n-k}} \oplus \boldsymbol{c}_2$	$\boldsymbol{v}_{2^{n-k}} \oplus \boldsymbol{c}_3$	$\cdots$	$\boldsymbol{v}_{2^{n-k}} \oplus \boldsymbol{c}_{2^k}$
		area between the balls belonging to $\boldsymbol{c}_3$		

etc. It is clear that this process will terminate. Since $\#(V_n/\boldsymbol{C}) = 2^{n-k}$, there are exactly 2^{n-k} rows in the standard array.

The decoding algorithm based on the standard array is simple. We have to find the received vector in the array, to determine the column where it lives, and to decode it to the code word which is on the top of the column.

THEOREM 4.49. *If we use the standard array for decoding the code $\boldsymbol{C}$, then for a received vector $\boldsymbol{y} = \boldsymbol{c} \oplus \boldsymbol{e}$ ($\boldsymbol{c} \in \boldsymbol{C}$):*

a) *the code vector $\boldsymbol{c}$ will always be correctly recovered if* $\mathrm{wt}(\boldsymbol{e}) \leq t$;

b) *the code vector $\boldsymbol{c}$ will be correctly recovered if* $\mathrm{wt}(\boldsymbol{e}) > t$ *and* $\boldsymbol{e} = \boldsymbol{c} \oplus \boldsymbol{y}$ *coincides with the coset leader.*

PROOF. Statement a) follows from the construction of the standard array.

Statement b) follows from the fact that if the vector $\boldsymbol{e} = \boldsymbol{c} \oplus \boldsymbol{y}$ for $\mathrm{wt}(\boldsymbol{e}) > t$ is not a coset leader and if $\boldsymbol{y}$ does not belong to the coset $\boldsymbol{v}_i \oplus \boldsymbol{C}$, then it is not in the column below the code word $\boldsymbol{c}$ as $\boldsymbol{v}_i \oplus \boldsymbol{c} \neq \boldsymbol{e} \oplus \boldsymbol{c}$. Therefore, it will be decoded incorrectly. □

There exist two basic types of decoders, which can be described with the help of standard array: *complete decoders* and *incomplete decoders*. A complete decoder concatenates the closest code word with the received vector (Theorem 4.49, a) and b)). An incomplete decoder concatenates the received vector with a code word at the distance at most t from it if such a vector exists; otherwise it refuses to decode (Theorem 4.49, a)). The received vector should be found in the standard array. If it is above the horizontal line, we decode it to the code word on the top of the column where the received vector is. If it is below the horizontal line, the decoder refuses to decode, and in this case there are more than t errors.

Search in an array requires much time. The search can be avoided if, for example, for the coset leaders the relations $H\boldsymbol{v}_{i_1}^{\top} \neq H\boldsymbol{v}_{i_2}^{\top}$ ($i_1 \neq i_2$) hold, where

H is the parity-check matrix of $\boldsymbol{C}$. Then, after receiving the vector $\boldsymbol{y} = \boldsymbol{c} \oplus \boldsymbol{e} = \boldsymbol{c} \oplus \boldsymbol{v}_j$ ($\boldsymbol{e} = \boldsymbol{v}_j$), we calculate its syndrome $S(\boldsymbol{y}) = H(\boldsymbol{c} \oplus \boldsymbol{v}_j)^\top = H\boldsymbol{v}_j^\top$. Using this syndrome, we uniquely determine the coset leader $\boldsymbol{v}_j$ and thus the code word $\boldsymbol{c} = \boldsymbol{y} \oplus \boldsymbol{v}_j$.

DEFINITION 4.50. The *word error probability* P_{err} for a given decoding scheme is the probability of producing an incorrect code word by the decoder provided all code words of the code $\boldsymbol{C}$ are used with equal probability 2^{-k}.

By $P\{\dots\}$ we will denote the probability of the event indicated in braces. One has

$$P_{\text{err}} = 2^{-k} \sum_{i=1}^{n} P\{\text{output of the decoder} \neq c_i \mid c_i \text{ was sent}\}.$$

If we decode using a standard array, then we get a decoding error only when the error vector $\boldsymbol{e}$ does not coincide with any of the coset leaders. Thus,

$$P_{\text{err}} = P\{\boldsymbol{e} \neq \text{coset leader}\}.$$

Suppose that γ_i of the coset leaders have weight i. Then

$$P_{\text{err}} = 1 - \sum_{i=0}^{n} \gamma_i p^i (1-p)^{n-i},$$

where p is the channel error (see Figure 4.1). As $d_{\boldsymbol{C}} = 2t+1$, we have

$$\gamma_i = \binom{n}{i}, \qquad 0 \leq i \leq t.$$

It is rather hard to calculate γ_i for $i > t$ if n and k are big. In order to overcome this difficulty, certain special codes, which will be considered in the next section, must be used.

Since standard arrays ensure decoding of the received vector to the closest code word, i.e., the maximum likelihood decoding, any other decoding scheme of the code will have error probability P_{err} no less than $1 - \sum_{i=0}^{n} \gamma_i p^i (1-p)^{n-i}$.

4.3. Cyclic codes

The most studied codes among all linear codes are the cyclic codes. As we will see below, their successful study is possible because they possess a rich structure. A mathematical apparatus for studying cyclic codes is the theory of finite fields, which is not very useful outside this class of codes. Most of the codes used in practice are based on ideas of finite field theory applied to cyclic codes. As was mentioned in [**34**], the importance of cyclic codes is determined by their rich algebraic structure, well-defined in terms of finite field theory, which leads to encoding and decoding procedures effectively from both algorithmic and computational points of view. The class of cyclic codes contains important subclasses, and its representatives are used in the construction of many well-known codes.

Let $\boldsymbol{C}$ be a linear $[n,k]$-code over the field $\mathbb{F}_2$. As before, we denote the code words by row vectors of length n with elements from $\mathbb{F}_2$: $\boldsymbol{c} = (c^{(0)}, c^{(1)}, \dots, c^{(n-1)})$.

DEFINITION 4.51. A linear $[n,k]$-code $\boldsymbol{C}$ over the field $\mathbb{F}_2$ is called *cyclic* if for any code word $\boldsymbol{c} = (c^{(0)}, c^{(1)}, \dots, c^{(n-1)}) \in \boldsymbol{C}$ the row vector $\boldsymbol{c}' = (c^{(n-1)}, c^{(0)}, c^{(1)}, \dots, c^{(n-2)})$ is also a code word in $\boldsymbol{C}$.

EXAMPLE 4.52. The linear code $\boldsymbol{C} = \{(0,0,0),(0,1,1),(1,1,0),(1,0,1)\}$ is a cyclic $[3,2]$-code.

There is a 1-1 correspondence between $\boldsymbol{c} = (c^{(0)}, c^{(1)}, \dots, c^{(n-1)}) \in V_n$ and the set of polynomials $c(\lambda) = c^{(0)} \oplus c^{(1)}\lambda \oplus \cdots \oplus c^{(n-1)}\lambda^{(n-1)} \in \mathbb{F}_2[\lambda]$ of degree $\leq n-1$. Later we will use both representations interchangeably.

Let $\mathfrak{R}_n = \mathbb{F}_2[\lambda]/(\lambda^n \oplus 1)$ be the ring of residue classes of the ring $\mathbb{F}_2[\lambda]$ modulo the polynomial $\lambda^n \oplus 1$. Since $\mathbb{F}_2[\lambda]$ is a principal ideal domain, $\mathfrak{R}_n$ is also a ring of principal ideals. We will choose polynomials of degree at most $n-1$ as representatives of residue classes in $\mathfrak{R}_n$. In order to distinguish ideals of the ring $\mathfrak{R}_n$ from the ideals of the ring $\mathbb{F}_2[\lambda]$, we will denote by $\langle g(\lambda) \rangle$ the ideal in $\mathfrak{R}_n$ generated by the polynomial $g(\lambda)$, as opposed to the notation $(g(\lambda)) \subset \mathbb{F}_2[\lambda]$ for ideals in the ring $\mathfrak{R}_n$ (we hope that there will be no confusion with scalar product). If $g(\lambda)$ and $g'(\lambda)$ are equal in the ring $\mathfrak{R}_n$, we will write $g(\lambda) \equiv g'(\lambda)$.

Since a cyclic code $\boldsymbol{C} \subset \mathfrak{R}_n$ is invariant under a cyclic shift, it follows from $c(\lambda) \in \boldsymbol{C}$ and $c'(\lambda) \equiv \lambda c(\lambda)$ that $c'(\lambda) \in \boldsymbol{C}$. In other words, multiplication of a polynomial $c(\lambda)$ by λ in the ring $\mathfrak{R}_n$ corresponds to the cyclic shift of its coefficients. Therefore, for any $t(\lambda) \in \mathfrak{R}_n$ and any $c(\lambda) \in \boldsymbol{C}$ we have $t(\lambda)c(\lambda) \in \boldsymbol{C}$, and this implies that $\boldsymbol{C}$ is an ideal of the ring $\mathfrak{R}_n$.

From all that was said above we can easily conclude that the study of cyclic codes in V_n is equivalent to the study of ideals in $\mathfrak{R}_n$. The corresponding mathematical statement is contained in the next theorem.

THEOREM 4.53. *Let $\boldsymbol{C}$ be a nonzero cyclic code in $\mathfrak{R}_n$ (i.e., a nonzero ideal of the ring $\mathfrak{R}_n$). Then there exists a polynomial $g(\lambda) \in \boldsymbol{C}$ such that:*

1. *$g(\lambda)$ is the unique minimal polynomial of minimum degree in $\boldsymbol{C}$.*
2. *$\boldsymbol{C} = \langle g(\lambda) \rangle$.*
3. *$g(\lambda)$ divides $\lambda^n \oplus 1$; if $k = n - \deg g(\lambda)$ and $g(\lambda) = \bigoplus_{i=0}^{n-1} g_i \lambda^i$, where $g_{n-k} = 1$ and $g_{n-k+1} = \cdots = g_{n-1} = 0$, then*
4. *The dimension of the code $\boldsymbol{C}$ is equal to k and $\{g(\lambda), \lambda g(\lambda), \dots, \lambda^{k-1} g(\lambda)\}$ is a basis of the code $\boldsymbol{C}$.*
5. *The generator matrix G of the code $\boldsymbol{C}$ is of the form*

$$G = \begin{pmatrix} g_0 & g_1 & \cdots & g_{n-k} & 0 & \cdots & 0 & 0 \\ 0 & g_0 & \cdots & g_{n-k-1} & g_{n-k} & \cdots & 0 & 0 \\ \cdots & \cdots & \cdots & \cdots & \cdots & \cdots & \cdots & \cdots \\ 0 & 0 & \cdots & 0 & 0 & \cdots & g_{n-k-1} & g_{n-k} \end{pmatrix}.$$

PROOF. 1. Let $g(\lambda)$ be a nonzero polynomial of minimum degree in $\boldsymbol{C}$. Suppose that $g_1(\lambda)$ is another polynomial in $\boldsymbol{C}$ such that $\deg g_1(\lambda) = \deg g(\lambda)$. Then the polynomial $g(\lambda) \oplus g_1(\lambda)$, which is nonzero, belongs to the code $\boldsymbol{C}$ and has degree less than $\deg g(\lambda)$. This contradicts the minimality of the degree of the polynomial $g(\lambda)$. Therefore, $g(\lambda)$ is unique.

2. Suppose that $c(\lambda) \in \boldsymbol{C}$. Since $\deg g(\lambda) \leq \deg c(\lambda)$, the Euclidean algorithm shows that $c(\lambda) = g(\lambda)q(\lambda) \oplus r(\lambda)$ in $\mathbb{F}_2[\lambda]$, where $\deg r(\lambda) < \deg g(\lambda)$, $0 \leq \deg q(\lambda) \leq n - 1 - \deg g(\lambda)$. It is clear that $c(\lambda) \equiv g(\lambda)q(\lambda) \oplus r(\lambda)$, whence the polynomial $r(\lambda) \equiv c(\lambda) \oplus g(\lambda)q(\lambda)$ belongs to the code $\boldsymbol{C}$. This is possible only if $r(\lambda) = 0$. Therefore, $g(\lambda)$ divides $c(\lambda)$ and $\boldsymbol{C} = \langle g(\lambda) \rangle$.

3. Suppose that $\lambda^n \oplus 1 = g(\lambda)h(\lambda) \oplus r(\lambda)$ in $\mathbb{F}_2[\lambda]$. An argument similar to that from the proof of part 2 shows that $r(\lambda) = 0$, whence $g(\lambda) \mid \lambda^n \oplus 1$.

4. In view of parts 2 and 3 the code $\boldsymbol{C}$ contains different polynomials of the type $a(\lambda)g(\lambda)$, where $a(\lambda) \in \mathbb{F}_2[\lambda]$ and $\deg a(\lambda) < n - \deg g(\lambda)$. On the other hand, suppose that $c(\lambda) \in \langle g(\lambda) \rangle$. Then $c(\lambda) \equiv f(\lambda)g(\lambda)$ in $\mathfrak{R}_n$. Hence, $c(\lambda) = f(\lambda)g(\lambda) \oplus d(\lambda)(\lambda^n \oplus 1)$ in $\mathbb{F}_2[\lambda]$ for some polynomial $d(\lambda)$. Since $g(\lambda) \mid \lambda^n \oplus 1$,

$$c(\lambda) = \Big(f(\lambda) \oplus \frac{(\lambda^n \oplus 1)d(\lambda)}{g(\lambda)}\Big)g(\lambda)$$

in the ring $\mathbb{F}_2[\lambda]$. Thus, the polynomial $c(\lambda)$ is represented in the ring $\mathbb{F}_2[\lambda]$ by a product of the polynomial $g(\lambda)$ and some other polynomial, which proves part 4.

5. Immediate from part 4. □

DEFINITION 4.54. If $\boldsymbol{C} = \langle g(\lambda) \rangle$ is a nonzero cyclic code in $\mathfrak{R}_n$ (i.e., a nonzero ideal in $\mathfrak{R}_n$), the polynomial $g(\lambda)$ is called the *generator polynomial of the cyclic code* $\boldsymbol{C}$.

From what was said it follows that a cyclic code of length n over the field $\mathbb{F}_2$ exists for any polynomial $g(\lambda) \in \mathbb{F}_2[\lambda]$ which divides $\lambda^n \oplus 1$ (we regard the polynomial $\lambda^n \oplus 1$, which is equal to 0 in $\mathfrak{R}_n$, as a generator polynomial of the zero cyclic code $\{0\}$). Therefore, it is necessary to study the generator polynomial in detail. The most natural approach is to factorize $\lambda^n \oplus 1$ into irreducible factors,

$$\lambda^n \oplus 1 = d_1(\lambda) \cdot d_2(\lambda) \cdots d_s(\lambda),$$

where s is the number of such factors. The product of an arbitrary subset of these factors gives the generator polynomial $g(\lambda)$ of a cyclic code. From now on in this chapter we will deal with a certain value of n called the primitive length of the code.

DEFINITION 4.55. If the length n of a cyclic code $\boldsymbol{C}$ over $\mathbb{F}_2$ is of the form $n = 2^m - 1$, it is called a *primitive*. A cyclic code with primitive length over $\mathbb{F}_2$ is called a *primitive cyclic code*.

The polynomial $\lambda^{2^m-1} - 1$ is coprime with its derivative as $\gcd(2^m - 1, 2) = 1$. Its roots are all the nonzero elements of the field $\mathbb{F}_{2^m}$, and its factorization into prime factors has, according to Lemma 1.93 and Theorem 1.123, the form

$$\lambda^{2^m-1} \oplus 1 = \prod_s m_s(\lambda),$$

where s runs over all representatives of 2-cyclotomic classes (Definition 1.146) modulo $n = 2^m - 1$, and $m_s(\lambda) = m_{\alpha^s}(\lambda)$ is the minimal polynomial of the element $b = \alpha^s$, where α is a primitive element of the field $\mathbb{F}_{2^m}$.

EXAMPLE 4.56. Let $m = 4$ and $n = 2^4 - 1 = 15$. The polynomial $\lambda^{15} \oplus 1$ factors as follows:

$$\lambda^{15} \oplus 1 = (\lambda \oplus 1)(\lambda^2 \oplus \lambda \oplus 1)(\lambda^4 \oplus \lambda \oplus 1)(\lambda^4 \oplus \lambda^3 \oplus 1)(\lambda^4 \oplus \lambda^3 \oplus \lambda^2 \oplus \lambda \oplus 1).$$

There are $2^5 = 32$ such factors and thus 32 generator polynomials of cyclic codes of length 15. Two of them ($g(\lambda) = \lambda^{15} \oplus 1$ with $k = 0$ and $g(\lambda) = 1$ with $k = n$) are trivial and 30 generate nontrivial cyclic codes.

PROBLEM 4.57. Let $\boldsymbol{C}_1$ and $\boldsymbol{C}_2$ be primitive cyclic codes of equal length with generator polynomials $g_1(\lambda)$ and $g_2(\lambda)$ respectively. Prove that $\boldsymbol{C}_1 \subseteq \boldsymbol{C}_2$ if and only if $g_2(\lambda)$ divides $g_1(\lambda)$.

PROBLEM 4.58. Suppose that $f(\lambda) = \prod_{s \in S}(\lambda - \alpha^s)$, where α is a primitive element of the field $\mathbb{F}_{2^m}$ and S is a subset of the set $\{0, 1, \dots, 2^m - 2\}$. Prove that the coefficients of $f(\lambda)$ lie in $\mathbb{F}_2$ if and only if $s \in S$ implies $2s \pmod{2^m - 1} \in S$ (i.e., S is a union of cyclotomic classes modulo $2^m - 1$).

PROBLEM 4.59. Prove that if $h(\lambda) = \prod_{s \in S}(\lambda - \alpha^s)$ divides $\lambda^{2^m-1} \oplus 1$ over $\mathbb{F}_2$, then

$$\lambda^{\deg h(\lambda)} h(\lambda^{-1}) = \prod_{s \in S} (\lambda - \alpha^{-s}).$$

DEFINITION 4.60. Suppose that $\boldsymbol{C}$ is a primitive cyclic code of length $n = 2^m - 1$ with generator polynomial

$$g(\lambda) = \prod_{s \in S} (\lambda - \alpha^s).$$

The elements $\{\alpha^s \mid s \in S\}$ from $\mathbb{F}_{2^m}$ are called the *zeros of the code* $\boldsymbol{C}$. The rest of the elements $\mathbb{F}^*_{2^m} \setminus \{\alpha^s \mid s \in S\}$ are called the *nonzeros* of the code $\boldsymbol{C}$.

It is easy to see that either the set of zeros of a code or the set of its nonzeros is a union of cyclotomic classes (see Problem 4.58). The set of zeros of the code $\boldsymbol{C}$ coincides with the set of roots of the polynomial

$$h(\lambda) = \frac{\lambda^{2^m-1} \oplus 1}{g(\lambda)}.$$

DEFINITION 4.61. Suppose that $\boldsymbol{C}$ is a primitive cyclic code of length $n = 2^m - 1$ with generator polynomial $g(\lambda)$. Then the primitive cyclic code $\boldsymbol{C}'$ of the same length with generator polynomial $h(\lambda) = (\lambda^{2^m-1} \oplus 1)/g(\lambda)$ is called *complementary* of the code $\boldsymbol{C}$. The polynomial $h(\lambda)$ is called the *check* polynomial of the code $\boldsymbol{C}$.

Now we can characterize the code words of a cyclic code in terms of the roots of its generator polynomial.

LEMMA 4.62. *Suppose that the elements $b_1, b_2, \dots, b_r \in \mathbb{F}_{2^m}$ are roots of the generator polynomial $g(\lambda)$ of a primitive cyclic code $\boldsymbol{C}$. A polynomial $c(\lambda) \in \mathbb{F}_2[\lambda]$, $\deg c(\lambda) < 2^m - 1$, is a code word if and only if*

$$c(b_1) = c(b_2) = \cdots = c(b_r) = 0.$$

PROOF. If $c(\lambda) \in \boldsymbol{C}$, then $c(\lambda) = a(\lambda)g(\lambda)$ and we have $c(b_j) = a(b_j)g(b_j) = 0$ for any $j = 1, 2, \dots, r$.

Suppose that $c(b_j) = 0$, $j = 1, 2, \dots, r$. Using the Euclidean algorithm we get

$$c(\lambda) = a(\lambda) m_{b_j}(\lambda) \oplus t(\lambda),$$

where the degree of the polynomial $t(\lambda)$ is smaller than that of the polynomial $m_{b_j}(\lambda)$ and $m_{b_j}(\lambda)$ is the minimal polynomial of the element b_j. Now it follows from the equalities

$$0 = c(b_j) = a(b_j) m_{b_j}(b_j) \oplus t(b_j) = t(b_j)$$

that $t(\lambda) = 0$. Thus, $c(\lambda)$ must be divisible by $m_{b_j}(\lambda)$ for all $j = 1, 2, \dots, r$ and also by $\operatorname{lcm}[m_{b_1}(\lambda), m_{b_2}(\lambda), \dots, m_{b_r}(\lambda)] = g(\lambda)$. □

LEMMA 4.63. *Suppose that* $\boldsymbol{C}$ *is a primitive cyclic code of length* $n = 2^m - 1$ *with generator polynomial* $g(\lambda)$ *and let*

$$h(\lambda) = \frac{\lambda^{2^m-1} \oplus 1}{g(\lambda)} = \bigoplus_{j=0}^{k} h_j \lambda^j, \quad h_k = 1,$$

be its check polynomial. Then the parity-check matrix of the code $\boldsymbol{C}$ *has the form*

$$H = \begin{pmatrix} 0 & \dots & 0 & h_k & \dots & h_2 & h_1 & h_0 \\ 0 & \dots & h_k & h_{k-1} & \dots & h_1 & h_0 & 0 \\ \dots & \dots & \dots & \dots & \dots & \dots & \dots & \dots \\ h_k & \dots & h_2 & h_1 & h_0 & 0 & \dots & 0 \end{pmatrix}. \tag{4.5}$$

PROOF. Suppose that $c(\lambda) = \bigoplus_{i=0}^{n-1} c_i \lambda^i = a(\lambda)g(\lambda)$ is an arbitrary code word of $\boldsymbol{C}$. Then

$$c(\lambda) \cdot h(\lambda) \equiv \left[\bigoplus_{i=0}^{n-1} c_i \lambda^i\right] \cdot \left[\bigoplus_{j=0}^{k} h_j \lambda^j\right] \equiv a(\lambda)g(\lambda)h(\lambda) \equiv 0.$$

The coefficient of λ^j in the last formula is

$$\bigoplus_{i=0} c_i h_{j-i} = 0, \quad j = 0, 1, \dots, n-1, \tag{4.6}$$

where the indices are taken modulo n. From (4.6) we get the check equations which must be satisfied by the code words of $\boldsymbol{C}$. It is clear from these equations that if $\boldsymbol{c} \in \boldsymbol{C}$, then for the matrix H (see (4.5)) one has $H\boldsymbol{c}^\top = 0$. Since $k = \deg h(\lambda) = n - \deg g(\lambda)$ is the dimension of the code $\boldsymbol{C}$ and since the rows of the matrix H are, obviously, linearly independent, the condition $H\boldsymbol{c}^\top = 0$ is also necessary for a word $\boldsymbol{c}$ to belong to the code $\boldsymbol{C}$. □

THEOREM 4.64. *Suppose that* $\boldsymbol{C}$ *is a primitive cyclic code with generator polynomial* $g(\lambda)$ *and check polynomial* $h(\lambda) = \frac{\lambda^{2^m-1}\oplus 1}{g(\lambda)}$. *Then the dual code* $\boldsymbol{C}^\perp$ *is also a primitive cyclic code with generator polynomial*

$$g^\perp(\lambda) = \lambda^{\deg h(\lambda)} h(\lambda^{-1}) = h^*(\lambda),$$

where $h^*(\lambda)$ *is reciprocal to* $h(\lambda)$ *(see Definition* 1.135*). The nonzeros of the code* $\boldsymbol{C}^\perp$ *are the elements* $\{b_1^{-1}, \dots, b_r^{-1}\}$, *where* $\{b_1, \dots, b_r\}$ *are the zeros of* $\boldsymbol{C}$.

PROOF. According to Theorem 4.16, the matrix H (see (4.5)) is a generator matrix for the code $\boldsymbol{C}^\perp$; since H looks exactly like matrices of cyclic codes, the code $\boldsymbol{C}^\perp$ is a primitive cyclic code. The coefficients of its generator polynomial $g^\perp(\lambda)$ are also coefficients of the check polynomial $h(\lambda)$ written in reverse order, i.e.,

$$g^\perp(\lambda) = \lambda^{\deg h(\lambda)} h(\lambda^{-1}).$$

Let b_i, $i = 1, \dots, r$, be zeros of the code $\boldsymbol{C}$. Then b_i, $i = 1, \dots, r$, are not roots of the polynomial $h(\lambda)$. According to the result of Problem 4.59, the elements b_i^{-1}, $i = 1, \dots, r$, are not roots of the polynomial $\lambda^{\deg h(\lambda)} h(\lambda^{-1}) = g^\perp(\lambda)$. Therefore, b_i^{-1}, $i = 1, \dots, r$, are nonzeros of the code $\boldsymbol{C}^\perp$. In other words, if $\{b_i\} = \{\alpha^{s_i}\}$ (where α is a primitive element of the field) are zeros of the code $\boldsymbol{C}$, where s_i runs over the cyclotomic classes $\mathcal{C}_{m,u_1}, \mathcal{C}_{m,u_2}, \dots$, then the nonzeros of the code $\boldsymbol{C}^\perp$ are the elements $\{b_i^{-1}\} = \{\alpha^{s'_i}\}$, where s'_i runs over the cyclotomic classes $\mathcal{C}_{m,-u_1}, \mathcal{C}_{m,-u_2}, \dots$. □

So far we have chosen as a generator polynomial $g(\lambda)$ of the cyclic code $\boldsymbol{C}$ the polynomial of minimum degree in $\boldsymbol{C}$. However, other choices of generator polynomials are possible.

LEMMA 4.65. *Suppose that $\boldsymbol{C}$ is a primitive cyclic code of length $n = 2^m - 1$, $m > 0$, over $\mathbb{F}_2$, with generator polynomial $g(\lambda)$. Then the polynomial $k(\lambda) \in \mathbb{F}_2[\lambda]$ is a generator polynomial for the code $\boldsymbol{C}$ iff* $\operatorname{lcm}(k(\lambda), \lambda^n \oplus 1) = g(\lambda)$ *(in other words, the polynomial $k(\lambda)$ generates the code $\boldsymbol{C}$ if and only if the roots of the polynomial $k(\lambda)$ coincide with the zeros of the code $\boldsymbol{C}$).*

PROOF. Suppose first that $\operatorname{lcm}(k(\lambda), \lambda^n \oplus 1) = g(\lambda)$. Then there exist polynomials $a(\lambda), b(\lambda) \in \mathbb{F}_2[\lambda]$ such that

$$g(\lambda) = a(\lambda)k(\lambda) \oplus b(\lambda)(\lambda^n \oplus 1).$$

Therefore, $g(\lambda) \equiv a(\lambda)k(\lambda)$ in $\mathfrak{R}_n$ and $\boldsymbol{C} = \langle k(\lambda) \rangle$.

Suppose now that $\boldsymbol{C} = \langle k(\lambda) \rangle$. Then there exist polynomials $c(\lambda), d(\lambda) \in \mathfrak{R}_n$ such that $g(\lambda) = c(\lambda)k(\lambda) \oplus d(\lambda)(\lambda^n \oplus 1)$ in $\mathbb{F}_2[\lambda]$. Therefore, any root of the polynomial $k(\lambda)$ is also a root of the polynomial $g(\lambda)$, i.e., a zero of the code $\boldsymbol{C}$. □

DEFINITION 4.66. An element $e(\lambda)$ of the ring $\mathfrak{R}_n$ is called an *idempotent* if $e^2(\lambda) \equiv e(\lambda)$ in the ring $\mathfrak{R}_n$. An idempotent $e(\lambda)$ of the ring $\mathfrak{R}_n$ is called *proper* if $e(\lambda) \not\equiv 0, 1$ (see Definition 1.117).

EXAMPLE 4.67. If $n = 2^3 - 1 = 7$, then the polynomials

$$e'(\lambda) = 1, \quad e''(\lambda) = \lambda \oplus \lambda^2 \oplus \lambda^4, \quad e'''(\lambda) = \lambda^3 \oplus \lambda^5 \oplus \lambda^6$$

are idempotents in the ring $\mathfrak{R}_n$.

LEMMA 4.68. *A polynomial $e(\lambda) \in \mathfrak{R}_n$ is idempotent if and only if*

$$e(\lambda) = \bigoplus_{s \in S} \bigoplus_{i \in \mathcal{C}_{n,s}} \lambda^i, \tag{4.7}$$

where S is a subset of the set of representatives of 2-cyclotomic classes modulo n.

PROOF. If $e(\lambda)$ has the form (4.7), it is easy to check that $e^2(\lambda) \equiv e(\lambda)$, so $e(\lambda)$ is an idempotent in the ring $\mathfrak{R}_n$.

Conversely, let $e(\lambda) = e_0 \oplus e_1\lambda \oplus \cdots \oplus e_{n-1}\lambda^{n-1}$ be an idempotent in the ring $\mathfrak{R}_n$. Then

$$e^2(\lambda) = [e_0 \oplus e_1\lambda \oplus \cdots \oplus e_{n-1}\lambda^{n-1}]^2 = e_0 \oplus e_1\lambda^2 \oplus \cdots \oplus e_{n-1}\lambda^{2(n-1)} = e(\lambda).$$

Comparing the coefficients at equal powers of λ in these polynomials for any representative s of 2-cyclotomic class, we see that $e_s = e_{2s} = \cdots$. Hence, $e(\lambda)$ has the form (4.7). □

LEMMA 4.69. *Let α be a primitive element of the field $\mathbb{F}_{2^m}$, $n = 2^m - 1$. A polynomial $e(\lambda) \in \mathfrak{R}_n$ is an idempotent if and only if $e(\alpha^j) = 0$ or 1 for all $j = 0, 1, \ldots, n-1$.*

PROOF. Suppose that $e(\lambda)$ is idempotent in $\mathfrak{R}_n$. Then for any $j = 0, 1, \ldots, n-1$ we have $e^2(\alpha^j) = e(\alpha^j)$, whence the value of $e(\alpha^j)$ belongs to the field $\mathbb{F}_2$.

Conversely, if for any $j = 0, 1, \ldots, n-1$ the value of $e(\alpha^j)$ is equal to 0 or 1, then $e(\alpha^{2j}) = e^2(\alpha^j) = e(\alpha^j)$. Look at the sum Q_i $(i = 0, 1, \ldots, n-1)$:

$$Q_i = \sum_{j=0}^{n-1} e(\alpha^j)\alpha^{-ij} = \sum_{j=0}^{n-1}\sum_{t=0}^{n-1} e_t \alpha^{j(t-i)} = \sum_{t=0}^{n-1} e_t \sum_{j=0}^{n-1} \alpha^{j(t-i)}. \tag{4.8}$$

According to Lemma 1.144, the inner sum in (4.8) is equal to zero if $t \neq i$, whence $Q_i = ne_i = e_i$. Since $\gcd(2, n) = 1$, one has

$$e_{2i} = Q_{2i} = \sum_{j=0}^{n-1} e(\alpha^j)\alpha^{-2ij} = \sum_{j=0}^{n-1} e(\alpha^{2j})\alpha^{-i(2j)} = \sum_{j'=0}^{n-1} e(\alpha^{j'})\alpha^{-ij'} = e_i.$$

Thus, $e(\lambda) = \bigoplus_{s\in S} \bigoplus_{j\in C_{n,s}} \lambda^j$. By virtue of Lemma 4.68, the polynomial $e(\lambda)$ is an idempotent in the ring $\mathfrak{R}_n$. □

PROBLEM 4.70. Prove that if $c(\lambda)$ is an idempotent, then $1 \oplus e(\lambda)$ is also an idempotent.

THEOREM 4.71. *Suppose that $\boldsymbol{C}$ is a primitive cyclic code in $\mathfrak{R}_n$. Then there exists a unique idempotent $e(\lambda) \in \boldsymbol{C}$ such that $\boldsymbol{C} = \langle e(\lambda) \rangle$. If $e(\lambda)$ is a nonzero idempotent, then $\boldsymbol{C} = \langle e(\lambda)\rangle$ if and only if $e(\lambda)$ is the identity in $\boldsymbol{C}$.*

PROOF. Suppose that $g(\lambda)$ is a generator polynomial of the code $\boldsymbol{C}$, $\lambda^n \oplus 1 = g(\lambda)h(\lambda)$, where $g(\lambda)$ and $h(\lambda)$ are coprime. Then there exist polynomials $a(\lambda)$ and $b(\lambda)$ in $\mathbb{F}_2[\lambda]$ such that

$$a(\lambda)g(\lambda) \oplus b(\lambda)h(\lambda) = 1. \tag{4.9}$$

Set $e(\lambda) = a(\lambda)g(\lambda)$. It follows from (4.9) that

$$a(\lambda)g(\lambda)[a(\lambda)g(\lambda) \oplus b(\lambda)h(\lambda)] \equiv a^2(\lambda)g^2(\lambda) \equiv a(\lambda)g(\lambda),$$

i.e., $e^2(\lambda) \equiv e(\lambda)$ in $\mathfrak{R}_n$ and $e(\lambda)$ is an idempotent.

Suppose that $\boldsymbol{C} = \langle e(\lambda)\rangle$ and $e(\lambda)$ is an idempotent. We can present any polynomial $c(\lambda) \in \boldsymbol{C}$ in the form $c(\lambda) = a(\lambda)e(\lambda)$. Then $e(\lambda) \cdot c(\lambda) \equiv e^2(\lambda)a(\lambda) \equiv e(\lambda)a(\lambda) \equiv c(\lambda)$, i.e., $e(\lambda)$ is the identity in $\boldsymbol{C}$.

The converse statement is obvious. □

For a polynomial $a(\lambda) = a_0 \oplus a_1\lambda \oplus \cdots \oplus a_{n-1}\lambda^{n-1} \in \mathfrak{R}_n$ we define the polynomial $\widetilde{a}(\lambda) = a_0 \oplus a_{n-1}\lambda \oplus \cdots \oplus a_1\lambda^{n-1}$ (all coefficients except the first are written in reverse order). It is clear that $a(\alpha^{-1}) = \widetilde{a}(\alpha)$, $\alpha \in \mathbb{F}_{2^m}$.

PROBLEM 4.72. Prove that if $e(\lambda)$ is an idempotent, then so is $\widetilde{e}(\lambda)$.

PROBLEM 4.73. Suppose that $\boldsymbol{C} = \langle g(\lambda)\rangle = \langle e(\lambda)\rangle$, $\lambda^n \oplus 1 = g(\lambda)h(\lambda)$, and $e(\lambda)$ is an idempotent. Prove that $1 \oplus e(\lambda)$ is an idempotent of the code with generator polynomial $h(\lambda)$.

PROBLEM 4.74. Suppose that $\boldsymbol{C}$ is a cyclic code with generator idempotent $e(\lambda)$. Prove that the code $\boldsymbol{C}^\perp$ has a generator idempotent $(\widetilde{1 \oplus e(\lambda)})$.

In Table 4.2 all nonzero primitive cyclic codes $\boldsymbol{C}_i$ of length 7 are listed.

Two of the codes of dimension 4 in the table turn out to be Hamming $[7, 4, 3]$-codes.

TABLE 4.2

i	dim	$g_i(\lambda)$	$e_i(\lambda)$
1	1	$1 \oplus \lambda \oplus \lambda^2 \oplus \cdots \oplus \lambda^6$	$1 \oplus \lambda \oplus \lambda^2 \oplus \cdots \oplus \lambda^6$
2	3	$1 \oplus \lambda^2 \oplus \lambda^3 \oplus \lambda^4$	$1 \oplus \lambda^3 \oplus \lambda^5 \oplus \lambda^6$
3	3	$1 \oplus \lambda \oplus \lambda^2 \oplus \lambda^4$	$1 \oplus \lambda \oplus \lambda^2 \oplus \lambda^4$
4	4	$1 \oplus \lambda \oplus \lambda^3$	$\lambda \oplus \lambda^2 \oplus \lambda^4$
5	4	$1 \oplus \lambda^2 \oplus \lambda^3$	$\lambda^3 \oplus \lambda^5 \oplus \lambda^6$
6	6	$1 \oplus \lambda$	$\lambda \oplus \lambda^2 \oplus \cdots \oplus \lambda^6$
7	7	1	1

LEMMA 4.75. *Suppose that $\boldsymbol{C}$ is a primitive cyclic $[n,k]$-code with generator idempotent $e(\lambda) = \bigoplus_{i=0}^{n-1} e_i\lambda^i$, $e(\lambda) \neq 0$. Then the $(k \times n)$ matrix*

$$\begin{pmatrix} e_0 & e_1 & e_2 & \cdots & e_{n-2} & e_{n-1} \\ e_{n-1} & e_0 & e_1 & \cdots & e_{n-3} & e_{n-2} \\ \cdots & \cdots & \cdots & \cdots & \cdots & \cdots \\ e_{n-k+1} & e_{n-k+2} & e_{n-k+3} & \cdots & e_{n-k-1} & e_{n-k} \end{pmatrix}$$

is a generator matrix of the code $\boldsymbol{C}$.

PROOF. It is enough to show that if $a(\lambda) \in \mathbb{F}_2[\lambda]$, $\deg a(\lambda) < k$ and $e(\lambda)a(\lambda) \equiv 0$, then $a(\lambda) = 0$. This follows immediately from Theorem 4.71. □

Suppose that $\boldsymbol{C}_1$ and $\boldsymbol{C}_2$ are linear codes. Then

$$\boldsymbol{C}_1 \cap \boldsymbol{C}_2 = \{\boldsymbol{c} \mid \boldsymbol{c} \in \boldsymbol{C}_1 \text{ and } \boldsymbol{c} \in \boldsymbol{C}_2\}$$

is the *intersection of codes* $\boldsymbol{C}_1$ and $\boldsymbol{C}_2$, and

$$\boldsymbol{C}_1 \oplus \boldsymbol{C}_2 = \{\boldsymbol{c} \oplus \boldsymbol{c}' \mid \boldsymbol{c} \in \boldsymbol{C}_1, \boldsymbol{c}' \in \boldsymbol{C}_2\}$$

is the *sum of codes* $\boldsymbol{C}_1$ and $\boldsymbol{C}_2$. The resulting codes are also linear: it is easy to show that they are linear subspaces.

PROBLEM 4.76. Suppose that $\boldsymbol{C}_1$ and $\boldsymbol{C}_2$ are primitive cyclic codes. Prove that $\boldsymbol{C}_1 \cap \boldsymbol{C}_2$ and $\boldsymbol{C}_1 \oplus \boldsymbol{C}_2$ are also cyclic codes.

THEOREM 4.77. *Let $\boldsymbol{C}_1$ and $\boldsymbol{C}_2$ be primitive cyclic codes of length $n = 2^m - 1$, $m > 0$ with generator polynomials $g_1(\lambda)$, $g_2(\lambda)$ and generator idempotents $e_1(\lambda)$, $e_2(\lambda)$ respectively. Then*

1. *The primitive cyclic code $\boldsymbol{C}_1 \cap \boldsymbol{C}_2$ has generator polynomial* $\gcd(g_1(\lambda), g_2(\lambda))$ *and generator idempotent $e_1(\lambda)e_2(\lambda)$.*
2. *The primitive cyclic code $\boldsymbol{C}_1 \oplus \boldsymbol{C}_2$ has generator polynomial* $\gcd(g_1(\lambda), g_2(\lambda))$ *and generator idempotent $e_1(\lambda) \oplus e_2(\lambda) \oplus e_1(\lambda)e_2(\lambda)$.*

PROOF. 1. Suppose that $l(\lambda) = \gcd(g_1(\lambda), g_2(\lambda))$ and $c(\lambda) \in \boldsymbol{C}_1 \cap \boldsymbol{C}_2$. Since $c(\lambda) \in \boldsymbol{C}_1$, one has $c(\lambda) \equiv a(\lambda)g_1(\lambda)$ in $\mathfrak{R}_n$ for some polynomial $a(\lambda)$. Similarly, since $c(\lambda) \in \boldsymbol{C}_2$, one has $c(\lambda) \equiv b(\lambda)g_2(\lambda)$ in $\mathfrak{R}_n$ for some polynomial $b(\lambda)$. Then $c(\lambda) = a(\lambda)g_1(\lambda) \oplus (\lambda^n \oplus 1)u(\lambda)$ and $c(\lambda) = b(\lambda)g_2(\lambda) \oplus (\lambda^n \oplus 1)v(\lambda)$ in $\mathbb{F}_2[\lambda]$. Thus, $c(\lambda)$ is divisible by $g_1(\lambda)$ and $g_2(\lambda)$. Therefore, $c(\lambda)$ is also divisible by $l(\lambda)$, i.e., $c(\lambda) \in \langle l(\lambda)\rangle$ and $\boldsymbol{C}_1 \cap \boldsymbol{C}_2 \subseteq \langle l(\lambda)\rangle$. On the other hand, $\langle l(\lambda)\rangle \subset \langle g_1(\lambda)\rangle$ and $\langle l(\lambda)\rangle \subset \langle g_2(\lambda)\rangle$. We get $\boldsymbol{C}_1 \cap \boldsymbol{C}_2 \supseteq \langle l(\lambda)\rangle$, whence $\boldsymbol{C}_1 \cap \boldsymbol{C}_2 = \langle l(\lambda)\rangle$.

Since $e_1(\lambda)$ and $e_2(\lambda)$ are idempotents in $\mathfrak{R}_n$, it is clear that $e_1(\lambda) \cdot e_2(\lambda)$ is also an idempotent in the same ring. Since $\boldsymbol{C}_1 = \langle e_1(\lambda)\rangle \supset \langle e_1(\lambda) \cdot e_2(\lambda)\rangle$ and

$\boldsymbol{C}_2 = \langle e_2(\lambda)\rangle \supset \langle e_1(\lambda)\cdot e_2(\lambda)\rangle$, one has $\boldsymbol{C}_1 \cap \boldsymbol{C}_2 \supset \langle e_1(\lambda)\cdot e_2(\lambda)\rangle$. Obviously $e_1(\lambda)\cdot e_2(\lambda)$ is the identity of the ideal $\boldsymbol{C}_1\cap\boldsymbol{C}_2$, and by Theorem 4.71, it generates the whole ideal, i.e., $\boldsymbol{C}_1 \cap \boldsymbol{C}_2 = \langle e_1(\lambda)\cdot e_2(\lambda)\rangle$.

2. Suppose that $g(\lambda) = \gcd(g_1(\lambda), g_2(\lambda))$. The Euclidean algorithm in $\mathbb{F}_2[\lambda]$ shows that there exist polynomials $u(\lambda)$ and $v(\lambda)$ such that

$$g(\lambda) = g_1(\lambda)u(\lambda) \oplus g_2(\lambda)v(\lambda).$$

Hence,

$$\langle g(\lambda)\rangle = \{g_1(\lambda)a(\lambda) \oplus g_2(\lambda)b(\lambda) \mid a(\lambda), b(\lambda) \in \mathfrak{R}_n\} = \boldsymbol{C}_1 \oplus \boldsymbol{C}_2.$$

Since $g(\lambda)$ divides $\lambda^n \oplus 1$, $g(\lambda)$ is a generator polynomial for the code $\boldsymbol{C}_1 \oplus \boldsymbol{C}_2$.

The idempotent $e_1(\lambda) \oplus e_2(\lambda) \oplus e_1(\lambda)e_2(\lambda)$ obviously belongs to $\boldsymbol{C}_1 \oplus \boldsymbol{C}_2$. Besides, if $c(\lambda) \equiv c_1(\lambda) \oplus c_2(\lambda)$, where $c_i(\lambda) \in \boldsymbol{C}_i$, $i = 1, 2$, then

$$\begin{aligned} &c(\lambda)[e_1(\lambda) \oplus e_2(\lambda) \oplus e_1(\lambda)e_2(\lambda)] \\ &\quad\equiv c_1(\lambda) \oplus c_1(\lambda)e_2(\lambda) \oplus c_1(\lambda)e_2(\lambda) \oplus c_2(\lambda) \oplus c_2(\lambda)e_1(\lambda) \oplus c_2(\lambda)e_1(\lambda) \\ &\quad\equiv c_1(\lambda) \oplus c_2(\lambda) \equiv c(\lambda); \end{aligned}$$

in view of Theorem 4.71 this implies the desired statement. □

According to Lemma 1.93 and Theorem 1.123, the factorization of the polynomial $\lambda^n \oplus 1 = \lambda^{2^m-1} \oplus 1$ into irreducible factors is

$$\lambda^{2^m-1} \oplus 1 = \prod_s m_s(\lambda),$$

where s runs over the set of representatives of 2-cyclotomic classes modulo $n = 2^m - 1$, and $m_s(\lambda) = m_{\alpha^s}(\lambda)$ is a minimal polynomial of $b = \alpha^s$ (α is a primitive element of the field $\mathbb{F}_{2^m}$). Put $\widehat{m}_s(\lambda) = (\lambda^{2^m-1} \oplus 1)/m_s(\lambda)$. In the following theorem we will show that the ideals $\langle \widehat{m}_s(\lambda)\rangle$ are minimal ideals in the ring $\mathfrak{R}_n$. Denote the generator idempotent of the code $\langle \widehat{m}_s(\lambda)\rangle$ by $\widehat{e}_s(\lambda)$.

DEFINITION 4.78. Primitive cyclic codes which are minimal ideals in $\mathfrak{R}_n$ are called *minimal (irreducible) primitive cyclic codes*. Their generator idempotents are called *primitive*.

Suppose that $\{s_1, s_2, \dots, s_t\}$ is a set of representatives of 2-cyclotomic classes modulo $n = 2^m - 1$.

THEOREM 4.79. *Primitive idempotents of the ring $\mathfrak{R}_n$ have the following properties.*

1. *Only the ideals $\langle \widehat{m}_{s_i}(\lambda)\rangle$, $i = 1, 2, \dots, t$, are minimal ideals of the ring $\mathfrak{R}_n$.*
2. *If $i \neq j$, then $\widehat{e}_{s_i}(\lambda)\cdot\widehat{e}_{s_j}(\lambda) \equiv 0$ in the ring $\mathfrak{R}_n$.*
3. *The ring $\mathfrak{R}_n$ regarded as a vector space over $\mathbb{F}_2$ is a direct sum of its minimal ideals.*
4. $\bigoplus_{i=1}^t \widehat{e}_{s_i}(\lambda) = 1$.

PROOF. 1. Suppose that $\langle g(\lambda)\rangle$ is a minimal ideal of the ring $\mathfrak{R}_n$ and $g(\lambda)$ is its generator. Then the polynomial $h(\lambda) = (\lambda^{2^m-1} \oplus 1)/g(\lambda)$ must be irreducible and the polynomial $\lambda^{2^m-1} \oplus 1$ coincides with $m_{s_i}(\lambda)$ for some i. Thus, $g(\lambda) = \widehat{m}_{s_i}(\lambda)$. Arguing in the reverse order, we obtain that $\langle \widehat{m}_{s_i}(\lambda)\rangle$ is a minimal ideal of the ring $\mathfrak{R}_n$.

2. It is clear that if $i \neq j$, then according to part 1, $\langle \widehat{m}_{s_i}(\lambda) \rangle \cap \langle \widehat{m}_{s_j}(\lambda) \rangle = \{0\}$. On the other hand, $\widehat{e}_{s_i}(\lambda)\widehat{e}_{s_j}(\lambda)$ belongs either to the ideal $\langle \widehat{m}_{s_i}(\lambda) \rangle$ or to the ideal $\langle \widehat{m}_{s_j}(\lambda) \rangle$. Therefore, $\widehat{e}_{s_i}(\lambda)\widehat{e}_{s_j}(\lambda) \equiv 0$ in the ring $\mathfrak{R}_n$.

3. Since $\gcd(\widehat{m}_{s_1}(\lambda), \widehat{m}_{s_2}(\lambda), \dots, \widehat{m}_{s_t}(\lambda)) = 1$, the Euclidean algorithm shows that there exist polynomials $a_i(\lambda)$, $i = 1, 2, \dots, t$, from $\mathbb{F}_2[\lambda]$ such that

$$\bigoplus_{i=1}^{t} a_i(\lambda)\widehat{m}_{s_i}(\lambda) = 1.$$

Therefore, $\mathfrak{R}_n$ is a sum of its minimal ideals $\langle \widehat{m}_{s_i}(\lambda) \rangle$. Theorem 4.77 shows that the polynomial generating the ideal $\bigoplus_{j \neq i} \langle \widehat{m}_{s_j}(\lambda) \rangle$ coincides with

$$\operatorname{lcm}(\widehat{m}_{s_1}(\lambda), \dots, \widehat{m}_{s_{i-1}}(\lambda), \widehat{m}_{s_{i+1}}(\lambda), \dots, \widehat{m}_{s_t}(\lambda)).$$

Since $\gcd(m_{s_i}(\lambda), \widehat{m}_{s_i}(\lambda)) = 1$, one has

$$\langle \widehat{m}_{s_i}(\lambda) \rangle \cap \bigoplus_{j \neq i} \langle \widehat{m}_{s_j}(\lambda) \rangle = \langle \widehat{m}_{s_i}(\lambda) \rangle \cap \{m_{s_i}(\lambda)\} = \{0\}.$$

Therefore, $\mathfrak{R}_n$ as a vector space is a direct sum of its minimal ideals.

4. From parts 2 and 3 it follows that $\bigoplus_{i=1}^{t} \widehat{e}_{s_i}(\lambda)$ is the identity in the ring $\mathfrak{R}_n$, i.e., that $\bigoplus_{i=1}^{t} \widehat{e}_{s_i}(\lambda) \equiv 1$. □

The polynomial $\widetilde{\widehat{e}}_s(\lambda)$ is also a primitive idempotent, so there exists an s' such that it is a representative of a 2-cyclotomic class modulo n, where $\widetilde{\widehat{e}}_s(\lambda) = \widehat{e}_{s'}(\lambda)$. Therefore, $s' \in \mathcal{C}_{n,-s}$. The set $\{\alpha^j \mid j \in \mathcal{C}_{n,-s}\}$ coincides with set of elements of the field $\mathbb{F}_{2^m}$ which are not roots of the polynomial $\widetilde{\widehat{e}}_s(\lambda)$.

In order to explain the relation between various primitive cyclic codes and its generator idempotents (polynomials) we will need a special mapping.

Put $n = 2^m - 1$, $m > 0$, and let d be an integer such that $1 \leq d \leq n$, $\gcd(n, d) = 1$. We define the mapping θ_d from the set $\{0, 1, \dots, n-1\}$ into itself by the formula

$$\theta_d(i) \equiv id \pmod{n}.$$

As d and n are coprime, the mapping θ_d is a permutation of the set $\{0, 1, \dots, n-1\}$. We extend the action of this mapping to the ring $\mathfrak{R}_n$ as follows. If $a(\lambda) \in \mathfrak{R}_n$, then

$$\theta_d(a(\lambda)) \equiv a(\lambda^d)$$

in the ring $\mathfrak{R}_n$. It is clear that $\theta_d(a(\lambda) \oplus b(\lambda)) \equiv \theta_d(a(\lambda)) \oplus \theta_d(b(\lambda))$ and $\theta_d(a(\lambda)b(\lambda)) \equiv \theta_d(a(\lambda))\theta_d(b(\lambda))$. Thus, the mapping θ_d is an endomorphism of the ring $\mathfrak{R}_n$. It is easy to show that the kernel of this endomorphism is trivial. Therefore, the mapping θ_d is an automorphism of $\mathfrak{R}_n$ and it maps idempotents to idempotents. The 2-cyclotomic classes modulo n are invariant under the mapping θ_2, so the order of θ_2 coincides with the multiplicative order of 2 modulo n.

THEOREM 4.80. *Let $\boldsymbol{C}$ be a primitive cyclic code of length $n = 2^m - 1$, $m > 0$, over the field $\mathbb{F}_2$, and let d be a positive integer, $\gcd(n, d) = 1$. Suppose that $e(\lambda)$ is a generator idempotent of this code. Then:*

1. $\theta_d \boldsymbol{C} = \langle \theta_d(e(\lambda)) \rangle$ and $\theta_d(e(\lambda))$ is a generator idempotent of the primitive cyclic code $\theta_d \boldsymbol{C}$ (i.e., θ_d maps primitive cyclic codes into primitive cyclic codes and generator idempotents into generator idempotents).

2. $\theta_2(e(\lambda)) \equiv e(\lambda)$ and $\theta_2 \in \operatorname{Aut}(\boldsymbol{C})$.

PROOF. 1. Apply the mapping θ_d to $\boldsymbol{C}$:

$$\begin{aligned}\theta_d\boldsymbol{C} &= \{\theta_d(e(\lambda)a(\lambda)) \mid a(\lambda) \in \mathfrak{R}_n\} = \{\theta_d(e(\lambda))\theta_d(a(\lambda)) \mid a(\lambda) \in \mathfrak{R}_n\} \\ &= \{\theta_d(e(\lambda))a(\lambda) \mid a(\lambda) \in \mathfrak{R}_n\} = \langle\theta_d(e(\lambda))\rangle\end{aligned}$$

since θ_d is an automorphism of the ring $\mathfrak{R}_n$. Therefore, $\theta_d\boldsymbol{C}$ is a primitive cyclic code and $\theta_d(e(\lambda))$ is its generator idempotent.

2. Put $\boldsymbol{C} = \langle g(\lambda)\rangle$, where $g(\lambda)$ is a generator of $\boldsymbol{C}$. The polynomial $g(\lambda)$ has roots $\{\alpha^s \mid s \in I\}$, where I is a union of 2-cyclotomic classes modulo n. Then $\theta_2(g(\lambda)) \equiv g(\lambda^2) \equiv [g(\lambda)]^2$ also has roots $\{\alpha^s \mid s \in I\}$. Thus, $\theta_2(g(\lambda)) \in \boldsymbol{C}$ and $\big\langle\theta_2(g(\lambda))\big\rangle \subset \boldsymbol{C}$. Since θ_2 is an automorphism of $\mathfrak{R}_n$, one has $\#\big\langle\theta_2(g(\lambda))\big\rangle = \#\boldsymbol{C}$. Therefore, $\big\langle\theta_2(g(\lambda))\big\rangle = \boldsymbol{C}$. Since the generator idempotent is unique, we obtain the second statement. □

LEMMA 4.81. *Let $\boldsymbol{C}$ be a primitive cyclic code of length n over $\mathbb{F}_2$. Let $\mathfrak{A}$ be the group of order n generated by the cyclic shift $i \mapsto i+1 \mod n$, and denote by $\mathfrak{B}$ the group generated by the mapping θ_2. Then the group $\mathfrak{G}$ generated by the groups $\mathfrak{A}$ and $\mathfrak{B}$ is a subgroup of* $\mathrm{Aut}(\boldsymbol{C})$ *of order nl, where l has multiplicative order 2 modulo n.*

PROOF. The lemma follows from the structure of the normalizer of the group $\mathfrak{A}$ in the symmetric group of degree n and from Theorem 4.80. □

LEMMA 4.82. *Let $\boldsymbol{C}$ be a primitive cyclic code of length n over $\mathbb{F}_2$; denote its zeros by $\{\alpha^i \mid i \in I\}$. Suppose that* $\gcd(n,d) = 1$. *Then $\{\alpha^{di} \mid i \in I\}$ are zeros of the code $\theta_d\boldsymbol{C}$.*

PROOF. Let $e(\lambda)$ be the generator idempotent of the code $\boldsymbol{C}$. According to Theorem 4.80, a generator idempotent of the code $\theta_d\boldsymbol{C}$ is the polynomial $\theta_d(e(\lambda))$. By virtue of Lemma 4.65, the zeros of the codes $\boldsymbol{C}$ and $\theta_d(\boldsymbol{C})$ correspond to the zeros of the polynomials $e(\lambda)$ and $\theta_d(e(\lambda))$. Hence, the lemma follows immediately from the congruence $\theta_d(e(\lambda)) \equiv e(\lambda^d)$. □

THEOREM 4.83. *Let $\boldsymbol{C}$ be a primitive cyclic code of length n over the field $\mathbb{F}_2$ with a generator polynomial $g(\lambda)$ and with generator idempotent $e(\lambda)$. Let $\boldsymbol{C}'$ be the dual of the code $\boldsymbol{C}$. Then*

1. $\boldsymbol{C}^\perp = \theta_{-1}\boldsymbol{C}'$.
2. $\boldsymbol{C}^\perp$ *has generator idempotent* $1 \oplus \theta_{-1}(e(\lambda))$.
3. $\boldsymbol{C}^\perp$ *has generator* $\lambda^k h(\lambda^{-1})$, *where* $h(\lambda) = \frac{\lambda^n \oplus 1}{g(\lambda)}$.

PROOF. 1. The code $\boldsymbol{C}'$, which is the dual of the code $\boldsymbol{C}$, is a sum of minimal ideals of $\mathfrak{R}_n$ that are not in $\boldsymbol{C}$. One has $\boldsymbol{C} \oplus \boldsymbol{C}' = \mathfrak{R}_n$ and $\boldsymbol{C} \cap \boldsymbol{C}' = \{0\}$. The generator idempotent of the code $\boldsymbol{C}'$ is $e'(\lambda) = 1 \oplus e(\lambda)$. The zeros of the code $\boldsymbol{C}'$ will be nonzeros of $\boldsymbol{C}$. Let $\{\alpha^j \mid j \in I\}$ be nonzeros of $\boldsymbol{C}$ and zeros of $\boldsymbol{C}'$. According to Lemma 4.82, $\{\alpha^{-j} \mid j \in I\}$ are zeros of the code $\theta_{-1}\boldsymbol{C}'$. Since $\{\alpha^j \mid j \in I\}$ are roots of the generator polynomial $\lambda^k h(\lambda^{-1})$ of the code $\boldsymbol{C}'$, one has $\lambda^k h(\lambda^{-1}) \in \theta_{-1}\boldsymbol{C}'$. Hence, $\boldsymbol{C}^\perp = \big\langle\lambda^k h(\lambda^{-1})\big\rangle \subset \theta_{-1}\boldsymbol{C}'$. However, as θ_{-1} is an automorphism of $\mathfrak{R}_n$, one has $\#\boldsymbol{C}^\perp = \theta_{-1}\boldsymbol{C}'$, whence the first statement.

2. Part 2 follows directly from Theorem 4.80 and the congruence $\theta_{-1}(e'(\lambda)) \equiv \theta_{-1}(1 \oplus e(\lambda)) \equiv 1 \oplus \theta_{-1}(e(\lambda))$.

3. Part 3 is proved in Theorem 4.71. □

4.4. Some classes of primitive cyclic codes

In this section some important classes of binary primitive cyclic codes are presented. We give here their detailed description in order to show by examples how the finite fields technique may work. Note that, since cyclic codes are not the main subject of this book, we do not have enough room to describe all important classes of cyclic codes (for example, Reed–Solomon codes, which are not binary).

Hamming codes and simplex codes. We recall (see Example 4.43) that the Hamming code $\boldsymbol{C}_m$ ($m \geq 2$) has length $n = 2^m - 1$ and dimension $k = 2^m - m - 1$. The rows of its parity-check matrix H are all $2^m - 1$ different nonzero m-columns (in Example 4.43 they are placed in lexicographic order). We use a more compact representation of this matrix based on the extension $\mathbb{F}_{2^m}$ of the field $\mathbb{F}_2$, which is isomorphic to the quotient ring $\mathbb{F}_2[\lambda]/(a(\lambda))$ (see Theorem 1.94), where $a(\lambda)$ is an irreducible polynomial over $\mathbb{F}_2$ of degree m. When writing polynomials of degree at most $m-1$ representing elements of the field $\mathbb{F}_{2^m}$ in the matrix form, we use the elements of the first row of the matrix H as coefficients of λ^0, those of the second row as coefficients of λ^1, etc.; finally, the elements of the mth row are coefficients of λ^{m-1}. Let α be a primitive element of the field $\mathbb{F}_{2^m}$. Then the elements $1, \alpha, \alpha^2, \dots, \alpha^{2^m-2}$ are distinct and can be represented in the way described above as nonzero binary m-columns.

The parity-check matrix H of the Hamming code $\boldsymbol{C}_m$ may be written in the vector form

$$H = (1, \alpha, \alpha^2, \dots, \alpha^{2^m-2}),$$

where any element of this $(1 \times (2^m - 1))$ matrix can be represented as the corresponding m-column. It is clear that in this representation the order of the m-columns may change. Recall that speaking about a code we have in mind the class of equivalent (up to permutation of the coordinates) codes with the same parameters.

THEOREM 4.84. *The Hamming code $\boldsymbol{C}_m$ is a primitive cyclic code of length $n = 2^m - 1$ with generator polynomial $g(\lambda) = m_1(\lambda) = m_\alpha(\lambda)$.*

PROOF. The row $\boldsymbol{c} = (c_0, c_1, \dots, c_{n-1})$ belongs to the code $\boldsymbol{C}_m$ if and only if $\boldsymbol{H}\boldsymbol{c}^\top = 0$. Therefore, $\bigoplus_{i=0}^{n-1} c_i\alpha^i = 0$. The last equality shows that $c(\alpha) = 0$, where $c(\lambda) = c_0 \oplus c_1\lambda \oplus \cdots \oplus c_{n-1}\lambda^{n-1}$. By virtue of the result of Problem 1.122, a polynomial $c(\lambda)$ belongs to the code $\boldsymbol{C}_m$ if and only if $c(\lambda)$ is divisible by $m_1(\lambda) = m_\alpha(\lambda)$, which is the minimal polynomial of the element $\alpha = \alpha^1$. Thus, the code $\boldsymbol{C}_m$ consists of polynomials of degree at most $n-1$ that are multiples of $m_1(\lambda)$. □

According to Theorem 4.53, the generator matrix of the code $\boldsymbol{C}_m$ is

$$G = \begin{pmatrix} m_1(\lambda) & & & 0 \\ & \lambda m_1(\lambda) & & \\ & & \ddots & \\ 0 & & & \lambda^{n-m-1} m_1(\lambda) \end{pmatrix}.$$

EXAMPLE 4.85. For the code $\boldsymbol{C}_3$,

$$H = (1, \alpha, \alpha^2, \alpha^3, \alpha^4, \alpha^5, \alpha^6) = \begin{pmatrix} 0 & 0 & 1 & 0 & 1 & 1 & 1 \\ 0 & 1 & 0 & 1 & 1 & 1 & 0 \\ 1 & 0 & 0 & 1 & 0 & 1 & 1 \end{pmatrix},$$

where $\alpha \in \mathbb{F}_{2^3}$ is a root of the irreducible polynomial $\lambda^3 \oplus \lambda \oplus 1$. Its generator matrix is

$$G = \begin{pmatrix} 1 & 1 & 0 & 1 & 0 & 0 & 0 \\ 0 & 1 & 1 & 0 & 1 & 0 & 0 \\ 0 & 0 & 1 & 1 & 0 & 1 & 0 \\ 0 & 0 & 0 & 1 & 1 & 0 & 1 \end{pmatrix}.$$

Let $\boldsymbol{C}'$ be the primitive cyclic code which is the dual of the Hamming code $\boldsymbol{C}_m$. Then, by Definition 4.61, the generator polynomial of the code $\boldsymbol{C}'$ is the polynomial $(\lambda^{2^m-1} \oplus 1)/m_1(\lambda)$. According to Theorem 4.79, the code $\boldsymbol{C}'_m$ is a minimal primitive cyclic code.

DEFINITION 4.86. The dual of the Hamming code $\boldsymbol{C}_m$ is called the *simplex code* $\boldsymbol{S}_m$, i.e.,

$$\boldsymbol{S}_m = \boldsymbol{C}_m^{\perp}.$$

LEMMA 4.87. *The simplex code $\boldsymbol{S}_m$ is a minimal primitive cyclic $[2^m - 1, m]$-code, and its set of zeros coincides with $\{\alpha^j \mid j \in \mathcal{C}_{m,-1}\}$.*

PROOF. It follows from Theorem 4.83 that $\boldsymbol{S}_m = \boldsymbol{C}_m^{\perp} = \theta_{-1}(\boldsymbol{C}'_m)$ and that a generator polynomial of this code has the form $\lambda^{2^m-m-1}h(\lambda^{-1})$, where $h(\lambda) = (\lambda^{2^m-1} \oplus 1)/m_1(\lambda)$. It is obvious that the dimension of the code $\boldsymbol{S}_m$ is m. The set $\{\alpha^j \mid j \in \{0, 1, \ldots, 2^m - 2\} \backslash \mathcal{C}_{m,1}\}$ is the set of roots of the polynomial $h(\lambda)$. Using the result of Problem 4.59 we get that the set $\{\alpha^{-j} \mid j \in \{0, 1, \ldots, 2^m - 2\} \backslash \mathcal{C}_{m,1}\}$ is the set of zeros of the polynomial $\lambda^{2^m-m-1}h(\lambda^{-1})$. Therefore, the nonzeros of the code $\boldsymbol{S}_m$ form the set $\{\alpha^j \mid j \in \mathcal{C}_{m,1}\}$, and the generator polynomial of this code is $(\lambda^{2^m-1} \oplus 1)/m_{-1}(\lambda)$. Thus, the simplex code $\boldsymbol{S}_m$ is a minimal primitive cyclic $[2^m - 1, m]$-code. $\square$

The check polynomial of the code $\boldsymbol{S}_m$ is $m_{-1}(\lambda) = 1 \oplus h_1\lambda \oplus \cdots \oplus h_{m-1}\lambda^{m-1} \oplus \lambda^m$. Hence, the parity-check $((2^m - m - 1) \times (2^m - 1))$ matrix of the code $\boldsymbol{S}_m$ is

$$H = \begin{pmatrix} & & & & 1 & h_{m-1} & h_{m-2} & \ldots & h_2 & h_1 & 1 \\ & 0 & & 1 & h_{m-1} & h_{m-2} & \ldots & h_2 & h_1 & 1 & \\ \cdots & \cdots & \cdots & \cdots & \cdots & \cdots & \cdots & \cdots & \cdots & \cdots & \cdots \\ 1 & h_{m-1} & h_{m-2} & \ldots & h_2 & h_1 & 1 & & & 0 & \end{pmatrix}.$$

Therefore, a word $\boldsymbol{c} = (c_0, c_1, \ldots, c_{2^m-2})$ belongs to the code $\boldsymbol{S}_m$ if and only if the following linear equations hold:

$$c_{i+m} = c_i \oplus c_{i+1}h_{m-1} \oplus \cdots \oplus c_{i+m-1}h_1, \quad i = 0, \ldots, 2^m - 2, \tag{4.10}$$

where addition in the indices is modulo $2^m - 1$.

These equations show that each bit of the word $\boldsymbol{c}$ is recurrently expressed via m previous bits (taking into account the addition modulo $2^n - 1$ in the indices).

Figure 4.2 shows a linear feedback shift register of length m that generates a linear recurrent sequence satisfying (4.10). The sequence of the states of this register is described by the following matrix equation:

$$\begin{pmatrix} c_{i+1} \\ c_{i+2} \\ \cdots \\ c_{i+m-1} \\ c_{i+m} \end{pmatrix} = \begin{pmatrix} 0 & 1 & 0 & \ldots & 0 \\ 0 & 0 & 1 & \ldots & 0 \\ \cdots & \cdots & \cdots & \cdots & \cdots \\ 0 & 0 & 0 & \ldots & 1 \\ 1 & h_{m-1} & h_{m-2} & \ldots & h_1 \end{pmatrix} \cdot \begin{pmatrix} c_i \\ c_{i+1} \\ \cdots \\ c_{i+m-2} \\ c_{i+m-1} \end{pmatrix} = A \cdot \begin{pmatrix} c_i \\ c_{i+1} \\ \cdots \\ c_{i+m-2} \\ c_{i+m-1} \end{pmatrix},$$

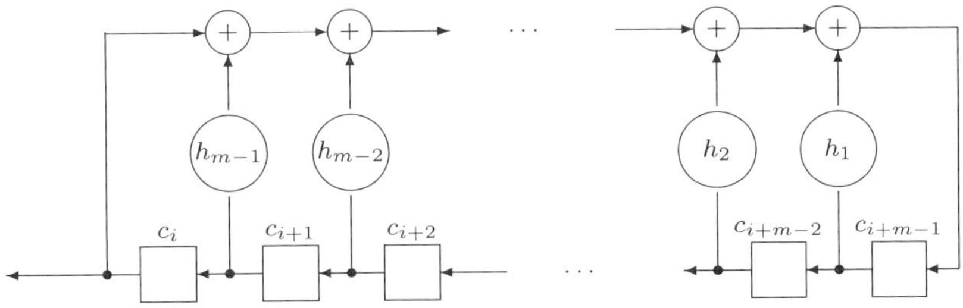

FIGURE 4.2

where $i = 0, 1, 2, \dots$ and the matrix A is the so-called attendant matrix (see [**225**]) of the polynomial $\lambda^m m_{-1}(\lambda^{-1}) = m_1(\lambda)$. From a well-known fact of linear algebra (see [**250, 317**]) it follows that the characteristic polynomial of the matrix A coincides with $m_1(\lambda)$, i.e.,

$$\det(A \oplus \lambda E_m) = m_1(\lambda) \quad \text{and} \quad m_1(A) = 0.$$

Since the polynomial $\lambda^{2^m-1} \oplus 1$ is divisible by $m_1(\lambda)$, one has $A^{2^m-1} = E_m$. The polynomial $m_1(\lambda)$ is a primitive polynomial, whence its period coincides with $2^m - 1$ and $A^r \neq E_m$ for $0 < r < 2^m - 1$ (see Lemma 1.133). Thus, the minimum period of the sequence of the shift register (see Figure 4.2), and hence of the output sequence of the register, is equal to $2^m - 1$. What was said above shows that all 2^m code words of the code $\boldsymbol{S}_m$ can be obtained by a linear feedback shift register of length m if one feeds to it all binary m-tuples.

Bose–Chaudhuri–Hocquenghem codes. Now we examine one of the most important classes of codes, Bose–Chaudhuri–Hocquenghem codes (BCH codes). These codes are interesting because, although their length is not too big, they provide good (but, as a rule, not the best) and simple constructive encoding and decoding methods. From the point of view of this book, a good understanding of the properties of BCH codes is useful for the study of other classes of codes.

We will start our acquaintance with BCH codes with the following statement.

THEOREM 4.88 (BCH bound). *Let $\boldsymbol{C}$ be a primitive cyclic $[n, k, d]$-code over the field $\mathbb{F}_2$ and let α be a primitive element of the field $\mathbb{F}_{2^m}$ $(n + 1 = 2^m)$. Suppose that there exist integers $r \geq 0$ and $d \geq 2$ such that the set $\{\alpha^i \mid r \leq i \leq r + d^* - 2\}$ is a subset of the set of zeros of the code $\boldsymbol{C}$. Then $d \geq d^*$.*

PROOF. Let $\boldsymbol{c}(\lambda)$ be a nonzero element of the code $\boldsymbol{C}$; put $\mathrm{wt}(\boldsymbol{c}) = w$, i.e.,

$$\boldsymbol{c}(\lambda) = \bigoplus_{j=1}^{w} c_{i_j} \lambda^{i_j}.$$

Suppose that $w < d^*$. Since $c(\alpha^i) = 0$ for $r \leq i \leq r + d^* - 2$, one has $M\boldsymbol{u}^\top = 0$, where

$$M = \begin{pmatrix} \alpha^{i_1 r} & \alpha^{i_2 r} & \dots & \alpha^{i_w r} \\ \alpha^{i_1(r+1)} & \alpha^{i_2(r+1)} & \dots & \alpha^{i_w(r+1)} \\ \dots & \dots & \dots & \dots \\ \alpha^{i_1(r+w-1)} & \alpha^{i_2(r+w-1)} & \dots & \alpha^{i_w(r+w-1)} \end{pmatrix}$$

and $\boldsymbol{u} = (c_{i_1}, c_{i_2}, \ldots, c_{i_w}) = \mathbf{1}$. Since $\boldsymbol{u} \neq \mathbf{0}$, the matrix M is degenerate, whence $\det M = 0$. It is easy to see that

$$\det M = \alpha^{(i_1+i_2+\cdots+i_w)r} \det M',$$

where M' is the Vandermonde matrix

$$M' = \begin{pmatrix} 1 & 1 & \ldots & 1 \\ \alpha^{i_1 r} & \alpha^{i_2 r} & \ldots & \alpha^{i_w r} \\ \cdots & \cdots & \cdots & \cdots \\ \alpha^{i_1(w-1)} & \alpha^{i_2(w-1)} & \ldots & \alpha^{i_w(w-1)} \end{pmatrix}.$$

As all α^{i_j} are different, $\det M' \neq 0$ and we get a contradiction. Therefore, $\mathrm{wt}(\boldsymbol{c}) \geq d^*$ for any $\boldsymbol{c} \neq \mathbf{0}$ from $\boldsymbol{C}$, whence $d \geq d^*$. □

EXAMPLE 4.89. Let $\boldsymbol{C}$ be a primitive cyclic $[31, 25, d]$-code with the generator polynomial

$$g(\lambda) = \prod_{j \in \mathcal{C}_{31,3} \cup \{0\}} (\lambda \oplus \alpha^j),$$

where α is a primitive element of the field $\mathbb{F}_{32}$ and $\mathcal{C}_{31,3}$ is a 2-cyclotomic class modulo 31. The set of zeros of the code $\boldsymbol{C}$ is $\{\alpha^j \mid j \in \{0, 3, 6, 12, 17, 24\}\}$. Applying the BCH bound to the code $\boldsymbol{C}$, we get $d \geq 2$. But, according to Lemma 4.82, the equivalent code $\boldsymbol{C}_1 = \theta_{21}\boldsymbol{C}$ has the set of zeros $\{\alpha^j \mid j \in \{0, 1, 2, 4, 8, 16\}\}$. Applying the BCH bound to the code $\boldsymbol{C}_1$, we get $d \geq 4$. In particular, $d = 4$, as $\boldsymbol{C}_1$ is the subcode of the Hamming code $\boldsymbol{C}_5$ consisting of code words of even weight.

This example shows that the BCH bound can evaluate the minimum distance of different codes having equal error correcting capabilities with different precision.

DEFINITION 4.90. The parameter d^* from Theorem 4.88 is called the *constructive code distance*.

According to Theorem 4.88, the constructive code distance is no greater than the ordinary code distance.

DEFINITION 4.91. Let d^* be an integer such that $2 \leq d^* \leq n$, $n = 2^m - 1$. The *primitive BCH code* $\boldsymbol{C}$ over the field $\mathbb{F}_2$ of length n with constructive distance d^* is the primitive cyclic code with generator polynomial

$$g(\lambda) = \mathrm{lcm}\{m_{\alpha^r}(\lambda), m_{\alpha^{r+1}}(\lambda), \ldots, m_{\alpha^{r+d^*-2}}(\lambda)\},$$

where α is a primitive element of the field $\mathbb{F}_{2^m}$.

In other words, the primitive BCH code $\boldsymbol{C}$ is a primitive cyclic code with maximum possible dimension for which the set $\{\alpha^j \mid r \leq j \leq r + d^* - 2\}$ is a subset of the set of zeros of the code.

If $r = 1$ in Definition 4.91, the corresponding code $\boldsymbol{C}$ is called a *narrow-sense primitive BCH code*. When r is fixed, BCH codes with constructive distance d_1^* contain codes with constructive distance d_2^* if and only if $d_1^* \leq d_2^*$.

Note that the dual of a BCH code is not a BCH code.

THEOREM 4.92. *Let $\boldsymbol{C}$ be a primitive BCH code of length $n = 2^m - 1$ with constructive distance d^*. Then:*

1) $d_{\boldsymbol{C}} \geq d^*$;

2) $k \geq n - m(d^* - 1)$.

PROOF. Part 1) follows immediately from the BCH bound (see Theorem 4.88).

2) As the generator polynomial of the code $\boldsymbol{C}$ has the form

$$g(\lambda) = \gcd\{m_{\alpha^r}(\lambda), m_{\alpha^{r+1}}(\lambda), \dots, m_{\alpha^{r+d^*-2}}(\lambda)\},$$

for any $c(\lambda) \in \boldsymbol{C}$ we have

$$c(\alpha^r) = c(\alpha^{r+1}) = \dots = c(\alpha^{r+d^*-2}) = 0.$$

Therefore, the matrix

$$H = \begin{pmatrix} 1 & \alpha^r & \alpha^{2r} & \dots & \alpha^{(n-1)r} \\ 1 & \alpha^{r+1} & \alpha^{2(r+1)} & \dots & \alpha^{(n-1)(r+1)} \\ \dots & \dots & \dots & \dots & \dots \\ 1 & \alpha^{r+d^*-2} & \alpha^{2(r+d^*-2)} & \dots & \alpha^{(n-1)(r+d^*-2)} \end{pmatrix}$$

defines the check equations for the code words of the code $\boldsymbol{C}$. Since every element of the matrix H belongs to the field $\mathbb{F}_{2^m}$, each of them can be replaced by a column of m elements of the field $\mathbb{F}_2$. The rows of the resulting matrix over $\mathbb{F}_2$ define check equations for the code $\boldsymbol{C}$. There will be $m(d^* - 1)$ rows in the resulting matrix; since they are not necessarily linearly independent, the dimension k of the code $\boldsymbol{C}$ is at least $n - (d^* - 1)m$. □

LEMMA 4.93. *Suppose that $\boldsymbol{C}$ is a narrow-sense primitive BCH code of length $n = 2^m - 1$ and with constructive distance $d^* = 2t + 1$. Then $k_{\boldsymbol{C}} \geq n - mt$.*

PROOF. The generator polynomial of the code is

$$g(\lambda) = \operatorname{lcm}\{m_\alpha(\lambda), \dots, m_{\alpha^{2t}}(\lambda)\}.$$

According to Theorem 1.129, for any i we have $m_{\alpha^{2i}}(\lambda) = m_{\alpha^i}(\lambda)$. Thus,

$$\begin{aligned} g(\lambda) &= \operatorname{lcm}\{m_\alpha(\lambda), m_{\alpha^2}(\lambda), m_{\alpha^3}(\lambda), \dots, m_{\alpha^{2t}}(\lambda)\} \\ &= \operatorname{lcm}\{m_\alpha(\lambda), m_{\alpha^3}(\lambda), m_{\alpha^5}(\lambda), \dots, m_{\alpha^{2t-1}}(\lambda)\}. \end{aligned}$$

Since, by virtue of Theorem 1.129, for any i one has $\deg m_{\alpha^i}(\lambda) \leq m$, it follows that $\deg g(\lambda) \leq mt$. Hence, $k_{\boldsymbol{C}} \geq n - mt$. □

Returning to the beginning of the previous section, recall that, starting with Definition 4.55, we have worked only with primitive cyclic codes. Nevertheless, most of what we proved above is true in the more general case when the length of the code n is not primitive, $\gcd(n, 2) = 1$, and m is the minimum natural number such that $2^m \equiv 1 \pmod{n}$, $\lambda^n \oplus 1$ divides $\lambda^{2^m-1} \oplus 1$ and does not divide $\lambda^{2^s-1} \oplus 1$ for $0 < s < m$.

In this case, instead of a primitive element of the field $\mathbb{F}_{2^m}$, one uses the nth root of unity which lies in $\mathbb{F}_{2^m}$ and the 2-cyclotomic classes modulo n (see [**315, 241**]).

Suppose that $n = 23$, $m = 11$. Then $2^m - 1 = 23 \cdot 89$. For any set $\boldsymbol{c} \in V_{23}$ the ball of radius 3 centered at $\boldsymbol{c}$ will have the volume equal to $\binom{23}{0} + \binom{23}{1} + \binom{23}{2} + \binom{23}{3}$. One can check that

$$\left[\binom{23}{0} + \binom{23}{1} + \binom{23}{2} + \binom{23}{3}\right] \cdot 2^{12} = 2^{23}. \tag{4.11}$$

The last equality implies that the sum of the volumes of 2^{12} balls of radius 3 equals the cardinality of V_{23}, which is a necessary (but not sufficient) condition for the existence of a perfect triple error correcting a $[23, 12, 7]$-code over the field $\mathbb{F}_2$. Such a code, called the *binary Golay code*, exists. Because of its small length, it does not have many practical applications, but it plays an important role in coding theory. The good sphere packing it provides is related to many mathematical objects. Besides, it is interesting as an example of a binary cyclic code which is not primitive. We will describe it without proofs.

Let α be a primitive element of the field $\mathbb{F}_{2^{11}}$. Then the element $\beta = \alpha^{89}$ is a 23rd root of unity. The minimal polynomial of β is

$$m_\beta(\lambda) = (\lambda - \beta)(\lambda - \beta^2)\cdots(\lambda - \beta^{2^{r-1}}),$$

where $\mathcal{C}_{11,1} = \{1, 2, \ldots, 2^{r-1}\}$ is the 2-cyclotomic class modulo 23 containing 1. Similarly,

$$m_{\beta^{-1}}(\lambda) = (\lambda - \beta^{-1})(\lambda - \beta^{-2})\cdots(\lambda - \beta^{-2^{r-1}}),$$

where $\mathcal{C}_{11,-1} = \{-1, -2, \ldots, -2^{r-1}\}$ is the 2-cyclotomic class modulo 23 containing $-1 \equiv 22 \pmod{23}$. It is easy to show directly that

$$(\lambda \oplus 1)m_\beta(\lambda)m_{\beta^{-1}}(\lambda) = \lambda^{23} \oplus 1,$$

where

$$m_\beta(\lambda) = \lambda^{11} \oplus \lambda^{10} \oplus \lambda^6 \oplus \lambda^5 \oplus \lambda^4 \oplus \lambda^2 \oplus 1,$$
$$m_{-\beta}(\lambda) = \lambda^{11} \oplus \lambda^9 \oplus \lambda^7 \oplus \lambda^6 \oplus \lambda^5 \oplus \lambda \oplus 1 = m_\beta^*(\lambda).$$

The *binary Golay code* is the cyclic code of length $n = 23$ over the field $\mathbb{F}_2$ with generator polynomial $m_\beta(\lambda)$, where β is a 23rd root of unity in the field $\mathbb{F}_{2^{11}}$. Therefore, the dimension of the Golay code is $k = 23 - \deg m_\beta(\lambda) = 12$. Proving consecutively the following statements (see, for example, [**35**]):

1) there are no code words of weight 4 or less in the Golay code,

2) there are no code words of weights 2, 6, 10, 14, 18, and 22 in the Golay code, and

3) there are no code words of weights 1, 5, 9, 13, 17, and 21 in the Golay code,

we obtain that the minimum weight of the Golay code is at least 7. On the other hand, (4.11) shows that the spheres of radius 3 around the 2^{12} code words do not intersect and cover the whole space $\mathbb{F}_2^{23}$. Thus, the minimum distance of the Golay code is equal to 7.

Comments to Chapter 4

We suggest that the reader will consult the remarkable book [**241**] while studying this chapter. All important aspects of coding theory are exposed there in a simple and accessible manner. From our point of view, it is very important that the authors of that book (as we do) often restrict their attention to binary codes, which significantly simplifies the theory. Besides, there are many useful and instructive examples in [**241**].

If necessary, the readers can also use the well-known books [**22, 309**]. As a natural supplement to [**241**], we recommend the book [**34**]. Convolutional codes, which have recently been used in cryptology for constructing efficient algorithms for the correlation method, are systematically exposed in [**187**].

Properties of cyclic codes and the corresponding decoding algorithms are presented well enough in the above-mentioned literature. Besides, we would like to draw the reader's attention to the book [**197**]. It is well known that cyclic codes have a simple decoding algorithm. It was designed by efficiently using the technique of finite fields, rings, and ideals. In [**197**] a detailed description of classes of cyclic codes, methods for their construction, and an analysis of their error-correcting capabilities and of decoding methods are given. The reader will also may find the Handbook of Coding Theory [**315**] very useful.

A detailed description of the algebraic approach to coding theory can be found in [**440**].

CHAPTER 5

Reed–Muller Codes

This chapter is devoted to the exposition of the properties of a particular class of block codes, namely the Reed–Muller codes. These codes play a significant role in different cryptographic investigations. The reason is the natural functional interpretation of the Reed–Muller codes. It allows us, on the one hand, to formulate important cryptographic properties with the help of the parameters of these codes and, on the other hand, to use their decoding algorithms in developing methods for cryptanalysis.

5.1. General properties of the Reed–Muller codes

Let $f \in \mathcal{F}_m$. Denote by Ω_f the row of values of the function f, i.e.,

$$(5.1) \quad \Omega_f = (f(\boldsymbol{u}_0), f(\boldsymbol{u}_1), \ldots, f(\boldsymbol{u}_{2^m-1})), \quad \boldsymbol{u}_i \in V_m, \quad i = 0, 1, \ldots, 2^m - 1.$$

We note that $\mathrm{wt}(f) = \mathrm{wt}(\Omega_f)$. The vectors $\boldsymbol{u}_i$ of the space V_m are ordered in some way. In this section we assume that they are ordered lexicographically. Later, we will consider some other orderings, but we will always make a note and comment on this.

DEFINITION 5.1. For arbitrary natural numbers m and r, $0 \le r \le m$, the *Reed–Muller* code $RM(r, m)$ of order r and length $n = 2^m$ is defined as the set of rows Ω_f for all Boolean functions $f \in \mathcal{F}_m$ with algebraic degree $\deg f$ not exceeding r, i.e.,

$$RM(r, m) = \{\Omega_f \mid f \in \mathcal{F}_m, \deg f \le r\}.$$

PROPOSITION 5.2. *The dimension k of the code $RM(r, m)$ equals to*

$$k = \dim RM(r, m) = 1 + \binom{m}{1} + \cdots + \binom{m}{r}.$$

PROOF. The assertion follows from the fact that the subspace of all functions of algebraic degree less than r is generated by monomials of algebraic degree less than r. □

An important parameter of a code $RM(r, m)$ is its minimum distance $d = d_{RM(r,m)}$. To find the minimum distance of $RM(r, m)$, the following statement will be proved.

PROPOSITION 5.3. *The algebraic degree of a function $f \in \mathcal{F}_m$ is equal to r if and only if the algebraic degree of all its derivatives is not greater than $r - 1$ and there exists a direction in which the derivative has algebraic degree exactly $r - 1$, i.e.,*

$$\deg f = r \iff \begin{cases} \deg D_{\boldsymbol{u}} f \le r - 1 \text{ for all } \boldsymbol{u} \in V_m; \\ \text{there exists } \boldsymbol{u}_0 \in V_m \colon \deg D_{\boldsymbol{u}_0} f = r - 1. \end{cases}$$

PROOF. Let $\deg f = r$. From the definition of a Boolean function it is clear that $\deg D_{\boldsymbol{u}} f \leq r-1$ for any $\boldsymbol{u} \in V_m$. Let i be the index of an arbitrary variable from a monomial of degree r in the algebraic normal form of the function f. Then

$$f(x^{(1)}, x^{(2)}, \ldots, x^{(m)}) = x^{(i)} f_1(x^{(1)}, \ldots, x^{(i-1)}, x^{(i+1)}, \ldots, x^{(m)}) \\ \oplus f_2(x^{(1)}, \ldots, x^{(i-1)}, x^{(i+1)}, \ldots, x^{(m)}),$$

where $\deg f_1 = r-1$. Put $\boldsymbol{u} = \boldsymbol{e}_i = (0, \ldots, 0, \underset{i}{1}, 0, \ldots, 0)^\top$. Then

$$D_{\boldsymbol{u}} f = f_1, \quad \deg D_{\boldsymbol{u}} f = \deg f_1 = r-1.$$

Let $\deg D_{\boldsymbol{u}_0} f = r-1$. Without loss of generality we can assume that $\boldsymbol{u}_0 = \boldsymbol{e}_1$. Then

$$f(x^{(1)}, x^{(2)}, \ldots, x^{(m)}) = x^{(1)} D_{\boldsymbol{e}_1} f(x^{(2)}, \ldots, x^{(m)}) \oplus f_2(x^{(2)}, \ldots, x^{(m)})$$

and thus $\deg f = r$. □

LEMMA 5.4. *Let $f \in \mathcal{F}_m$. Then for any vector $\boldsymbol{u} \in V_m$ the following inequality holds:*

$$\mathrm{wt}(f) \geq \frac{1}{2} \mathrm{wt}(D_{\boldsymbol{u}} f).$$

PROOF. As $D_{\boldsymbol{u}} f(\boldsymbol{x}) = f(\boldsymbol{x}) \oplus f(\boldsymbol{x} \oplus \boldsymbol{u})$ and $\mathrm{wt}(f(\boldsymbol{x} \oplus \boldsymbol{u})) = \mathrm{wt}(f(\boldsymbol{x}))$, we have

$$\mathrm{wt}(D_{\boldsymbol{u}} f) \leq \mathrm{wt}(f(\boldsymbol{x} \oplus \boldsymbol{u})) + \mathrm{wt}(f(\boldsymbol{x})) \leq 2 \cdot \mathrm{wt}(f). \qquad \square$$

LEMMA 5.5. *Let $f \in \mathcal{F}_m$, $f \not\equiv 0$, and $\deg f = d$. Then $\mathrm{wt}(f) \geq 2^{m-d}$.*

PROOF. We use induction on the degree of d.

There exists a unique function f for which $\deg f = 0$, namely the constant function 1: $f(\boldsymbol{x}) = 1$ for all $\boldsymbol{x} \in V_m$. In this case $\mathrm{wt}(f) = 2^m$ and the lemma holds for $d = 0$.

Suppose the lemma holds for all d: $0 \leq d \leq d_0 \leq n-1$. We will prove that it holds for $d = d_0 + 1$. Let $\deg f = d_0 + 1$. According to Proposition 5.3, there exists a direction $\boldsymbol{u} \in V_m$ such that $\deg D_{\boldsymbol{u}} f = d_0$. By Lemma 5.4 the following inequality holds:

$$\mathrm{wt}(f) \geq \frac{1}{2} \mathrm{wt}(D_{\boldsymbol{u}} f). \tag{5.2}$$

Since $\deg D_{\boldsymbol{u}} f = d_0$, the induction hypothesis implies that

$$\mathrm{wt}(D_{\boldsymbol{u}} f) \geq 2^{m-d_0}.$$

Substituting the last inequality in (5.2) we obtain

$$\mathrm{wt}(f) \geq \frac{1}{2} \mathrm{wt}(D_{\boldsymbol{u}} f) \geq 2^{m-d_0-1} = 2^{m-(d_0+1)}.$$

Thus the lemma holds also for the function f. □

For completeness, we will give one more statement for lower and upper bounds for the weight of a Boolean function without using the properties of its derivative.

LEMMA 5.6. *Let $f \in \mathcal{F}_m$ and $\deg f = d$, $d \geq 1$. Then*

$$2^{m-d} \leq \mathrm{wt}(f) \leq 2^m - 2^{m-d}.$$

PROOF. Let $x^{(i_1)}x^{(i_2)}\cdots x^{(i_d)}$ be a monomial of length d of the algebraic normal form of the function f. Substituting the values 0 and 1 for each of the other $m-d$ variables, we obtain 2^{m-d} subfunctions $f_1,\dots,f_{2^{m-d}}\in\mathcal{F}_d$. Each subfunction contains the monomial $x^{(i_1)}x^{(i_2)}\cdots x^{(i_d)}$ in its algebraic normal form. Thus $1\le \mathrm{wt}(f_i)\le 2^d-1$, $i=1,2,\dots,2^{m-d}$. Obviously $\mathrm{wt}(f)=\sum_{i=1}^{2^{m-d}}\mathrm{wt}(f_i)$ and thus $2^{m-d}\le\mathrm{wt}(f)\le 2^m-2^{m-d}$. □

Now we can formulate a result concerning the minimum distance of a code $RM(r,m)$.

THEOREM 5.7. *The minimum distance of a code $RM(r,m)$ is equal to 2^{m-r}.*

PROOF. According to Lemma 5.5, for any code word of a code $RM(r,m)$ the following holds: $\mathrm{wt}(\Omega_f)=\mathrm{wt}(f)\ge 2^{m-\deg f}$. Since $\deg f\le r$ for the vectors Ω_f from $RM(r,m)$, we have $\mathrm{wt}(\Omega_f)\ge 2^{m-r}$. Thus the minimum distance of a code $RM(r,m)$ is not less than 2^{m-r}.

Consider the function $f_0(x^{(1)},x^{(2)},\dots,x^{(m)})=x^{(1)}x^{(2)}\cdots x^{(r)}$ of algebraic degree r. The vector Ω_{f_0} belongs to the code $RM(r,m)$ and $\mathrm{wt}(f_0)=\mathrm{wt}(\Omega_{f_0})=2^{m-r}$. Thus $d_{RM(r,m)}=2^{m-r}$. □

Proposition 5.2 shows that the generator matrix G of the code $RM(r,m)$ has dimensions $\left[1+\binom{m}{1}+\cdots+\binom{m}{r}\right]\times 2^m$ and the following structure:

$$G=\begin{pmatrix}G_0\\ G_1\\ \cdots\\ G_r\end{pmatrix}, \tag{5.3}$$

where $G_0=(1,1,\dots,1)=\Omega_1$,

$$G_1=\begin{pmatrix}\Omega_{x^{(m)}}\\ \cdots\\ \Omega_{x^{(1)}}\end{pmatrix},\quad G_2=\begin{pmatrix}\Omega_{x^{(m-1)}x^{(m)}}\\ \cdots\cdots\cdots\\ \Omega_{x^{(1)}x^{(3)}}\\ \Omega_{x^{(1)}x^{(2)}}\end{pmatrix},\quad\dots,\quad G_r=\begin{pmatrix}\Omega_{x^{(m-r+1)}x^{(m-r+2)}\cdot\ldots\cdot x^{(m)}}\\ \cdots\cdots\cdots\\ \Omega_{x^{(1)}\cdot\ldots\cdot x^{(r-1)}x^{(r+1)}}\\ \Omega_{x^{(1)}\cdot\ldots\cdot x^{(r-1)}x^{(r)}}\end{pmatrix}.$$

Note that the code $RM(m,m)$ consists of all vectors of length 2^m, the code $RM(m-1,m)$ consists of all vectors of length 2^m and even weight, and the code $RM(0,m)$ consists of just the all-one and the all-zero vectors.

THEOREM 5.8. *For each r, $0\le r\le m-1$, the following equality holds:*

$$RM(r,m)^\perp=RM(m-r-1,m).$$

PROOF. Let $\Omega_f\in RM(m-r-1)$ and $\Omega_g\in RM(r,m)$. Then $\deg f\le m-r-1$, $\deg g\le r$. Thus $\deg(f\cdot g)\le m-1$. Therefore, $\Omega_{f\cdot g}\in RM(m-1,m)$ and the weight of the vector $\Omega_{f\cdot g}$ is even. Thus $\langle\Omega_f,\Omega_g\rangle\equiv\mathrm{wt}(\Omega_{f\cdot g})\equiv 0 \pmod 2$ and $RM(m-r-1,m)\subseteq RM(r,m)^\perp$. On the other hand,

$$\begin{aligned}&\dim RM(m-r-1,m)+\dim RM(r,m)\\ &\quad=1+\binom{m}{1}+\cdots+\binom{m}{m-r-1}+1+\binom{m}{1}+\cdots+\binom{m}{r}=2^m,\end{aligned}$$

and thus $RM(m-r-1,m)=RM(r,m)^\perp$. □

DEFINITION 5.9. For each r, $0 \le r \le m-1$, the *punctured* code $RM^*(r,m)$ is the code obtained from $RM(r,m)$ by deleting all code words with coordinates corresponding to the value of the Boolean function at the zero vector $\mathbf{0}$.

The code words of a code $RM^*(r,m)$ will be denoted by Ω_f^*, $\deg f \le r$, where (see (5.1)) $\Omega_f = (f(\boldsymbol{u}_0), \Omega_f^*)$, assuming that $\boldsymbol{u}_0 = \mathbf{0}$.

Obviously, the length of a code $RM^*(r,m)$ is equal to $2^m - 1$.

PROBLEM 5.10. Prove that a code $RM^*(r,m)$ has minimum distance $2^{m-r}-1$ and dimension $1 + \binom{m}{1} + \cdots + \binom{m}{r}$.

Now we will need the representation of Boolean functions in terms of traces introduced in Section 2.2. According to Lemma 2.67, for each Boolean function $f \in \mathcal{F}_m$ there exists a polynomial $P(\boldsymbol{x})$ in the set $\mathcal{O}_m \subset \mathbb{F}_{2^m}[\boldsymbol{x}]$ such that $f(\boldsymbol{x}) = f_P(\boldsymbol{x}) = \operatorname{Tr}(P(\boldsymbol{x}))$ for any $\boldsymbol{x} \in \mathbb{F}_{2^m}$. Recall that $\mathcal{O}_m$ is the set of polynomials $P' \in \mathbb{F}_{2^m}[\boldsymbol{x}]$ of the form

$$P'(\boldsymbol{x}) = c_0\boldsymbol{\alpha} \oplus P(\boldsymbol{x}) \oplus c_{2^m-1}\boldsymbol{\alpha}\boldsymbol{x}^{2^m-1},$$

where $P(\boldsymbol{x})$ is an arbitrary cr-polynomial, $\boldsymbol{\alpha} \in \mathbb{F}_{2^m}$ is such that $\operatorname{Tr}_m(\boldsymbol{\alpha}) = 1$, and c_0 and c_{2^m-1} are elements of the field $\mathbb{F}_2$.

Let $\mathcal{O}$ be an additive subgroup of the ring $\mathbb{F}_{2^m}[\boldsymbol{x}]$ (for example, $\mathcal{O}_m$ is such a subgroup). We define the following subspaces of the spaces V_{n+1} and V_n, $n = 2^m-1$:

$$\Omega(\mathcal{O}) = \{\Omega_{f_P} \mid P \in \mathcal{O}\} \quad \text{and} \quad \Omega^*(\mathcal{O}) = \{\Omega^*_{f_P} \mid P \in \mathcal{O}\}.$$

We will prove the following useful description of the Reed–Muller codes.

THEOREM 5.11. *Let* $\mathcal{O} = \left\{\bigoplus_{i=1}^{l} a_i\boldsymbol{x}^{s_i} \;\middle|\; a_i \in \mathbb{F}_{2^m}\right\}$ *be an additive subgroup of the ring* $\mathbb{F}_{2^m}[\boldsymbol{x}]$ *and* α *a primitive element of the field* $\mathbb{F}_{2^m}$. *Then the dual of the cyclic code* $\boldsymbol{B}$ *of length* $n = 2^m - 1$ *with zeros* $\alpha^{s_1}, \dots, \alpha^{s_l}$ *is the code* $\Omega^*(\mathcal{O})$.

PROOF. Suppose $\boldsymbol{b} \in V_n$, $\boldsymbol{b} = (b^{(0)}, b^{(1)}, \dots, b^{(n-1)})^\top$, and $\boldsymbol{c} = \Omega^*_{f_P}$, $P(\boldsymbol{x}) = \bigoplus_{i=1}^{l} a_i\boldsymbol{x}^{s_i}$, $a_i \in \mathbb{F}_{2^m}$. We let $b(\boldsymbol{x}) = \bigoplus_{j=0}^{n-1} b^{(j)}\boldsymbol{x}^j$. Then

$$\begin{aligned}
\langle \boldsymbol{b}, \boldsymbol{c}\rangle &= \bigoplus_{j=0}^{n-1} b^{(j)} \operatorname{Tr}(P(\alpha^j)) = \bigoplus_{j=0}^{n-1} b^{(j)} \operatorname{Tr}\Bigl(\bigoplus_{i=1}^{l} a_i\alpha^{js_i}\Bigr) \\
&= \bigoplus_{j=0}^{n-1} b^{(j)} \bigoplus_{i=1}^{l}\bigoplus_{k=0}^{m-1} a_i^{2^k}\alpha^{js_i\cdot 2^k} = \bigoplus_{i=1}^{l}\bigoplus_{k=0}^{m-1} a_i^{2^k}\Bigl(\bigoplus_{j=0}^{n-1} b^{(j)}\alpha^{js_i}\Bigr)^{2^k} \qquad (5.4)\\
&= \bigoplus_{i=1}^{l}\bigoplus_{k=0}^{m-1} a_i^{2^k}(b(\alpha^{s_i}))^{2^k} = \bigoplus_{i=1}^{l} \operatorname{Tr}(a_i b(\alpha^{s_i})).
\end{aligned}$$

If $\boldsymbol{b} \in \boldsymbol{B}$, then $b(\alpha^{s_i}) = 0$ for $i = 1, 2, \dots, l$ and thus $\langle \boldsymbol{b}, \boldsymbol{c}\rangle = 0$ for any $\boldsymbol{c} \in \Omega^*(\mathcal{O})$.

We will prove the converse. Let $\langle \boldsymbol{b}, \boldsymbol{c}\rangle = 0$ for any $\boldsymbol{c} \in \Omega^*(\mathcal{O})$. If we put $P(\boldsymbol{x}) = a_1\boldsymbol{x}^{s_1}$, we obtain from (5.4) that $\operatorname{Tr}(a_1 b(\alpha^{s_1})) = 0$ for any $a_1 \in \mathbb{F}_{2^m}$, and therefore $b(\alpha^{s_1}) = 0$. Similarly $b(\alpha^{s_i}) = 0$, $i = 2, \dots, l$. Thus, $\boldsymbol{b} \in \boldsymbol{B}$. □

Let us put

$$\mathcal{A}_r = \Big\{P(\boldsymbol{x}) = \bigoplus_{i\in\Pi'_m} a_i\boldsymbol{x}^i \in \mathcal{O}_m \mid a_i = 0 \text{ if } w_2(i) > r\Big\}, \quad 0 \le r \le m,$$

$$\mathcal{B} = \Big\{P(\boldsymbol{x}) = \bigoplus_{i\in\Pi'_m} a_i\boldsymbol{x}^i \in \mathbb{F}_{2^m}[\boldsymbol{x}] \mid a_i = 0 \text{ if } w_2(i) > r\Big\}.$$

According to Problem 2.66 we have

$$\Omega(\mathcal{A}_r) = \Omega(\mathcal{B}) \quad\text{and}\quad \Omega^*(\mathcal{A}_r) = \Omega^*(\mathcal{B}). \tag{5.5}$$

THEOREM 5.12. *The following equality holds:* $RM(r,m) = \Omega(\mathcal{A}_r) = \{\Omega_{f_P} \mid P \in \mathcal{A}_r\}$.

PROOF. Let $\boldsymbol{\theta}_1,\dots,\boldsymbol{\theta}_m$ be some basis of the field $\mathbb{F}_{2^m}$ over $\mathbb{F}_2$. For a polynomial $P\in\mathcal{A}_r$, as before, we put

$$f_P(x^{(1)},\dots,x^{(m)}) = \mathrm{Tr}(P(x^{(1)}\boldsymbol{\theta}_1 \oplus\cdots\oplus x^{(m)}\boldsymbol{\theta}_m)).$$

Each polynomial $P\in\mathcal{A}_r$ is a sum of monomials $a_i\boldsymbol{x}^i$, where $w_2(i)\le r$. If $Q(\boldsymbol{x}) = a_i\boldsymbol{x}^i$, $i = 2^{i_1}+\cdots+2^{i_s}$ $(s\le r)$, then we have

$$\begin{aligned} Q(\boldsymbol{x}) &= Q(x^{(1)}\boldsymbol{\theta}_1 \oplus\cdots\oplus x^{(m)}\boldsymbol{\theta}_m) \\ &= a_i(x^{(1)}\boldsymbol{\theta}_1 \oplus\cdots\oplus x^{(m)}\boldsymbol{\theta}_m)^{2^{i_1}}\cdots(x^{(1)}\boldsymbol{\theta}_1 \oplus\cdots\oplus x^{(m)}\boldsymbol{\theta}_m)^{2^{i_s}} \\ &= a_i(x^{(1)}\boldsymbol{\theta}_1^{2^{i_1}} \oplus\cdots\oplus x^{(m)}\boldsymbol{\theta}_m^{2^{i_1}})\cdots(x^{(1)}\boldsymbol{\theta}_1^{2^{i_s}} \oplus\cdots\oplus x^{(m)}\boldsymbol{\theta}_m^{2^{i_s}}). \end{aligned}$$

Thus the polynomial $Q(\boldsymbol{x})$ can be written as a sum of monomials $ax^{(j_1)}\cdots x^{(j_t)}$, where $a\in\mathbb{F}_{2^m}$, $t\le s\le r$. The same holds for any polynomial $P\in\mathcal{B}$. Therefore, $f_P(x^{(1)},\dots,x^{(m)}) = \mathrm{Tr}(P(\boldsymbol{x})) = \mathrm{Tr}(P(x^{(1)}\boldsymbol{\theta}_1\oplus\cdots\oplus x^{(m)}\boldsymbol{\theta}_m))$ is a sum of monomials of the form $\mathrm{Tr}(a)x^{(j_1)}\cdots x^{(j_t)}$, $t\le r$, i.e., $\deg f_P\le r$ and thus $\Omega_{f_P}\in RM(r,m)$. This gives the inclusion $\Omega(\mathcal{A}_r)\subset RM(r,m)$.

Let us prove the converse. Denote $\mathcal{D}_r = \Omega(\mathcal{A}_r) = \Omega(\mathcal{B})$ and $\mathcal{D}_r^* = \Omega^*(\mathcal{A}_r) = \Omega^*(\mathcal{B})$ (see (5.5)). According to Theorem 5.11, $\mathcal{D}_r^{*\perp}$ is a cyclic code with zeros $\boldsymbol{\alpha}^j$, $w_2(j)\le r$ $(0\le j\le 2^m-2)$.

There are exactly $1+\binom{m}{1}+\cdots+\binom{m}{r}$ such elements. Therefore, $\dim\mathcal{D}_r^* = \sum_{i=0}^r\binom{m}{i}$. Then $\dim\mathcal{D}_r \ge \dim\mathcal{D}_r^* = \dim RM(r,m)$. Together with the inclusion $\Omega(\mathcal{A}_r)\subset RM(r,m)$ we get $\Omega(\mathcal{A}_r) = RM(r,m)$. □

COROLLARY 5.13. *The punctured Reed–Muller code $RM^*(r,m)$ of order r is a cyclic code of length 2^m-1 and its zeros are the elements $\boldsymbol{\alpha}^s$ satisfying the following conditions: $1\le w_2(s)\le m-r-1$ and $1\le s\le 2^m-2$.*

The generator and check polynomials of the code $RM^*(r,m)$ for $0\le r\le m-1$ are

$$g(\boldsymbol{x}) = \prod_{\substack{s\in\Pi'_m \\ 1\le w_2(s)\le m-r-1}} m_{\boldsymbol{\alpha}^s}(\boldsymbol{x}), \qquad h(\boldsymbol{x}) = (\boldsymbol{x}+1)\prod_{\substack{s\in\Pi'_m \\ m-r\le w_2(s)\le m-1}} m_{\boldsymbol{\alpha}^s}(\boldsymbol{x})$$

(recall that $m_{\boldsymbol{\alpha}^t}(\boldsymbol{x})$ is the minimal polynomial of the element $\boldsymbol{\alpha}^t$).

PROOF. The corollary follows immediately from Theorem 5.12 and the structure of the sets of roots of the cyclic code and of its dual. □

We now proceed with the description of words of minimum weight of a Reed–Muller code. As a rule, this description uses methods of finite geometry. One of the advantages of the geometric language is the possibility to precisely determine the minimum weight words in $RM(r,m)$: they are the incidence vectors of $(m-r)$-dimensional planes of a finite geometry of dimension m over the field $\mathbb{F}_2$. We will not give this description since it requires many preliminary results of finite geometry. To those who are interested we recommend the book [**241**]. We will give a description of code words of minimum weight of a Reed–Muller code using functional language.

Recall (see Chapter 3) that $\mathfrak{T}_m$ is the shift group and $\mathfrak{I}_{\mathfrak{T}_m}(f)$ the moment group of a Boolean function f in the shift group:

$$\mathfrak{I}_{\mathfrak{T}_m}(f) = \{\boldsymbol{u} \in V_m \mid D_{\boldsymbol{u}} f(\boldsymbol{x}) = f(\boldsymbol{x}) \oplus f(\boldsymbol{x} \oplus \boldsymbol{u}) = 0 \text{ for all } \boldsymbol{x} \in V_m\}.$$

PROBLEM 5.14. Let $f \in \mathcal{F}_m$, $\dim \mathfrak{I}_{\mathfrak{T}_m}(f) = s \le m$. Prove that there exist $A \in \mathfrak{GL}(V_m)$ and $f^- \in \mathcal{F}_{m-s}$ such that if $\boldsymbol{y} = A\boldsymbol{x}$, then

$$f(A\boldsymbol{x}) = f^-(y^{(s+1)}, \dots, y^{(m)}).$$

Also, $\mathfrak{I}_{\mathfrak{T}_{m-s}}(f^-) = \{\boldsymbol{0}\}$, $\operatorname{wt}(f) = 2^s \operatorname{wt}(f^-)$, and $\deg f^- = \deg f$.

THEOREM 5.15. *The vector* Ω_f, $f \in \mathcal{F}_m$, $\deg f \le r$, *is a code word of minimum weight* $\operatorname{wt}(f) = 2^{m-r}$ *in a code* $RM(r,m)$ *if and only if the following conditions hold:*

1) $\dim \mathfrak{I}_{\mathfrak{T}_m}(f) = m - r$;
2) $\operatorname{wt}(f^-) = 1$.

PROOF. Sufficiency follows from Problem 5.14.

We will prove the necessity. Let $f \in \mathcal{F}_m$, $\deg f \le r$, and $\operatorname{wt}(f) = 2^{m-r}$. Then $\deg f = r$, and according to Problem 5.14, $\dim \mathfrak{I}_{\mathfrak{T}_m}(f) = m - r - t \le m - r$, $\operatorname{wt}(f^-) = 2^t$, $\mathfrak{I}_{\mathfrak{T}_{r+t}}(f^-) = \{\boldsymbol{0}\}$, $\deg f^- = \deg f = r$. As $\mathfrak{I}_{\mathfrak{T}_{r+t}}(f^-) = \{\boldsymbol{0}\}$, then $\operatorname{wt}(D_{\boldsymbol{u}} f^-) \ne 0$ for any $\boldsymbol{u} \in V_{r+t}$, $\boldsymbol{u} \ne \boldsymbol{0}$. By Lemma 5.4, $\operatorname{wt}(D_{\boldsymbol{u}} f^-) \le 2 \operatorname{wt}(f^-) = 2^{t+1}$. On the other hand, since $D_{\boldsymbol{u}} f^- \not\equiv 0$, $D_{\boldsymbol{u}} f^- \in \mathcal{F}_{r+t}$, $\deg D_{\boldsymbol{u}} f^- \le r-1$, it follows from Lemma 5.5 that $\operatorname{wt}(D_{\boldsymbol{u}} f^-) \ge 2^{(r+t)-(r-1)} = 2^{t+1} = 2\operatorname{wt}(f^-)$ for any $\boldsymbol{u} \ne \boldsymbol{0}$. This is possible only if

$$\{\boldsymbol{x} \in V_{r+t} \mid f^-(\boldsymbol{x}) = 1\} \cup \{\boldsymbol{x} \in V_{r+t} \mid f^-(\boldsymbol{x} \oplus \boldsymbol{u}) = 1\} = \varnothing$$

for every $\boldsymbol{u} \in V_{r+t}$, $\boldsymbol{u} \ne \boldsymbol{0}$. But then $2^{r+t} \operatorname{wt}(f^-) \le 2^{r+t}$, hence $\operatorname{wt}(f^-) = 1$ and $t = 0$. □

COROLLARY 5.16. *The minimum weight words of a code* $RM(r,m)$ *and only they are incidence vectors of the cosets of a space* V_m *modulo subspace of dimension* $m-r$. *In other words, the minimum weight words in* $RM(r,m)$ *and only they are incidence vectors of the set of solutions of a system of linear Boolean equations of the type*

$$A\boldsymbol{x} = \boldsymbol{b},$$

where $\boldsymbol{b} \in V_m$, A *is a Boolean* $(m \times m)$ *matrix, and* $\operatorname{rank} A = r$.

PROBLEM 5.17. Prove that a product of words of minimum weight of an arbitrary Reed–Muller code is again a word of minimum weight of a certain Reed–Muller code. In the language of functions, prove that if $f_1, f_2 \in \mathcal{F}_m$, $\deg f_1 = r_1$, $\deg f_2 = r_2$, $\operatorname{wt}(f_1) = 2^{m-r_1}$, and $\operatorname{wt}(f_2) = 2^{m-r_2}$, then $\deg f_1 \cdot f_2 = r$ and $\operatorname{wt}(f_1 \cdot f_2) = 2^{m-r}$ for some $r \le r_1 + r_2$.

Consider now automorphism groups of the Reed–Muller codes. Recall some constructions of Chapter 3. As before, let $\mathfrak{S}_{2^m}$ be the symmetric group of degree 2^m of permutations of elements of the space V_m that induce on the set $\mathcal{F}_m$ the action

$$f^{\mathfrak{g}}(\boldsymbol{x}) = f(\mathfrak{g}\boldsymbol{x}),$$

where $f \in \mathcal{F}_m$, $\boldsymbol{x} \in V_m$, and $\mathfrak{g}\boldsymbol{x}$ is the image of the vector $\boldsymbol{x}$ under a permutation $\mathfrak{g} \in \mathfrak{S}_{2^m}$. It is easy to see that this action satisfies the following conditions:

$$(f_1 \oplus f_2)^{\mathfrak{g}} = f_1^{\mathfrak{g}} \oplus f_2^{\mathfrak{g}}, \tag{5.6}$$

$$(f_1 \cdot f_2)^{\mathfrak{g}} = f_1^{\mathfrak{g}} \cdot f_2^{\mathfrak{g}} \tag{5.7}$$

for any $f_1, f_2 \in \mathcal{F}_m$, $\mathfrak{g} \in \mathfrak{S}_{2^m}$.

It is clear that the automorphism groups of Reed–Muller codes are subgroups of the group $\mathfrak{S}_{2^m}$ and the automorphism groups of punctured Reed–Muller codes belong to the subgroup $\mathfrak{S}_{2^m}^{(0)}$ of permutations that fix the vector $\mathbf{0} \in V_m$, and this subgroup is isomorphic to the symmetric group $\mathfrak{S}_{2^m-1}$ of degree $2^m - 1$ that consists of permutations of elements in $V_m \setminus \{\mathbf{0}\}$.

The following statement is almost obvious.

Lemma 5.18. *Let $r = 0$ or $r = m - 1$. Then*

$$\mathrm{Aut}(RM^*(r,m)) = \mathfrak{S}_{2^m-1}, \quad \mathrm{Aut}(RM(r,m)) = \mathfrak{S}_{2^m}.$$

Before we proceed to the statement characterizing the automorphism group of the codes $RM(r,m)$ and $RM^*(r,m)$ in a more meaningful case, we will consider some auxiliary statements.

Lemma 5.19. *Let $\mathfrak{g} \in \mathrm{Aut}(RM(r,m))$, $1 \le r \le m-1$. Then $\mathfrak{g}$ transforms every minimum weight word from $RM(s,m)$, $s \ge r$ into a minimum weight word from the same code.*

Proof. Let $\Omega_f \in RM(s,m)$, $s \ge r$, be a minimum weight word, i.e., $\mathrm{wt}(f) = 2^{m-s}$. Then $\mathrm{wt}(f^{\mathfrak{g}}) = 2^{m-s}$. It is obvious from Corollary 5.16 that there exist words $\Omega_{f_1}, \dots, \Omega_{f_l} \in RM(r,m)$ of minimum weight such that $f = f_1 f_2 \cdots f_l$. Then $\Omega_{f_1^{\mathfrak{g}}}, \dots, \Omega_{f_l^{\mathfrak{g}}}$ are also minimum weight words. According to (5.7), $f^{\mathfrak{g}} = f_1^{\mathfrak{g}} f_2^{\mathfrak{g}} \cdots f_l^{\mathfrak{g}}$ and $\mathrm{wt}(f^{\mathfrak{g}}) = 2^{m-s}$, and by Problem 5.17, $\Omega_{f^{\mathfrak{g}}}$ is a minimum weight word of a code $RM(s,m)$. ☐

Problem 5.20. Prove that for $1 \le r \le m-2$ the following inclusions hold:

$$\mathfrak{GA}(V_m) \subset \mathrm{Aut}(RM(r,m)), \quad \mathfrak{GL}(V_m) \subset \mathrm{Aut}(RM^*(r,m)).$$

Remark 5.21. The groups $\mathfrak{GL}(V_m)$ and $\mathfrak{GA}(V_m)$ are two and three times transitive respectively (for definitions of t-transitivity of permutation groups see, for example, [**410**]).

Theorem 5.22. *For $1 \le r \le m-1$ the following inclusions hold:*
1) $\mathrm{Aut}(RM(r,m)) \subset \mathrm{Aut}(RM(r+1,m))$;
2) $\mathrm{Aut}(RM^*(r,m)) \subset \mathrm{Aut}(RM^*(r+1,m))$.

Proof. 1) Let $\mathfrak{g} \in \mathrm{Aut}(RM(r,m))$ and $\Omega_f \in RM(r+1,m)$. If $\deg f \le r$, i.e., $\Omega_f \in RM(r,m)$, then $\Omega_{f^{\mathfrak{g}}} \in RM(r,m) \subset RM(r+1,m)$. Let $\deg f = r+1$. Then a function f can be represented as follows:

$$f = f_0 \oplus f_1 \oplus \cdots \oplus f_t,$$

where $\deg f_0 \le r$, $t \ge 1$, and $f_1 \oplus \cdots \oplus f_t$ is a sum of all t monomials of length $r+1$ in the algebraic normal form of function f. Using (5.6) for the function f, we get

$$f^{\mathfrak{g}} = f_0^{\mathfrak{g}} \oplus f_1^{\mathfrak{g}} \oplus \cdots \oplus f_t^{\mathfrak{g}}. \tag{5.8}$$

Since $\deg f_0 \le r$ and $\mathfrak{g} \in \mathrm{Aut}(RM(r,m))$, we have $\deg f_0^{\mathfrak{g}} \le r$. Consider now the remaining summands. Let, for example, $f_1(\boldsymbol{x}) = x^{(j_1)}x^{(j_2)}\dots x^{(j_r)}x^{(j_{r+1})}$. We represent the function f_1 as $f_1 = f_{10} \cdot f_{11}$, where $f_{10}(\boldsymbol{x}) = x^{(j_1)}x^{(j_2)}\dots x^{(j_r)}$ and $f_{11}(\boldsymbol{x}) = x^{(j_2)}\cdots x^{(j_{r+1})}$, $\deg f_{10} = \deg f_{11} = r$, $\mathrm{wt}(f_{10}) = \mathrm{wt}(f_{11}) = 2^{m-r}$. Also, $\mathrm{wt}(f_1) = \mathrm{wt}(f_1^{\mathfrak{g}}) = 2^{m-r-1}$ and $\mathrm{wt}(f_{10}^{\mathfrak{g}}) = \mathrm{wt}(f_{11}^{\mathfrak{g}}) = 2^{m-r}$, $\Omega_{f_{10}^{\mathfrak{g}}} \in RM(r,m)$, $\Omega_{f_{11}^{\mathfrak{g}}} \in RM(r,m)$. Since $f_1^{\mathfrak{g}} = f_{10}^{\mathfrak{g}} \cdot f_{11}^{\mathfrak{g}}$, the functions $f_{10}^{\mathfrak{g}}$, $f_{11}^{\mathfrak{g}}$, and $f_1^{\mathfrak{g}}$ satisfy the assumptions of Problem 5.17 and $\deg f_1^{\mathfrak{g}} \le r+1$, i.e., $\Omega_{f_1} \in RM(r+1,m)$. Using similar arguments for $f_2^{\mathfrak{g}}, \dots, f_t^{\mathfrak{g}}$, we get from (5.8) that $\deg f^{\mathfrak{g}} \le r+1$, i.e., $\Omega_{f^{\mathfrak{g}}} \in RM(r+1,m)$ and $\mathfrak{g} \in \mathrm{Aut}(RM(r+1,m))$.

2) Part 2) can be proved in a similar way. □

THEOREM 5.23. *For $1 \le r \le m-2$ the following equalities hold:*
1) $\mathrm{Aut}(RM(r,m)) = \mathfrak{GA}(V_m)$;
2) $\mathrm{Aut}(RM^*(r,m)) = \mathfrak{GL}(V_m)$.

PROOF. The code $\boldsymbol{C}_m^{\perp}$ (see §4.4) is the simplex code of maximum length $n = 2^m - 1$. In other words, $\boldsymbol{C}_m^{\perp} = \{\Omega_f^* \mid f \in \mathcal{L}_m\}$. If we add to the code $\boldsymbol{C}_m^{\perp}$ the all-one code word, we obtain $RM^*(1,m)$. The code $RM^*(1,m)$ itself is obtained from the code $RM(1,m)$ by puncturing the components corresponding to the vector $\mathbf{0}$. According to Problem 4.22, we have $\mathrm{Aut}(\boldsymbol{C}_m^{\perp}) = \mathrm{Aut}(RM^*(1,m))$, and according to Problem 5.20, the following inclusion holds: $\mathfrak{GL}(V_m) \subset \mathrm{Aut}(RM^*(1,m))$. Then by Lemma 4.23 and Problem 4.24, we get $\mathrm{Aut}(\boldsymbol{C}_m^{\perp}) = \mathrm{Aut}(RM^*(1,m)) = \mathfrak{GL}(V_m)$.

Let $\mathfrak{G}_1 = \mathrm{Aut}(RM^*(1,m))$, $\mathfrak{G}_2 = \mathrm{Aut}(RM(1,m))$. It is clear that $\mathfrak{G}_1$ is a subgroup of the group $\mathfrak{G}_2$ that fixes the component of the code word which corresponds to the vector $\mathbf{0}$. Since the group $\mathfrak{GA}(V_m)$ is transitive, the group $\mathfrak{G}_2$ is also transitive. Since different cosets of the subgroup $\mathfrak{G}_2$ in the group $\mathfrak{G}_1$ transform the vector $\mathbf{0}$ into different vectors of the space V_m, we have $\#\mathfrak{G}_2 = 2^m \cdot \#\mathfrak{G}_1$. Therefore, $\mathfrak{G}_2 = \mathfrak{GA}(V_m)$. Using Problem 4.21, we get $\mathrm{Aut}(RM(m-2,m)) = \mathrm{Aut}(RM(1,m)) = \mathfrak{GA}(V_m)$.

From everything we have said so far it follows that

$$\begin{aligned}
\mathfrak{GA}(V_m) &\subseteq \mathrm{Aut}(RM(1,m)) \subseteq \mathrm{Aut}(RM(2,m)) \subseteq \cdots \\
&\subseteq \mathrm{Aut}(RM(m-2,m)) = \mathfrak{GA}(V_m), \\
\mathfrak{GL}(V_m) &\subseteq \mathrm{Aut}(RM^*(1,m)) \subseteq \mathrm{Aut}(RM^*(2,m)) \subseteq \cdots \\
&\subseteq \mathrm{Aut}(RM^*(m-2,m)) = \mathfrak{GL}(V_m).
\end{aligned}$$
□

5.2. Reed's decoding algorithm

Now we continue with Reed's decoding algorithm for Reed–Muller codes. These codes are linear over the field of two elements $\mathbb{F}_2$, and they have a simple description and an easy decoding algorithm (the simple majority algorithm). As a rule, in any book devoted to coding theory, the practical significance of Reed–Muller codes is illustrated by its application in deep space communication.

We will study a decoding algorithm that was specially designed for Reed–Muller codes. This algorithm is also interesting because in essence it is a method for determining the minimum distance of the code. It differs from most decoding algorithms because it allows for recovering of the information symbols directly from the received word without finding the exact value of the error. There is also no need for intermediate variables, syndromes for example.

The decoding algorithm for a code $RM(r,m)$ can correct $\frac{1}{2}2^{m-r}-1$ errors and recover $k=\sum_{i=0}^{r}\binom{m}{i}$ information symbols. From this it follows that the minimum distance (see Theorem 5.7) of a code $RM(r,m)$ is at least 2^{m-r}.

Suppose that we have a decoder (device realizing the decoding algorithm) of the Reed–Muller code of $(r-1)$th order correcting $\frac{1}{2}2^{m-(r-1)}-1$ errors. We will build a decoder for the Reed–Muller code of rth order correcting $\frac{1}{2}2^{m-r}-1$ errors by reducing this case to the previous one.

The code $RM(0,m)$ with code words $\boldsymbol{c}_0=\mathbf{0}$ and $\boldsymbol{c}_1=\mathbf{1}$ can obviously correct $\frac{1}{2}2^{m}-1$ errors by majority logic decoding (i.e., by voting). This will be the base of induction.

Let us see how the induction step is done. Divide the information vector $\boldsymbol{u}$ of length $k=\sum_{i=0}^{r}\binom{m}{i}$ into $r+1$ parts $\boldsymbol{u}=(\boldsymbol{u}_0,\boldsymbol{u}_1,\dots,\boldsymbol{u}_r)$, where $\boldsymbol{u}_i$, $i=0,\dots,r$ contains $\binom{m}{i}$ information bits. The coding can be represented as multiplication of a vector and a matrix

$$\boldsymbol{c}=\boldsymbol{u}G=(\boldsymbol{u}_0,\boldsymbol{u}_1,\dots,\boldsymbol{u}_r)\begin{pmatrix}G_0\\G_1\\\cdots\\G_r\end{pmatrix},$$

where to any information part corresponds one of the $r+1$ blocks of the generator matrix (see (5.3)) of $RM(r,m)$. If we can recover the information bits of the part $\boldsymbol{u}_r$, we can calculate its contribution to the received word and subtract it from this word. Then the problem will be reduced to decoding a code $RM(r-1,m)$. The decoding procedure is a sequence of majority decodings, and it begins with finding the information symbols in the rth part by the majority method.

We write the received word as

$$\boldsymbol{v}=(\boldsymbol{u}_0,\boldsymbol{u}_1,\dots,\boldsymbol{u}_r)\begin{pmatrix}G_0\\G_1\\\cdots\\G_r\end{pmatrix}\oplus\boldsymbol{e},$$

where $\boldsymbol{e}\in V_{2^m}$ is the channel error. At the beginning, the decoding algorithm recovers $\boldsymbol{u}_r$ from the vector $\boldsymbol{v}$ and then calculates the difference

$$\boldsymbol{v}'=\boldsymbol{v}\oplus\boldsymbol{u}_rG_r=(\boldsymbol{u}_0,\boldsymbol{u}_1,\dots,\boldsymbol{u}_{r-1})\begin{pmatrix}G_0\\G_1\\\cdots\\G_{r-1}\end{pmatrix}\oplus\boldsymbol{e},$$

which is a corrupted code word of $RM(r-1,m)$.

For convenience we index the components of the information vector $\boldsymbol{u}$ of the code word $\boldsymbol{c}$ and the received word $\boldsymbol{v}=\boldsymbol{c}\oplus\boldsymbol{e}$ as follows:

$$\boldsymbol{c}=(c(\mathbf{0}),\dots,c(\mathbf{1})),\quad \boldsymbol{v}=(v(\mathbf{0}),\dots,v(\mathbf{1})),$$
$$\boldsymbol{u}_l=(u_l^{(0,\dots,0,1,\dots,1)},\dots,u_l^{(1,\dots,1,0,\dots,0)}),$$

$l = 0, 1, \dots, r$, for any $u_l^{(\sigma)}$, $\mathrm{wt}(\sigma) = l$, $\sigma \in V_m$. The part $u_l^{(\sigma)}$ corresponds to the row $\Omega_{g_\sigma^l(x)}$ of the matrix G_l, where $g_\sigma^l(\boldsymbol{x}) = (x^{(1)})^{\sigma^{(1)}} \cdots (x^{(m)})^{\sigma^{(m)}}$ (see (5.3)) and

$$c(\boldsymbol{x}) = \bigoplus_{l=0}^{r} \bigoplus_{\substack{\sigma \in V_m \\ \mathrm{wt}(\sigma)=l}} u_l^{(\sigma)} g_\sigma^l(\boldsymbol{x}). \tag{5.9}$$

Consider decoding of the information bit $u_r^{(1,\dots,1,0,\dots,0)}$, which is multiplied by the last row of the generator matrix G (see (5.3)) of the form $\Omega_{g^r_{(1,\dots,1,0,\dots,0)}(\boldsymbol{x})}$, where $g^r_{(1,\dots,1,0,\dots,0)}(\boldsymbol{x}) = x^{(1)} \cdots x^{(r)}$. We decode this bit by calculating 2^{m-r} check sums of 2^r bits from 2^m bits of the received word $\boldsymbol{v}$ in such a way that any received bit is contained in only one sum. The check sums are such that the symbol $u_r^{(1,\dots,1,0,\dots,0)}$ contributes only one bit and all the rest contribute an even number of bits to any check sum.

For any $\boldsymbol{b} \in V_{m-r}$ we define the check sum

$$\theta_{\boldsymbol{b},r}^{(1,\dots,1,0,\dots,0)}(\boldsymbol{v}) = \bigoplus_{\substack{\boldsymbol{x}\in V_m,\, \boldsymbol{x}^{(r+i)}=\boldsymbol{b}^{(i)} \\ 1\le i\le m-r}} v(\boldsymbol{x}). \tag{5.10}$$

If there are no errors in the communication channel, the equality $\boldsymbol{v} = \boldsymbol{c}$ holds, and according to (5.9) and (5.10) one has

$$\begin{aligned} \theta_{\boldsymbol{b},r}^{(1,\dots,1,0,\dots,0)}(\boldsymbol{v}) &= \bigoplus_{\substack{\boldsymbol{x}\in V_m,\, \boldsymbol{x}^{(r+i)}=\boldsymbol{b}^{(i)} \\ 1\le i\le m-r}} \bigoplus_{l=0}^{r} \bigoplus_{\boldsymbol{\sigma}\in V_m,\, \mathrm{wt}(\boldsymbol{\sigma})=l} u_l^{(\boldsymbol{\sigma})} g_{\boldsymbol{\sigma}}^l(\boldsymbol{x}) \\ &= \bigoplus_{l=0}^{r} \bigoplus_{\substack{\boldsymbol{\sigma}\in V_m \\ \mathrm{wt}(\boldsymbol{\sigma})=l}} u_l^{(\boldsymbol{\sigma})} \bigoplus_{\substack{\boldsymbol{x}\in V_m,\, \boldsymbol{x}^{(r+i)}=\boldsymbol{b}^{(i)} \\ 1\le i\le m-r}} g_{\boldsymbol{\sigma}}^l(\boldsymbol{x}). \end{aligned} \tag{5.11}$$

The inner sum in (5.11) can be written as

$$\begin{aligned} \bigoplus_{\substack{\boldsymbol{x}\in V_m,\, \boldsymbol{x}^{(r+i)}=\boldsymbol{b}^{(i)} \\ 1\le i\le m-r}} g_{\boldsymbol{\sigma}}^l(\boldsymbol{x}) &= (b^{(1)})^{\sigma^{(r+1)}} \cdots (b^{(m-r)})^{\sigma^{(m)}} \\ &\oplus \Bigl(\bigoplus_{x^{(i)}\in\mathbb{F}_2,\, 1\le i\le r} (x^{(1)})^{\sigma^{(1)}} \cdots (x^{(r)})^{\sigma^{(r)}} \Bigr). \end{aligned} \tag{5.12}$$

The expression $\bigoplus_{x^{(i)}\in\mathbb{F}_2,\,1\le i\le r}(x^{(1)})^{\sigma^{(1)}} \cdots (x^{(r)})^{\sigma^{(r)}}$ is equal to 1 if and only if $\sigma^{(1)} = \cdots = \sigma^{(r)} = 1$; thus, the sum (5.12) is also equal to 1 if $\boldsymbol{\sigma} = (\underbrace{1,\dots,1}_{r}, 0, \dots, 0)^\top$ and equal to 0 if $\boldsymbol{\sigma} \ne (\underbrace{1,\dots,1}_{r}, 0, \dots, 0)^\top$. Therefore, $\theta_{\boldsymbol{b},r}^{(1,\dots,1,0,\dots,0)}(\boldsymbol{u}) = u_r^{(1,\dots,1,0,\dots,0)}$ for any $\boldsymbol{b} \in V_{m-r}$. Thus, if there are no errors, then all the 2^{m-r} check sums are equal to $u_r^{(1,\dots,1,0,\dots,0)}$. But if there are no more than $\frac{1}{2}2^{m-r} - 1$ errors, most of the check sums will be equal to $u_r^{(1,\dots,1,0,\dots,0)}$.

Similarly, any information symbol which is multiplied by one row of the generator matrix G can be decoded. This is possible because the rows of the matrix G_r are indistinguishable from each other. Permuting the columns, we can make any row of the matrix G to look like the last one, i.e., G_r. Therefore, the same check sums can be used, only the indexes included in these sums have to be permuted. After constructing 2^{m-r} check sums, any bit is decoded by the majority method. After obtaining the information part $\boldsymbol{u}_r$ and undoing its effect on the received word, we proceed with the decoding of the code $RM(r-1, m)$.

One continues with these computations until all the information parts are found.

To conclude this section we now present a useful fact concerning divisibility of weights of Reed–Muller codes.

THEOREM 5.24 (McEliece). *The weight of any code word of $RM(r, m)$ is divisible by $2^{\lceil m/r \rceil - 1}$.*

PROOF. Suppose that $f \in \mathcal{F}_m$, $0 < \deg f = d \leq r$; let

$$\begin{aligned} f(\boldsymbol{x}) = f(x^{(1)}, \dots, x^{(m)}) &= \bigoplus_{\boldsymbol{u} \in V_m} g_f(\boldsymbol{u}) \boldsymbol{x}^{\boldsymbol{u}} \\ &= \bigoplus_{(u^{(1)}, \dots, u^{(m)}) \in V_m} g_f(u^{(1)}, \dots, u^{(m)}) (x^{(1)})^{u^{(1)}} \cdots (x^{(m)})^{u^{(m)}} \end{aligned}$$

be its algebraic normal form. Put

$$U_f = \{\boldsymbol{u} \in V_m \mid g_f(\boldsymbol{u}) = 1\}.$$

Then $f(\boldsymbol{x}) = \bigoplus_{\boldsymbol{u} \in U_f} \boldsymbol{x}^{\boldsymbol{u}}$, and $\mathrm{wt}(\boldsymbol{u}) \leq d$ for any $\boldsymbol{u} \in U_f$. Consider the Walsh–Hadamard coefficient of the function f at zero: $W_f(\mathbf{0}) = \sum_{\boldsymbol{x} \in V_m} (-1)^{f(\boldsymbol{x})}$. As $(-1)^{\bigoplus_{\boldsymbol{u} \in U_f} \boldsymbol{x}^{\boldsymbol{u}}} = (-1)^{\sum_{\boldsymbol{u} \in U_f} \boldsymbol{x}^{\boldsymbol{u}}}$, in the proof of this theorem we will use real sums. We have

$$W_f(\mathbf{0}) = \sum_{\boldsymbol{x} \in V_m} (-1)^{\sum_{\boldsymbol{u} \in U_f} \boldsymbol{x}^{\boldsymbol{u}}} = \sum_{\boldsymbol{x} \in V_m} \prod_{\boldsymbol{u} \in U_f} (1 - 2\boldsymbol{x}^{\boldsymbol{u}}).$$

We denote by A all images of the mapping $\alpha \colon U_f \to \{0, 1\}$. One has

$$W_f(\mathbf{0}) = \sum_{\boldsymbol{x} \in V_m} \sum_{\alpha \in A} \prod_{\boldsymbol{u} \in U_f} (-2\boldsymbol{x}^{\boldsymbol{u}})^{\alpha(\boldsymbol{u})} = \sum_{\alpha \in A} (-2)^{\sum_{\boldsymbol{u} \in U_f} \alpha(\boldsymbol{u})} \sum_{\boldsymbol{x} \in V_m} \boldsymbol{x}^{\sum_{\boldsymbol{u} \in U_f} \alpha(\boldsymbol{u}) \boldsymbol{u}}.$$

Let A_s, $0 \leq s \leq m$, be the set of those images $\alpha \in A$ for which the integer vector $\sum_{\boldsymbol{u} \in U_f} \alpha(\boldsymbol{u}) \boldsymbol{u}$ has exactly s nonzero coordinates. If $\alpha \in A_s$, it is clear that

$$\sum_{\boldsymbol{x} \in V_m} \boldsymbol{x}^{\sum_{\boldsymbol{u} \in U_f} \alpha(\boldsymbol{u}) \boldsymbol{u}} = 2^{m-s}.$$

Thus,

$$W_f(\mathbf{0}) = \sum_{0 \leq s \leq m} \sum_{\alpha \in A_s} (-1)^{\sum_{\boldsymbol{u} \in U_f} \alpha(\boldsymbol{u})} 2^{m - s + \sum_{\boldsymbol{u} \in U_f} \alpha(\boldsymbol{u})}. \tag{5.13}$$

We note that any term in the sum (5.13) is divisible by $2^{m-s+\sum_{\boldsymbol{u} \in U_f} \alpha(\boldsymbol{u})}$, $\alpha \in A_s$, $0 \leq s \leq m$. We will find the minimum among all these powers of 2. As $\sum_{\boldsymbol{u} \in U_f} \alpha(\boldsymbol{u}) \boldsymbol{u}$ has exactly s nonzero coordinates for any $\alpha \in A_s$ and $\mathrm{wt}(\boldsymbol{u}) \leq d$ for any $\boldsymbol{u} \in U_f$, it is easy to see that $\sum_{\boldsymbol{u} \in U_f} \alpha(\boldsymbol{u}) \geq \left\lceil \frac{s}{d} \right\rceil$ for any $\alpha \in A_s$. It is obvious that

$$\min_{\substack{0 \leq s \leq m \\ 0 < d \leq r}} \left(m - s + \left\lceil \frac{s}{d} \right\rceil\right) = \left\lceil \frac{m}{r} \right\rceil.$$

Thus, $W_f(\mathbf{0})$ is divisible by $2^{\lceil m/r \rceil}$ for any $\Omega_f \in RM(r,m)$, whence $\mathrm{wt}(\Omega_f)$ is divisible by $2^{\lceil m/r \rceil - 1}$. $\square$

COROLLARY 5.25. *Suppose that $f \in \mathcal{F}_m$ and $\deg f \leq r$. Then $W_f(\boldsymbol{u})$ is divisible by $2^{\lceil m/r \rceil - 1}$ for any $\boldsymbol{u} \in V_m$.*

In connection with this statement we will note that the minimum power of 2 dividing all Walsh–Hadamard coefficients of a given Boolean function is an interesting but less studied affine invariant of Boolean functions. This motivates the following problem.

PROBLEM 5.26 (Open problem). Find an effective algorithm for calculation of the above-mentioned affine invariant for some classes of Boolean functions important in practice and describe its connection with other cryptographic parameters.

5.3. First order Reed–Muller codes and connections with other codes

In this section we will consider Reed– Muller codes of the first order, $RM(1,m)$, in detail. A code $RM(1,m)$ is an $[n,k,d]$-code with the following basic parameters: $n = 2^m$, $k = m+1$, $d = 2^{m-1}$. Its generator matrix G is

$$G = \begin{pmatrix} \Omega_1 \\ \Omega_{x^{(m)}} \\ \cdots \\ \Omega_{x^{(1)}} \end{pmatrix}.$$

A punctured code $RM^*(1,m)$ is a $[2^m - 1, m+1, 2^{m-1} - 1]$-cyclic code with generator polynomial $g(\lambda) = (\lambda^{2^m - 1} \oplus 1)/m_{\alpha^{-1}}(\lambda)(\lambda \oplus 1)$, where α is a primitive element of the field $\mathbb{F}_{2^m}$.

The weight spectrum of a code $RM(1,m)$ is

$$A_0 = A_{2^m} = 1, \quad A_{2^{m-1}} = 2^{m+1} - 2.$$

In various practical situations the problem of calculating weight spectra for cosets of Reed–Muller codes of first order (i.e., $\Omega_f \oplus RM(1,m)$, $f \in \mathcal{F}_m$) arises. For example, the problem of finding the weight spectrum for the coset containing the vector Ω_f, $f \in \mathcal{F}_m$, is reduced to the problem of finding the Walsh–Hadamard transform for this vector.

PROPOSITION 5.27. *The weight spectrum of a coset of $RM(1,m)$ containing Ω_f is equal to*

$$\left\{ 2^{m-1} \pm \frac{1}{2} W_f(\boldsymbol{u}) \mid \boldsymbol{u} \in V_m \right\}.$$

PROOF. The expression for distances between functions f and $l_{\boldsymbol{u},a}$, $\boldsymbol{u} \in V_m$, $a \in \mathbb{F}_2$ immediately follows from the properties of the Walsh–Hadamard transform. $\square$

PROPOSITION 5.28. *Suppose that functions f and h are $\mathfrak{GA}(V_m)$-equivalent. Then the weight spectrum of the coset of $RM(1,m)$ containing Ω_f coincides with the weight spectrum of the coset of $RM(1,m)$ containing Ω_g.*

PROOF. It follows from the hypothesis that there exists a permutation $\mathfrak{g} \in \mathfrak{GA}(V_m)$, $\mathfrak{g}\boldsymbol{x} = A\boldsymbol{x} \oplus \boldsymbol{b}$, $A \in \mathfrak{GL}(V_m)$, $\boldsymbol{b} \in V_m$ such that $f^{\mathfrak{g}} = h$. Taking into account the properties of the Walsh–Hadamard coefficients, we find that the sets $\left\{ 2^{m-1} \pm \frac{1}{2} W_f(\boldsymbol{u}) \,\middle|\, \boldsymbol{u} \in V_m \right\}$ and $\left\{ 2^{m-1} \pm \frac{1}{2} W_h(\boldsymbol{u}) \,\middle|\, \boldsymbol{u} \in V_m \right\}$ coincide. $\square$

We proceed with the problem of decoding codes $RM(1,m)$. Maximum likelihood decoding is reduced to comparing the received (from the communication channel) vector Ω_f with all code words of the code $RM(1,m)$ and choosing the closest one. According to Proposition 2.51, this is equivalent to finding the greatest value of $|W_f(\boldsymbol{u})|$ for $\boldsymbol{u} \in V_m$. Suppose that the greatest value is $|W_f(\boldsymbol{u})|$ for some $\boldsymbol{u} = (u^{(1)}, \dots, u^{(m)})^\top \in V_m$. Then if $W_f(\boldsymbol{u}) \geq 0$, then the vector Ω_f is decoded into the code word $\Omega_{l_{\boldsymbol{u}},0}$, where $l_{\boldsymbol{u},0}(\boldsymbol{x}) = \langle \boldsymbol{u}, \boldsymbol{x} \rangle = \bigoplus_{i=1}^m u^{(i)} x^{(i)}$. If $W_f(\boldsymbol{u}) < 0$, then the vector Ω_f is decoded to the code word $\Omega_{l_{\boldsymbol{u},1}}$, where $l_{\boldsymbol{u},1}(\boldsymbol{x}) = \langle \boldsymbol{u}, \boldsymbol{x} \rangle \oplus 1 = \bigoplus_{i=1}^m u^{(i)} x^{(i)} \oplus 1$. Thus, the central part of the procedure of decoding $RM(1,m)$ for codes is calculating the Walsh–Hadamard transform.

Put

$$\exp \Omega_f = \bar{\Omega}_f = ((-1)^{f(\boldsymbol{u}_0)}, (-1)^{f(\boldsymbol{u}_1)}, \dots, (-1)^{f(\boldsymbol{u}_{2^m-1})}).$$

Definition 1.17 shows that the direct calculation of

$$W_f = (W_f(\boldsymbol{u}_0), W_f(\boldsymbol{u}_1), \dots, W_f(\boldsymbol{u}_{2^m-1})) = \bar{\Omega}_f H_m \tag{5.14}$$

by multiplying the row $\bar{\Omega}_f$ by the Sylvester-type Hadamard matrix H_m requires $\mathcal{O}(2^{2m})$ additions and subtractions. There is a fast method for calculating W based on the so-called Hadamard fast transform. According to this method, H_m is represented as product of m $(2^m \times 2^m)$ matrices, both containing only two nonzero elements ± 1 in each column. With this method the row $W = \bar{\Omega}_f H_m$ can be found by $m \cdot 2^m$ additions and subtractions.

In the calculation of the Hadamard matrix H_m, the Kronecker (tensor) product of matrices is used. For a detailed description of this factorization of Hadamard matrices we will need some properties of the Kronecker product. We will state these properties as (not too hard) problems for a special case of the Kronecker product: product of square matrices.

DEFINITION 5.29. Suppose that $A = (a_{ij})$ is a $(t \times t)$ matrix over an arbitrary field $\mathbb{F}$ and $B = (b_{kl})$ is an $(s \times s)$ matrix over the same field. The *Kronecker product* $A \otimes B$ of these matrices is the $(ts \times ts)$ matrix over the same field $\mathbb{F}$ obtained by replacing each element a_{ij} of the matrix A with the matrix $a_{ij}B$, i.e.,

$$A \otimes B = \begin{pmatrix} a_{11}B & a_{12}B & \dots & a_{1t}B \\ a_{21}B & a_{22}B & \dots & a_{2t}B \\ \multicolumn{4}{c}{\dotfill} \\ a_{t1}B & a_{t2}B & \dots & a_{tt}B \end{pmatrix}.$$

The Kronecker product is not commutative.

PROBLEM 5.30. Prove the following properties of the Kronecker product of square matrices over an arbitrary field:

(1) $A \otimes (B \otimes C) = (A \otimes B) \otimes C$;
(2) $(A + B) \otimes C = (A \otimes C) + (B \otimes C)$;
(3) $(A \otimes B)(C \otimes D) = (AC) \otimes (BD)$;
(4) $(A \otimes B)^\top = A^\top \otimes B^\top$.

THEOREM 5.31 (Fast Hadamard transform). *The Hadamard matrix H_m for an arbitrary natural number $m \in \mathbb{N}$ can be represented as*

$$H_m = M_m^{(1)} M_m^{(2)} \cdots M_m^{(m)}, \tag{5.15}$$

where

$$M_m^{(j)} = I_{m-j} \otimes H_1 \otimes I_{j-1}, \quad 1 \le j \le m,$$

and I_t *is the identity* $(2^t \times 2^t)$ *matrix.*

PROOF. We will use induction on m. For $m = 1$ the theorem is obvious. Suppose that the theorem holds for all $m \le k$. Then for $1 \le j \le k$ one has

$$M_{k+1}^{(j)} = I_{k+1-j} \otimes H_1 \otimes I_{j-1} = I_1 \otimes I_{k-j} \otimes H_1 \otimes I_{j-1} = I_1 \otimes M_k^{(j)}, \tag{5.16}$$

and for $j = k+1$ one has

$$M_{k+1}^{(k+1)} = H_1 \otimes I_k. \tag{5.17}$$

According to (5.15)–(5.17), the induction hypothesis, and property (3) from Problem 5.30, we get

$$\begin{aligned} M_{k+1}^{(1)} M_{k+1}^{(2)} \cdots M_{k+1}^{(k)} M_{k+1}^{(k+1)} &= (I_1 \otimes M_k^{(2)}) \cdots (I_1 \otimes M_k^{(k)})(H_1 \otimes I_k) \\ &= H_1 \otimes (M_k^{(1)} M_k^{(2)} \cdots M_k^{(k)}) = H_1 \otimes H_k = H_{k+1}. \end{aligned}$$

Therefore, (5.15) holds also for $k+1$ and thus for an arbitrary m. □

EXAMPLE 5.32. Suppose that $m = 2$. Then $M_2^{(1)} = I_1 \otimes H_1$, $M_2^{(2)} = H_1 \otimes I_1$, and

$$\begin{aligned} H_2 = H_1 \otimes H_1 &= M_2^{(1)} M_2^{(2)} \\ &= \begin{pmatrix} 1 & 1 & 0 & 0 \\ 1 & -1 & 0 & 0 \\ 0 & 0 & 1 & 1 \\ 0 & 0 & 1 & -1 \end{pmatrix} \begin{pmatrix} 1 & 0 & 1 & 0 \\ 0 & 1 & 0 & 1 \\ 1 & 0 & -1 & 0 \\ 0 & 1 & 0 & -1 \end{pmatrix} = \begin{pmatrix} 1 & 1 & 1 & 1 \\ 1 & -1 & 1 & -1 \\ 1 & 1 & -1 & -1 \\ 1 & -1 & -1 & 1 \end{pmatrix}. \end{aligned}$$

COROLLARY 5.33. *The matrix* $M_m^{(j)}$, $1 \le j \le m$, *from* (5.15) *has exactly two nonzero elements in any column (and in any row); they lie in the set* $\{+1, -1\}$.

PROOF. The assertion is immediate from the expression for the matrix $M_m^{(j)}$ and the properties of the Kronecker product. □

COROLLARY 5.34. *The row* $W = \bar{\Omega}_f H_m$ *can be found by* $m2^m$ *additions and subtractions.*

PROOF. Multiplying consecutively by matrices $M_m^{(j)}$, $j = 1, \dots, m$, we need exactly 2^m additions and subtractions for multiplication by any matrix $M_m^{(j)}$; hence, the total number of additions and subtractions is $\underbrace{2^m + \cdots + 2^m}_{m} = m2^m$. □

A scheme based on Theorem 5.31 and called the correlation decoding of the code $RM(1,m)$ was suggested by R. Green. Green's scheme needs $\mathcal{O}(m2^m)$ operations for obtaining the vector W and finding the maximum modulus of its components. It has the following nice property: the decoding scheme for $RM(1,m+1)$ differs from the decoding scheme of $RM(1,m)$ by adding one more step, the $(m+1)$th decoding step, and making a small change in the mth step.

There also exist other factorizations of Hadamard matrices different from the one given in Theorem 5.31; the matrix factors used in these schemes are completely different. It is possible to make a correlation decoding faster by using subcodes (linear or nonlinear) of the code $RM(1,m)$.

Suppose that the code words used for information transmission over a binary symmetric channel are taken from the following set: $\mathcal{L}_{r,m} = \{\Omega_{l_{\boldsymbol{a},0}} \in RM(1,m) \mid l_{\boldsymbol{a},0}(\boldsymbol{x}) = \langle \boldsymbol{a}, \boldsymbol{x} \rangle,\ \boldsymbol{a} \in V_m,\ \mathrm{wt}(\boldsymbol{a}) = r\}$; this is a nonlinear subcode of $RM(1,m)$, $\mathcal{L}_{r,m} \subset RM(1,m)$. Denote by $\mathfrak{g}_1$ the element of the group $\mathfrak{T}_m$ acting on V_m by the formula

$$\mathfrak{g}_1(\boldsymbol{x}) = \mathfrak{g}_1(x^{(1)}, \ldots, x^{(m)}) = (x^{(1)} \oplus 1, \ldots, x^{(m)} \oplus 1)$$

for any $\boldsymbol{x} \in V_m$.

We will set $\mathfrak{g}_1(\boldsymbol{x}) = \widetilde{\boldsymbol{x}}$. On the receiving end of a binary symmetric channel we get a corrupted code word Ω_f for some Boolean function $f \in \mathcal{F}_m$. Now we have to perform the Hadamard transform of the corrupted code word and to obtain $W = \bar{\Omega}_f \cdot H_m$. As we have already done it many times before, we assume that the components of the vectors $\bar{\Omega}_f$, W and the elements of the matrix H_m are indexed by vectors of the space V_m ordered lexicographically:

$$\bar{\Omega}_f = ((-1)^{f(\boldsymbol{u}_0)}, (-1)^{f(\boldsymbol{u}_1)}, \ldots, (-1)^{f(\boldsymbol{u}_{2^m-1})}),$$
$$W = (w_{\boldsymbol{u}_0}, w_{\boldsymbol{u}_1}, \ldots, w_{\boldsymbol{u}_{2^m-1}}), \quad w_i = W_f(\boldsymbol{u}_i), \quad i = 0, \ldots, 2^m - 1,$$
$$H_m = ((-1)^{\langle \boldsymbol{u}_i, \boldsymbol{v}_j \rangle}), \quad i = 0, \ldots, 2^m - 1; \quad j = 0, \ldots, 2^m - 1.$$

As a complexity measure for decoding of $\mathcal{L}_{r,m}$ we choose the maximum of the number of additions and the number of subtractions; the complexity will be denoted by $\mathcal{E}(r,m)$.

The component $w_{\boldsymbol{v}}$ of W has the form

$$w_{\boldsymbol{v}} = \sum_{\boldsymbol{u} \in V_n} (-1)^{f(\boldsymbol{u}) \oplus \langle \boldsymbol{v}, \boldsymbol{u} \rangle}. \tag{5.18}$$

For decoding we need to calculate the components of the row W and choose from the components of the weight index r those of maximum weight. We estimate the complexity of the recursive algorithm for calculation of the components of the row W denoted by $\mathcal{B}(r,m)$. If $r = 0$ or $r = m$, then $w_{\mathbf{0}}$ and $w_{\mathbf{1}}$ are calculated according to (5.18) by summation:

$$w_{\mathbf{0}} = \sum_{\boldsymbol{u} \in V_m} (-1)^{f(\boldsymbol{u})}, \quad w_{\mathbf{1}} = \sum_{\boldsymbol{u} \in V_m} (-1)^{f(\boldsymbol{u})} (-1)^{\bigoplus_{i=1}^m u^{(i)}}. \tag{5.19}$$

To compute one of the sums of (5.19), we need $2^m - 1$ additions and subtractions of two numbers. Suppose that $0 < r < m$. The recursive algorithm suggested below is based on the following relation, which holds for the element $\mathfrak{g}_1 \in \mathfrak{T}_m$:

$$(-1)^{\langle \mathfrak{g}_1(\boldsymbol{u}), \boldsymbol{v} \rangle} = (-1)^{\langle \tilde{\boldsymbol{u}}, \boldsymbol{v} \rangle} = (-1)^{\mathrm{wt}(\boldsymbol{v})} (-1)^{\langle \boldsymbol{u}, \boldsymbol{v} \rangle} = (-1)^r (-1)^{\langle \boldsymbol{u}, \boldsymbol{v} \rangle}, \tag{5.20}$$

where $\boldsymbol{v} \in V_m$, $\mathrm{wt}(\boldsymbol{v}) = r$.

The vector space V_m is a disjoint union of the equivalence classes $\{\boldsymbol{u}, \widetilde{\boldsymbol{u}}\}$ for the permutation $\mathfrak{g}_1$. We set up a one-to-one correspondence between the equivalence classes $\{\boldsymbol{u}, \widetilde{\boldsymbol{u}}\}$ and the vectors $\boldsymbol{u}^* = (u^{(1)}, \ldots, u^{(m-1)})^\top \in V_{m-1}$ as follows:

$$\boldsymbol{u}^* \longleftrightarrow \{\boldsymbol{u}, \widetilde{\boldsymbol{u}}\},$$

where $\boldsymbol{u} = (\boldsymbol{u}^*, 0)^\top = (u^{(1)}, \ldots, u^{(m-1)}, 0)^\top$. Later, in the description of the recursive algorithm, we will assume that the vectors $\boldsymbol{u}^*$ and $\boldsymbol{u}$ are connected by this relation, unless stipulated otherwise. We split the sum (5.18) into two sums as follows:

$$
(5.21)\qquad \begin{aligned} w_{\boldsymbol{v}} &= \sum_{\boldsymbol{u}\in V_m} (-1)^{f(\boldsymbol{u})}(-1)^{\langle \boldsymbol{v},\boldsymbol{u}\rangle} \\ &= \sum_{\boldsymbol{u}^*\in V_{m-1}} (-1)^{f(\boldsymbol{u})}(-1)^{\langle \boldsymbol{v},\boldsymbol{u}\rangle} + \sum_{\boldsymbol{u}^*\in V_{m-1}} (-1)^{f(\widetilde{\boldsymbol{u}})}(-1)^{\langle \boldsymbol{v},\widetilde{\boldsymbol{u}}\rangle}. \end{aligned}
$$

If we substitute (5.20) in (5.21), taking into account that $\mathrm{wt}(\boldsymbol{v}) = r$, we get

$$
(5.22)\qquad w_{\boldsymbol{v}} = \sum_{\boldsymbol{u}^*\in V_{m-1}} (-1)^{\langle \boldsymbol{v},\boldsymbol{u}\rangle}[(-1)^{f(\boldsymbol{u})} + (-1)^{f(\widetilde{\boldsymbol{u}})}\cdot(-1)^r].
$$

Introduce the following notation:

$$
(5.23)\qquad \begin{aligned} q^0_{\boldsymbol{u}^*} &= (-1)^{f(\boldsymbol{u})} + (-1)^{f(\widetilde{\boldsymbol{u}})}\cdot(-1)^r, \\ w^1_{\boldsymbol{v}^*} = w_{\boldsymbol{v}} \quad (v^{(m)} &= 1), \qquad w^0_{\boldsymbol{v}^*} = w_{\boldsymbol{v}} \quad (v^{(m)} = 0), \end{aligned}
$$

where $\boldsymbol{v}^* = (v^{(1)},\dots,v^{(m-1)})^\top$ for the vector $\boldsymbol{v} = (v^{(1)},\dots,v^{(m-1)},v^{(m)})^\top$. For vectors $\boldsymbol{u}$ in (5.22) we have $u^{(m)} = 0$, whence

$$
(5.24)\qquad (-1)^{\langle \boldsymbol{u},\boldsymbol{v}\rangle} = (-1)^{\langle \boldsymbol{u}^*,\boldsymbol{v}^*\rangle}.
$$

Note that the weight of the reduced vector $\boldsymbol{v}^*$ does not change if $v^{(m)} = 0$, i.e., $\mathrm{wt}(\boldsymbol{v}^*) = r$, and decreases by one if $v^{(m)} = 1$, i.e., $\mathrm{wt}(\boldsymbol{v}^*) = r-1$. Formulas (5.22) and (5.24) allow us to reduce the problem to a computational problem of a smaller dimension. After substituting (5.24) into (5.22) and taking into account the notation (5.23) and the equalities $v^{(m)} = 0$ or $v^{(m)} = 1$, we obtain:

$$
(5.25)\qquad \begin{aligned} w^0_{\boldsymbol{v}^*} &= \sum_{\boldsymbol{u}^*\in V_{m-1}} (-1)^{\langle \boldsymbol{u}^*,\boldsymbol{v}^*\rangle}\cdot q^0_{\boldsymbol{u}^*}, \quad v^{(m)} = 0; \\ w^1_{\boldsymbol{v}^*} &= \sum_{\boldsymbol{u}^*\in V_{m-1}} (-1)^{\langle \boldsymbol{u}^*,\boldsymbol{v}^*\rangle}\cdot q^0_{\boldsymbol{u}^*}, \quad v^{(m)} = 1. \end{aligned}
$$

We can now see that the sums (5.25) have the form (5.18). Since $\mathrm{wt}(\boldsymbol{v}) = r$, we get $\mathrm{wt}(\boldsymbol{v}^*) = r$ for $w^0_{\boldsymbol{v}^*}$ $(v^{(m)} = 0)$ and $\mathrm{wt}(\boldsymbol{v}^*) = r-1$ for $w^1_{\boldsymbol{v}^*}$ $(v^{(m)} = 1)$. Thus, the calculation of $w_{\boldsymbol{v}}$ for all $\boldsymbol{v}\in V_m$ such that $\mathrm{wt}(\boldsymbol{v}) = r$ is reduced to the calculation of $q^0_{\boldsymbol{u}^*}$ using (5.23) (the complexity of the algorithm for this calculation equals 2^{m-1} additions and subtractions of numbers) and the calculation of the quantities of the type (5.18): $w^0_{\boldsymbol{v}^*}$ for $\boldsymbol{v}^*\in V_{m-1}$, $\mathrm{wt}(\boldsymbol{v}^*) = r$ and $w^1_{\boldsymbol{v}^*}$ for $\boldsymbol{v}^*\in V_{m-1}$, $\mathrm{wt}(\boldsymbol{v}^*) = r-1$. In other words, the original problem must be solved for the smaller dimension $m-1$. Thus, the suggested algorithm is applied recursively for the calculation of $w^0_{\boldsymbol{v}^*}$ and $w^1_{\boldsymbol{v}^*}$ for the vector $\boldsymbol{v}^*\in V_{m-1}$. Hence, the complexity of the calculation of $\{w_{\boldsymbol{v}} \mid \boldsymbol{v}\in V_m,\ \mathrm{wt}(\boldsymbol{v}) = r\}$ satisfies the following recursive formula:

$$
(5.26)\qquad \mathcal{B}(r,m) = 2^{m-1} + \mathcal{B}(r,m-1) + \mathcal{B}(r-1,m-1).
$$

This implies the following estimate.

LEMMA 5.35. $\mathcal{B}(r,m) < (r+1)2^m$.

PROOF. It is not hard to show that $\mathcal{B}(r,m) = \mathcal{B}(m-r,m)$, which allows us to treat only the case $r \le \lceil n/2\rceil$. We will perform induction on m. If $m = 1$, then $\mathcal{B}(0,1) = 1 < 2$ and $\mathcal{B}(1,1) = 1 < 2$. Suppose that the lemma holds for $m-1$ and for all r, $0\le r\le m-1$. We will show that the lemma holds for m and for all r, $0 \le r \le m$. For $r = 0$ or $r = m$, this is obvious. Suppose that $0 < r < m$. Formula (5.26) and the induction hypothesis imply that

$$\begin{aligned}\mathcal{B}(r,m) &= 2^{m-1} + \mathcal{B}(r,m-1) + \mathcal{B}(r-1,m-1)\\ &< 2^{m-1} + (r+1)2^{m-1} + r2^{m-1} = (r+1)2^m.\end{aligned}$$ □

LEMMA 5.36.

$$\mathcal{B}(r,m) = (r+1)2^m - 2\sum_{i=0}^{r-1}\binom{m}{i}(r-i) - \binom{m}{r}.$$

PROOF. Using the well-known identity $\binom{m}{r} = \binom{m-1}{r} + \binom{m-1}{r-1}$, where $t \le r$ and $(m-t) \ge 0$, we get

$$\mathcal{B}(r,m) = t2^{m-1} + \sum_{i=0}^{t}\binom{m}{i}\mathcal{B}(r-i,m-t).$$

To complete the proof, use induction. □

COROLLARY 5.37.

$$\mathcal{E}(r,m) = (r+1)2^m - 2\sum_{i=0}^{r-1}\binom{m}{i}(r-i).$$

PROOF. The assertion immediately follows from the equality

$$\mathcal{E}(r,m) = \mathcal{B}(r,m) + \binom{m}{r}.$$ □

We now look at another important characteristic of $RM(1,m)$: its covering radius $r_{RM(1,m)}$. For brevity, in this section we will denote it by $r_m = r_{RM(1,m)}$. The values of r_m are known for small m: $r_0 = r_1 = 0$, $r_2 = 1$, $r_3 = 2$, $r_4 = 6$, $r_5 = 12$, $r_6 = 28$. No general formula for r_m is known, although one can write down the exact value of r_m for infinitely many m. In general, only upper and lower bounds for r_m are known. The covering radius of $RM(1,m)$ is an important cryptographic parameter. We refer the reader to Chapter 7, where it will be studied in more detail; for the time being we will prove the following upper bound.

LEMMA 5.38.

$$r_m \le 2^{m-1} - 2^{(m-2)/2}.$$

PROOF. Definition 4.7 shows that

$$r_m = \max_{f \in \mathcal{F}_m} \operatorname{dist}(f, \mathcal{A}_m). \tag{5.27}$$

Since

$$\operatorname{dist}(f,\mathcal{A}_m) = 2^{m-1} - \frac{1}{2}\max_{\boldsymbol{u}\in V_m}|W_f(\boldsymbol{u})|,$$

Parseval's formula implies the inequality $\max_{\boldsymbol{u}\in V_m}|W_f(\boldsymbol{u})| \ge 2^{m/2}$, whence

$$\operatorname{dist}(f,\mathcal{A}_m) \le 2^{m-1} - 2^{(m-2)/2}$$

for any function $f \in \mathcal{F}_m$. Now it follows from (5.27) that

$$r_m \le 2^{m-1} - 2^{(m-2)/2}.$$ □

COROLLARY 5.39. *If $m \in \mathbb{N}$ is even, $m > 2$, then*

$$r_m = 2^{m-1} - 2^{(m-2)/2}.$$

PROOF. Suppose that $m = 2t$, $t = 2, 3, \dots$. Define the following function on $\mathcal{F}_m$:

$$f(x^{(1)}, \dots, x^{(m)}) = x^{(1)}x^{(2)} \oplus x^{(3)}x^{(4)} \oplus \dots \oplus x^{(m-1)}x^{(m)}.$$

It is easy to prove that $|W_f(\boldsymbol{u})| = 2^{m/2}$ for any $\boldsymbol{u} \in V_m$. In other words, $\operatorname{dist}(f, \mathcal{A}_m) = 2^{m-1} - 2^{(m-2)/2}$. The bound of Lemma 5.38 is attained for the function f, which proves the corollary. □

LEMMA 5.40. *If $m \geq 2$, then $r_m \geq 2r_{m-2} + 2^{m-2}$.*

PROOF. Let $f \in \mathcal{F}_m$ be an arbitrary Boolean function of m variables. We represent it as

$$f(x^{(1)}, \dots, x^{(m-2)}, x^{(m-1)}, x^{(m)}) = \bigoplus_{u^{(m-1)}, u^{(m)} \in \mathbb{F}_2} (x^{(m-1)} \oplus u^{(m-1)} \oplus 1)(x^{(m)} \oplus u^{(m)} \oplus 1) f_{u^{(m-1)}u^{(m)}}(x^{(1)}, \dots, x^{(m-2)}),$$

where $f_{u^{(m-1)}u^{(m)}} \in \mathcal{F}_{m-2}$, $u^{(m-1)}, u^{(m)} \in \mathbb{F}_2$. Define $\widetilde{\boldsymbol{x}}$ and $\widetilde{\boldsymbol{u}}$ by the formulas $(\widetilde{\boldsymbol{x}}, x^{(m-1)}, x^{(m)}) \in V_m$, $\boldsymbol{u} = (\widetilde{\boldsymbol{u}}, u^{(m-1)}, u^{(m)}) \in V_m$, where $\widetilde{\boldsymbol{x}} \in V_{m-2}$, $\widetilde{\boldsymbol{u}} \in V_{m-2}$. Then

$$W_f(\boldsymbol{u}) = W_f(\widetilde{\boldsymbol{u}}, u^{(m-1)}, u^{(m)}) = \sum_{\boldsymbol{x} \in V_m} (-1)^{f(\boldsymbol{x}) \oplus \langle \boldsymbol{u}, \boldsymbol{x} \rangle} = \sum_{\substack{\widetilde{\boldsymbol{x}} \in V_{m-2} \\ x^{(m-1)}, x^{(m)} \in \mathbb{F}_2}} (-1)^A$$

$$\Big[\text{where } A = \bigoplus_{u^{(m-1)}, u^{(m)} \in \mathbb{F}_2} (x^{(m-1)} \oplus u^{(m-1)} \oplus 1)(x^{(m)} \oplus u^{(m)} \oplus 1) f_{u^{(m-1)}u^{(m)}} \oplus \langle \widetilde{\boldsymbol{u}}, \widetilde{\boldsymbol{x}} \rangle \oplus u^{(m-1)}x^{(m-1)} \oplus u^{(m)}x^{(m)}\Big]$$

$$\begin{aligned}
&= \sum_{\widetilde{\boldsymbol{x}} \in V_{m-2}} (-1)^{f_{00}(\boldsymbol{x}) \oplus \langle \widetilde{\boldsymbol{u}}, \widetilde{\boldsymbol{x}} \rangle} + (-1)^{u^{(m)}} \sum_{\widetilde{\boldsymbol{x}} \in V_{m-2}} (-1)^{f_{01}(\boldsymbol{x}) \oplus \langle \widetilde{\boldsymbol{u}}, \widetilde{\boldsymbol{x}} \rangle} \\
&\quad + (-1)^{u^{(m-1)}} \sum_{\widetilde{\boldsymbol{x}} \in V_{m-2}} (-1)^{f_{10}(\boldsymbol{x}) \oplus \langle \widetilde{\boldsymbol{u}}, \widetilde{\boldsymbol{x}} \rangle} \\
&\quad + (-1)^{u^{(m-1)} \oplus u^{(m)}} \sum_{\widetilde{\boldsymbol{x}} \in V_{m-2}} (-1)^{f_{11}(\boldsymbol{x}) \oplus \langle \widetilde{\boldsymbol{u}}, \widetilde{\boldsymbol{x}} \rangle} \\
&= W_{f_{00}}(\widetilde{\boldsymbol{u}}) + (-1)^{u^{(m)}} W_{f_{01}}(\widetilde{\boldsymbol{u}}) + (-1)^{u^{(m-1)}} W_{f_{10}}(\widetilde{\boldsymbol{u}}) \\
&\quad + (-1)^{u^{(m-1)} \oplus u^{(m)}} W_{f_{11}}(\widetilde{\boldsymbol{u}}).
\end{aligned}$$

Since $\operatorname{dist}(g, l_{\widetilde{\boldsymbol{u}},1}) = \operatorname{dist}(g \oplus 1, l_{\widetilde{\boldsymbol{u}},0})$ for any function $g \in \mathcal{F}_{m-2}$, there exist a function $g_{\max}$ and a linear function $l_{\widetilde{\boldsymbol{u}},0}$ such that $\operatorname{dist}(g_{\max}, l_{\widetilde{\boldsymbol{u}},0}) = r_{m-2}$ and $\operatorname{dist}(g_{\max}, l_{\widetilde{\boldsymbol{u}},b}) \geq r_{m-2}$ for any $\widetilde{\boldsymbol{u}} \in V_{m-2}$, $b \in \mathbb{F}_2$. Let the initial function f be such that $f_{00} = f_{01} = f_{10} = g_{\max}$, $f_{11} = g_{\max} \oplus 1$. Then

$$\begin{aligned}
\operatorname{dist}(f_{\max}, l_{\widetilde{\boldsymbol{u}},b}) &= 2^{m-1} - \frac{(-1)^b}{2} W_f(\boldsymbol{u}) \\
&= 2^{m-1} - \frac{(-1)^b}{2} \Big[W_{g_{\max}}(\widetilde{\boldsymbol{u}}) + (-1)^{u^{(m)}} W_{g_{\max}}(\widetilde{\boldsymbol{u}}) \\
&\quad + (-1)^{u^{(m-1)}} W_{g_{\max}}(\widetilde{\boldsymbol{u}}) + (-1)^{u^{(m-1)} \oplus u^{(m)}} W_{g_{\max} \oplus 1}(\widetilde{\boldsymbol{u}}) \Big] \\
&= \left[2^{m-3} - \frac{(-1)^b}{2} W_{g_{\max}}(\widetilde{\boldsymbol{u}}) \right] + \left[2^{m-3} - \frac{(-1)^{b \oplus u^{(m)}}}{2} W_{g_{\max}}(\widetilde{\boldsymbol{u}}) \right] \\
&\quad + \left[2^{m-3} - \frac{(-1)^{b \oplus u^{(m-1)}}}{2} W_{g_{\max}}(\widetilde{\boldsymbol{u}}) \right] \\
&\quad + \left[2^{m-3} - \frac{(-1)^{b \oplus u^{(m-1)} \oplus u^{(m)}}}{2} W_{g_{\max} \oplus 1}(\widetilde{\boldsymbol{u}}) \right] \\
&= \operatorname{dist}(g_{\max}, l_{\widetilde{\boldsymbol{u}},b}) + \operatorname{dist}(g_{\max}, l_{\widetilde{\boldsymbol{u}},b \oplus u^{(m)}}) \\
&\quad + \operatorname{dist}(g_{\max}, l_{\widetilde{\boldsymbol{u}},b \oplus u^{(m-1)}}) + \operatorname{dist}(g_{\max} \oplus 1, l_{\widetilde{\boldsymbol{u}},b \oplus u^{(m)} \oplus u^{(m-1)}}) \\
&= \operatorname{dist}(g_{\max}, l_{\widetilde{\boldsymbol{u}},b}) + \operatorname{dist}(g_{\max}, l_{\widetilde{\boldsymbol{u}},b \oplus u^{(m)}}) \\
&\quad + \operatorname{dist}(g_{\max}, l_{\widetilde{\boldsymbol{u}},b \oplus u^{(m-1)}}) + (2^{m-2} - \operatorname{dist}(g_{\max}, l_{\widetilde{\boldsymbol{u}},b \oplus u^{(m)} \oplus u^{(m-1)}})).
\end{aligned}$$

If $(u^{(m-1)}, u^{(m)})$ is of type $(0,0)$, $(0,1)$, $(1,0)$, then

$$\operatorname{dist}(g_{\max}, l_{\boldsymbol{u},b}) = 2^{m-2} + 2 \operatorname{dist}(g_{\max}, l_{\widetilde{\boldsymbol{u}},b}),$$

and for $(u^{(m-1)}, u^{(m)})$ of type $(1,1)$ one has

$$\operatorname{dist}(g_{\max}, l_{\boldsymbol{u},b}) = 2^{m-2} + 2 \operatorname{dist}(g_{\max}, l_{\widetilde{\boldsymbol{u}},b \oplus 1}).$$

Since $\operatorname{dist}(g_{\max}, l_{\widetilde{\boldsymbol{u}},b}) \geq r_{m-2}$ for any $\widetilde{\boldsymbol{u}} \in V_{m-2}$, $b \in \mathbb{F}_2$, from the latter equality it follows that

$$r_m \geq \operatorname{dist}(g_{\max}, l_{\widetilde{\boldsymbol{u}},b}) \geq 2^{m-2} + 2r_{m-2}. \qquad \square$$

PROBLEM 5.41. Prove that for odd $m \geq 3$ the following inequality holds:

$$r_m \geq 2^{m-1} - 2^{(m-1)/2}.$$

PROBLEM 5.42 (Open problem). Find the exact value of r_m (or at least an asymptotic expression for it) for odd $m \geq 3$.

5.4. Reed–Muller codes of second order and related codes

Suppose that $\deg f \leq 2$ and $\Omega_f \in RM(2,m)$. As was noted in §3.3, a quadratic function f can be represented in the matrix form as follows:

$$f(\boldsymbol{x}) = f(x^{(1)}, \ldots, x^{(m)}) = \boldsymbol{x}^\top Q_f \boldsymbol{x} \oplus \boldsymbol{b}_f^\top \boldsymbol{x} \oplus \varepsilon, \tag{5.28}$$

where Q_f is an upper triangular $(m \times m)$ matrix with zeros on the main diagonal and $\boldsymbol{b}_f \in V_m$, $\varepsilon \in \mathbb{F}_2$. Since

$$\Omega_f \oplus RM(1,m) = \Omega_{\boldsymbol{x}^\top Q_f \boldsymbol{x}} \oplus RM(1,m), \tag{5.29}$$

there is a one-to-one correspondence between symplectic matrices $\widetilde{Q}_f = Q_f \oplus Q_f^\top$ and cosets of $RM(1,m)$ in $RM(2,m)$. Therefore, calculation of the weight spectrum of the code $RM(2,m)$ is naturally reduced to calculation of the weight spectra of the cosets of $RM(1,m)$ in $RM(2,m)$. Now, using Dickson's theorem (see Chapter 3), it is easy to prove the following statement.

LEMMA 5.43. *Let f be a quadratic function of the form* (5.29), $\operatorname{rank}\widetilde{Q}_f = 2s$, $0 \le s \le [m/2]$. *Then the weight spectrum of the coset $\Omega_f \oplus RM(1,m)$ coincides with the weight spectrum of the coset $\Omega_{\bigoplus_{i=1}^{s} x^{(2i-1)}x^{(2i)}} \oplus RM(1,m)$.*

PROBLEM 5.44. Prove that

$$\operatorname{wt}\left(\bigoplus_{i=1}^{s} x^{(2i-1)}x^{(2i)} \oplus \varepsilon\right) = 2^{2s-1} + (-1)^{\varepsilon}2^{s-1}.$$

COROLLARY 5.45. *If we regard the function $\bigoplus_{i=1}^{s} x^{(2i-1)}x^{(2i)} \oplus \varepsilon$ as an element of the space of functions $\mathcal{F}_m$, where $m \ge 2s$, i.e., as a function of m variables, then*

$$wt\left(\bigoplus_{i=1}^{s} x^{(2i-1)}x^{(2i)} \oplus \varepsilon\right) = 2^{m-1} + (-1)^{\varepsilon}2^{m-s-1}.$$

PROBLEM 5.46. Suppose that

$$f(x^{(1)},\dots,x^{(m)}) = \bigoplus_{i=1}^{s} x^{(2i-1)}x^{(2i)} \oplus \bigoplus_{t=2s+1}^{m} b^{(t)}x^{(t)},$$

where $s \le [m/2]$, $(b^{(2s+1)},\dots,b^{(m)}) \ne \mathbf{0}$, $\varepsilon \in \mathbb{F}_2$. Prove that $\operatorname{wt}(f) = 2^{m-1}$.

THEOREM 5.47. *Suppose that f is a quadratic function of type* (5.29), $\operatorname{rank}\widetilde{Q}_f = 2s$, $0 \le s \le [m/2]$. *Then the weight spectrum of the coset $\Omega_f \oplus RM(1,m)$ has the form*

$$A'_{2^{m-1}-2^{m-s-1}} = A'_{2^{m-1}+2^{m-s-1}} = 2^{2s}, \quad A'_{2^{m-1}} = 2^{m+1} - 2^{2s+1}.$$

PROOF. For $s = 0$ we have $A'_0 = A'_{2^m} = 1$, $A'_{2^{m-1}} = 2^{m+1} - 2$, which coincides with the spectrum of $RM(1,m)$. Suppose now that $s > 0$. Lemma 5.43 shows that the weight spectrum of the coset $\Omega_f \oplus RM(1,m)$ coincides with the weight spectrum of the coset $\Omega_{\bigoplus_{i=1}^{s} x^{(2i-1)}x^{(2i)}} \oplus RM(1,m)$. For a row $\Omega_{\bigoplus_{i=1}^{s} x^{(2i-1)}x^{(2i)}} \oplus \boldsymbol{b}^\top\boldsymbol{x} \oplus \varepsilon$, $\boldsymbol{b} = (b^{(1)},\dots,b^{(2s)},b^{(2s+1)},\dots,b^{(m)})^\top \in V_m$, the following cases are possible. If $b^{(2s+1)} = b^{(2s+2)} = \cdots = b^{(m)} = 0$, then there are exactly 2^{2s+1} such rows (since $\varepsilon \in \mathbb{F}_2$). For the function $\bigoplus_{i=1}^{s} x^{(2i-1)}x^{(2i)} \oplus \bigoplus_{t=1}^{2s} b^{(t)}x^{(t)} \oplus \varepsilon$, Theorem 3.34 shows that there exists a nondegenerate affine transformation of the variables which transforms it into the function $\bigoplus_{i=1}^{s} x^{(2i-1)}x^{(2i)} \oplus \varepsilon'$. According to Corollary 5.45, in the initial coset we get 2^{2s} rows of weight $2^{m-1} + 2^{m-s-1}$ and 2^{2s} rows of weight $2^{m-1} - 2^{m-s-1}$.

If $(b^{(2s+1)},\dots,b^{(m)}) \ne \mathbf{0}$, then by Theorem 3.34 there is a nondegenerate affine transformation of the variables that transforms the function $\bigoplus_{i=1}^{s} x^{(2i-1)}x^{(2i)} \oplus \boldsymbol{b}^\top\boldsymbol{x} \oplus \varepsilon$ into the function $\bigoplus_{i=1}^{s} x^{(2i-1)}x^{(2i)} \oplus \bigoplus_{t=2s+1}^{m} b^{(t)}x^{(t)} \oplus \varepsilon'$. There will be exactly $2^{m+1} - 2^{2s+1}$ rows corresponding to these functions in the initial coset. According to Problem 5.46, the weight of these rows will be 2^{m-1}. $\square$

THEOREM 5.48. *The weight spectrum of $RM(2,m)$ has the form $A_0 = A_{2^m} = 1$ and*

$$A_{2^{m-1}-2^{m-s-1}} = A_{2^{m-1}+2^{m-s-1}} = 2^{s(s+1)} \frac{(2^m-1)(2^{m-1}-1)\cdots(2^{m-2s+1}-1)}{(2^{2s}-1)(2^{2s-2}-1)\cdots(2^2-1)}$$

for all $1 \le s \le [m/2]$. The rest of the A_i are equal to zero except $A_{2^{m-1}}$, which can be found from the relation

$$\sum_{i=0}^{2^m} A_i = 2^{m^2/2+m/2+1}.$$

PROOF. The assertion follows from Theorem 5.47 if we take into account the formula for the number $\Gamma(m,2s)$ of symplectic $(m \times m)$ matrices of rank $2s$, $s \ge 1$, obtained in Theorem 3.32. □

For a symplectic $(m \times m)$ matrix B we put

$$\ker B = \{\boldsymbol{y} \in V_m \mid \boldsymbol{x}^\top B\boldsymbol{y} = 0 \text{ for any } x \in V_m\}. \tag{5.30}$$

The set $\ker B$ will be called the *kernel of the matrix* B (and of the corresponding bilinear form). It is obvious that $\ker B$ is a subspace in the space V_m. The following statement is well known.

THEOREM 5.49. *Suppose that B is a symplectic $(m \times m)$ matrix over $\mathbb{F}_2$, $\operatorname{rank} B = 2s$, $0 \le s \le [m/2]$. Then*

$$\dim(\ker B) = m - \operatorname{rank} B.$$

Now we will briefly describe classes of nonlinear codes related to $RM(2,m)$ (see [**241, 16, 56**]).

For convenience we will represent the Boolean function f as a mapping $f\colon V_m \to \mathbb{F}_2$. Decompose V_m as $V_m = V_{m-1} \times \mathbb{F}_2$: for any $\boldsymbol{x} \in V_m$ we have $\boldsymbol{x} = (\boldsymbol{x}', \varepsilon)$, $\boldsymbol{x}' \in \mathbb{F}_{2^{m-1}}$, $\varepsilon \in \mathbb{F}_2$. For arbitrary $\boldsymbol{x} = (\boldsymbol{x}', \varepsilon)$, $\boldsymbol{y} = (\boldsymbol{y}', \eta) \in V_m = \mathbb{F}_{2^{m-1}} \times \mathbb{F}_2$, and $\boldsymbol{u} \in \mathbb{F}_{2^{m-1}}$ we define the mapping

$$\Phi_{\boldsymbol{u}}\colon (\mathbb{F}_{2^{m-1}} \times \mathbb{F}_2) \times (\mathbb{F}_{2^{m-1}} \times \mathbb{F}_2) \to \mathbb{F}_2$$

by the formula

$$\begin{aligned} \Phi_{\boldsymbol{u}}(\boldsymbol{x},\boldsymbol{y}) &= \Phi_{\boldsymbol{u}}((\boldsymbol{x}',\varepsilon),(\boldsymbol{y}',\eta)) \\ &= \operatorname{Tr}_{m-1}(\boldsymbol{u}^2\boldsymbol{x}'\boldsymbol{y}') \oplus \operatorname{Tr}_{m-1}(\boldsymbol{u}\boldsymbol{x}')\operatorname{Tr}_{m-1}(\boldsymbol{u}\boldsymbol{y}') \\ &\quad \oplus \varepsilon \operatorname{Tr}_{m-1}(\boldsymbol{u}\boldsymbol{y}') \oplus \eta \operatorname{Tr}_{m-1}(\boldsymbol{u}\boldsymbol{x}'). \end{aligned} \tag{5.31}$$

PROBLEM 5.50. Prove that the mapping $\Phi_{\boldsymbol{u}}$ is a symplectic form.

PROBLEM 5.51. Prove that for $\boldsymbol{u} \in \mathbb{F}_{2^{m-1}}$, $\boldsymbol{u} \ne \boldsymbol{0}$, the mapping $\Phi_{\boldsymbol{u}}$ defined in (5.31) satisfies the equation

$$\dim(\ker \Phi_{\boldsymbol{u}}) = 0.$$

PROBLEM 5.52. Suppose that $\boldsymbol{u}_1, \boldsymbol{u}_2 \in \mathbb{F}_{2^{m-1}}$, $\boldsymbol{u}_1 \ne \boldsymbol{u}_2$. Prove that

$$\dim(\ker(\Phi_{\boldsymbol{u}_1} \oplus \Phi_{\boldsymbol{u}_2})) = 0.$$

DEFINITION 5.53. Suppose that m is even, $m \ge 4$. By definition, the *Kerdock code* $\boldsymbol{K}_m$ is the union of 2^{m-1} cosets of $RM(1,m)$ in $RM(2,m)$ whose associate symplectic forms are given by (5.31).

It follows from the definition that the length of the Kerdock code is $n_{\boldsymbol{K}_m} = 2^m$ and its cardinality is $\#\boldsymbol{K}_m = 2^{2m}$.

PROBLEM 5.54. Prove that the minimum distance $r_{\boldsymbol{K}_m}$ for the Kerdok code $\boldsymbol{K}_m$ is equal to $2^{m-1} - 2^{(m-2)/2}$.

PROBLEM 5.55. Prove that the weight spectrum of the Kerdok code $\boldsymbol{K}_m$ is

i	A_i
0	1
$2^{m-1} - 2^{(m-2)/2}$	$2^m(2^{m-1} - 1)$
2^{m-1}	$2^{m+1} - 2$
$2^{m-1} + 2^{(m-2)/2}$	$2^m(2^{m-1} - 1)$
2^m	1

It is easy to see that $RM(1, m) \subset \boldsymbol{K}_m \subset RM(2, m)$.

Choose an ordering of the elements of the field $\mathbb{F}_{2^{m-1}}$ (see, for example, §5.1) and suppose that $M \subset \mathbb{F}_{2^{m-1}}$, $L \subset \mathbb{F}_{2^{m-1}}$. Consider the rows

$$(\Omega_{I_M}, \Omega_{I_L}), \tag{5.32}$$

where I_M (resp. I_L) is the indicator function of the set M (resp. L).

DEFINITION 5.56. Let m be even, $m \geq 4$. The *Preparata code* $\boldsymbol{P}_m$ is defined to be the set of rows of type (5.32) of length 2^m for which the sets M and L satisfy the following conditions:

1) $\#M = \#L \equiv 0 \pmod 2$;
2) $\bigoplus_{\boldsymbol{x} \in M} \boldsymbol{x} = \bigoplus_{\boldsymbol{y} \in L} \boldsymbol{y}$;
3) $\bigoplus_{\boldsymbol{x} \in M} \boldsymbol{x}^3 \oplus \bigoplus_{\boldsymbol{y} \in L} \boldsymbol{y}^3 = \left(\bigoplus_{\boldsymbol{x} \in M} \boldsymbol{x}\right)^3$.

For $m = 4$ we have $\boldsymbol{P}_4 = \boldsymbol{K}_4$. The Preparata code is (see [**56**]) the union of a linear subcode $\boldsymbol{C}$ of $RM(m-2, m)$ and $2^m - 1$ cosets of $\boldsymbol{C}$ in $RM(m-2, m)$. Besides, $RM(m-3, m) \subset \boldsymbol{P}_m \subset RM(m-2, m)$, the length is $n_{\boldsymbol{P}_m} = 2^m$, and the cardinality is $\#\boldsymbol{P}_m = 2^{2^m - 2m}$. The weight spectrum of the Preparata code $\boldsymbol{P}_m$ is equal to the MacWilliams transform of the weight spectrum of the Kerdock code $\boldsymbol{K}_m$ (see [**56**]).

5.5. Classification of Boolean functions and Reed–Muller codes of the 3rd order

This section is closely related to §3.4. The classification of Boolean (and not only Boolean) functions with respect to the transformation group $\mathfrak{GA}(V_m)$ (or $\mathfrak{GL}(V_m)$) was completed quite a long time ago. This classification allows one to describe the weight spectra of the Reed–Muller codes of the second order. This situation significantly differs from what is known about Boolean functions of degree greater than 2. The study of group equivalence of Boolean functions of degree 3 provided only certain partial results. Similarly, in coding theory, the study of weight spectra of Reed–Muller codes of third order is far from completion. All this demonstrates the principal difference in the complexity of these problems for Boolean functions of degrees 2 and 3.

PROBLEM 5.57 (Open problem). Find the weight spectrum of the code $RM(r, m)$ for $r \geq 3$.

In §3.4 some results on classification of particular sets of Boolean functions of degree greater than 2 were given. In this section we will present results that show the complexity of the group equivalence and weight spectrum problems for cubic Boolean functions from the coding-theoretic point of view. These results suggest that once these problems are solved for cubic Boolean functions, they will also be solved for functions of greater degrees.

LEMMA 5.58. *Suppose that the functions* $f_1, f_2, g \in \mathcal{F}_m$ *and* $f \in \mathcal{F}_m$ *are defined as follows:*

$$f(\boldsymbol{x}) = f_1(\boldsymbol{x}) \cdot f_2(\boldsymbol{x}) \oplus g(\boldsymbol{x}) \tag{5.33}$$

for any $\boldsymbol{x} \in V_m$. *Then*

$$\sum_{\boldsymbol{x}\in V_m} (-1)^{f(\boldsymbol{x})} = \frac{1}{2} \sum_{\boldsymbol{x}\in V_m} \sum_{a,b\in\mathbb{F}_2} (-1)^{ab\oplus af_1(\boldsymbol{x})\oplus bf_2(\boldsymbol{x})\oplus g(\boldsymbol{x})}. \tag{5.34}$$

PROOF. It is easy to see that

$$\sum_{\boldsymbol{x}\in V_m} (-1)^{f(\boldsymbol{x})} = \frac{1}{2} \sum_{\boldsymbol{x}\in V_m} \sum_{a,b\in\mathbb{F}_2} (-1)^{ab\oplus f(\boldsymbol{x})}, \tag{5.35}$$

as the products $a \cdot b$ assume the value 0 three times and the value 1 only once. In addition,

$$\sum_{a,b\in\mathbb{F}_2} (-1)^{a\cdot b} = \sum_{a,b\in\mathbb{F}_2} (-1)^{(a\oplus f_2(\boldsymbol{x}))\cdot(b\oplus\ f_1(\boldsymbol{x}))} \tag{5.36}$$

for any fixed $\boldsymbol{x} \in V_m$. In view of (5.35) and (5.36), the following identities hold for the function (5.33):

$$\begin{aligned}\sum_{\boldsymbol{x}\in V_m} (-1)^{f(\boldsymbol{x})} &= \frac{1}{2} \sum_{\boldsymbol{x}\in V_m} \sum_{a,b\in\mathbb{F}_2} (-1)^{(a\oplus f_2(\boldsymbol{x}))\cdot(b\oplus f_1(\boldsymbol{x}))\oplus f(\boldsymbol{x})} \\ &= \frac{1}{2} \sum_{\boldsymbol{x}\in V_m} \sum_{a,b\in\mathbb{F}_2} (-1)^{ab\oplus af_1(\boldsymbol{x})\oplus bf_2(\boldsymbol{x})\oplus g(\boldsymbol{x})}. \qquad \square\end{aligned}$$

COROLLARY 5.59. *For the functions* $f, f_1, f_2, g \in \mathcal{F}_m$ *connected by the relation* (5.33), *one has*

$$\operatorname{wt}(f) = \frac{1}{2}\operatorname{wt}(f^{[1]}) - 2^{m-1},$$

where

$$f^{[1]}(\boldsymbol{x}, y^{(1)}, y^{(2)}) = y^{(1)}y^{(2)} \oplus y^{(1)}f_1(\boldsymbol{x}) \oplus y^{(2)}f_2(\boldsymbol{x}) \oplus g(\boldsymbol{x}).$$

PROOF. For any Boolean function from $\mathcal{F}_m$ one has

$$\operatorname{wt}(f) = 2^{m-1} - \frac{1}{2}\sum_{\boldsymbol{x}\in V_m} (-1)^{f(\boldsymbol{x})}.$$

Then, using Lemma 5.58, we get

$$\begin{aligned}\operatorname{wt}(f) &= 2^{m-1} - \frac{1}{2}\Big(\frac{1}{2}\sum_{\boldsymbol{x}\in V_m}\sum_{a,b\in\mathbb{F}_2} (-1)^{f^{[1]}(\boldsymbol{x},a,b)}\Big) \\ &= 2^{m-1} - \frac{1}{4}(2^{m+2} - 2\operatorname{wt}(f^{[1]})) = \frac{1}{2}\operatorname{wt}(f^{[1]}) - 2^{m-1}. \qquad \square\end{aligned}$$

REMARK 5.60. It is clear that

$$\deg f^{[1]} = \max(2, \deg f_1 + 1, \deg f_2 + 1, \deg g).$$

LEMMA 5.61. *For any $f \in \mathcal{F}_m$ there exist a nonnegative integer and a function $f^{[t]} \in \mathcal{F}_{m+2t}$ such that $\deg f^{[t]} \leq 3$ and*

$$\sum_{\boldsymbol{x}\in V_m} (-1)^{f(\boldsymbol{x})} = \frac{1}{2^t} \sum_{\boldsymbol{x}\in V_m} \sum_{\boldsymbol{y}\in V_{2t}} (-1)^{f^{[t]}(\boldsymbol{x},\boldsymbol{y})},$$

$$\mathrm{wt}(f) = 2^{m-1} - 2^{m+t-1} + \frac{1}{2^t}\,\mathrm{wt}(f^{[t]}).$$

PROOF. Suppose that $f \in \mathcal{F}_m$. If $\deg f \leq 3$, we put $t = 0$ and $f^{[0]} = f$. Suppose now that $\deg f > 3$. Consider the algebraic normal form of the function f, i.e.,

$$f(x^{(1)}, \ldots, x^{(m)}) = \bigoplus_{i=1}^{t} f_i,$$

where f_i are monomials and $\deg f_i \leq \deg f$. Every monomial f_i can be written as a product of two monomials $f_{i,1}$ and $f_{i,2}$, where $\deg f_{i,1} \leq \lceil \frac{\deg f_i}{2} \rceil$ and $\deg f_{i,2} \leq \lceil \frac{\deg f_i}{2} \rceil$, i.e.,

$$f = f_{1,1} \cdot f_{1,2} \oplus \cdots \oplus f_{t,1} \cdot f_{t,2}.$$

Using Lemma 5.58 for functions f and $g = f_{2,1} \cdot f_{2,2} \oplus \cdots \oplus f_{t,1} \cdot f_{t,2}$, we get

$$\sum_{\boldsymbol{x}\in V_m} (-1)^{f(\boldsymbol{x})} = \frac{1}{2} \sum_{\boldsymbol{x}\in V_m} \sum_{y^{(1,1)},y^{(1,2)}\in\mathbb{F}_2} (-1)^{f^{[1]}(\boldsymbol{x},y^{(1,1)},y^{(1,2)})},$$

where

$$f^{[1]}(\boldsymbol{x}, y^{(1,1)}, y^{(1,2)}) = y^{(1,1)}y^{(1,2)} \oplus y^{(1,1)} f_{1,1}(\boldsymbol{x}) \oplus y^{(1,2)} f_{1,2}(\boldsymbol{x}) \oplus \bigoplus_{i=2}^{t} f_{i,1}(\boldsymbol{x}) f_{i,2}(\boldsymbol{x}).$$

Applying Lemma 5.58 to the function $f^{[1]}$ from $\mathcal{F}_{m+2}$ and the identity

$$g = y^{(1,1)}y^{(1,2)} \oplus y^{(1,1)} f_{1,1}(\boldsymbol{x}) \oplus y^{(1,2)} f_{1,2}(\boldsymbol{x}) \oplus \bigoplus_{i=3}^{t} f_{i,1}(\boldsymbol{x}) f_{i,2}(\boldsymbol{x}),$$

we get

$$\begin{aligned}\sum_{\boldsymbol{x}\in V_m} \sum_{y^{(1,1)},y^{(1,2)}\in\mathbb{F}_2} &(-1)^{f^{[1]}(\boldsymbol{x},y^{(1,1)},y^{(1,2)})} \\ &= \frac{1}{2} \sum_{\boldsymbol{x}\in V_m} \sum_{y^{(1,1)},y^{(1,2)}\in\mathbb{F}_2} \sum_{y^{(2,1)},y^{(2,2)}\in\mathbb{F}_2} (-1)^{f^{[2]}(\boldsymbol{x},y^{(1,1)},y^{(1,2)},y^{(2,1)},y^{(2,2)})},\end{aligned}$$

where

$$\begin{aligned}f^{[2]}&(\boldsymbol{x}, y^{(1,1)}, y^{(1,2)}, y^{(2,1)}, y^{(2,2)}) \\ &= y^{(1,1)}y^{(1,2)} \oplus y^{(1,1)} f_{1,1}(\boldsymbol{x}) \oplus y^{(1,2)} f_{1,2}(\boldsymbol{x}) \\ &\quad \oplus y^{(2,1)}y^{(2,2)} \oplus y^{(2,1)} f_{2,1}(\boldsymbol{x}) \oplus y^{(2,2)} f_{2,2}(\boldsymbol{x}) \oplus \bigoplus_{i=3}^{t} f_{i,1}(\boldsymbol{x}) f_{i,2}(\boldsymbol{x}).\end{aligned}$$

Applying $t-2$ more times Lemma 5.58, we obtain the function $f^{[t]} \in \mathcal{F}_{m+2t}$ for which $\deg f^{[t]} \leq 1 + \left\lceil \frac{\deg f}{2} \right\rceil$ and

$$\sum_{\boldsymbol{x} \in V_m} (-1)^{f(\boldsymbol{x})} = \frac{1}{2^t} \sum_{\boldsymbol{x} \in V_m} \sum_{\boldsymbol{y} \in V_{2t}} (-1)^{f^{[t]}(\boldsymbol{x},\boldsymbol{y})}.$$

Continue this process until $\deg f^{[t]} = \left\lceil \frac{\deg f^{[t]}}{2} \right\rceil$, i.e., $\deg f^{[t]} = 3$. Therefore, there exists t such that $f^{[t]}$ satisfies the following equality:

$$\operatorname{wt}(f) = 2^{m-1} - \frac{1}{2^{t+1}} \sum_{\boldsymbol{x} \in V_m} \sum_{\boldsymbol{y} \in V_{2t}} (-1)^{f^{[t]}(\boldsymbol{x},\boldsymbol{y})} = 2^{m-1} - 2^{m+t-1} + \frac{1}{2^t} \operatorname{wt}(f^{[t]}). \quad \square$$

Lemma 5.61 shows that describing the weight structure of $RM(3, m)$ is no harder than that of Reed–Muller codes of higher orders.

COROLLARY 5.62. *For any $k \in \mathbb{Z}$ there exist natural numbers (t, s) and a function $h \in \mathcal{F}_s$, $\deg h \leq 3$ (i.e., $\Omega_h \in RM(3, s)$), such that $k = \frac{1}{2^t}(2^s - 2\operatorname{wt}(h))$.*

PROOF. Let m be an integer for which $-2^{m-1} \leq k \leq 2^{m-1}$. As $0 \leq 2^{m-1} - k \leq 2^m$, there exists a Boolean function $f \in \mathcal{F}_m$ for which $\operatorname{wt}(f) = 2^{m-1} - k$, i.e., $k = \sum_{\boldsymbol{x} \in V_m} (-1)^{f(\boldsymbol{x})}$. Using Lemma 5.61 and putting $s = m + 2(t-1)$ and $h = f^{[t-1]}$, we finish the proof. $\square$

Corollary 5.62 shows how $RM(2, m)$ differs from $RM(3, m)$: for $RM(2, m)$ the difference $2^m - 2\operatorname{wt}(h)$ for $h \in RM(2, m)$ assumes only the values $\pm 2^i$ or 0.

Comments to Chapter 5

Reed–Muller codes [**34, 241, 279, 323**], if described in terms of Boolean functions (the so-called polynomial representation; other descriptions, such as Euclidean geometrical, projective geometrical, etc., are also possible) form a class of linear codes over $\mathbb{F}_2$ with a simple description, and they can be decoded using majority algorithms. Most of the parameters and properties of these codes are well studied. Their Boolean function description assign to these codes a special place in cryptographic investigations: e.g., many of the parameters of these codes became also cryptographic parameters (see Chapters 5 and 6).

There is a generalization of Reed–Muller codes to the case of arbitrary finite fields, which is also in the class of polynomial codes. These results can be found in the works mentioned in the concluding remarks to Chapter 4. We would like to mention one additional reference, a very well written work [**101**]. Information about cyclic codes (primitive and nonprimitive), constructed on the base of Reed–Muller codes, can be found in [**181, 408**].

The results presented in Chapters 3 and 5 show that there are deep mutually connected difficulties in the classification problems for Boolean functions and the description of the weight spectra of Reed–Muller codes. It might be useful to refer the reader to the works [**182, 186, 374, 398**], having to do with these problems.

CHAPTER 6

Nonlinearity

The notion of nonlinearity for Boolean functions, which was introduced in §2.2, is basic for cryptography. In this chapter we will describe this notion by using examples of well-known classes of Boolean functions having extremal (or close to extremal) nonlinearity. Besides, some generalizations of the notion of nonlinearity will be touched upon.

6.1. Nonlinearity as a measure of cryptographic quality

It is known from mathematical investigations that the linearity (in the mathematical sense) of an object allows us to obtain a complete description of its properties, parameters, or behavior in almost all cases. Linearity is successfully used in algebra, system theory, mathematical cybernetics, and other fields of mathematics. Besides, if an object is in a sense close to linearity, it simplifies its study. However, from the cryptographic point of view, nonlinearity is important for constructing reliable systems, and closeness to linearity is regarded as weakness in cryptographic functions. The presence of linearity contradicts the fundamental principles (see §2.5) of constructing cryptographic functions. For the needs of practical cryptanalysis of various cryptographic systems, many properties and parameters of cryptographic functions were distinguished which have to do with nonlinearity. The nonlinearity (N_f) of a Boolean function f gives a numerical expression for the distance (in the Hamming metric) from a given function to the set of affine functions (Definition 2.50). Recall that it can be expressed via the Walsh–Hadamard coefficients as follows:

$$N_f = \operatorname{dist}(f, \mathcal{A}_n) = 2^{n-1} - \frac{1}{2} \max_{\boldsymbol{\alpha} \in V_n} |W_f(\boldsymbol{\alpha})|. \tag{6.1}$$

The Walsh–Hadamard coefficients of an arbitrary Boolean function $f \in \mathcal{F}_n$ satisfy Parseval's equation:

$$\sum_{\boldsymbol{\alpha} \in V_n} W_f^2(\boldsymbol{\alpha}) = 2^{2n}.$$

Since the number of the Walsh–Hadamard coefficients of a Boolean function in $\mathcal{F}_n$ is equal to 2^n, the maximal modulus of the Walsh–Hadamard coefficients satisfies the inequality

$$\max_{\boldsymbol{\alpha} \in V_n} |W_f(\boldsymbol{\alpha})| \geq 2^{n/2}. \tag{6.2}$$

Thus, for the nonlinearity N_f of an arbitrary Boolean function f the following upper bound holds:

$$N_f \leq 2^{n-1} - 2^{n/2-1}. \tag{6.3}$$

In the case of a Boolean mapping $\Phi = (f_1, \ldots, f_m) \in \mathcal{F}_{n,m}$, the nonlinearity is characterized by the parameter

$$N_\Phi = \min_{\substack{\boldsymbol{c}\in V_m \\ \boldsymbol{c}\neq \boldsymbol{0}}} \{N_{\langle \boldsymbol{c},\Phi\rangle}\}, \tag{6.4}$$

where $N_{\langle \boldsymbol{c},\Phi\rangle}$ is the nonlinearity of the Boolean function $\langle \boldsymbol{c}, \Phi\rangle = c^{(1)} f_1 \oplus \cdots \oplus c^{(m)} f_m$.

From the coding-theoretic point of view, the set of affine functions $\mathcal{A}_n$ represents the first order Reed–Muller code $RM(1,n)$ and the upper bound for N_f in inequality (6.3) is an upper bound for the covering radius of the code $RM(1,n)$. If n is even, it coincides with the covering radius of the code $RM(1,n)$. If n is odd, the exact value of the covering radius of $RM(1,n)$ is known only for some values of n (see §5.3), and for the rest, only upper and lower bounds are known. These bounds determine the interval in which the maximum possible value of nonlinearity of the function lies.

The choice of affine functions as "bad" from the cryptographic point of view is natural, but other choices of the class of "bad" functions are possible. They will give rise to various definitions of "generalized nonlinearity". It is natural to choose a class of "bad" functions that contains the class $\mathcal{A}_n$. One example is the following class of Boolean functions: $\{\mathrm{Tr}(\boldsymbol{a}x^d) \oplus \varepsilon,\ \boldsymbol{a} \in V_n,\ (d, 2^n - 1) = 1,\ \varepsilon = 0, 1\}$.

Various applications of maximally nonlinear functions to coding theory and cryptography, their generalizations and modifications lead to a number of problems related to partial Boolean functions, and this gives rise to problems of the approximation of partial functions and of generalizing the nonlinearity theory to the case of partial functions.

6.2. Maximum-nonlinear bent functions and their properties

DEFINITION 6.1. A function $f \in \mathcal{F}_n$ is called *maximum-nonlinear* (*mn*-) function if its nonlinearity N_f attains the maximum possible value for a function in $\mathcal{F}_n$.

Since, by Proposition 2.51, $N_f = 2^{n-1} - \frac{1}{2}\max_{\boldsymbol{\alpha}\in V_n}|W_f(\boldsymbol{\alpha})|$, a function $f \in \mathcal{F}_n$ is maximum-nonlinear if and only if

$$\max_{\boldsymbol{\alpha}\in V_n}|W_f(\boldsymbol{\alpha})| = \min_{\varphi\in\mathcal{F}_n}\max_{\boldsymbol{\alpha}\in V_n}|W_\varphi(\boldsymbol{\alpha})|.$$

The given definition describes a class of functions that are maximally distant from the set $\mathcal{A}_n$. Note that the value of this distance is known only for some values of n (see §§5.3, 6.1). The functions that are maximally distant from a given class $\mathcal{R}$ are called *deep holes* for the set $\mathcal{R}$. This name is connected with the following geometrical interpretation. In the metric space $\mathcal{F}_n$, choose each of the points of the set $\mathcal{R}$ as the center of a ball and make the radii of these balls grow uniformly; the deep holes of $\mathcal{R}$ are the points of $\mathcal{F}_n$ that are the last to be covered by these growing balls. For example, maximum-nonlinear functions are deep holes for the class of affine functions.

DEFINITION 6.2. A function $f \in \mathcal{F}_n$ is called a *bent function* if all its Walsh–Hadamard coefficients are equal to $\pm 2^{n/2}$.

We will denote by $\mathcal{B}_n$ the set of all bent functions of n variables. Since Walsh–Hadamard coefficients are integers, bent functions do not exist for odd n. Parseval's identity implies the following proposition.

PROPOSITION 6.3. *For even n, the set of maximum-nonlinear functions coincides with the set of bent functions.*

EXAMPLE 6.4. The Boolean function $f(x^{(1)}, x^{(2)}) = x^{(1)}x^{(2)}$ is a maximum-nonlinear bent function. Indeed, its Walsh–Hadamard coefficients are

$$W_f(0,0) = (-1)^{f(0,0)} + (-1)^{f(0,1)} + (-1)^{f(1,0)} + (-1)^{f(1,1)} = 1+1+1-1 = 2;$$
$$W_f(0,1) = (-1)^{f(0,0)} - (-1)^{f(0,1)} + (-1)^{f(1,0)} - (-1)^{f(1,1)} = 1-1+1+1 = 2;$$
$$W_f(1,0) = (-1)^{f(0,0)} + (-1)^{f(0,1)} - (-1)^{f(1,0)} - (-1)^{f(1,1)} = 1+1-1+1 = 2;$$
$$W_f(1,1) = (-1)^{f(0,0)} - (-1)^{f(0,1)} - (-1)^{f(1,0)} + (-1)^{f(1,1)} = 1-1-1-1 = -2.$$

Obviously, for any function $f \in \mathcal{B}_{2n}$,

$$\operatorname{wt}(f) \in \{2^{2n-1} + 2^{n-1},\ 2^{2n-1} - 2^{n-1}\}. \tag{6.5}$$

Our next criterion for maximal nonlinearity is related to the notion of Hadamard matrix, which is studied in combinatorial analysis and coding theory.

DEFINITION 6.5. An *Hadamard matrix* of order N, $N \in \mathbb{N}$, is an $(N \times N)$ matrix H with elements $+1$ and -1 such that

$$HH^{\top} = NE,$$

where E is the identity matrix.

An important example of an Hadamard matrix is the Sylvester–Hadamard matrix (see Definition 2.77):

$$H_N = \begin{pmatrix} 1 & 1 \\ 1 & -1 \end{pmatrix}^{[N]}.$$

REMARK 6.6. It is known that if an Hadamard matrix of order N exists, then either $N = 1$, or $N = 2$, or N is divisible by 4. It is also conjectured that Hadamard matrices of order N exist for any N divisible by 4, but as of now this has not been proved. (For more details about Hadamard matrices see [**6**].)

PROBLEM 6.7 (Open problem). Describe all Hadamard matrices and/or find their number.

THEOREM 6.8. *A Boolean function $f \in \mathcal{F}_{2n}$ is maximum-nonlinear if and only if the $(2^{2n} \times 2^{2n})$ matrix*

$$\frac{1}{2^n}\left(W_f(\boldsymbol{\alpha} \oplus \boldsymbol{\beta})\right), \quad \boldsymbol{\alpha} \in V_{2n}, \quad \boldsymbol{\beta} \in V_{2n},$$

is an Hadamard matrix.

PROOF. The theorem follows immediately from the definitions of bent functions and Hadamard matrices, taking into account the orthogonality correlation for the Walsh–Hadamard coefficients:

$$\sum_{\boldsymbol{\alpha} \in V_{2n}} W_f(\boldsymbol{\alpha})W_f(\boldsymbol{\alpha} \oplus \boldsymbol{\beta}) = \begin{cases} 0 & \text{if } \boldsymbol{\beta} \neq \mathbf{0}, \\ 2^{4n} & \text{if } \boldsymbol{\beta} = \mathbf{0}. \end{cases} \qquad \square$$

If $f \in \mathcal{B}_{2n}$, then obviously there exists a Boolean function $\tilde{f} \in \mathcal{F}_{2n}$ such that

$$W_f(\boldsymbol{\alpha}) = 2^n(-1)^{\tilde{f}(\boldsymbol{\alpha})}. \tag{6.6}$$

The Boolean function $\tilde{f} \in \mathcal{F}_{2n}$ defined by the equation (6.6) is called the *dual function* of the function f.

PROPOSITION 6.9. *The function $\tilde{f} \in \mathcal{F}_{2n}$ is also a maximum-nonlinear function.*

PROOF. From (6.6) it follows that

$$\begin{aligned} W_{\tilde{f}}(\boldsymbol{\alpha}) &= \sum_{\boldsymbol{x} \in V_{2n}} (-1)^{\tilde{f}(\boldsymbol{x}) \oplus \langle \boldsymbol{\alpha}, \boldsymbol{x} \rangle} \\ &= 2^{-n} \sum_{\boldsymbol{x} \in V_{2n}} W_f(\boldsymbol{x})(-1)^{\langle \boldsymbol{\alpha}, \boldsymbol{x} \rangle} = 2^n(-1)^{f(\boldsymbol{\alpha})}. \end{aligned} \tag{6.7}$$

□

The definition of dual functions and equation (6.7) imply the following.

COROLLARY 6.10. 1) *For any $f \in \mathcal{B}_{2n}$ the equality $\tilde{\tilde{f}} = f$ holds.*

2) *If $f \in \mathcal{B}_{2n}$, $\boldsymbol{a}, \boldsymbol{b} \in V_{2n}$, $\varepsilon \in \mathbb{F}_2$, and $g(\boldsymbol{x}) = f(\boldsymbol{x} \oplus \boldsymbol{\alpha}) \oplus \langle \boldsymbol{b}, \boldsymbol{x} \rangle \oplus \varepsilon$, then $g \in \mathcal{B}_{2n}$ and*

$$\widetilde{g}(\boldsymbol{\alpha}) = \tilde{f}(\boldsymbol{\alpha} \oplus \boldsymbol{b}) \oplus \langle \boldsymbol{a}, \boldsymbol{\alpha} \rangle \oplus \varepsilon \oplus \langle \boldsymbol{a}, \boldsymbol{b} \rangle.$$

PROBLEM 6.11. Suppose that a function $h \in \mathcal{F}_{2n+2m}$ is representable as a sum of two functions $f \in \mathcal{F}_{2n}$, $g \in \mathcal{F}_{2m}$:

$$h(\boldsymbol{x}, \boldsymbol{y}) = f(\boldsymbol{x}) \oplus g(\boldsymbol{y}), \quad \boldsymbol{x} \in V_{2n}, \quad \boldsymbol{y} \in V_{2m}.$$

Prove that $h \in \mathcal{B}_{2n+2m}$ if and only if $f \in \mathcal{B}_{2n}$, $g \in \mathcal{B}_{2m}$, and

$$\widetilde{h}(\boldsymbol{\alpha}, \boldsymbol{\beta}) = \tilde{f}(\boldsymbol{\alpha}) \oplus \widetilde{g}(\boldsymbol{\beta}), \quad \boldsymbol{\alpha} \in V_{2n}, \quad \boldsymbol{\beta} \in V_{2m}.$$

If for a function f one has $f = \tilde{f}$, the function is called *self-dual*.

EXAMPLE 6.12. It follows from Example 6.4 that the function $f(x^{(1)}, x^{(2)}) = x^{(1)}x^{(2)}$ is self-dual. According to Problem 6.11, the function

$$f(x^{(1)}, x^{(2)}, \ldots, x^{(2n-1)}, x^{(2n)}) = x^{(1)}x^{(2)} \oplus \cdots \oplus x^{(2n-1)}x^{(2n)}$$

is self-dual for any $n \in \mathbb{N}$.

LEMMA 6.13. *If $A \in \mathfrak{GL}(V_{2n})$ and $A^* = (A^{-1})^\top \in \mathfrak{GL}(V_{2n})$, then $f^A \in \mathcal{B}_{2n}$ and*

$$\widetilde{f^A} = (\tilde{f})^{A^*}$$

for any $f \in \mathcal{B}_{2n}$.

PROOF. Since A is a nondegenerate linear transformation, the following chain of equalities holds:

$$\begin{aligned} \sum_{\boldsymbol{x} \in V_{2n}} (-1)^{f^A(\boldsymbol{x}) \oplus \langle \boldsymbol{\alpha}, \boldsymbol{x} \rangle} &= \sum_{\boldsymbol{x} \in V_{2n}} (-1)^{f(A\boldsymbol{x}) \oplus \langle \boldsymbol{\alpha}, \boldsymbol{x} \rangle} = \sum_{\boldsymbol{y} \in V_{2n}} (-1)^{f(\boldsymbol{y}) \oplus \langle \boldsymbol{\alpha}, A^{-1}\boldsymbol{y} \rangle} \\ &= \sum_{\boldsymbol{y} \in V_{2n}} (-1)^{f(\boldsymbol{y}) \oplus \langle (A^{-1})^\top \boldsymbol{\alpha}, \boldsymbol{y} \rangle} = \sum_{\boldsymbol{y} \in V_{2n}} (-1)^{f(\boldsymbol{y}) \oplus \langle A^* \boldsymbol{\alpha}, \boldsymbol{y} \rangle}. \end{aligned}$$

Therefore,

$$2^n(-1)^{\widetilde{f^A}(\boldsymbol{\alpha})} = 2^n(-1)^{(\tilde{f})^{A^*}(\boldsymbol{\alpha})}$$

for any $\boldsymbol{\alpha} \in V_{2n}$. □

The following corollary is implied by Theorem 6.8 and Proposition 6.9.

COROLLARY 6.14. *A Boolean function $f \in \mathcal{F}_{2n}$ is maximum-nonlinear if and only if the matrix*

$$\left((-1)^{f(\boldsymbol{\alpha}\oplus\boldsymbol{\beta})}\right), \quad \boldsymbol{\alpha} \in V_{2n}, \quad \boldsymbol{\beta} \in V_{2n},$$

is an Hadamard matrix.

Most often, Corollary 6.14 is used in the following equivalent form:

THEOREM 6.15 (Rothaus criterion). *A Boolean function $f \in \mathcal{F}_{2n}$ is maximum-nonlinear if and only if its derivative $D_{\boldsymbol{\sigma}} f(\boldsymbol{x})$ in any nonzero direction $\boldsymbol{\sigma} \in V_{2n}$, $\boldsymbol{\sigma} \neq \mathbf{0}$ is balanced.*

A criterion for being a bent function may be stated in terms of the propagation criterion PC.

COROLLARY 6.16. *A Boolean function $f \in \mathcal{F}_{2n}$ is maximum-nonlinear if and only if*

$$\mathrm{PC}_f = V_{2n} \setminus \{\mathbf{0}\}.$$

The distance between the vector $(\Delta_f(\boldsymbol{\alpha}))$, $\boldsymbol{\alpha} \in V_{2n} \setminus \{\mathbf{0}\}$, and the zero vector is called the *global avalanche characteristic (GAC)* of the function f. As a numerical measure of this distance two characteristics (see Chapter 8) are used:

$$\sigma_f = \sum_{\boldsymbol{\alpha} \in V_n} \Delta_f^2(\boldsymbol{\alpha}), \quad \Delta_f = \max_{\substack{\boldsymbol{\alpha} \in V_n \\ \boldsymbol{\alpha} \neq \mathbf{0}}} |\Delta_f(\boldsymbol{\alpha})|.$$

The following theorem is a corollary of the Rothaus criterion.

THEOREM 6.17. *A Boolean function $f \in \mathcal{F}_{2n}$ is maximum-nonlinear if and only if one of the two equivalent conditions is satisfied:*

1. $\Delta_f = 0$.
2. $\sigma_f = 2^{4n}$.

The notion of bent function is related to the combinatorial notion of a difference set.

DEFINITION 6.18. Let G be a finite group of order $v = \#G$. A subset $D \subset G$ of order $k = \#D$ is called a *difference set* with parameters (v, k, λ, n) if any nonzero element $g \in G$ can be represented in λ different ways as

$$g = d^{-1} d',$$

where $d, d' \in D$ and $n = k - \lambda$.

The definition of difference sets and the Rothaus criterion imply the following.

PROPOSITION 6.19. *A Boolean function $f \in \mathcal{F}_{2n}$ is maximum-nonlinear if and only if the set*

$$\{(\boldsymbol{x}, f(\boldsymbol{x})) \mid \boldsymbol{x} \in V_n\} \subseteq V_{2n+1}$$

is a difference set with parameters $(2^{2n+1}, 2^{2n}, 2^{2n-1}, 2^{2n} - 2^{2n-1})$ in the additive group of the space V_{2n+1} (such difference sets are called simple Hadamard sets).

We now state without proof two more criteria for maximum nonlinearity of Boolean functions. The following notation for the indicator function of the set $M \subset V_n$ will be used:

$$I_M(\boldsymbol{x}) = \begin{cases} 1 & \text{if } \boldsymbol{x} \in M, \\ 0 & \text{otherwise.} \end{cases}$$

THEOREM 6.20. *A Boolean function $f \in \mathcal{F}_{2n}$ is a bent function if and only if there exist subspaces $L_1, \dots, L_k$ of the space V_{2n} (all of dimension n) and integers $m_1, \dots, m_k$ such that for every $\boldsymbol{x} \in V_{2n}$ the following holds:*

$$\sum_{i=1}^{k} m_i I_{L_i}(\boldsymbol{x}) \equiv 2^{n-1} I_{\mathbf{0}}(\boldsymbol{x}) + f(\boldsymbol{x}) \pmod{2^n}.$$

THEOREM 6.21. *A Boolean function $f \in \mathcal{F}_{2n}$ is a bent function if and only if there exist subspaces $L_1, \dots, L_k$ of the space V_{2n} of dimensions n or $n+1$ and integers $m_1, \dots, m_k$ such that for every $\boldsymbol{x} \in V_{2n}$ the following holds:*

$$\sum_{i=1}^{k} m_i I_{L_i}(\boldsymbol{x}) = \pm 2^{n-1} I_{\mathbf{0}}(\boldsymbol{x}) + f(\boldsymbol{x}).$$

For maximum-nonlinear functions, the statement of Lemma 2.40 becomes more symmetric because of the existence of dual functions and equation (6.6).

LEMMA 6.22. *Let $f \in \mathcal{B}_{2n}$ and let $L \subseteq V_{2n}$ be an arbitrary subset and $\boldsymbol{a}, \boldsymbol{b} \in V_{2n}$ arbitrary vectors. Then*

$$(6.8) \qquad \sum_{\boldsymbol{x} \in \boldsymbol{a} \oplus L} (-1)^{f(\boldsymbol{x}) \oplus \langle \boldsymbol{b}, \boldsymbol{x} \rangle} = 2^{\dim L - n} (-1)^{\langle \boldsymbol{a}, \boldsymbol{b} \rangle} \sum_{\boldsymbol{\alpha} \in \boldsymbol{b} \oplus L^{\perp}} (-1)^{\tilde{f}(\boldsymbol{\alpha}) \oplus \langle \boldsymbol{a}, \boldsymbol{\alpha} \rangle}.$$

It is sometimes more convenient to write the relation (6.8) in the following more symmetric form:

$$(6.9) \qquad \frac{\left| \sum_{\boldsymbol{x} \in \boldsymbol{a} \oplus L} (-1)^{f(\boldsymbol{x}) \oplus \langle \boldsymbol{b}, \boldsymbol{x} \rangle} \right|}{2^{\dim L / 2}} = \frac{\left| \sum_{\boldsymbol{\alpha} \in \boldsymbol{b} \oplus L^{\perp}} (-1)^{\tilde{f}(\boldsymbol{\alpha}) \oplus \langle \boldsymbol{a}, \boldsymbol{\alpha} \rangle} \right|}{2^{\dim L^{\perp} / 2}}.$$

Now it is natural to give the following definition.

DEFINITION 6.23. We say that a subspace $L \subseteq V_{2n}$ is *stable* for a bent function $f \in \mathcal{B}_{2n}$ if

$$\left| \sum_{\boldsymbol{x} \in \boldsymbol{a} \oplus L} (-1)^{f(\boldsymbol{x}) \oplus \langle \boldsymbol{b}, \boldsymbol{x} \rangle} \right| = 2^{\dim L / 2}$$

for $\boldsymbol{a}, \boldsymbol{b} \in V_{2n}$.

REMARK 6.24. The notion of stability introduced above is an implementation of a general concept of inheriting properties under restriction of Boolean functions; we will study it in more detail in § 7.2.

Formula (6.9) implies the following useful proposition.

PROPOSITION 6.25. *A subspace $L \subseteq V_{2n}$ is stable for an mn-bent function $f \in \mathcal{B}_{2n}$ if and only if $L^{\perp} \subseteq V_{2n}$ is stable for the mn-bent function $\tilde{f} \in \mathcal{B}_{2n}$.*

There exists a simple criterion of stability for a function $f \in \mathcal{B}_{2n}$; it is stated in terms of the properties of the autocorrelation function.

LEMMA 6.26. *A subspace $L \subseteq V_{2n}$ is stable for an mn-bent function $f \in \mathcal{B}_{2n}$ if and only if the equality*

$$\sum_{\boldsymbol{x} \in \boldsymbol{a} \oplus L} (-1)^{f(\boldsymbol{x}) \oplus f(\boldsymbol{x} \oplus \boldsymbol{u})} = 0$$

holds for any nonzero $\boldsymbol{u} \in L$ and any $\boldsymbol{a} \in V_{2n}$.

To prove the lemma it is enough to slightly modify the approach used in the proof of the mutual correlation Theorem 2.80.

Since the restrictions of a function $f \in \mathcal{B}_{2n}$ to cosets of a subspace L of dimension 2 can be viewed as Boolean functions of two variables, it is easy to derive a criterion for a subspace L of dimension 2 to be stable for $f \in \mathcal{B}_{2n}$.

PROBLEM 6.27. Choose a basis in V_{2n} such that

$$L = \{(x^{(1)}, x^{(2)}, 0, \ldots, 0)^\top \in V_{2n} \mid x^{(1)}, x^{(2)} \in \mathbb{F}_2\}.$$

Prove that L is stable for a function $f \in \mathcal{B}_{2n}$ if and only if the algebraic normal form of the function f is

$$\begin{aligned} &f(x^{(1)}, x^{(2)}, x^{(3)}, \ldots, x^{(2n)}) \\ &\quad = x^{(1)}x^{(2)} \oplus x^{(1)} f_1(x^{(3)}, \ldots, x^{(2n)}) \oplus x^{(2)} f_2(x^{(3)}, \ldots, x^{(2n)}) \oplus f_3(x^{(3)}, \ldots, x^{(2n)}) \end{aligned}$$

for some $f_1, f_2, f_3 \in \mathcal{F}_{2n-2}$.

The correlation (6.8) is even more symmetric and useful if $\boldsymbol{a}$ (and/or $\boldsymbol{b}$) is equal to the zero vector and $\dim L = n$:

$$(6.10) \qquad \sum_{\boldsymbol{x} \in \boldsymbol{a} \oplus L} (-1)^{f(\boldsymbol{x})} = \sum_{\boldsymbol{\alpha} \in L^\perp} (-1)^{\tilde{f}(\boldsymbol{\alpha}) \oplus \langle \boldsymbol{a}, \boldsymbol{\alpha} \rangle}, \quad \sum_{\boldsymbol{x} \in L} (-1)^{f(\boldsymbol{x})} = \sum_{\boldsymbol{\alpha} \in L^\perp} (-1)^{\tilde{f}(\boldsymbol{\alpha})}.$$

An important consequence of the equations obtained is the following theorem.

THEOREM 6.28. *If $f \in \mathcal{B}_{2n}$, $n \geq 2$, then* $\deg f \leq n$.

PROOF. Suppose that the Zhegalkin polynomial of the function f is

$$f(\boldsymbol{x}) = \bigoplus_{\boldsymbol{\alpha} \in V_{2n}} g_f(\boldsymbol{\alpha}) \boldsymbol{x}^{\boldsymbol{\alpha}}.$$

According to Theorem 2.10, the coefficients $g_f(\boldsymbol{\alpha})$ can be found from the formula

$$g_f(\boldsymbol{\alpha}) = \bigoplus_{\substack{\boldsymbol{x} \in V_{2n} \\ \boldsymbol{x} \preccurlyeq \boldsymbol{\alpha}}} f(\boldsymbol{x}).$$

Since $f \in \mathcal{B}_{2n}$, we can use (6.10), which implies that, for $L = L_{\boldsymbol{\alpha}} = \{\boldsymbol{x} \in V_{2n} \mid \boldsymbol{x} \preccurlyeq \boldsymbol{\alpha}\}$ with $\dim L = \mathrm{wt}(\boldsymbol{\alpha})$, taking into account that $(-1)^{f(\boldsymbol{x})} = 1 - 2f(\boldsymbol{x})$, we have the following relation:

$$g_f(\boldsymbol{\alpha}) = 2^{\mathrm{wt}(\boldsymbol{\alpha})-1} - 2^{n-1} + 2^{\mathrm{wt}(\boldsymbol{\alpha})-n} \sum_{\boldsymbol{u} \in L_{\boldsymbol{\alpha}}^\perp} \tilde{f}(\boldsymbol{u}) \pmod 2.$$

If $n \geq 2$ and $\mathrm{wt}(\boldsymbol{\alpha}) \geq n+1$, this gives $g_f(\boldsymbol{\alpha}) = 0$. □

PROBLEM 6.29. Suppose that $f \in \mathcal{B}_{2n}$. Prove that $g_f(\boldsymbol{\alpha}) = g_{\tilde{f}}(\boldsymbol{\alpha} \oplus \mathbf{1})$ for all $\boldsymbol{\alpha} \in V_{2n}$ such that $\mathrm{wt}(\boldsymbol{\alpha}) = n$. In particular, $\deg f = n$ if and only if $\deg \tilde{f} = n$.

PROBLEM 6.30. Suppose that $f \in \mathcal{B}_{2n}$, $\deg f = n$. Prove that f cannot be represented as a sum of two functions with disjoint sets of variables (see Problem 6.11).

In Theorem 6.9 and Lemma 6.13 it was shown that the set $\mathcal{B}_{2n}$ is closed under nondegenerate affine substitutions of variables, addition of affine functions, and replacement of functions by their duals. Less trivial transformations preserving the class of maximum-nonlinear functions are described in the following proposition.

PROPOSITION 6.31. *Let $f \in \mathcal{B}_{2n}$ be a function such that*

$$f|_{\boldsymbol{a}_0 \oplus L_0}(\boldsymbol{x}) = \langle \boldsymbol{b}_0, \boldsymbol{x} \rangle \oplus \varepsilon$$

for each $\boldsymbol{x} \in V_{2n}$ and for some $\boldsymbol{a}_0, \boldsymbol{b}_0 \in V_{2n}$, $\varepsilon \in \mathbb{F}_2$, where $L_0 \subseteq V_{2n}$ is a subspace of dimension n. Then the function $f'(\boldsymbol{x}) = f(\boldsymbol{x}) \oplus I_{\boldsymbol{a}_0 \oplus L_0}(\boldsymbol{x})$ is also maximum-nonlinear. Besides,

$$\widetilde{f'}(\boldsymbol{x}) = \tilde{f}(\boldsymbol{x}) + I_{\boldsymbol{b}_0 \oplus L_0^{\perp}}(\boldsymbol{x}).$$

PROOF. Since $\dim L_0 = n$, it follows from Lemma 6.22 that $f|_{\boldsymbol{a}_0 \oplus L_0}(\boldsymbol{x}) = \langle \boldsymbol{b}_0, \boldsymbol{x} \rangle \oplus \varepsilon$ if and only if $\tilde{f}|_{\boldsymbol{b}_0 \oplus L_0^{\perp}}(\boldsymbol{x}) = \langle \boldsymbol{a}_0, \boldsymbol{\alpha} \rangle \oplus \langle \boldsymbol{a}_0, \boldsymbol{b}_0 \rangle \oplus \varepsilon$. Hence, if we simultaneously inverting the values of the function f on $\boldsymbol{a}_0 \oplus L_0$ and the values of the function $\tilde{f}$ on $\boldsymbol{b}_0 \oplus L_0^{\perp}$, the equalities

$$\sum_{\boldsymbol{x} \in \boldsymbol{a} \oplus L_0} (-1)^{f(\boldsymbol{x}) \oplus \langle \boldsymbol{b}, \boldsymbol{x} \rangle} = (-1)^{\langle \boldsymbol{a}, \boldsymbol{b} \rangle} \sum_{\boldsymbol{\alpha} \in \boldsymbol{b} \oplus L_0^{\perp}} (-1)^{\tilde{f}(\boldsymbol{\alpha}) \oplus \langle \boldsymbol{a}, \boldsymbol{\alpha} \rangle}$$

will still hold for all $\boldsymbol{a}, \boldsymbol{b} \in V_{2n}$. Now to finish the proof it suffices to use Problem 2.42. □

PROBLEM 6.32. Prove that if $f \in \mathcal{B}_{2n}$, $\dim L_0 > n$, then there exist no $\boldsymbol{a}_0, \boldsymbol{b}_0 \in V_{2n}$, $\varepsilon \in \mathbb{F}_2$ such that the relation

$$f|_{\boldsymbol{a}_0 \oplus L_0}(\boldsymbol{x}) = \langle \boldsymbol{b}_0, \boldsymbol{x} \rangle \oplus \varepsilon$$

can be fulfilled simultaneously for all $\boldsymbol{x} \in V_{2n}$.

EXAMPLE 6.33. It was shown by experiments (see [**326**]) that any maximum-nonlinear function of degree 3 in six variables is equivalent, up to a nondegenerate affine substitution of variables and addition of an affine function, to one of the following three functions:

1) $x^{(1)}x^{(2)}x^{(3)} \oplus x^{(1)}x^{(4)} \oplus x^{(2)}x^{(5)} \oplus x^{(3)}x^{(6)}$;

2) $x^{(1)}x^{(2)}x^{(3)} \oplus x^{(2)}x^{(4)}x^{(5)} \oplus x^{(1)}x^{(2)} \oplus x^{(1)}x^{(4)} \oplus x^{(2)}x^{(6)} \oplus x^{(3)}x^{(5)} \oplus x^{(4)}x^{(5)}$;

3) $x^{(1)}x^{(2)}x^{(3)} \oplus x^{(2)}x^{(4)}x^{(5)} \oplus x^{(3)}x^{(4)}x^{(6)} \oplus x^{(1)}x^{(4)} \oplus x^{(2)}x^{(6)} \oplus x^{(3)}x^{(4)} \oplus x^{(3)}x^{(5)} \oplus x^{(3)}x^{(6)} \oplus x^{(4)}x^{(5)} \oplus x^{(4)}x^{(6)}$.

PROBLEM 6.34 (open). Find $\#\mathcal{B}_{2n}$ for all n or at least find the asymptotic behavior of the number $\#\mathcal{B}_{2n}$ as $n \to \infty$.

6.3. Some classes of maximum-nonlinear bent functions

The class $\mathcal{X}$ of maximum-nonlinear functions is called *complete* if it is closed with respect to nondegenerate affine substitutions of variables and addition of affine functions (see Lemma 6.2). If the class $\mathcal{X}$ of maximum-nonlinear functions is not complete, we will denote by $\bar{\mathcal{X}}$ its "completion", i.e., the smallest class of functions which contains $\mathcal{X}$ and is closed under nondegenerate affine substitutions of variables and addition of affine functions.

Maiorana–McFarland family $\mathcal{M}$.

THEOREM 6.35. *Let $\pi\colon V_n \to V_n$ be an arbitrary permutation of the space V_n and $\psi \in \mathcal{F}_n$ an arbitrary Boolean function of n variables. Then the Boolean function $f \in \mathcal{F}_{2n}$ of $2n$ variables given by the equality*

$$f(\boldsymbol{x}, \boldsymbol{y}) = \langle \pi(\boldsymbol{y}), \boldsymbol{x} \rangle \oplus \psi(\boldsymbol{y}), \quad \boldsymbol{x}, \boldsymbol{y} \in V_n, \tag{6.11}$$

is a maximum-nonlinear function.

PROOF. It suffices to invoke the Rothaus criterion. □

The class of maximum-nonlinear functions described in Theorem 6.35 is called the Maiorana–McFarland class and is denoted by $\mathcal{M}$. Clearly the class $\mathcal{M}$ is not complete.

PROBLEM 6.36. Check that (for a fixed decomposition of the space V_{2n} into a direct sum of two spaces of dimension n) the cardinality of the class $\mathcal{M}$ of maximum-nonlinear functions is given by the formula

$$\#\mathcal{M} = 2^n! \cdot 2^{2^n}.$$

PROBLEM 6.37. Let f be a maximum-nonlinear function given by (6.11). Prove that its dual function $\widetilde{f}$ is defined by the equality

$$\widetilde{f}(\boldsymbol{x}, \boldsymbol{y}) = \langle \boldsymbol{y}, \pi^{-1}(\boldsymbol{x}) \rangle \oplus \psi(\pi^{-1}(\boldsymbol{x})), \quad \boldsymbol{x}, \boldsymbol{y} \in V_n.$$

PROBLEM 6.38. Prove that the algebraic degree of an mn-bent Boolean function of $2n$ variables can be any of the numbers $2, 3, \dots, n$. (*Hint*: Use Maiorana–McFarland construction.)

Maximal nonlinearity of functions defined by linear branching. Elements of the family $\mathcal{M}$ are not the only examples of mn-bent functions defined by a linear branching. It is possible to attain maximal nonlinearity for an arbitrary function defined by a linear branching. This enables us to construct new families of mn-bent functions. Recall that a Boolean function $f \in \mathcal{F}_n$ is represented by a linear branching if there exist t, s, $n = t + s$, $\Phi \in \mathcal{F}_{t,s}$, $h \in \mathcal{F}_s$ such that

$$f(\boldsymbol{z}) = f(\boldsymbol{x}, \boldsymbol{y}) = \langle \boldsymbol{x}, \Phi(\boldsymbol{y}) \rangle \oplus h(\boldsymbol{y}) \tag{6.12}$$

for each $\boldsymbol{z} = (\boldsymbol{x}, \boldsymbol{y}) \in V_n$, $\boldsymbol{x} \in V_t$, $\boldsymbol{y} \in V_s$.

To simplify what follows we give two definitions. A set $\boldsymbol{C} \subseteq V_n$ will be called a *bent set* if the following two conditions hold:

1) $\#C = 2^{2l}$ for some integer l;

2) for each $\boldsymbol{v} \in V_n \setminus \{\boldsymbol{0}\}$, the set $\boldsymbol{C_v} = \{\boldsymbol{y} \in \boldsymbol{C} \mid \boldsymbol{y} \oplus \boldsymbol{v} \in \boldsymbol{C}\}$ is either empty or has even cardinality.

Any one-element set $\boldsymbol{C}$ is bent since $\boldsymbol{C_v} = \varnothing$ for $\boldsymbol{v} \in V_n \setminus \{\boldsymbol{0}\}$. If $\boldsymbol{C}$ is a coset of a subspace L of even dimension, then $\boldsymbol{C}$ is bent since

$$\boldsymbol{C_v} = \begin{cases} \varnothing & \text{if } \boldsymbol{v} \notin L, \\ \boldsymbol{C} & \text{if } \boldsymbol{v} \in L. \end{cases}$$

The space V_n is a bent set for even n.

DEFINITION 6.39. Suppose that $h(\boldsymbol{x})$ is a Boolean function and $\boldsymbol{C}$ is a bent set. We say that the pair $(h; \boldsymbol{C})$ is a *partially defined mn-bent function*, or a *partial mn-bent function*, if for $\boldsymbol{v} \in V_n \setminus \{\boldsymbol{0}\}$ the weight $\mathrm{wt}(D_{\boldsymbol{v}} h(\boldsymbol{x})\big|_{\boldsymbol{C_v}})$ of the restriction of the function $D_{\boldsymbol{v}} h(\boldsymbol{x})$ to the set $\boldsymbol{C}_{(\boldsymbol{v})}$ is equal to $\frac{1}{2}\#\boldsymbol{C}_{(\boldsymbol{v})}$ (compare with Definition 6.74).

It is clear that a partial mn-bent function $(h; V_n)$ is an ordinary mn-bent function. Any Boolean function is a partial mn-bent function over a one-element set.

THEOREM 6.40. *A Boolean function $f(\boldsymbol{x}, \boldsymbol{y})$ of type* (6.12) *belongs to $\mathcal{B}_n$, $n = 2m$, if and only if for all $\boldsymbol{\alpha} \in V_t$ the following conditions are satisfied:*

1) $t \le \frac{n}{2}$;

2) $\#\Phi^{-1}(\boldsymbol{\alpha}) = 2^{n-2t}$;

3) $\Phi^{-1}(\boldsymbol{\alpha})$ *is a bent set;*
4) $(h; \Phi^{-1}(\boldsymbol{\alpha}))$ *is a partial mn-bent function.*

PROOF. We will use the criterion

$$f(\boldsymbol{x}, \boldsymbol{y}) \in \mathcal{B}_n \iff PC_f = V_n \setminus \{\boldsymbol{0}\}.$$

Let us find the derivative of the function $f(\boldsymbol{x}, \boldsymbol{y})$ of type (6.12) in an arbitrary direction $(\boldsymbol{u}, \boldsymbol{v}) \in V_n$. We have

(6.13)
$$\begin{aligned} D_{(\boldsymbol{u},\boldsymbol{v})} f(\boldsymbol{x}, \boldsymbol{y}) &= f(\boldsymbol{x} \oplus \boldsymbol{u}, \boldsymbol{y} \oplus \boldsymbol{v}) \oplus f(\boldsymbol{x}, \boldsymbol{y}) \\ &= \big\langle \Phi(\boldsymbol{y} \oplus \boldsymbol{v}), \boldsymbol{x} \big\rangle \oplus \big\langle \Phi(\boldsymbol{y} \oplus \boldsymbol{v}), \boldsymbol{u} \big\rangle \oplus h(\boldsymbol{y} \oplus \boldsymbol{v}) \oplus \big\langle \Phi(\boldsymbol{y}), \boldsymbol{x} \big\rangle \oplus h(\boldsymbol{y}) \\ &= \langle D_{\boldsymbol{v}} \Phi(\boldsymbol{y}), \boldsymbol{x} \rangle \oplus \langle \Phi(\boldsymbol{y} \oplus \boldsymbol{v}), \boldsymbol{u} \rangle \oplus D_{\boldsymbol{v}} h(\boldsymbol{y}). \end{aligned}$$

Now two cases are possible:

I. $\boldsymbol{v} = \boldsymbol{0}$, $\boldsymbol{u} \neq \boldsymbol{0}$.
II. $\boldsymbol{v} \neq \boldsymbol{0}$.

For each of the cases we will find necessary and sufficient conditions for the inclusion $(\boldsymbol{u}, \boldsymbol{v}) \in PC_f$. All the conditions together give the statement of the theorem.

I. Formula (6.13) becomes

$$D_{(\boldsymbol{u},\boldsymbol{0})} f(\boldsymbol{x}, \boldsymbol{y}) = \big\langle \Phi(\boldsymbol{y}), \boldsymbol{u} \big\rangle.$$

The requirement $(\boldsymbol{u}, \boldsymbol{0}) \in PC_f$ for all $\boldsymbol{u} \in V_t \setminus \{\boldsymbol{0}\}$ is equivalent to the condition that the Boolean function $\big\langle \Phi(\boldsymbol{y}), \boldsymbol{u} \big\rangle$ should be balanced for any $\boldsymbol{u} \in V_t \setminus \{\boldsymbol{0}\}$. The latter condition is well known: it holds when the mapping $\Phi \in \mathcal{F}_{s,t}$ is regular. It is also known that a mapping Φ is regular if and only if the following two conditions hold:

1) $t \leq s$;
2) $\#\Phi^{-1}(\boldsymbol{\alpha}) = 2^{n-2t}$ for any $\boldsymbol{\alpha} \in V_t$.

II. We calculate the weight of the function (6.13) as a sum of the weights of all its subfunctions for all possible values of $\boldsymbol{y} \in V_s$. The following cases are possible:

a) $D_{\boldsymbol{v}} \Phi(\boldsymbol{y}) \neq \boldsymbol{0}$; in this case $D_{(\boldsymbol{u},\boldsymbol{v})} f(\boldsymbol{x}, \boldsymbol{y})$ as a function of $\boldsymbol{x}$ is affine and for a given fixed $\boldsymbol{y}$ its weight is equal to 2^{t-1};

b) $D_{\boldsymbol{v}} \Phi(\boldsymbol{y}) = \boldsymbol{0}$, $\Phi(\boldsymbol{y}) = \boldsymbol{\alpha}$, $\boldsymbol{\alpha} \in V_t$; in this case $D_{(\boldsymbol{u},\boldsymbol{v})} f(\boldsymbol{x}, \boldsymbol{y})$ does not depend on $\boldsymbol{x}$ and its weight for a given fixed $\boldsymbol{y}$ is equal to

$$2^t(\langle \boldsymbol{u}, \boldsymbol{\alpha} \rangle \oplus D_{\boldsymbol{v}} h(\boldsymbol{y})).$$

We sum all the weights of the subfunctions for all $\boldsymbol{y} \in V_s$, $\boldsymbol{\alpha} \in V_t$ obtained in the cases a) and b); after simple transformations we obtain

$$\begin{aligned} \operatorname{wt}(D_{(\boldsymbol{u},\boldsymbol{v})} f(\boldsymbol{x}, \boldsymbol{y})) &= 2^{t-1}(2^s - \#T_{\boldsymbol{v}}) + 2^t \sum_{\boldsymbol{\alpha} \in V_t} \sum_{\boldsymbol{y} \in T_{\boldsymbol{v},\boldsymbol{\alpha}}} \Big(\langle \boldsymbol{u}, \boldsymbol{\alpha} \rangle \oplus D_{\boldsymbol{v}} h(\boldsymbol{y}) \Big) \\ &= 2^{n-1} + 2^{t-1} \Big[2 \sum_{\boldsymbol{\alpha} \in V_t} \operatorname{wt}(\langle \boldsymbol{u}, \boldsymbol{\alpha} \rangle \oplus D_{\boldsymbol{v}} h(\boldsymbol{y})|_{T_{\boldsymbol{v},\boldsymbol{\alpha}}}) - \#T_{\boldsymbol{v}} \Big], \end{aligned}$$

$$T_{\boldsymbol{v}} = (D_{\boldsymbol{v}} \Phi)^{-1}(\boldsymbol{0}), \quad T_{\boldsymbol{v},\boldsymbol{\alpha}} = T_{\boldsymbol{v}} \cap \Phi^{-1}(\boldsymbol{\alpha}).$$

It is easy to see that

$$\mathrm{wt}(\langle \boldsymbol{u}, \boldsymbol{\alpha}\rangle \oplus D_{\boldsymbol{v}}h(\boldsymbol{y})|_{T_{\boldsymbol{v},\boldsymbol{\alpha}}}) = \mathrm{wt}(D_{\boldsymbol{v}}h(\boldsymbol{y})|_{T_{\boldsymbol{v},\boldsymbol{\alpha}}}) + \langle \boldsymbol{u}, \boldsymbol{\alpha}\rangle \bigl(\#T_{\boldsymbol{v},\boldsymbol{\alpha}} - 2\,\mathrm{wt}(D_{\boldsymbol{v}}h(\boldsymbol{y})|_{T_{\boldsymbol{v},\boldsymbol{\alpha}}})\bigr),$$

$$\sum_{\boldsymbol{\alpha}\in V_t} \mathrm{wt}(D_{\boldsymbol{v}}h(\boldsymbol{y})|_{T_{\boldsymbol{v},\boldsymbol{\alpha}}}) = \mathrm{wt}(D_{\boldsymbol{v}}h(\boldsymbol{y})|_{T_{\boldsymbol{v}}}).$$

Therefore,

$$\mathrm{wt}(D_{(\boldsymbol{u},\boldsymbol{v})}f(\boldsymbol{x},\boldsymbol{y})) = 2^{n-1} + 2^{t-1}\Bigl[2\,\mathrm{wt}(D_{\boldsymbol{v}}h(\boldsymbol{y})|_{T_{\boldsymbol{v}}}) - \#T_{\boldsymbol{v}} + 2\sum_{\boldsymbol{\alpha}}\langle \boldsymbol{u}, \boldsymbol{\alpha}\rangle \bigl(\#T_{\boldsymbol{v},\boldsymbol{\alpha}} - 2\,\mathrm{wt}(D_{\boldsymbol{v}}h(\boldsymbol{y})\big|_{T_{\boldsymbol{v},\boldsymbol{\alpha}}})\bigr)\Bigr].$$

Thus, the condition $(\boldsymbol{u},\boldsymbol{v}) \in PC_f$ for all $\boldsymbol{v} \in V_s \setminus \{\boldsymbol{0}\}$, $\boldsymbol{u} \in V_t$ is equivalent to the condition

$$2\,\mathrm{wt}(D_{\boldsymbol{v}}h(\boldsymbol{y})|_{T_{\boldsymbol{v}}}) - \#T_{\boldsymbol{v}} + 2\sum_{\boldsymbol{\alpha}}\langle \boldsymbol{u}, \boldsymbol{\alpha}\rangle \bigl(\#T_{\boldsymbol{v},\boldsymbol{\alpha}} - 2\,\mathrm{wt}(D_{\boldsymbol{v}}h(\boldsymbol{y})\big|_{T_{\boldsymbol{v},\boldsymbol{\alpha}}})\bigr) = 0 \tag{6.14}$$

for all $\boldsymbol{v} \in V_s \setminus \{\boldsymbol{0}\}$, $\boldsymbol{u} \in V_t$.

Inspecting the cases $\boldsymbol{u} = \boldsymbol{0}$ and $\boldsymbol{u} \neq \boldsymbol{0}$ separately, we obtain that the system (6.14) is equivalent to the following nondegenerate linear system of equations:

$$\begin{cases} \sum_{\boldsymbol{\alpha}}\langle \boldsymbol{u}, \boldsymbol{\alpha}\rangle \bigl(\#T_{\boldsymbol{v},\boldsymbol{\alpha}} - 2\,\mathrm{wt}(D_{\boldsymbol{v}}h(\boldsymbol{y})|_{T_{\boldsymbol{v},\boldsymbol{\alpha}}})\bigr) = 0 & \text{for all } \boldsymbol{v} \in V_s \setminus \{\boldsymbol{0}\}, \\ & \boldsymbol{u} \in V_t \setminus \{\boldsymbol{0}\}, \\ \sum_{\boldsymbol{\alpha}} \bigl(\#T_{\boldsymbol{v},\boldsymbol{\alpha}} - 2\,\mathrm{wt}(D_{\boldsymbol{v}}h(\boldsymbol{y})|_{T_{\boldsymbol{v},\boldsymbol{\alpha}}})\bigr) = 0 & \text{for all } \boldsymbol{v} \in V_s \setminus \{\boldsymbol{0}\}. \end{cases} \tag{6.15}$$

This system has a unique solution

$$\#T_{\boldsymbol{v},\boldsymbol{\alpha}} - 2\,\mathrm{wt}(D_{\boldsymbol{v}}h(\boldsymbol{y})\big|_{T_{\boldsymbol{v},\boldsymbol{\alpha}}}) = 0 \quad \text{for all } \boldsymbol{v} \in V_s \setminus \{\boldsymbol{0}\},\ \boldsymbol{\alpha} \in V_t,$$

and conditions (6.15) are equivalent to the following: $(\boldsymbol{u},\boldsymbol{v}) \in PC_f$ for all $\boldsymbol{u} \in V_t$, $\boldsymbol{v} \in V_s \setminus \{\boldsymbol{0}\}$.

If we return to the definitions of bent sets and partial mn-bent functions, we see that the system of equations (6.15) is equivalent to conditions 3) and 4) of Theorem 6.40. □

To illustrate the approach suggested by Theorem 6.40, we present two simple examples.

EXAMPLE 6.41. Suppose that $t = n/2$. In this case condition 2) of the theorem means that the map $\Phi \in \mathcal{F}_{n/2,n/2}$ is a bijection. Furthermore, all sets $\Phi^{-1}(\boldsymbol{\alpha})$ are one-element sets, and conditions 3) and 4) of the theorem are satisfied.

Thus, the case $t = n/2$ yields a class $\mathcal{M}$ of mn-bent functions.

EXAMPLE 6.42. Let $t \le n/2$ be an arbitrary linear mapping $\Phi \in \mathcal{F}_{s,t}$. In this case condition 2) of the theorem is satisfied, $\Phi^{-1}(\boldsymbol{0})$ is a subspace of dimension $n-2t$, and $\Phi^{-1}(\boldsymbol{\alpha})$, $\boldsymbol{\alpha} \neq \boldsymbol{0}$, are its cosets. Thus, condition 3) of the theorem is also satisfied.

It is convenient to consider condition 4) first in the special case

$$\Phi(\boldsymbol{y}) = (y_1, \dots, y_t),$$

and then to make the necessary substitution of variables. Thus,

$$\Phi^{-1}(\boldsymbol{\alpha}) = (\alpha_1, \dots, \alpha_t, y_{t+1}, \dots, y_s),$$

the y_i are arbitrary, and so condition 4) of the theorem means that all 2^t Boolean functions $h(\alpha_1, \dots, \alpha_t, y_1, \dots, y_s)$ of $n-2t$ variables are mn-bent functions.

Thus, the example (up to an affine substitution of variables) yields the following class of mn-bent functions:

$$f(x_1,\dots,x_t,y_1,\dots,y_s) = x_1y_1 \oplus \cdots \oplus x_ty_t \oplus \bigoplus_{\varepsilon\in V_t} y_1^{\varepsilon_1}\cdot\dots\cdot y_t^{\varepsilon_t} h_\varepsilon(y_{t+1},\dots,y_s),$$

where $t \le \frac{n}{2}$, $y_i^{\varepsilon_i} = y_i \oplus \varepsilon_i \oplus 1$, and $h_\varepsilon(y_{t+1},\dots,y_s)$ is an arbitrary set of 2^t mn-bent functions.

PROBLEM 6.43 (Open problem). Let $f \in \mathcal{B}_{2n}$, $\operatorname{ill}(f) \le s$, and let f be representable as a linear branching. Find (or estimate) $\operatorname{ill}(\tilde{f})$ and a representation of $\tilde{f}$ as a linear branching (compare with Problems 6.37, 6.45, 6.47).

The class $\mathcal{PS}$.

PROPOSITION 6.44. *Suppose that $L_1,\dots,L_{2^{n-1}}$ is a family of 2^{n-1} subspaces of the space V_{2n} for which the following two conditions hold:*
1) $\dim L_i = n$, $i = 1,\dots,2^{n-1}$;
2) $L_i \cap L_j = \{\mathbf{0}\}$ *for* $i \ne j$, $i,j = 1,\dots,2^{n-1}$.
Then the function

$$f(\boldsymbol{x}) = \bigoplus_{i=1}^{2^{n-1}} I_{L_i}(\boldsymbol{x}) \tag{6.16}$$

is maximum-nonlinear.

The family of maximum-nonlinear functions obtained by the construction of Proposition 6.44 is denoted by $\mathcal{PS}^-$.

PROBLEM 6.45. Suppose that a maximum-nonlinear function f is given by the equality (6.16). Prove that:
1. The family of subspaces $L_1^\perp,\dots,L_{2^{n-1}}^\perp$, which are dual to the subspaces $L_1,\dots,L_{2^{n-1}}$, satisfies conditions 1) and 2) of Proposition 6.44.
2. The function $\widetilde{f}$ (the dual of the function f) satisfies the equation

$$\widetilde{f}(\boldsymbol{x}) = \bigoplus_{i=1}^{2^{n-1}} I_{L_i^\perp}(\boldsymbol{x}).$$

PROPOSITION 6.46. *Let $L_1,\dots,L_{2^{n-1}+1}$ be a family of $2^{n-1}+1$ subspaces of the space V_{2n} for which the following two conditions hold:*
1) $\dim L_i = n$, $i = 1,\dots,2^{n-1}+1$;
2) $L_i \cap L_j = \{\mathbf{0}\}$ *for* $i \ne j$, $i,j = 1,\dots,2^{n-1}+1$.
Then the function

$$f(\boldsymbol{x}) = \bigoplus_{i=1}^{2^{n-1}+1} I_{L_i}(\boldsymbol{x}) \tag{6.17}$$

is maximum-nonlinear.

The family of maximum-nonlinear functions obtained by the construction of Proposition 6.46 is denoted by $\mathcal{PS}^+$.

PROBLEM 6.47. Let the maximum-nonlinear function f be defined by the equality (6.17). Prove that

1. The family of subspaces $L_1^\perp, \dots, L_{2^{n-1}+1}^\perp$ (duals of the subspaces $L_1, \dots, L_{2^{n-1}+1}$) satisfies conditions 1) and 2) of Proposition 6.46.

2. The function $\widetilde{f}$ (dual of f) may be defined by the equation

$$\widetilde{f}(\boldsymbol{x}) = \bigoplus_{i=1}^{2^{n-1}+1} I_{L_i^\perp}(\boldsymbol{x}).$$

The union of the classes $\mathcal{PS}^-$ and $\mathcal{PS}^+$ of maximum-nonlinear functions forms the class $\mathcal{PS}$ (of Partial Spreads). The class $\mathcal{PS}$ (as well as each of the classes $\mathcal{PS}^-$ and $\mathcal{PS}^+$) is not complete.

The class $\mathcal{M}$ of maximum-nonlinear functions is not contained in the complement $\overline{\mathcal{PS}}$ of the class $\mathcal{PS}$.

The class $\mathcal{D}$.

PROPOSITION 6.48. *Suppose that L_1 and L_2 are subspaces of V_n and π is a permutation of V_n such that*

1) $\dim L_1 + \dim L_2 = n$;

2) $\pi(L_1) = L_2^\perp$.

Then the Boolean function of $2n$ variables

$$f(\boldsymbol{x}, \boldsymbol{y}) = \langle \boldsymbol{x}, \pi(\boldsymbol{y}) \rangle \oplus I_{L_1}(\boldsymbol{x}) I_{L_2}(\boldsymbol{y}), \quad \boldsymbol{x}, \boldsymbol{y} \in V_n, \tag{6.18}$$

is maximum-nonlinear.

PROOF. It suffices to apply Proposition 6.31 to the family $\mathcal{M}$ of maximum-nonlinear functions. □

COROLLARY 6.49. *The dual function of a maximum-nonlinear function of type* (6.18) *is the function*

$$\widetilde{f}(\boldsymbol{x}, \boldsymbol{y}) = \langle \boldsymbol{y}, \pi^{-1}(\boldsymbol{x}) \rangle \oplus I_{L_1^\perp}(\boldsymbol{x}) I_{L_2^\perp}(\boldsymbol{y}), \quad \boldsymbol{x}, \boldsymbol{y} \in V_n.$$

The class of maximum-nonlinear functions of type (6.18) is called the class $\mathcal{D}$.

EXAMPLE 6.50. If we put $L_1 = V_n$ and $L_2 = \{\boldsymbol{0}\}$, then, for any permutation π, the function

$$f(\boldsymbol{x}, \boldsymbol{y}) = \langle \boldsymbol{x}, \pi(\boldsymbol{y}) \rangle \oplus \prod_{i=1}^{n} (x^{(i)} \oplus 1), \quad \boldsymbol{x} = (x^{(1)}, \dots, x^{(n)})^\top, \ \boldsymbol{y} \in V_n,$$

is maximum-nonlinear. The class of such functions is denoted by $\mathcal{D}_0$.

All maximum-nonlinear functions of degree 3 in six variables belong to the class $\overline{\mathcal{D}_0}$.

The class of maximum-nonlinear functions $\mathcal{D}_0$ (hence, the class $\mathcal{D}$ as well) is not contained in the complement $\overline{\mathcal{M}}$ of the class $\mathcal{M}$. Besides, the class $\mathcal{D}$ is not contained in the complement $\overline{\mathcal{PS}}$ of the class $\mathcal{PS}$.

6.4. Partially maximum-nonlinear (partially bent) functions and their properties

Suppose that $f \in \mathcal{F}_n$ and E is a subset of the space V_n, $E \subset V_n$. Recall that the mapping $f|_E$ from E to $\mathbb{F}_2$ is called the restriction of the function f to the subset E. It is defined as follows: $f|_E(\boldsymbol{x}) = f(\boldsymbol{x})$ for $\boldsymbol{x} \in E$.

DEFINITION 6.51. Suppose that $f \in \mathcal{F}_n$ and L is a linear subspace of the space V_n, $\dim L = 2t \leq n$. We say that the function f is *maximum-nonlinear bent for the subspace* L if the restriction $f|_L$ is an mn-bent function, i.e., if

$$\sum_{\boldsymbol{x} \in L} (-1)^{f|_L(\boldsymbol{x}) \oplus \langle \boldsymbol{u}, \boldsymbol{x} \rangle} = \sum_{\boldsymbol{x} \in L} (-1)^{f(\boldsymbol{x}) \oplus \langle \boldsymbol{u}, \boldsymbol{x} \rangle} = \pm\sqrt{\#L} = \pm 2^t$$

for every $\boldsymbol{u} \in V_n$.

As before, for an arbitrary function $f \in \mathcal{F}_n$ we will use the following notation:

$$N\Delta_f = 2^n - \#\{\boldsymbol{u} \in V_n \mid \Delta_f(\boldsymbol{u}) = 0\}, \quad \Delta_f = \max_{\boldsymbol{u} \in V_n \setminus \{\boldsymbol{0}\}} |\Delta_f(\boldsymbol{u})|;$$
$$NW_f = 2^n - \#\{\boldsymbol{u} \in V_n \mid W_f(\boldsymbol{u}) = 0\}.$$

Recall that according to Corollary 2.25, for each function $f \in \mathcal{F}_n$ the following inequality holds:

$$N\Delta_f \cdot NW_f \geq 2^n.$$

DEFINITION 6.52. A Boolean function $f \in \mathcal{F}_n$ for which

$$N\Delta_f \cdot NW_f = 2^n, \tag{6.19}$$

is called a *partially mn-bent function*.

THEOREM 6.53. *A Boolean function $f \in \mathcal{F}_n$ is a partially mn-bent function if and only if one of the following equivalent conditions is satisfied:*

1. *There exists a vector $\boldsymbol{u}_0 \in V_n$ such that for any $\boldsymbol{s} \in V_n$ the value of $\Delta_f(\boldsymbol{s})$ is equal to 0 or $(-1)^{\langle \boldsymbol{u}_0, \boldsymbol{s} \rangle} 2^n$.*

2. *The space V_n can be represented as a direct sum of two subspaces E_f and E', $\dim E' = 2t$, such that*
   - (a) *$f(\boldsymbol{x}, \boldsymbol{y}) = \langle \boldsymbol{u}_0, \boldsymbol{x} \rangle \oplus f|_{E'}(\boldsymbol{y})$ for some $\boldsymbol{u}_0 \in V_n$ and for all $\boldsymbol{x} \in E_f$, $\boldsymbol{y} \in E'$;*
   - (b) *$f|_{E'}$ is a maximum-nonlinear bent function.*

PROOF. From the proof of Theorem 2.90 it follows that (6.19) holds if and only if

$$N\Delta_f = 2^{-n} \max_{\boldsymbol{u} \in V_n} W_f^2(\boldsymbol{u}) = 2^{-n} W_f^2(\boldsymbol{u}_0)$$

for some $\boldsymbol{u}_0 \in V_n$. Put $f'(\boldsymbol{x}) = f(\boldsymbol{x}) \oplus \langle \boldsymbol{u}_0, \boldsymbol{x} \rangle$. Then

$$W_{f'}^2(\boldsymbol{0}) = W_f^2(\boldsymbol{u}_0),$$

whence

$$\sum_{\boldsymbol{s} \in V_n} \Delta_{f'}(\boldsymbol{s}) = W_{f'}^2(\boldsymbol{0}) = W_f^2(\boldsymbol{u}_0) = 2^n N\Delta_f = \sum_{\substack{\boldsymbol{s} \in V_n \\ \Delta_{f'}(\boldsymbol{s}) \neq 0}} 2^n.$$

Therefore, for any $\boldsymbol{s} \in V_n$, the value of $\Delta_{f'}(\boldsymbol{s})$ equals either zero or 2^n. Since $\Delta_{f'}(\boldsymbol{s}) = (-1)^{\langle \boldsymbol{u}_0, \boldsymbol{s} \rangle} \Delta_f(\boldsymbol{s})$, the value of $\Delta_f(\boldsymbol{s})$ is either zero or $(-1)^{\langle \boldsymbol{u}_0, \boldsymbol{s} \rangle} 2^n$. Thus condition 1 is equivalent to (6.19).

Suppose now that condition 1 is satisfied. We put

$$E_f = \{\boldsymbol{x} \in V_n \mid \Delta_f(\boldsymbol{x}) = (-1)^{\langle \boldsymbol{u}_0, \boldsymbol{x} \rangle} 2^n\},$$

i.e., E_f is the space of linear structures of the function f.

Let E' be a linear subspace of V_n such that $V_n = E \dot{+} E'$. Then for any $\boldsymbol{v} \in E'$, $\boldsymbol{v} \neq \boldsymbol{0}$, we have $\Delta_f(\boldsymbol{v}) = 0$ and, according to Problem 6.17, $f|_{E'}$ is an mn-bent function. Thus, condition 2 is satisfied, taking into account the results from §2.5.

Suppose now that condition 2 is satisfied and the function $f(\boldsymbol{x}, \boldsymbol{y})$ is represented as a sum of two functions of disjoint variables, of which one is linear and the other is mn-bent with respect to the subspace. Calculating $\Delta_f(\boldsymbol{s})$ and using Problem 2.82, we prove that condition 1 is satisfied. □

PROBLEM 6.54. Suppose that f is a partially mn-bent function in $\mathcal{F}_n$, A is a nonsingular $(n \times n)$ matrix over the field $\mathbb{F}_2$, and $\boldsymbol{u}$ is an arbitrary vector in V_n. Prove that the function $f'(\boldsymbol{x}) = f(A\boldsymbol{x} \oplus \boldsymbol{u})$ is also a partially mn-bent function in $\mathcal{F}_n$.

PROBLEM 6.55. Prove that the set of partially mn-bent functions contains

- any affine function;
- any mn-bent function;
- any quadratic function.

PROBLEM 6.56. Prove that if f is a partially mn-bent function, then all nonzero squares of its Walsh–Hadamard coefficients are the same and $W_f(\boldsymbol{\alpha}) = 0$ if and only if $\boldsymbol{\alpha} \notin \boldsymbol{u}_0 \oplus E_f^{\perp}$, i.e.,

$$V_n \setminus CI_f = \boldsymbol{u}_0 \oplus E_f^{\perp}.$$

PROBLEM 6.57. Prove that a partially mn-bent function satisfies the PC(k) condition, $1 \leq k \leq n$, if and only if the weight of any vector from the space of linear structures of the function f is either greater than k or equal to zero. (*Hint*: If f is a partially mn-bent function, then $\{\boldsymbol{z} \in V_n \mid \Delta_f(\boldsymbol{z}) = 0\} = V_n \setminus E_f$.)

PROBLEM 6.58. Prove that a partially mn-bent function $f \in \mathcal{F}_n$ is balanced if and only if one of the following two equivalent conditions is satisfied:
1. $f|_{E_f} \neq \text{const}$.
2. There exists a $\boldsymbol{v} \in E_f$ such that $f(\boldsymbol{z} \oplus \boldsymbol{v}) = f(\boldsymbol{z}) \oplus 1$ for any $\boldsymbol{z} \in V_n$.
Otherwise $\text{wt}(f) = 2^{n-1} \pm 2^{n-1-t}$, where $t \leq n/2$.

PROPOSITION 6.59. *A partially mn-bent function $f \in \mathcal{F}_n$ is correlation-immune of order k (resp., balanced correlation-immune of order k) if and only if the coset $\boldsymbol{u}_0 \oplus E_f^{\perp}$ consists only of vectors of weight greater than k or equal to zero (resp., consists only of vectors of weight greater than k).*

PROOF. According to Problem 6.56, if f is a partially mn-bent function, then $W_f(\boldsymbol{\alpha}) \neq 0$ if and only if $\boldsymbol{\alpha} \in \boldsymbol{u}_0 \oplus E_f^{\perp}$.

Suppose that f is a correlation-immune function of order k (resp., balanced correlation-immune function of order k). Then $W_f(\boldsymbol{\alpha}) = 0$ for all $\boldsymbol{\alpha} \in V_n$, $1 \leq \text{wt}(\boldsymbol{\alpha}) \leq k$ (resp., for all $\boldsymbol{\alpha} \in V_n$, $0 \leq \text{wt}(\boldsymbol{\alpha}) \leq k$). Hence, these vectors do not belong to the coset $\boldsymbol{u}_0 \oplus E_f^{\perp}$, i.e., the condition of the theorem is satisfied.

Conversely, suppose that for any $\boldsymbol{\alpha} \in V_n$, $1 \leq \text{wt}(\boldsymbol{\alpha}) \leq k$ (resp., $0 \leq \text{wt}(\boldsymbol{\alpha}) \leq k$) it is true that $\boldsymbol{\alpha} \notin \boldsymbol{u}_0 \oplus E_f^{\perp}$. Then for such an $\boldsymbol{\alpha}$ one has $W_f(\boldsymbol{\alpha}) = 0$, whence f is a correlation-immune function of order k (resp., a balanced correlation-immune function of order k). □

6.5. Plateaued functions and partially defined mn-bent functions

If all nonzero squares of Walsh–Hadamard coefficients are equal, then it follows from Parseval's identity that they are equal to 2^{2n-2r} for some r.

DEFINITION 6.60. Suppose that $f \in \mathcal{F}_n$. If there exists a natural number $r \in \mathbb{N}$, $0 \le r \le n$, such that the square of any Walsh–Hadamard coefficient is equal either to 2^{2n-2r} or to 0, then the function f is called a *plateaued function of order* $2r$.

If we are not interested in the exact value of r, we will just say a *plateaued function*. According to Problem 6.56, any partial mn-bent function is plateaued.

PROBLEM 6.61. Prove that $f \in \mathcal{F}_n$ is a plateaued function of order n if and only if f is an mn-bent function.

PROBLEM 6.62. Prove that $f \in \mathcal{F}_n$ is a plateaued function of order 0 if and only if f is an affine function.

PROBLEM 6.63. Prove that the sum of a plateaued function and a linear function is a plateaued function of the same order.

PROBLEM 6.64. Prove that for any plateaued function there exists a linear function such that their sum is balanced.

PROBLEM 6.65. Prove that if f is a plateaued function of order $2r$, then $NW_f = 2^{2r}$.

For any plateaued function f of order $2r$, its support is defined as the set $S_f \subseteq V_n$ such that $\# S_f = 2^{2r}$ and

$$W_f^2(\boldsymbol{u}) = \begin{cases} 2^{2n-2r} & \text{if } \boldsymbol{u} \in S_f; \\ 0 & \text{otherwise.} \end{cases}$$

The set of plateaued functions with a fixed support $S \subseteq V_n$ will be denoted by $\mathcal{TF}_n(S)$.

PROBLEM 6.66. Prove that $f \in \mathcal{TF}_n(S)$ if and only if for any $\boldsymbol{\alpha} \in V_n$ the following condition holds:

$$\Delta_f(\boldsymbol{\alpha}) = \frac{2^n}{\# S} \sum_{\boldsymbol{u} \in S} (-1)^{\langle \boldsymbol{\alpha}, \boldsymbol{u} \rangle}.$$

Note that according to Problem 6.66 all Boolean functions in the set $\mathcal{TF}_n(S)$ have the same autocorrelation function, which is defined by the projection of the first order Reed–Muller code on the set S.

PROBLEM 6.67. Prove that all Boolean functions from $\mathcal{TF}_n(S)$ have the same space of linear structures L_S, which is defined by the following conditions:

$$L_S = \{\boldsymbol{x} \in V_n \mid \langle \boldsymbol{x}, \boldsymbol{u} \rangle = \text{const} \ \text{ for each } \boldsymbol{u} \in S\}.$$

Now we define Boolean functions dual to plateaued functions; they will be defined only on the set S.

DEFINITION 6.68. For $f \in \mathcal{TF}_n(S)$, the Boolean function $\tilde{f}$ defined on the set S by the formula

$$W_f(\boldsymbol{u}) = 2^{n-r}(-1)^{\tilde{f}(\boldsymbol{u})},$$

is called the *dual function to the plateaued function* f.

In order to study the properties of duals to plateaued functions, it is necessary to introduce, for partial Boolean functions, the same apparatus as for everywhere defined Boolean functions.

DEFINITION 6.69. A mapping $f_S : S \subseteq V_n \to \{0,1\}$ is called a *partially defined (partial, p.d.) Boolean function* with domain S. If $S = V_n$, we will say that f_S is everywhere defined.

We will denote by $\mathcal{F}_n(S)$ the set of all p.d. Boolean functions of n variables with domain S.

DEFINITION 6.70. The *distance between two p.d. Boolean functions* f_S and g_S is equal to the number of vectors in S for which the functions assume different values; it will be denoted by

$$\operatorname{dist}(f_S, g_S) = \#\{\boldsymbol{x} \in S \mid f_S(\boldsymbol{x}) \neq g_S(\boldsymbol{x})\}.$$

The distance from a p.d. function f_S to the set $H \subset \mathcal{F}_n(S)$ is equal to the smallest distance from f_S to a p.d. function in H; it is denoted by

$$\operatorname{dist}(f_S, H) = \min_{g_S \in H} \operatorname{dist}(f_S, g_S).$$

DEFINITION 6.71. The *incomplete Walsh–Hadamard transformation of a p.d. function* $f_S \in \mathcal{F}_n(S)$ is the mapping

$$W_{f_S}(\boldsymbol{u}) = \sum_{\boldsymbol{x} \in S} (-1)^{f_S(\boldsymbol{x}) \oplus \langle \boldsymbol{u}, \boldsymbol{x} \rangle}, \qquad \boldsymbol{u} \in V_n,$$

defined on the whole space V_n.

Using standard techniques (see Chapter 2), it is easy to prove the following important statement.

LEMMA 6.72. *For the incomplete Walsh–Hadamard transformation the following inversion formula holds:*

$$2^{-n} \sum_{\boldsymbol{u} \in V_n} W_{f_S}(\boldsymbol{u})(-1)^{\langle \boldsymbol{u}, \boldsymbol{x} \rangle} = \begin{cases} (-1)^{f_S(\boldsymbol{x})} & \text{if } x \in S; \\ 0 & \text{if } x \notin S. \end{cases}$$

PROBLEM 6.73. Prove the following analog of Parseval's identity for partial Boolean functions:

$$\sum_{\boldsymbol{u} \in V_n} W_{f_S}^2(\boldsymbol{u}) = \#S \cdot 2^n.$$

A notion similar to the notion of mn-bent functions for the case of everywhere defined functions can be introduced for partial functions (cf. Problem 6.73).

DEFINITION 6.74. A partial Boolean function $f_S \in \mathcal{F}_n(S)$ is called a *partial mn-bent function* if for each $\boldsymbol{u} \in V_n$ the following equality holds: $|W_{f_S}(\boldsymbol{u})| = \sqrt{\#S}$ (cf. Definition 6.39).

PROBLEM 6.75. Prove that Definitions 6.74 and 6.39 are equivalent.

We will denote by $\mathcal{B}_n(S)$ the set of all partial mn-bent functions with a fixed domain S. It is natural to ask for which sets $S \subseteq V_n$ there exist partial mn-bent functions, i.e., when the set $\mathcal{B}_n(S)$ is nonempty. It is clear from Definition 6.74 that there are no restrictions on n and that the cardinality $\#S$ must be a square. From the same definition it is clear that if S is a coset of a linear subspace V_n of dimension $2t$, then $\mathcal{B}_n(S) \neq \varnothing$ and $\mathcal{B}_n(S)$ can be identified with the set of all bent functions of $2t$ variables. It is also interesting to study the set $\mathcal{B}_n(S)$ for "worse" subsets S of the space V_n.

PROBLEM 6.76. Prove that the mapping $f(\boldsymbol{x}) \mapsto f(\boldsymbol{x} \oplus \boldsymbol{z})$ defines a one-to-one correspondence between $\mathcal{B}_n(S)$ and $\mathcal{B}_n(S \oplus \boldsymbol{z})$ for any $S \subseteq V_n$ and any $\boldsymbol{z} \in V_n$.

PROBLEM 6.77 (Open problem). Find constructive necessary and sufficient conditions for nonemptiness of $\mathcal{B}_n(S)$, in terms of the set $S \subseteq V_n$.

We observe that any function from $\mathcal{B}_n(L)$, where $L \subset V_n$ is a linear subspace with $\dim L = 2t$, can be extended to the whole space V_n in such a way that it will become a partially mn-bent function (see Theorem 6.53). Since the restriction of a Boolean function $f \in \mathcal{F}_n$ to a subset $L \subset V_n$ can be regarded as a partial Boolean function, any maximum-nonlinear bent function on L belongs to $\mathcal{B}_n(L)$.

For a partial mn-bent function the notion of a dual function can be introduced in a standard way. It turns out that this dual function is defined everywhere.

DEFINITION 6.78. For $f_S \in \mathcal{B}_n(S)$, the Boolean function $\tilde{f}_S$ defined on V_n by the formula

$$W_{f_S}(\boldsymbol{u}) = (-1)^{\tilde{f}_S(\boldsymbol{u})}\sqrt{\#S}$$

will be called the *dual function to the partial* mn-bent function f_S.

THEOREM 6.79. *Taking the duals of functions defines a one-to-one correspondence between the classes $\mathcal{B}_n(S)$ and $\mathcal{TF}_n(S)$, $\#S = 2^{2r}$.*

PROOF. Suppose that $f \in \mathcal{TF}_n(S)$. Then

$$\begin{aligned} W_{\tilde{f}}(\boldsymbol{u}) &= \sum_{\boldsymbol{x} \in S} (-1)^{\tilde{f}(\boldsymbol{x}) \oplus \langle \boldsymbol{u}, \boldsymbol{x} \rangle} = \frac{1}{2^{n-r}} \sum_{\boldsymbol{x} \in S} W_f(\boldsymbol{x})(-1)^{\langle \boldsymbol{u}, \boldsymbol{x} \rangle} \\ &= \frac{1}{2^{n-r}} \sum_{\boldsymbol{x} \in V_n} W_f(\boldsymbol{x})(-1)^{\langle \boldsymbol{u}, \boldsymbol{x} \rangle} = 2^r \cdot (-1)^{f(\boldsymbol{u})} \end{aligned}$$

for any $\boldsymbol{u} \in V_n$. Hence, $\tilde{f} \in \mathcal{B}_n(S)$ and $\tilde{\tilde{f}} = f$. The argument for the reverse implication is similar. □

PROBLEM 6.80. Prove that a partially defined function f_S belongs to $\mathcal{B}_n(S)$ if and only if for any $\boldsymbol{u} \in V_n$, $\boldsymbol{u} \neq \boldsymbol{0}$ the equality

$$\sum_{\boldsymbol{x} \in S \cap S \oplus \boldsymbol{u}} (-1)^{f_S(\boldsymbol{x}) \oplus f_S(\boldsymbol{x} \oplus \boldsymbol{u})} = 0$$

holds, where $S \oplus \boldsymbol{u} = \{\boldsymbol{y} \in V_n \mid \boldsymbol{y} \oplus \boldsymbol{u} \in S\}$ and the sum of empty sets is equal to zero by definition.

In other words, the result of Problem 6.80 can be stated as follows: For any $\boldsymbol{u} \in V_n$, $\boldsymbol{u} \neq \boldsymbol{0}$ the Boolean function $f_S(\boldsymbol{x}) \oplus f_S(\boldsymbol{x} \oplus \boldsymbol{u})$, which is defined on $S_{\boldsymbol{u}} = S \cap S \oplus \boldsymbol{u}$, is balanced.

PROPOSITION 6.81. *Suppose that $f_S \in \mathcal{B}_n(S) \neq \varnothing$, $\#S = 2^{2r}$. Then for any $\boldsymbol{u} \in V_n$, $\boldsymbol{u} \neq \boldsymbol{0}$ the cardinality $\#S_{\boldsymbol{u}}$ is divisible by 4 whenever $S_{\boldsymbol{u}} \neq \varnothing$.*

PROOF. The function $f_S(\boldsymbol{x}) \oplus f_S(\boldsymbol{x} \oplus \boldsymbol{u})$ over the set $S_{\boldsymbol{u}}$ assumes each of the values 0 and 1 an even number of times. □

PROBLEM 6.82. Suppose that $\mathcal{B}_n(S) \neq \varnothing$, $\#S = 4$. Prove that S is a linear subspace of dimension 2.

This result naturally suggests the following conjecture: If the set $\mathcal{B}_n(S)$ is nonempty, $\mathcal{B}_n(S) \neq \varnothing$, and $\#S = 2^{2r}$, then S is a linear subspace of dimension $2r$. It turns out that this conjecture is not true. This can be shown by the following counterexample.

EXAMPLE 6.83 ([**210**]). Let the function $f \in \mathcal{F}_5$ be defined as follows:

$$f(x^{(1)}, \dots, x^{(5)}) = \bigoplus_{i,j,k:\ 1 \le i<j<k \le 5} x^{(i)}x^{(j)}x^{(k)} \oplus \bigoplus_{i=1}^{3} x^{(i)}x^{(i+1)} \oplus x^{(1)}x^{(5)}.$$

A direct calculation shows that $W_f(\boldsymbol{u}) = 0$ for $\mathrm{wt}(\boldsymbol{u}) \in \{0,3,4\}$, $|W_f(\boldsymbol{u})| = 8$ for $\mathrm{wt}(\boldsymbol{u}) \in \{1,2,5\}$. In particular, $W_f(\boldsymbol{u}) = -8$ for $\boldsymbol{u} \in \{(1,1,1,1,1)^\top,\ (1,1,0,0,0)^\top,\ (0,1,1,0,0)^\top,\ (0,0,1,1,0)^\top,\ (0,0,0,1,1)^\top,\ (1,0,0,0,1)^\top\}$ and $W_f(\boldsymbol{u}) = 8$ for $\boldsymbol{u} \in \{(1,0,0,0,0)^\top,\ (0,1,0,0,0)^\top,\ (0,0,1,0,0)^\top,\ (0,0,0,1,0)^\top,\ (0,0,0,0,1)^\top,\ (1,0,1,0,0)^\top,\ (0,1,0,1,0)^\top,\ (0,0,1,0,1)^\top,\ (1,0,0,1,0)^\top,\ (0,1,0,0,1)^\top\}$. Thus $f \in \mathcal{TF}_n(S)$, where $S = \{\boldsymbol{u} \in V_5 \mid \mathrm{wt}(\boldsymbol{u}) \in \{1,2,5\}\}$. Then $\tilde{f}_S \in \mathcal{B}_n(S)$. It is easy to see that S is not a linear subspace.

The definition of p.d. bent functions suggests the natural question about inheritance of the property "of being an mn-bent function" when some of the variables are fixed.

DEFINITION 6.84. Suppose that $n \geq 4$ is even and that k is even, $2 \leq k \leq n-2$. A Boolean function $f \in \mathcal{F}_n$ is called an (n,k)*-maximum-nonlinear bent function*, if for any subset $J = \{j_1, \dots, j_k\} \subset \{1, \dots, n\}$ and for any $\boldsymbol{b} = (b^{(1)}, \dots, b^{(k)}) \in V_k$ the subfunction $f_{j_1,\dots,j_k}^{b^{(1)},\dots,b^{(k)}}$ is an mn-bent function in $\mathcal{F}_{n-k}$.

The following theorem, which we give without a proof, describes the set of (n,k)-maximum-nonlinear bent functions (see [**65**]).

THEOREM 6.85. *Any (n,k)-mn-bent function f is a quadratic function of the form*

$$f(\boldsymbol{x}) = \bigoplus_{1 \le i<j \le n} x^{(i)}x^{(j)} \oplus h(\boldsymbol{x}), \quad h \in \mathcal{A}_n.$$

Now we are going to characterize the plateaued functions as Boolean functions having some extremal properties. Put $W_{f,M} = \max_{\boldsymbol{u} \in V_n} |W_f(\boldsymbol{u})|$.

THEOREM 6.86. *Suppose $f \in \mathcal{F}_n$. Then the following statements are equivalent:*
1) *f is a plateaued function;*
2) $W_{f,M} = \frac{2^n}{\sqrt{NW_f}}, \quad N_f = 2^{n-1} - \frac{2^{n-1}}{\sqrt{NW_f}}$;
3) $W_{f,M} = \frac{\sqrt{\sigma_f}}{2^{n/2}}, \quad N_f = 2^{n-1} - \frac{\sqrt{\sigma_f}}{2^{n/2+1}}$.

Theorem 6.86 is implied by the following lemma, which is interesting in its own right.

LEMMA 6.87. *For any function $f \in \mathcal{F}_n$ the following equivalent inequalities hold:*

$$W_{f,M} \geq \frac{2^n}{\sqrt{NW_f}}, \quad N_f \leq 2^{n-1} - \frac{2^{n-1}}{\sqrt{NW_f}}.$$

These inequalities become equalities if and only if f is a plateaued function.

PROOF. Using Parseval's identity we get

$$2^{2n} = \sum_{\boldsymbol{u} \in V_n} W_f^2(\boldsymbol{u}) \leq W_{f,M}^2 NW_f,$$

whence

$$W_{f,M} \geq \frac{2^n}{\sqrt{NW_f}}.$$

The inequality becomes an equality if and only if all nonzero squares of Walsh–Hadamard coefficients are equal, i.e., f is plateaued. □

LEMMA 6.88. *For any function $f \in \mathcal{F}_n$ the following equivalent inequalities hold:*

$$W_{f,M} \geq \frac{\sqrt{\sigma_f}}{2^{n/2}}, \quad N_f \leq 2^{n-1} - \frac{\sqrt{\sigma_f}}{2^{n/2+1}}.$$

These inequalities become equalities if and only if f is a plateaued function.

PROOF. Using the mutual correlation theorem and Pareseval's identity, we see that

$$2^n \sum_{\boldsymbol{u} \in V_n} \Delta_f^2(\boldsymbol{u}) = \sum_{\boldsymbol{u} \in V_n} W_f^4(\boldsymbol{u}) \leq W_{f,M}^2 \cdot 2^{2n},$$

whence

$$W_{f,M} \geq \frac{\sqrt{\sigma_f}}{2^{n/2}}.$$

The conclusion of the proof is similar to that of Lemma 6.87. □

PROBLEM 6.89. Prove that if f is a plateaued function of order $2r$, then $\deg f \leq r$.

PROBLEM 6.90. Suppose that $f \in \mathcal{F}_n$ is a plateaued function of order $2r$, that A is a nonsingular $(n \times n)$ matrix over $\mathbb{F}_2$, and that $\boldsymbol{u} \in V_n$ is an arbitrary vector. Prove that the function $f'(\boldsymbol{x}) = f(A\boldsymbol{x} \oplus \boldsymbol{u})$ is plateaued of order $2r$.

THEOREM 6.91. *Let f be a plateaued function of order $2r$ in $\mathcal{F}_n$, $l = \dim L_f$. Then f is affine equivalent to a Boolean function f', which is represented as a sum of two functions of disjoint variables,*

$$f'(\boldsymbol{y}, \boldsymbol{z}) = g(\boldsymbol{y}) \oplus h(\boldsymbol{z}), \quad \boldsymbol{y} \in V_k, \quad \boldsymbol{z} \in V_l, \quad k + l = n. \tag{6.20}$$

In this expression, $2r \leq n - l$, $g \in \mathcal{F}_k$ is a plateaued function of order $2r$ without linear structures, and $h \in \mathcal{F}_l$ is a linear function.

In particular, $2r = n - l$ if and only if f is a partially mn-bent function.

PROOF. The affine equivalence of the functions f and f' of the form (6.20) is proved in §2.5. Problem 6.90 implies that the function f' is plateaued of order $2r$; hence, Problem 6.63 implies that the function $g \in \mathcal{F}_k$ is plateaued of order $2r \leq k$. According to Problem 6.61, $2r = k = n - l$ if and only if g is an mn-bent function, i.e., f is a partially mn-bent function. □

Using Problem 6.67, we can find the space of linear structures for any Boolean function in $\mathcal{TF}_n(S)$. Then with the help of Theorem 6.91 we can continue studying the plateaued functions without linear structures.

As any partial mn-bent function is a plateaued function, it is interesting to characterize partial mn-bent functions inside the class of plateaued functions.

PROBLEM 6.92. Let $f \in \mathcal{F}_n$ be a plateaued function of order $2r < n$. Prove that the following conditions are equivalent:
1) f is a partial mn-bent function;
2) $N\Delta_f = 2^{n-2r}$;
3) $\dim L_f = n - 2r$.

The following problem is related to Problem 6.92.

PROBLEM 6.93 (Open problem). For any given n, $2r < n$, $0 \le l \le n - 2r$, construct a plateaued function $f \in \mathcal{F}_n$ of order $2r$ for which

$$\dim L_f = l.$$

Obviously the most promising candidates for the required examples are Boolean functions defined by linear branching. We will recall the construction. One says that a Boolean function $f \in \mathcal{F}_n$ is defined by linear branching if

$$f(\boldsymbol{z}) = f(\boldsymbol{x}, \boldsymbol{y}) = \big\langle \boldsymbol{x}, \Phi(\boldsymbol{y}) \big\rangle \oplus h(\boldsymbol{y}), \tag{6.21}$$

where $\boldsymbol{z} = (\boldsymbol{x}, \boldsymbol{y}) \in V_n$, $\boldsymbol{x} \in V_t$, $\boldsymbol{y} \in V_s$, $n = t + s$, $\Phi \in \mathcal{F}_{s,t}$, $h \in \mathcal{F}_s$.

Suppose that $\boldsymbol{u} = (\boldsymbol{v}, \boldsymbol{w}) \in V_n$, $\boldsymbol{v} \in V_t$, $\boldsymbol{w} \in V_s$. According to Problem 2.135,

$$W_f(\boldsymbol{u}) = W_f(\boldsymbol{v}, \boldsymbol{w}) = 2^t \sum_{\boldsymbol{y} \in \Phi^{-1}(\boldsymbol{v})} (-1)^{h(\boldsymbol{y}) \oplus \langle \boldsymbol{y}, \boldsymbol{w} \rangle}. \tag{6.22}$$

We formulate without proof the following statement, which repeats the criteria for maximal nonlinearity of a Boolean function defined by a linear branching (see §6.3).

THEOREM 6.94. *Suppose that a function $f \in \mathcal{F}_n$ is defined by a linear branching satisfying the following conditions:*
1) *for any $\boldsymbol{\alpha} \in V_t$, either $\#\Phi^{-1}(\boldsymbol{\alpha}) = 2^{2l}$ or $\#\Phi^{-1}(\boldsymbol{\alpha}) = 0$;*
2) *for any $\boldsymbol{\alpha} \in V_t$, the functions $h|_{\Phi^{-1}(\boldsymbol{\alpha})}$ are partially mn-bent functions.*
Then the function f is a plateaued function of order $2(n - t - l)$.

We continue the study of properties of plateaued functions in $\mathcal{F}_n$ of order $n-1$ and their relations to mn-bent functions.

DEFINITION 6.95. Functions $g_1, g_2 \in \mathcal{F}_n$, where n is odd, are called *complementary $(n-1)$th order plateaued functions* if for each $\boldsymbol{u} \in V_n$, either $W_{g_1}(\boldsymbol{u}) = 0$, $W_{g_2}^2(\boldsymbol{u}) = 2^{n+1}$ or $W_{g_1}^2(\boldsymbol{u}) = 2^{n+1}$, $W_{g_2}(\boldsymbol{u}) = 0$.

We need the following number-theoretic statement.

LEMMA 6.96. *Suppose that k is a natural number, $k \ge 2$, and a, b are nonnegative integers such that $2^k = a^2 + b^2$. Then*

$$\begin{cases} a^2 = 2^k,\ b = 0 & \textit{if } k \textit{ is even;} \\ a = b = 2^{(k-1)/2} & \textit{if } k \textit{ is odd.} \end{cases}$$

PROOF. If $b = 0$, the statement is obvious.

Suppose that $b \neq 0$. If we write a and b in the form $a = 2^s\alpha$ and $b = 2^t\beta$, where $s \ge 0$, $t \ge 0$, and α, β are odd integers, then

$$2^k = 2^{2s}\alpha^2 + 2^{2t}\beta^2. \tag{6.23}$$

Since α and β are odd, their squares α^2 and β^2 are also odd.

Suppose that $s < t$. In this case the equation (6.23) can be rewritten as follows:

$$2^k = 2^{2s}(\alpha^2 + 2^{2(t-s)}\beta^2) \quad \text{or} \quad 2^{k-2s} = \alpha^2 + 2^{2(t-s)}\beta^2.$$

As α^2 is odd and $t > s$, we have an odd number on the right-hand side, which leads to a contradiction.

Thus, $s \geq t$. In this case (6.23) can be rewritten as follows:

$$2^k = 2^{2t}(2^{2(s-t)}\alpha^2 + \beta^2) \quad \text{or} \quad 2^{k-2t} = 2^{2(s-t)}\alpha^2 + \beta^2.$$

Since β^2 is odd, $s = t$. Thus the last equation becomes

$$2^{k-2t} = \alpha^2 + \beta^2. \tag{6.24}$$

Put $\alpha = 2A + 1$ and $\beta = 2B + 1$, where A and B are integers. Then

$$2^{k-2t} = (2A+1)^2 + (2B+1)^2 = 4(A^2 + B^2 + A + B) + 2.$$

Since the right-hand side of this equation is divisible by 4, this is possible only if $k - 2t = 1$, $\alpha = \beta = 1$. Thus, for $b \neq 0$ we see that $k = 2t + 1$, $s = t$, $a = b = 2^{(k-1)/2}$. □

THEOREM 6.97. *The functions $g_1, g_2 \in \mathcal{F}_n$ are complementary plateaued functions of order $n - 1$ if and only if one of the following equivalent conditions holds:*

1) $W_{g_1}^2(\boldsymbol{u}) + W_{g_2}^2(\boldsymbol{u}) = 2^{n+1}$ *for any* $\boldsymbol{u} \in V_n$;

2) $\Delta_{g_1}(\boldsymbol{u}) + \Delta_{g_2}(\boldsymbol{u}) = 0$ *for any nonzero* $\boldsymbol{u} \in V_n$.

PROOF. Since $n + 1$ is even, Lemma 6.96 implies that condition 1) is satisfied if and only if, for a $\boldsymbol{u} \in V_n$, either $W_{g_1}(\boldsymbol{u}) = 0$, $W_{g_2}^2(\boldsymbol{u}) = 2^{n+1}$ or $W_{g_1}^2(\boldsymbol{u}) = 2^{n+1}$, $W_{g_2}(\boldsymbol{u}) = 0$, i.e., g_1, g_2 are complementary plateaued functions of order $n-1$. Using Corollary 2.81, we see that

$$\begin{aligned} &(\Delta_{g_1}(\boldsymbol{u}_0) + \Delta_{g_2}(\boldsymbol{u}_0), \Delta_{g_1}(\boldsymbol{u}_1) + \Delta_{g_2}(\boldsymbol{u}_1), \ldots, \Delta_{g_1}(\boldsymbol{u}_{2^n-1}) + \Delta_{g_2}(\boldsymbol{u}_{2^n-1}))H_n \\ &= (W_{g_1}^2(\boldsymbol{u}_0) + W_{g_2}^2(\boldsymbol{u}_0), W_{g_1}^2(\boldsymbol{u}_1) + W_{g_2}^2(\boldsymbol{u}_1), \ldots, W_{g_1}^2(\boldsymbol{u}_{2^n-1}) + W_{g_2}^2(\boldsymbol{u}_{2^n-1})). \end{aligned} \tag{6.25}$$

From the well-known properties of the matrix H_n we conclude that conditions 1) and 2) are equivalent. □

PROBLEM 6.98. Prove that the property of being a pair of complementary plateaued functions of order $n - 1$ is stable under:

1) a simultaneous nondegenerate linear substitution of variables in both functions;

2) adding the same linear function to both functions;

3) an arbitrary shift of variables.

PROBLEM 6.99. Prove that if g_1, g_2 are complementary plateaued functions of order $n - 1$, then

1) $L_{g_1} = L_{g_2}$;

2) one and only one of these functions is balanced.

If g_1, g_2 is a complementary pair of plateaued functions of order $n-1$, it follows immediately from the definition that

$$W_{g_1}(\boldsymbol{u}) \pm W_{g_2}(\boldsymbol{u}) = \pm 2^{(n+1)/2} \text{ for all } \boldsymbol{u} \in V_n.$$

This gives us the opportunity to define, similarly to the general case of bent functions, the duality functions $\widetilde{g}_1(\boldsymbol{u})$, $\widetilde{g}_2(\boldsymbol{u})$:

$$\begin{aligned} W_{g_1}(\boldsymbol{u}) + W_{g_2}(\boldsymbol{u}) &= (-1)^{\widetilde{g}_1(\boldsymbol{u})} \cdot 2^{(n+1)/2}, \\ W_{g_1}(\boldsymbol{u}) - W_{g_2}(\boldsymbol{u}) &= (-1)^{\widetilde{g}_2(\boldsymbol{u})} \cdot 2^{(n+1)/2}. \end{aligned} \tag{6.26}$$

THEOREM 6.100. *Passing to dual functions preserves the property of being a pair of complementary plateaued functions of order $n-1$; besides,*

$$\widetilde{\widetilde{g}}_1 = g_1, \quad \widetilde{\widetilde{g}}_2 = g_2.$$

PROOF. Using (6.26) we get

$$\begin{aligned} W_{\widetilde{g}_1}(\boldsymbol{v}) = \sum_{\boldsymbol{u} \in V_n} (-1)^{\widetilde{g}_1(\boldsymbol{u}) \oplus \langle \boldsymbol{u}, \boldsymbol{v} \rangle} &= 2^{-\frac{n+1}{2}} \sum_{\boldsymbol{u} \in V_n} [W_{g_1}(\boldsymbol{u}) + W_{g_2}(\boldsymbol{u})](-1)^{\langle \boldsymbol{u}, \boldsymbol{v} \rangle} \\ &= 2^{\frac{n-1}{2}} [(-1)^{g_1(\boldsymbol{v})} + (-1)^{g_2(\boldsymbol{v})}], \\ W_{\widetilde{g}_2}(\boldsymbol{v}) = \sum_{\boldsymbol{u} \in V_n} (-1)^{\widetilde{g}_2(\boldsymbol{u}) \oplus \langle \boldsymbol{u}, \boldsymbol{v} \rangle} &= 2^{-\frac{n+1}{2}} \sum_{\boldsymbol{u} \in V_n} [W_{g_1}(\boldsymbol{u}) - W_{g_2}(\boldsymbol{u})](-1)^{\langle \boldsymbol{u}, \boldsymbol{v} \rangle} \\ &= 2^{\frac{n-1}{2}} [(-1)^{g_1(\boldsymbol{v})} - (-1)^{g_2(\boldsymbol{v})}]. \end{aligned}$$

It is easy to infer from these equations that $W^2_{\widetilde{g}_1}(\boldsymbol{v}) + W^2_{\widetilde{g}_2}(\boldsymbol{v}) = 2^{n+1}$ for any $\boldsymbol{v} \in V_n$,

$$\begin{aligned} W_{\widetilde{g}_1}(\boldsymbol{v}) + W_{\widetilde{g}_2}(\boldsymbol{v}) &= (-1)^{g_1(\boldsymbol{v})} \cdot 2^{(n+1)/2}, \\ W_{\widetilde{g}_1}(\boldsymbol{v}) - W_{\widetilde{g}_2}(\boldsymbol{v}) &= (-1)^{g_2(\boldsymbol{v})} \cdot 2^{(n+1)/2}. \end{aligned}$$

□

Now we proceed with the study of connections between plateaued functions and mn-bent functions. For an arbitrary function $f \in \mathcal{F}_n$, restrict it to a hyperplane and to one of its cosets. Let $L \subset V_n$ be a hyperplane ($\dim L = n-1$) and let $\boldsymbol{e} \in (V_n \setminus L) \cup L^{\perp}$ be a vector such that $\langle \boldsymbol{e}, \boldsymbol{e} \rangle = 1$. Then it is clear that any vector $\boldsymbol{x} \in V_n$ is uniquely represented as

$$\boldsymbol{x} = \boldsymbol{x}' \oplus \varepsilon \boldsymbol{e}, \quad \boldsymbol{x}' \in L, \quad \varepsilon \in \mathbb{F}_2,$$

and any function $f \in \mathcal{F}_n$ is uniquely represented as

$$f(\boldsymbol{x}' \oplus \varepsilon \boldsymbol{e}) = \varepsilon f|_L(\boldsymbol{x}') \oplus (\varepsilon \oplus 1) f|_{V_n \setminus L}(\boldsymbol{x}'). \tag{6.27}$$

This representation is a generalization of the well-known expansion of a Boolean function with respect to one variable. We also put $f|_L = f_1 \in \mathcal{F}_{n-1}$, $f|_{V_n \setminus L} = f_2 \in \mathcal{F}_{n-1}$. Using (6.27), it is easy to obtain by simple calculations that for any $\boldsymbol{u}' \in L$ the following equalities hold:

$$W_f(\boldsymbol{u}') = W_{f_1}(\boldsymbol{u}') + W_{f_2}(\boldsymbol{u}'), \quad W_f(\boldsymbol{u}' \oplus \boldsymbol{e}) = W_{f_1}(\boldsymbol{u}') - W_{f_2}(\boldsymbol{u}'),$$

whence

$$W^2_f(\boldsymbol{u}') + W^2_f(\boldsymbol{u}' \oplus \boldsymbol{e}) = 2(W^2_{f_1}(\boldsymbol{u}') + W^2_{f_2}(\boldsymbol{u}')). \tag{6.28}$$

THEOREM 6.101. *A function $f \in \mathcal{F}_n$ is mn-bent if and only if $f_1, f_2 \in \mathcal{F}_{n-1}$ is a pair of complementary plateaued functions of order $n-2$.*

PROOF. Since $n+1$ is odd, Lemma 6.96 implies that the equality

$$W^2_f(\boldsymbol{u}') + W^2_f(\boldsymbol{u}' \oplus \boldsymbol{e}) = 2^{n+1}$$

is equivalent to f being an mn-bent function, and according to (6.28), it is also equivalent to f_1, f_2 being a pair of complementary plateaued functions of order $n-2$. □

PROBLEM 6.102. Suppose that n is even, $f \in \mathcal{F}_n$, $i \in \{1, 2, \ldots, n\}$, and

$$\boldsymbol{x} = (x^{(1)}, \ldots, x^{(n)})^\top \in V_n, \quad \boldsymbol{y} = (x^{(1)}, \ldots, x^{(i-1)}, x^{(i+1)}, \ldots, x^{(n)})^\top \in V_{n-1}.$$

Put

$$f_1(x^{(1)}, \ldots, x^{(i-1)}, x^{(i+1)}, \ldots, x^{(n)}) = f(x^{(1)}, \ldots, x^{(i-1)}, 0, x^{(i+1)}, \ldots, x^{(n)}),$$
$$f_2(x^{(1)}, \ldots, x^{(i-1)}, x^{(i+1)}, \ldots, x^{(n)}) = f(x^{(1)}, \ldots, x^{(i-1)}, 1, x^{(i+1)}, \ldots, x^{(n)}).$$

Show that f is an mn-bent function if and only if

$$g(\boldsymbol{y}) = (x^{(i)} \oplus 1) f_1(\boldsymbol{y} \oplus \boldsymbol{u}) \oplus x^{(i)} f_2(\boldsymbol{y} \oplus \boldsymbol{v})$$

is an mn-bent function for $\boldsymbol{u}, \boldsymbol{v} \in V_{n-1}$.

6.6. Hyperbent functions

We will use the trace representation of Boolean functions in order to find the extended Walsh–Hadamard transformation. We identify the space V_n with the field $\mathbb{F}_{2^n}$ (cf. §2.2). Everywhere in this section, n will be even, and we put $n = 2m$.

Any affine function $l_{\boldsymbol{u},\varepsilon}(\boldsymbol{x}) = \langle \boldsymbol{u}, \boldsymbol{x} \rangle \oplus \varepsilon$, $\boldsymbol{u} \in V_n$, $\varepsilon \in \mathbb{F}_2$ can be represented as $l_{\boldsymbol{u},\varepsilon}(\boldsymbol{x}) = \mathrm{Tr}(\boldsymbol{a_u x}) \oplus \varepsilon$, where $\boldsymbol{a_u}$ is an appropriate element of the field $\mathbb{F}_{2^n}$. Besides affine functions, we will work with Boolean function in $\mathcal{F}_n$ defined using traces of monomials of the form $\boldsymbol{a}\boldsymbol{x}^s$, $\boldsymbol{a} \in \mathbb{F}_{2^n}$, $s > 1$, $s \in \Pi_n$ (see §2.2). We put

$$\mathcal{EA}_n = \{\mathrm{Tr}(\boldsymbol{a}\boldsymbol{x}^s) \oplus \varepsilon \mid \boldsymbol{a} \in \mathbb{F}_{2^n},\ s \in \Pi_n,\ \varepsilon \in \mathbb{F}_2\}. \tag{6.29}$$

We recall that s is a minimal representative of $\mathcal{C}_{n,s}$ modulo $2^n - 1$ and that $(s, 2^n - 1) = 1$.

DEFINITION 6.103. Suppose that $f \in \mathcal{F}_n$. We put

$$W_f(\boldsymbol{a}, s) = \sum_{\boldsymbol{x} \in \mathbb{F}_{2^n}} (-1)^{f(\boldsymbol{x}) \oplus \mathrm{Tr}(\boldsymbol{a}\boldsymbol{x}^s)}, \tag{6.30}$$

where $\boldsymbol{a} \in \mathbb{F}_{2^n}$, $s \in \Pi_n$. We will call $W_f(\boldsymbol{a}, s)$ the *extended Walsh–Hadamard transform*, EWHT, of the function f.

Note that the ordinary Walsh–Hadamard transform coincides with EWHT $W_f(\boldsymbol{a}, 1)$.

PROBLEM 6.104. Extend to the case of EWHT the known properties of the ordinary Walsh–Hadamard transform.

Using EWHT, we can generalize the notion of nonlinearity as follows.

DEFINITION 6.105. The *generalized nonlinearity* GN_f of a Boolean function $f \in \mathcal{F}_n$ is the minimum distance from the function f to the set $\mathcal{EA}_n$, which obviously can be calculated by

$$GN_f = 2^{n-1} - \frac{1}{2} \max_{\substack{\boldsymbol{a} \in \mathbb{F}_{2^n} \\ s \in \Pi_n}} |W_f(\boldsymbol{a}, s)|. \tag{6.31}$$

Thus, if we regard all the functions in $\mathcal{EA}_n$, as 'bad' from the cryptographic point of view, we define the cryptographic quality of a Boolean function as the distance to the set $\mathcal{EA}_n$. Continuing the analogy with the ordinary Walsh–Hadamard transformation and taking into account the generalized Parseval identity, we give the following definition.

DEFINITION 6.106. A Boolean function $f \in \mathcal{F}_n$ is called a *hyperbent function* if

$$W_f(\boldsymbol{a}, s) = \sum_{\boldsymbol{x} \in \mathbb{F}_{2^n}} (-1)^{f(\boldsymbol{x}) \oplus \mathrm{Tr}(\boldsymbol{a}\boldsymbol{x}^s)} = \pm 2^{n/2},$$

for any $\boldsymbol{a} \in \mathbb{F}_{2^n}$, $s \in \Pi_n$.

We will denote the set of hyperbent functions by $\mathcal{HB}_n$. Obviously, $\mathcal{HB}_n \subset \mathcal{B}_n$.

Now we will briefly (without proofs) describe only one class of hyperbent functions. The reader can find more details in [**421**]. We will use the definitions from Chapter 5. For a Boolean function $f \in \mathcal{F}_n$ we denote the shortened row of its values by $\Omega_f^* = (f(1), f(\boldsymbol{\alpha}), \ldots, f(\boldsymbol{\alpha}^{2^n-2}))$, where $\boldsymbol{\alpha}$ is a primitive element of the field $\mathbb{F}_{2^n}$. By its period we will mean the minimum period of the sequence $\Omega_f^*, \Omega_f^*, \Omega_f^*, \ldots$.

EXAMPLE 6.107. Suppose that $n = 2m$, $f \in \mathcal{F}_n$, $f(\mathbf{0}) = 0$, and the period of Ω_f^* is $2^m + 1$. Then f is a hyperbent function if and only if

$$\mathrm{wt}(f(1), f(\boldsymbol{\alpha}), \ldots, f(\boldsymbol{\alpha}^{2^m+1})) = 2^{m-1}.$$

Obviously, the cardinality of this class of hyperbent functions equals $\binom{2^m+1}{2^{m-1}}$. This class is interesting because it coincides with the class $\mathcal{PS}$ (see §6.3) of bent functions constructed earlier by Dillon [**108, 109**].

6.7. Biorthogonal bases

DEFINITION 6.108. We will say that a system of 2^n Boolean functions $e_{\boldsymbol{\alpha}}(\boldsymbol{x})$, $\boldsymbol{\alpha}, \boldsymbol{x} \in V_n$, is a *biorthogonal basis* if the following conditions are satisfied:

$$\sum_{\boldsymbol{x} \in V_n} (-1)^{e_{\boldsymbol{\alpha}}(\boldsymbol{x})}(-1)^{e_{\boldsymbol{\beta}}(\boldsymbol{x})} = 0 \quad \text{for } \boldsymbol{\alpha} \neq \boldsymbol{\beta}, \tag{6.32}$$

$$\sum_{\boldsymbol{\alpha} \in V_n} (-1)^{e_{\boldsymbol{\alpha}}(\boldsymbol{x})}(-1)^{e_{\boldsymbol{\alpha}}(\boldsymbol{y})} = 0 \quad \text{for } \boldsymbol{x} \neq \boldsymbol{y}. \tag{6.33}$$

This means that the system of functions $e_{\boldsymbol{\alpha}}(\boldsymbol{x})$ defines two orthogonal bases in a 2^n-dimensional real vector space. The vectors in the first basis are of the form

$$((-1)^{e_{\boldsymbol{\alpha}}(\boldsymbol{x}_1)}, \ldots, (-1)^{e_{\boldsymbol{\alpha}}(\boldsymbol{x}_{2^n})}) \quad \text{for all } \boldsymbol{\alpha} \in V_n,$$

and the vectors of the second basis are of the form

$$((-1)^{e_{\boldsymbol{\alpha}_1}(\boldsymbol{x})}, \ldots, (-1)^{e_{\boldsymbol{\alpha}_{2^n}}(\boldsymbol{x})}) \quad \text{for all } \boldsymbol{x} \in V_n,$$

where $\{\boldsymbol{x}_1, \ldots, \boldsymbol{x}_{2^n}\}$, $\{\boldsymbol{\alpha}_1, \ldots, \boldsymbol{\alpha}_{2^n}\}$ are two enumerations of the elements of the space V_n. In the matrix language, the conditions (6.32) and (6.33) mean that the $(2^n \times 2^n)$ matrix $\big((-1)^{e_{\boldsymbol{\alpha}}(\boldsymbol{x})}\big)$, $\boldsymbol{\alpha}, \boldsymbol{x} \in V_n$, is an Hadamard matrix by rows (condition (6.32)) and by columns (condition (6.33)). It is well known from linear algebra that any vector can be uniquely represented as a linear combination of elements of an orthogonal basis and that the coefficients of this linear combination are easily expressible in terms of scalar products. For the case of Boolean functions, this representation is as follows:

$$B_f(\boldsymbol{\alpha}) = \sum_{\boldsymbol{x} \in V_n} (-1)^{f(\boldsymbol{x})}(-1)^{e_{\boldsymbol{\alpha}}(\boldsymbol{x})} \qquad \text{for all } \boldsymbol{\alpha} \in V_n, \tag{6.34}$$

$$(-1)^{f(\boldsymbol{x})} = \frac{1}{2^n} \sum_{\boldsymbol{\alpha} \in V_n} B_f(\boldsymbol{\alpha})(-1)^{e_{\boldsymbol{\alpha}}(\boldsymbol{x})} \qquad \text{for all } \boldsymbol{x} \in V_n. \tag{6.35}$$

The coefficients and $B_f(\boldsymbol{\alpha})$ are chosen in such a way that the case $e_{\boldsymbol{\alpha}}(\boldsymbol{x}) = \langle \boldsymbol{\alpha}, \boldsymbol{x} \rangle$ yields $B_f(\boldsymbol{\alpha}) = W_f(\boldsymbol{\alpha})$, thus reducing to the case of Walsh–Hadamard transform. Except for the simplest case $e_{\boldsymbol{\alpha}}(\boldsymbol{x}) = \langle \boldsymbol{\alpha}, \boldsymbol{x} \rangle$, the following important class of biorthogonal bases is useful.

EXAMPLE 6.109. Suppose that $n = k + 2l$, so that any vector $\boldsymbol{x} \in V_n$ can be represented as $\boldsymbol{x} = (\boldsymbol{x}^{(1)}, \boldsymbol{x}^{(2)})$, where $\boldsymbol{x}^{(1)} \in V_k$, $\boldsymbol{x}^{(2)} \in V_{2l}$. Let $\varphi(\boldsymbol{x}^{(2)})$ be a bent function. Put

$$e_{\boldsymbol{\alpha}}(\boldsymbol{x}) = e_{(\boldsymbol{\alpha}^{(1)}, \boldsymbol{\alpha}^{(2)})}(\boldsymbol{x}^{(1)}, \boldsymbol{x}^{(2)}) = \langle \boldsymbol{\alpha}^{(1)}, \boldsymbol{x}^{(1)} \rangle \oplus \varphi(\boldsymbol{x}^{(2)} \oplus \boldsymbol{\alpha}^{(2)}).$$

The system $e_{\boldsymbol{\alpha}}(\boldsymbol{x})$ is a biorthogonal basis. It is easy to check conditions (6.32) and (6.33) taking into account that $\varphi(\boldsymbol{x}^{(2)})$ is a bent function:

$$\begin{aligned}&\sum_{\boldsymbol{x} \in V_n} (-1)^{e_{\boldsymbol{\alpha}}(\boldsymbol{x})} (-1)^{e_{\boldsymbol{\beta}}(\boldsymbol{x})} \\ &\qquad = \sum_{\boldsymbol{x}^{(1)} \in V_k} (-1)^{\langle \boldsymbol{\alpha}^{(1)} \oplus \boldsymbol{\beta}^{(1)}, \boldsymbol{x}^{(1)} \rangle} \cdot \sum_{\boldsymbol{x}^{(2)} \in V_{2l}} (-1)^{\varphi(\boldsymbol{\alpha}^{(2)} \oplus \boldsymbol{x}^{(2)}) \oplus \varphi(\boldsymbol{\beta}^{(2)} \oplus \boldsymbol{x}^{(2)})} = 0\end{aligned}$$

if $(\boldsymbol{\alpha}^{(1)}, \boldsymbol{\alpha}^{(2)}) \neq (\boldsymbol{\beta}^{(1)}, \boldsymbol{\beta}^{(2)})$, and

$$\begin{aligned}&\sum_{\boldsymbol{\alpha} \in V_n} (-1)^{e_{\boldsymbol{\alpha}}(\boldsymbol{x})} (-1)^{e_{\boldsymbol{\alpha}}(\boldsymbol{y})} \\ &\qquad = \sum_{\boldsymbol{\alpha}^{(1)} \in V_k} (-1)^{\langle \boldsymbol{x}^{(1)} \oplus \boldsymbol{y}^{(1)}, \boldsymbol{\alpha}^{(1)} \rangle} \cdot \sum_{\boldsymbol{\alpha}^{(2)} \in V_{2l}} (-1)^{\varphi(\boldsymbol{\alpha}^{(2)} \oplus \boldsymbol{x}^{(2)}) \oplus \varphi(\boldsymbol{\alpha}^{(2)} \oplus \boldsymbol{y}^{(2)})} = 0\end{aligned}$$

if $(\boldsymbol{x}^{(1)}, \boldsymbol{x}^{(2)}) \neq (\boldsymbol{y}^{(1)}, \boldsymbol{y}^{(2)})$.

Observe that using coding theory language and (6.34), it is easy to express the Hamming distance between the vector of values of the Boolean function $f(\boldsymbol{x})$ and the vector of values of the Boolean function $e_{\boldsymbol{\alpha}}(\boldsymbol{x})$:

$$\operatorname{dist}(f(\boldsymbol{x}), e_{\boldsymbol{\alpha}}(\boldsymbol{x})) = 2^{n-1} - \frac{1}{2} B_f(\boldsymbol{\alpha}). \tag{6.36}$$

In particular, it follows that a function in the system $e_{\boldsymbol{\alpha}}(\boldsymbol{x})$ that is the closest (in the sense of the Hamming distance) to the function $f(\boldsymbol{x})$ is the function $e_{\boldsymbol{\alpha}_0}(\boldsymbol{x})$ with $\boldsymbol{\alpha}_0 \in V_n$ for which the number $B_f(\boldsymbol{\alpha}_0)$ is maximal among all $B_f(\boldsymbol{\alpha})$. Thus, the problem of approximation of Boolean functions by elements of a biorthogonal basis is reduced to the problem of finding $\max_{\boldsymbol{\alpha} \in V_n} B_f(\boldsymbol{\alpha})$.

It is convenient to restate some of the above results using the language of coding theory. To this end, we need one more definition.

DEFINITION 6.110. An *Hadamard code* $H(e_{\boldsymbol{\alpha}}(\boldsymbol{x}))$ is the set of rows which are vectors of values of the Boolean functions $e_{\boldsymbol{\alpha}}(\boldsymbol{x})$, $e_{\boldsymbol{\alpha}}(\boldsymbol{x}) \oplus 1$, for $\boldsymbol{\alpha} \in V_n$, where $e_{\boldsymbol{\alpha}}(\boldsymbol{x})$ is a biorthogonal basis.

An Hadamard code is nonlinear except for the case $e_{\boldsymbol{\alpha}}(\boldsymbol{x}) = \langle \boldsymbol{\alpha}, \boldsymbol{x} \rangle$, when it coincides with the first order Reed–Muller code. The values of basic parameters of an Hadamard code will be given in the following theorem.

THEOREM 6.111. *An Hadamard code has the following parameters:*
1) *length* 2^n*;*
2) *cardinality* 2^{n+1}*;*
3) *code distance* 2^{n-1}*;*
4) *covering radius* $\leq 2^{n-1} - 2^{\frac{n}{2}-1}$.

PROOF. Parts 1) and 2) are obvious; part 3) follows from (6.32). For part 4) we will prove an analog of Parseval's identity for the numbers $B_f(\boldsymbol{\alpha})$, $f \in \mathcal{F}_n$. From (6.34) and (6.33) we have

$$
\begin{aligned}
\sum_{\boldsymbol{\alpha}\in V_n} B_f^2(\boldsymbol{\alpha}) &= \sum_{\boldsymbol{\alpha}\in V_n} \sum_{\boldsymbol{x},\boldsymbol{y}\in V_n} (-1)^{f(\boldsymbol{x})}(-1)^{f(\boldsymbol{y})}(-1)^{e_{\boldsymbol{\alpha}}(\boldsymbol{x})}(-1)^{e_{\boldsymbol{\alpha}}(\boldsymbol{y})} \\
&= \sum_{\boldsymbol{x},\boldsymbol{y}\in V_n} (-1)^{f(\boldsymbol{x})}(-1)^{f(\boldsymbol{y})} \sum_{\boldsymbol{\alpha}\in V_n} (-1)^{e_{\boldsymbol{\alpha}}(\boldsymbol{x})}(-1)^{e_{\boldsymbol{\alpha}}(\boldsymbol{y})} = 2^{2n}.
\end{aligned}
\tag{6.37}
$$

Whence,

$$
\min_f \max_{\boldsymbol{\alpha}\in V_n} |B_f(\boldsymbol{\alpha})| \le 2^{n/2}. \tag{6.38}
$$

Taking into account (6.36), we obtain part 4). □

THEOREM 6.112. *If $n = 2m + 2l$, $\varphi(\boldsymbol{x}^{(2)})$ is a bent function, and*

$$
e_{\boldsymbol{\alpha}}(\boldsymbol{x}) = e_{(\boldsymbol{\alpha}^{(1)},\boldsymbol{\alpha}^{(2)})}(\boldsymbol{x}^{(1)},\boldsymbol{x}^{(2)}) = \langle \boldsymbol{\alpha}^{(1)}, \boldsymbol{x}^{(1)}\rangle \oplus \varphi(\boldsymbol{x}^{(2)} \oplus \boldsymbol{\alpha}^{(2)}),
$$

where $\boldsymbol{\alpha}^{(1)}, \boldsymbol{x}^{(1)} \in V_{2m}$ and $\boldsymbol{\alpha}^{(2)}, \boldsymbol{x}^{(2)} \in V_{2l}$, then the covering radius of an Hadamard code $H(e_{\boldsymbol{\alpha}}(\boldsymbol{x}))$ is equal to $2^{n-1} - 2^{\frac{n}{2}-1}$.

PROOF. Taking into account the proof of Theorem 6.111, it is enough to prove that for some Boolean function $f(\boldsymbol{x})$ the inequalities (6.38) are sharp. Put

$$
f(\boldsymbol{x}) = f(\boldsymbol{x}^{(1)}, \boldsymbol{x}^{(2)}) = \psi(\boldsymbol{x}^{(1)}) \oplus \langle \boldsymbol{a}, \boldsymbol{x}^{(2)}\rangle,
$$

$\boldsymbol{x}^{(1)} \in V_{2m}$, $\boldsymbol{a}, \boldsymbol{x}^{(2)} \in V_{2l}$, where $\psi(\boldsymbol{x}^{(1)})$ is some bent function.

Formula (6.34) implies that

$$
\begin{aligned}
B_f(\boldsymbol{\alpha}) &= \sum_{\boldsymbol{x}\in V_n} (-1)^{f(\boldsymbol{x})}(-1)^{e_{\boldsymbol{\alpha}}(\boldsymbol{x})} \\
&= \sum_{\boldsymbol{x}^{(1)}\in V_{2m}} (-1)^{\psi(\boldsymbol{x}^{(1)})}(-1)^{\langle \boldsymbol{\alpha}^{(1)},\boldsymbol{x}^{(1)}\rangle} \sum_{\boldsymbol{x}^{(2)}\in V_{2l}} (-1)^{\varphi(\boldsymbol{x}^{(2)}\oplus\boldsymbol{\alpha}^{(2)})}(-1)^{\langle \boldsymbol{a},\boldsymbol{x}^{(2)}\rangle} \\
&= (\pm 2^m)(\pm 2^l) = \pm\, 2^{n/2}
\end{aligned}
$$

for all $\boldsymbol{\alpha} = (\boldsymbol{\alpha}^{(1)}, \boldsymbol{\alpha}^{(2)}) \in V_n$, as $\varphi(\boldsymbol{x}^{(2)})$, $\psi(\boldsymbol{x}^{(1)})$ are bent functions. □

Thus, the basic parameters of an arbitrary Hadamard code coincide with the parameters of a first order Reed–Muller code.

Compare now the possibilities for approximating Boolean functions by elements of two biorthogonal bases: affine functions $\langle \boldsymbol{\alpha}, \boldsymbol{x}\rangle \oplus \varepsilon$ and functions of the type $\langle \boldsymbol{\alpha}^{(1)}, \boldsymbol{x}^{(1)}\rangle \oplus \varphi(x^{(2)} \oplus \boldsymbol{\alpha}^{(2)}) \oplus \varepsilon$. From the argument above it follows that in the general case, in order to choose the best approximation, it is enough to compare the numbers $\max_{\boldsymbol{\alpha}\in V_n} |W_f(\boldsymbol{\alpha})|$ and $\max_{\boldsymbol{\alpha}\in V_n} |B_f(\boldsymbol{\alpha})|$. The larger of these determines which base approximates the functions. Finding $\max_{\boldsymbol{\alpha}\in V_n} |W_f(\boldsymbol{\alpha})|$ and $\max_{\boldsymbol{\alpha}\in V_n} |B_f(\boldsymbol{\alpha})|$ is computationally expensive. However, it turns out that some preliminary conclusions can be made by analyzing a useful analytical relation between $W_f(\boldsymbol{\alpha})$ and $B_f(\boldsymbol{\alpha})$.

LEMMA 6.113. *For any Boolean function $f(\boldsymbol{x})$ and any system of functions $e_{\boldsymbol{\alpha}}(\boldsymbol{x}) = e_{(\boldsymbol{\alpha}^{(1)},\boldsymbol{\alpha}^{(2)})}(\boldsymbol{x}^{(1)}, \boldsymbol{x}^{(2)}) = \langle \boldsymbol{\alpha}^{(1)}, \boldsymbol{x}^{(1)}\rangle \oplus \varphi(x^{(2)} \oplus \boldsymbol{\alpha}^{(2)}) \oplus \varepsilon$ satisfying the* *hypotheses of Example* 6.109, *the following relation holds:*

$$B_f(\boldsymbol{\alpha}) = \frac{1}{2^l} \sum_{\boldsymbol{\beta}^{(2)}\in V_{2l}} (-1)^{\langle \boldsymbol{\beta}^{(2)}, \boldsymbol{\alpha}^{(2)}\rangle} (-1)^{\widetilde{\varphi}(\boldsymbol{\beta}^{(2)})} W_f(\boldsymbol{\alpha}^{(1)}, \boldsymbol{\beta}^{(2)}), \tag{6.39}$$

where $\widetilde{\varphi}(\boldsymbol{\beta}^{(2)})$ *is the dual bent function to the function* $\varphi(\boldsymbol{x}^{(2)})$.

PROOF. First we write down the following inverse Walsh–Hadamard transformations:

$$(-1)^{f(\boldsymbol{x})} = \frac{1}{2^n} \sum_{\boldsymbol{\beta}\in V_n} W_f(\boldsymbol{\beta})(-1)^{\langle \boldsymbol{\beta}, \boldsymbol{x}\rangle},$$

$$(-1)^{e_{\boldsymbol{\alpha}}(\boldsymbol{x})} = \frac{1}{2^n} \sum_{\boldsymbol{\gamma}\in V_n} W_{e_{\boldsymbol{\alpha}}}(\boldsymbol{\gamma})(-1)^{\langle \boldsymbol{\gamma}, \boldsymbol{x}\rangle}.$$

Now we substitute the above expressions into the definition (6.34) of $B_f(\boldsymbol{\alpha})$:

$$\begin{aligned} B_f(\boldsymbol{\alpha}) &= \sum_{\boldsymbol{x}\in V_n} (-1)^{f(\boldsymbol{x})}(-1)^{e_{\boldsymbol{\alpha}}(\boldsymbol{x})} \\ &= \sum_{\boldsymbol{x}\in V_n} \Big(\frac{1}{2^n} \sum_{\boldsymbol{\beta}\in V_n} W_f(\boldsymbol{\beta})(-1)^{\langle \boldsymbol{\beta}, \boldsymbol{x}\rangle}\Big)\Big(\frac{1}{2^n} \sum_{\boldsymbol{\gamma}\in V_n} W_{e_{\boldsymbol{\alpha}}}(\boldsymbol{\gamma})(-1)^{\langle \boldsymbol{\gamma}, \boldsymbol{x}\rangle}\Big) \\ &= \frac{1}{2^{2n}} \sum_{\boldsymbol{\beta},\boldsymbol{\gamma}\in V_n} W_f(\boldsymbol{\beta})W_{e_{\boldsymbol{\alpha}}}(\boldsymbol{\gamma})\Big(\sum_{\boldsymbol{x}\in V_n} (-1)^{\langle \boldsymbol{\beta}, \boldsymbol{x}\rangle}(-1)^{\langle \boldsymbol{\gamma}, \boldsymbol{x}\rangle}\Big) \\ &= \frac{1}{2^n} \sum_{\boldsymbol{\beta}\in V_n} W_f(\boldsymbol{\beta})W_{e_{\boldsymbol{\alpha}}}(\boldsymbol{\beta}). \end{aligned} \tag{6.40}$$

The value of $W_{e_{\boldsymbol{\alpha}}}(\boldsymbol{\beta})$ can be computed as follows:

$$\begin{aligned} W_{e_{\boldsymbol{\alpha}}}(\boldsymbol{\beta}) &= \sum_{\substack{\boldsymbol{x}^{(1)}\in V_k \\ \boldsymbol{x}^{(2)}\in V_{2l}}} (-1)^{\langle \boldsymbol{\alpha}^{(1)}, \boldsymbol{x}^{(1)}\rangle}(-1)^{\varphi(\boldsymbol{\alpha}^{(2)}\oplus \boldsymbol{x}^{(2)})}(-1)^{\langle \boldsymbol{\beta}^{(1)}, \boldsymbol{x}^{(1)}\rangle}(-1)^{\langle \boldsymbol{\beta}^{(2)}, \boldsymbol{x}^{(2)}\rangle} \\ &= \sum_{\boldsymbol{x}^{(1)}\in V_k} (-1)^{\langle \boldsymbol{\alpha}^{(1)}\oplus\boldsymbol{\beta}^{(1)}, \boldsymbol{x}^{(1)}\rangle} \sum_{\boldsymbol{x}^{(2)}\in V_{2l}} (-1)^{\varphi(\boldsymbol{\alpha}^{(2)}\oplus \boldsymbol{x}^{(2)})}(-1)^{\langle \boldsymbol{\beta}^{(2)}, \boldsymbol{x}^{(2)}\rangle} \\ &= \begin{cases} 0 & \text{if } \boldsymbol{\beta}^{(1)} \neq \boldsymbol{\alpha}^{(1)}, \\ (-1)^{\langle \boldsymbol{\beta}^{(2)}, \boldsymbol{\alpha}^{(2)}\rangle}(-1)^{\widetilde{\varphi}(\boldsymbol{\beta}^{(2)})} \cdot 2^{k+l} & \text{if } \boldsymbol{\beta}^{(1)} = \boldsymbol{\alpha}^{(1)}. \end{cases} \end{aligned}$$

If we substitute this expression into (6.40), we obtain (6.39). □

Comments to Chapter 6

The problem of enumerating all Boolean (bent) functions on $\mathcal{F}_{2n}$ is not yet solved. Thus, it is interesting to obtain estimates for $\#\mathcal{B}_{2n}$. Using the result of Theorem 6.28 (see [**326**]), it is easy to obtain the following upper bound for $\#\mathcal{B}_{2n}$:

$$\#\mathcal{B}_{2n} \le 2^{1+2n+\cdots+\binom{2n}{n}} = 2^{2^{2n-1}+\frac{1}{2}\binom{2n}{n}}.$$

For the case $2n = 6$ (the maximum value of $2n$ for which the exact number of (bent) functions is known can be found [**320**]) we have $\#\mathcal{B}_6 = 2^{32}$, which is much less than $2^{2^5+\frac{1}{2}\binom{6}{3}} = 2^{42}$.

A lower bound for $\#\mathcal{B}_n$ can be obtained using estimates for cardinalities of various classes of (bent) functions (see §6.3). In [**7**], the following lower bound for $2n = 8$ was obtained:

$$\#\mathcal{B}_8 \geq 1\,559\,994\,535\,674\,013\,286\,400 \simeq 2^{70.4}.$$

The upper bound

$$\#\mathcal{B}_{2n} \leq 2^{2^{2n-1}+\frac{1}{2}\binom{2n}{n}-2^n+n+1}(1+\varepsilon) + 2^{2^{2n-1}-\frac{1}{2}\binom{2n}{n}}$$

for $2n \geq 6$ can be found in [**74**]. Representations of (bent) functions as sums of indicators of linear spaces (Theorems 6.20, 6.21) are studied in [**62, 63**]. Various classes (families) of Boolean (bent) functions were studied by a number of authors. We mention only some of them: [**24, 59, 60, 108, 109, 114, 265, 326**]. For partial (partial bent) functions we refer the reader to [**57, 58**].

Plateaued functions (see [**72, 438**]) are naturally connected with partial functions; their properties are described in [**210, 211**]. We would like to emphasize the close relationship between (bent) functions and plateaued functions (see [**439**]).

Various possible generalizations of the notion of nonlinearity are possible. As was discussed in §6.1, different classes of functions can be regarded as "bad". In particular, a class such as $\mathcal{LS}_n$ contains functions in $\mathcal{F}_n$ having nontrivial space of linear structures. It is clear that $\mathcal{LS}_n \supset \mathcal{A}_n$. It was shown in [**267**] that $\operatorname{dist}(f, \mathcal{LS}_n) \leq 2^{n-2}$ for any function $f \in \mathcal{F}_n$ and that the distance to $\mathcal{LS}_n$ has the affine invariance property. Another possible "bad" class of functions is the class of Boolean functions of type $\{\operatorname{Tr}(ax^s) \oplus \varepsilon\}$ (see §6.6). Functions that are maximally distant from this class are called "hyperbent functions" in [**421**]. This notion has been used earlier in [**65**] for the description of a totally different property of Boolean functions.

It is necessary to note one special feature in connection with the attempts for generalization of the notions of "maximum-nonlinear function" and "bent function" to mappings of various algebraic structures (finite fields, rings, groups, etc.). In the Boolean case these these two notions are "luckily" equivalent, but in general this is not the case, which is ignored by some authors. Results in this direction can be found in [**11, 234, 89, 257, 289, 300**].

CHAPTER 7

Correlation Immunity and Resiliency

The properties of correlation immunity of Boolean functions (see [**362, 365**]) and resiliency of Boolean mappings (see [**21, 87**]) originated in cryptographic problems. However, certain (mathematically essential) features of these properties can be expressed in terms of equivalent and easily formulated quantitative characteristics of Boolean functions and mappings. This chapter is devoted to the study of these properties and their connections with other cryptographic properties.

7.1. Main definitions and properties

The notion of correlation immunity (see [**362, 365**]) of a Boolean function represents its ability to resist correlation analysis (see §9.3). It can be said that this property measures the difficulty, given a known value of a Boolean function, of finding information about the values the function assumes on certain subsets, or for certain functions of its arguments. Otherwise, if such information existed, it would be possible to try to use it "to find the key". Of course, this argument makes sense once we have chosen a probabilistic model.

In order to completely and correctly formulate the property of correlation immunity, we recall certain concepts from probability and information theory.

Let X be a random variable taking values in a finite set U and let $\mathcal{P}\{X = u\}$, $u \in U$, be its probability distribution.

DEFINITION 7.1. The *self-information of the event* $X = u$ (see [**198, 131**]) is defined by the formula

$$I(X = u) = \log \frac{1}{\mathcal{P}\{X = u\}} = -\log \mathcal{P}\{X = u\}.$$

As of now, we will not fix the base of the logarithm since it only gives a scale for measuring the information. Depending on the applications, binary, natural, and physical scales are possible (see [**198, 373, 131**]).

Self-information is always a nonnegative number which increases as $\mathcal{P}\{X = u\}$ decreases. Note that if $\mathcal{P}\{X = u\} = 1$, then the self-information $I(X = u) = \log \frac{1}{\mathcal{P}\{X=u\}} = 0$, i.e., the event $X = u$ surely will happen and there is no self-information. If $\mathcal{P}\{X = u\} = 0$, then the self-information $I(X = u) = \log \frac{1}{\mathcal{P}\{X=u\}} = \infty$, which corresponds to the certainty that the event $X = u$ surely will not happen. Obviously, I is a numerical random variable assuming the value $I(X = u)$ with probability $\mathcal{P}\{X = u\}$, $u \in U$.

DEFINITION 7.2. The *entropy* $H(X)$ of a random variable X is defined as the mean of the self-information I:

$$H(X) = \sum_{u \in U} \mathcal{P}(X = u) \log \frac{1}{\mathcal{P}\{X = u\}} = -\sum_{u \in U} \mathcal{P}(X = u) \log \mathcal{P}\{X = u\}$$

(here we assume that $0 \log 0 = 0$).

EXAMPLE 7.3. Let X be a binary random variable taking value 1 with probability p and value 0 with probability $1 - p$, $0 \le p \le 1$. In this case, we take the base of the logarithm to be 2. Then the entropy of X is

$$H(X) = H(p) = -p \log_2 p - (1 - p) \log_2 (1 - p).$$

The function $H(p)$ satisfies the conditions

$$H(0) = H(1) = 0, \quad 0 \le H(p) \le 1.$$

These conditions show that in the cases $p = 0$ or 1 the entropy is minimal, i.e., we know in advance the value of the random variable. Obviously, the entropy is maximal when $p = 1 - p = \frac{1}{2}$. In this case $H\left(\frac{1}{2}\right) = 1$.

Let Z be another random variable taking values in the set V and let $\mathcal{P}\{Z = v\}$, $v \in V$, be its distribution. We denote by $\mathcal{P}\{X = u, Z = v\}$, $u \in U$, $v \in V$, the joint distribution of the random variables X and Z. We have

$$\sum_{v \in V} \mathcal{P}\{X = u, Z = v\} = \mathcal{P}\{X = u\}, \quad \sum_{u \in U} \mathcal{P}\{X = u, Z = v\} = \mathcal{P}\{Z = v\}.$$

DEFINITION 7.4. The random variables X and Z are called *independent* if for all $u \in U$ and $v \in V$ the following equality holds:

$$\mathcal{P}\{X = u, Z = v\} = \mathcal{P}\{X = u\} \cdot \mathcal{P}\{Z = v\}. \tag{7.1}$$

The conditional distribution of a random variable$Z$ given the occurrence of the event $X = u$ has the following form:

$$\mathcal{P}\{Z = v \mid X = u\} = \frac{\mathcal{P}\{X = u, Z = v\}}{\mathcal{P}\{X = u\}}, \quad v \in V,$$

$$\sum_{v \in V} \mathcal{P}\{Z = v \mid X = u\} = \frac{1}{\mathcal{P}\{X = u\}} \sum_{v \in V} \mathcal{P}\{X = u, Z = v\} = 1.$$

The conditional entropy of a random variable Z given the occurrence of the event $X = u$ is given by

$$H(Z \mid X = u) = -\sum_{v \in V} \mathcal{P}\{Z = v \mid X = u\} \log \mathcal{P}\{Z = v \mid X = u\}.$$

The conditional entropyof a random variable Z with respect to a random variable X is defined as the mean value of the entropy of the random variable Z given the occurrence of the event $X = u$:

$$\begin{aligned}
H(Z \mid X) &= \sum_{u \in U} \mathcal{P}\{X = u\} H(Z \mid X = u) \\
&= -\sum_{u \in U} \mathcal{P}\{X = u\} \sum_{v \in V} \mathcal{P}\{Z = v \mid X = u\} \log \mathcal{P}\{Z = v \mid X = u\} \\
&= -\sum_{u \in U} \sum_{v \in V} \mathcal{P}\{X = u, Z = v\} \log \mathcal{P}\{Z = v \mid X = u\}.
\end{aligned}$$

It is easy to show that

$$H(X, Z) = H(X) + H(Z \mid X) = H(Z) + H(X \mid Z).$$

DEFINITION 7.5. The *mutual information* $I(X, Z)$ of the random variables X and Z is defined by the expression

$$I(X, Z) = \sum_{u \in U} \sum_{v \in V} \mathcal{P}\{X = u, Z = v\} \log \frac{\mathcal{P}\{X = u, Z = v\}}{\mathcal{P}\{X = u\}\mathcal{P}\{Z = v\}}. \tag{7.2}$$

It is clear that the mutual information is positive; a simple calculation shows that

$$I(X, Z) = H(X) - H(X \mid Z) = H(Z) - H(Z \mid X).$$

In other words, the mutual information $I(X, Z)$ is equal to the average reduction of the self-information of the random variable Z given the condition that the value of the random variable X is known. By analogy, the mutual information $I(X, Z)$ is equal to the average decrement of the self-information of the random variable X given the condition that the value of the random variable Z is known. Thus, the decrement of the self-information is the same in both cases, which means that the random variable Z contains as much information about the random variable X as X contains about Z, namely, $I(X, Z)$.

Later, we will need the following statement.

LEMMA 7.6. *The mutual information of random variables is equal to zero if and only if these variables are independent.*

PROOF. If the random variables X and Z are independent, i.e., if for any $u \in U$ and $v \in V$ equality (7.1) holds, then obviously $I(X, Z) = 0$.

We will prove the converse. Suppose that $I(X, Z) = 0$. Since

$$\sum_{\substack{u \in U \\ v \in V}} \mathcal{P}\{X = u\}\mathcal{P}\{Z = v\} = 1,$$

the well-known inequality (see **[254]**) between arithmetic and geometric means yields

$$\prod_{\substack{u \in U \\ v \in V}} \Big(\frac{\mathcal{P}\{X = u\}\mathcal{P}\{Z = v\}}{\mathcal{P}\{X = u, Z = v\}}\Big)^{\mathcal{P}\{X = u, Z = v\}} \leq 1,$$

and the equality holds if and only if, for any $u \in U$, $v \in V$, equality (7.1) is true. Applying logarithms to both sides of the inequality (without loss of generality we may assume that this is the natural logarithm) and multiplying both sides of the result by -1, we get

$$I(X, Z) = \sum_{\substack{u \in U \\ v \in V}} \mathcal{P}\{X = u, Z = v\} \log \frac{\mathcal{P}\{X = u\}\mathcal{P}\{Z = v\}}{\mathcal{P}\{X = u, Z = v\}} \geq 0,$$

where the equality holds if and only if, for any $u \in U$, $v \in V$, equality (7.1) holds. Since by assumption we have $I(X, Z) = 0$, equality (7.1) holds for any $u \in U$, $v \in V$. Therefore, the random variables X and Z are independent. □

Let $f \in \mathcal{F}_n$, $X^{(1)}, X^{(2)}, \ldots, X^{(n)}$ be independent uniformly distributed binary random variables with the distribution $\mathcal{P}\{X^{(i)} = 0\} = \mathcal{P}\{X^{(i)} = 1\} = 1/2$ for

$i = 1, 2, \ldots, n$. The random variable $Z = f(X^{(1)}, X^{(2)}, \ldots, X^{(n)})$ is distributed as follows:

$$\mathcal{P}\{Z = 1\} = \frac{\mathrm{wt}(f)}{2^n}, \quad \mathcal{P}\{Z = 0\} = 1 - \frac{\mathrm{wt}(f)}{2^n}.$$

DEFINITION 7.7. We say that a Boolean function $f(x^{(1)}, x^{(2)}, \ldots, x^{(n)}) \in \mathcal{F}_n$ is *correlation-immune of order* m, $0 < m \le n$, if for any $1 \le i_1 < i_2 < \cdots < i_m \le n$ one has

$$I(X^{(i_1,\ldots,i_m)}, Z) = 0,$$

where $X^{(i_1,\ldots,i_m)} = (X^{(i_1)}, X^{(i_2)}, \ldots, X^{(i_m)})$.

In other words, a Boolean function $f(x^{(1)}, x^{(2)}, \ldots, x^{(n)}) \in \mathcal{F}_n$ is correlation-immune of order m if for any $1 \le i_1 < i_2 < \cdots < i_m \le n$ the random variables $X^{(i_1,\ldots,i_m)} = (X^{(i_1)}, X^{(i_2)}, \ldots, X^{(i_m)})$ and $Z = f(X^{(1)}, X^{(2)}, \ldots, X^{(n)})$ are independent.

It is easy to see that the only correlation-immune functions of order m for $m = n$ in $\mathcal{F}_n$ are the constant functions: $f = 0$, $f = 1$.

LEMMA 7.8. *A Boolean function $f(x^{(1)}, x^{(2)}, \ldots, x^{(n)}) \in \mathcal{F}_n$ is correlation-immune of order m, $0 < m \le n$, if and only if for any $1 \le i_1 < i_2 < \cdots < i_m \le n$, $a^{(j)} \in \mathbb{F}_2$, $j = 1, \ldots, m$, the following equality holds:*

$$\mathrm{wt}\big(f_{i_1,\ldots,i_m}^{a^{(1)},\ldots,a^{(m)}}\big) = \frac{\mathrm{wt}(f)}{2^m}. \tag{7.3}$$

PROOF. Let the Boolean function $f(x^{(1)}, x^{(2)}, \ldots, x^{(n)}) \in \mathcal{F}_n$ be correlation-immune of order m. Then according to Lemma 7.6, for any $1 \le i_1 < i_2 < \cdots < i_m \le n$ the random variables $X^{(i_1,\ldots,i_m)}$ and $Z = f(X^{(1)}, X^{(2)}, \ldots, X^{(n)})$ are independent, i.e.,

$$\begin{aligned} \mathcal{P}\{X^{(i_1,\ldots,i_m)} &= (a^{(1)}, \ldots, a^{(m)}) \mid Z = b\} \\ &= \frac{\mathcal{P}\{X^{(i_1,\ldots,i_m)} = (a^{(1)}, \ldots, a^{(m)}), Z = b\}}{\mathcal{P}\{Z = b\}} \\ &= \mathcal{P}\{X^{(i_1,\ldots,i_m)} = (a^{(1)}, \ldots, a^{(m)})\} \end{aligned} \tag{7.4}$$

for any $a^{(1)}, a^{(2)}, \ldots, a^{(m)}, b \in \mathbb{F}_2$. Let us express these probabilities for $b = 1$ via the parameters of the function f. It is obvious that

$$\mathcal{P}\{Z = 1\} = \frac{\mathrm{wt}(f)}{2^n},$$

$$\mathcal{P}\{X^{(i_1,\ldots,i_m)} = (a^{(1)}, \ldots, a^{(m)}), Z = 1\} = \frac{\mathrm{wt}\big(f_{i_1,\ldots,i_m}^{a^{(1)},\ldots,a^{(m)}}\big)}{2^n},$$

$$\mathcal{P}\{X^{(i_1,\ldots,i_m)} = (a^{(1)}, \ldots, a^{(m)})\} = 2^{-m}.$$

If we substitute these expressions in (7.4), we get

$$\mathrm{wt}\big(f_{i_1,\ldots,i_m}^{a^{(1)},\ldots,a^{(m)}}\big) = \frac{\mathrm{wt}(f)}{2^m}.$$

Now we will prove the converse. Suppose that for any $1 \le i_1 < i_2 < \cdots < i_m \le n$, $a^{(1)}, a^{(2)}, \ldots, a^{(m)}, b \in \mathbb{F}_2$ the relation (7.3) holds. Arguing in the opposite direction, we obtain that (7.4) holds for $b = 1$. Let us show that (7.4) holds for $b = 0$ as well. It is easy to see that the following chain of equalities holds:

$$
\begin{aligned}
&\mathcal{P}\{X^{(i_1,\dots,i_m)} = (a^{(1)},\dots,a^{(m)}) \mid Z = 0\} \\
&\quad= \frac{\mathcal{P}\{X^{(i_1,\dots,i_m)} = (a^{(1)},\dots,a^{(m)}), Z = 0\}}{\mathcal{P}\{Z = 0\}} \\
&\quad= \frac{\mathcal{P}\{X^{(i_1,\dots,i_m)} = (a^{(1)},\dots,a^{(m)})\}}{1 - \mathcal{P}\{Z = 1\}} - \frac{\mathcal{P}\{X^{(i_1,\dots,i_m)} = (a^{(1)},\dots,a^{(m)}), Z = 1\}}{1 - \mathcal{P}\{Z = 1\}} \\
&\quad= \frac{1}{1 - \frac{\mathrm{wt}(f)}{2^n}} \left(2^{-m} - \frac{\mathrm{wt}(f_{i_1,\dots,i_m}^{a^{(1)},\dots,a^{(m)}})}{2^n} \right) \\
&\quad= \frac{1}{1 - \frac{\mathrm{wt}(f)}{2^n}} \left(2^{-m} - \frac{2^{-m} \cdot \mathrm{wt}(f)}{2^n} \right) \\
&\quad= 2^{-m} = \mathcal{P}\{X^{(i_1,\dots,i_m)} = (a^{(1)},\dots,a^{(m)})\}.
\end{aligned}
$$

Therefore, for any $1 \leq i_1 < i_2 < \cdots < i_m \leq n$ the random variables $X^{(i_1,\dots,i_m)}$ and Z are independent. By virtue of Lemma 7.6 the function f is correlation-immune of order m. □

PROBLEM 7.9. Prove that a correlation-immune function on $\mathcal{F}_n$ of order m is also correlation-immune of any smaller order.

Taking into account Problem 7.9, it is natural to introduce the following quantity:

$$\operatorname{cor} f = \max\{m \in \mathbb{N} \mid f \text{ is correlation-immune of order } m\}.$$

It is possible to classify correlation-immune functions in terms of the Walsh–Hadamard coefficients as follows.

THEOREM 7.10. *A Boolean function $f \in \mathcal{F}_n$ is correlation-immune of order m if and only if $W_f(\boldsymbol{u}) = 0$ for all vectors $\boldsymbol{u} \in V_n$ such that $1 \leq \mathrm{wt}(\boldsymbol{u}) \leq m$.*

PROOF. Suppose that $f(x^{(1)}, x^{(2)}, \dots, x^{(n)})$ is a correlation-immune function of order m, $\boldsymbol{u} \in V_n$, $\mathrm{wt}(\boldsymbol{u}) = m$. Denote by $i_1, \dots, i_m$ ($i_1 < i_2 < \cdots < i_m$) the indices of coordinates of the vector $\boldsymbol{u}$ that are equal to 1. Then the expression $W_f(\boldsymbol{u})$ for the Walsh–Hadamard coefficients is

$$W_f(\boldsymbol{u}) = \sum_{\boldsymbol{x} \in V_n} (-1)^{f(\boldsymbol{x}) \oplus \langle \boldsymbol{u}, \boldsymbol{x} \rangle} = \sum_{\boldsymbol{x} \in V_n} (-1)^{f(\boldsymbol{x}) \oplus x^{(i_1)} \oplus \cdots \oplus x^{(i_m)}}. \tag{7.5}$$

Splitting the sum (7.5) into 2^m subsums with fixed values of the coordinates $x^{(i_1)}, \dots, x^{(i_m)}$ and using Lemma 7.8, we obtain:

$$\begin{aligned}
W_f(\boldsymbol{u}) &= \sum_{\boldsymbol{x}\in V_n} (-1)^{f(\boldsymbol{x})\oplus x^{(i_1)}\oplus\cdots\oplus x^{(i_m)}} \\
&= \sum_{a^{(1)},\dots,a^{(m)}\in\mathbb{F}_2} (-1)^{\bigoplus_{s=1}^m a^{(s)}} \\
&\quad\times \sum_{\substack{x^{(j)}\in\mathbb{F}_2\\ j\neq i_1,\dots,i_m}} (-1)^{f_{i_1,\dots,i_m}^{a^{(1)},\dots,a^{(m)}}(x^{(1)},\dots,x^{(i_1-1)},x^{(i_1+1)},\dots,x^{(i_m-1)},x^{(i_m+1)},\dots,x^{(n)})} \\
&= \sum_{a^{(1)},\dots,a^{(m)}\in\mathbb{F}_2} (-1)^{\bigoplus_{s=1}^m a^{(s)}} \sum_{\substack{x^{(j)}\in\mathbb{F}_2\\ j\neq i_1,\dots,i_m}} \big(1 - 2f_{i_1,\dots,i_m}^{a^{(1)},\dots,a^{(m)}}(x^{(1)},\dots, \\
&\qquad\qquad x^{(i_1-1)}, x^{(i_1+1)}, \dots, x^{(i_m-1)}, x^{(i_m+1)}, \dots, x^{(n)})\big) \\
&= \left[2^{n-m} - \frac{\mathrm{wt}(f)}{2^{m-1}}\right] \sum_{a^{(1)},\dots,a^{(m)}\in\mathbb{F}_2} (-1)^{\bigoplus_{s=1}^m a^{(s)}} = 0.
\end{aligned}$$

Using now the result of Problem 7.9, we see that $W_f(\boldsymbol{u}) = 0$, $\mathrm{wt}(\boldsymbol{u}) = 1, 2, \dots, m-1$.

Let us prove the converse. Suppose that $W_f(\boldsymbol{u}) = 0$, $1 \le \mathrm{wt}(\boldsymbol{u}) \le m$. According to Problem 2.41, for an arbitrary vector $\boldsymbol{u}$, $\mathrm{wt}(\boldsymbol{u}) = m$, we have

$$\sum_{\substack{\boldsymbol{x}\in V_n\\ \boldsymbol{x}\preccurlyeq\boldsymbol{u}}} W_f(\boldsymbol{x}) = 2^n - 2^{m+1}\,\mathrm{wt}(f_{i_1,\dots,i_m}^{0,\dots,0}), \tag{7.6}$$

where $i_1, \dots, i_m$ are the indices of coordinates in the vector $\boldsymbol{u}$ that are equal to 1. By hypothesis, the sum on the left-hand side of the latter equation is equal to the zeroth Walsh–Hadamard coefficient $W_f(\boldsymbol{0})$, which can be expressed in terms of the weight of the function f as follows: $W_f(\boldsymbol{0}) = 2^n - 2\cdot\mathrm{wt}(f)$. If we substitute this expression in (7.6), we obtain

$$\mathrm{wt}\big(f_{i_1,\dots,i_m}^{0,\dots,0}\big) = \frac{\mathrm{wt}(f)}{2^m}.$$

Suppose that $\boldsymbol{b} = (0, \dots, 0, \underset{i_1}{a^{(1)}}, 0, \dots, 0, \underset{i_m}{a^{(m)}}, 0, \dots, 0) \in V_n$. Define a Boolean function $\varphi(\boldsymbol{x}) = f(\boldsymbol{x}\oplus\boldsymbol{b})$. It is obvious that

$$W_\varphi(\boldsymbol{x}) = (-1)^{\langle\boldsymbol{b},\boldsymbol{x}\rangle} W_f(\boldsymbol{x}), \qquad \mathrm{wt}(\varphi_{i_1,\dots,i_m}^{0,\dots,0}) = \mathrm{wt}(f_{i_1,\dots,i_m}^{a^{(1)},\dots,a^{(m)}}.$$

Using now equality (7.6) for the function φ and applying similar argument, we arrive at the equality

$$\mathrm{wt}\big(f_{i_1,\dots,i_m}^{a^{(1)},\dots,a^{(m)}}\big) = \frac{\mathrm{wt}(f)}{2^m} \tag{7.7}$$

for any vector $(a^{(1)}, \dots, a^{(m)})$. That means that the function f is correlation-immune of order m. □

COROLLARY 7.11. *We have*

$$\operatorname{cor} f = \max\big\{m\in\mathbb{N} \mid \{\boldsymbol{u}\in V_n, \boldsymbol{u}\neq\boldsymbol{0} \mid W_f(\boldsymbol{u}) = 0\} \supseteq \{\boldsymbol{u}\in V_n \mid 1\le \mathrm{wt}(\boldsymbol{u})\le m\}\big\}.$$

The definitions of correlation-immunity and the proposition we proved demonstrate the cryptographic motivation and the origins of this notion. From the pure mathematical point of view, it is possible to develop the notion of correlation-immunity starting with a "local" definition. We will briefly describe this approach.

Suppose that $f, g \in \mathcal{F}_n$, $X^{(1)}, \dots, X^{(n)}$ are, as above, random equiprobable independent entries of these functions and $p_\varepsilon = \mathcal{P}\{f(X) = \varepsilon\}$, $p'_\varepsilon = \mathcal{P}\{g(X) = \varepsilon\}$, $p_{\varepsilon,\delta} = \mathcal{P}\{f(X) = \varepsilon, g(X) = \delta\}$.

PROBLEM 7.12. Prove that for any $\varepsilon, \delta \in \mathbb{F}_2$ the following equality holds:

$$p_\varepsilon p'_\delta - p_{\varepsilon,\delta} = \Big(p_\epsilon - \frac{1}{2}\Big)\Big(p'_\delta - \frac{1}{2}\Big) + \frac{1}{2}\Big[\frac{1}{2} - (p_{\varepsilon,\delta\oplus 1} + p_{\varepsilon\oplus 1,\delta})\Big].$$

COROLLARY 7.13. *The two properties of Boolean functions, "independence of the random variables $f(X)$, $g(X)$" and "zero mutual correlation $\Delta_{f,g}$", are equivalent if and only if one of the random variables $f(X)$, $g(X)$ is equiprobable.*

DEFINITION 7.14. A Boolean function $f \in \mathcal{F}_n$ is called *correlation-immune in the direction $\boldsymbol{u} \in V_n$* if the random variables $f(X)$ and $\langle \boldsymbol{u}, X\rangle$ are independent.

LEMMA 7.15. *A Boolean function $f \in \mathcal{F}_n$ is correlation-immune in the direction $\boldsymbol{u} \in V_n$ if and only if $W_f(\boldsymbol{u}) = 0$.*

DEFINITION 7.16. A balanced Boolean function $f \in \mathcal{F}_n$ satisfies *propagation criterion in the direction $\boldsymbol{u} \in V_n$* if the random variables $f(X)$ and $f(\boldsymbol{u} \oplus X)$ are independent (cf. Definitions 2.110 and 8.68).

LEMMA 7.17. *A balanced Boolean function $f \in \mathcal{F}_n$ satisfies propagation criterion in the direction $\boldsymbol{u} \in V_n$ if and only if $\Delta_f(\boldsymbol{u}) = 0$.*

The above-given statements add "local" cryptographic meaning to the zeros of the Walsh–Hadamard coefficients of the function $f \in \mathcal{F}_n$ and to the zeros of the autocorrelation function. Actually, these "local" properties allow us to study "overall" the correlation-immunity of a Boolean function and propagation criteria. This is why the study of the following sets connected with a given Boolean function $f \in \mathcal{F}_n$ is of great interest for cryptography:

$$\mathrm{PC}_f = \{\boldsymbol{u} \in V_n \mid \Delta_f(\boldsymbol{u}) = 0\}, \quad \mathrm{CI}_f = \{\boldsymbol{u} \in V_n \mid W_f(\boldsymbol{u}) = 0\}.$$

PROBLEM 7.18 (Open problem). Suppose that $M \subseteq V_n$ is a subset. Find efficient necessary and sufficient conditions for the existence of a Boolean function $f \in \mathcal{F}_n$ such that

(1) $\mathrm{PC}_f = M$;
(2) $\mathrm{CI}_f = M$.

PROBLEM 7.19. Prove that a Boolean function $f \in \mathcal{F}_n$ is correlation-immune of order m if and only if $\overline{W}_f(\boldsymbol{u}) = 0$ for all vectors $\boldsymbol{u} \in V_n$ such that $1 \le \mathrm{wt}(\boldsymbol{u}) \le m$.

PROBLEM 7.20. Prove that

$$\operatorname{cor} f^{\mathfrak{g}} = \operatorname{cor} f$$

for any $f \in \mathcal{F}_n$ and any $\mathfrak{g} \in \mathfrak{D}_n$.

PROBLEM 7.21. Suppose that $f \in \mathcal{F}_n$, $l \in \mathcal{F}_k$ are functions of disjoint variables and l is an affine function. Prove that

$$\operatorname{cor}(f \oplus l) = \begin{cases} \operatorname{cor} f & \text{if } f \text{ is nonbalanced;} \\ k + \operatorname{cor} f & \text{if } f \text{ is balanced.} \end{cases}$$

THEOREM 7.22. 1. *If* $f \in \mathcal{F}_n$, $\operatorname{cor} f = m \le n-1$, *then for any vector* $\boldsymbol{u} \in V_n$ *one has* $W_f(\boldsymbol{u}) \equiv 0 \pmod{2^{m+1}}$.

2. *If* $f \in \mathcal{F}_n$ *is balanced,* $\operatorname{cor} f = m \le n-2$, *then for any vector* $\boldsymbol{u} \in V_n$ *one has* $W_f(\boldsymbol{u}) \equiv 0 \pmod{2^{m+2}}$.

PROOF. Let $\operatorname{cor} f = m \le n-1$. Then

$$\operatorname{wt}(f) = 2^m \operatorname{wt}(f^{a^{(1)},\dots,a^{(m)}}_{i_1,\dots,i_m}),$$

i.e., $\operatorname{wt}(f)$ is divisible by 2^m. We will prove the theorem by induction on $\operatorname{wt}(\boldsymbol{u})$. If $\boldsymbol{u} = \mathbf{0}$, then $W_f(\mathbf{0}) = 2^n - 2 \cdot \operatorname{wt}(f)$. The hypothesis of the theorem implies that $n \ge m+1$ and 2^n is divisible by 2^{m+1}. Hence, $W_f(\mathbf{0}) \equiv 0 \pmod{2^{m+1}}$.

For $1 \le \operatorname{wt}(\boldsymbol{u}) \le m$ all Walsh–Hadamard coefficients $W_f(\boldsymbol{u})$ are zero according to Theorem 7.10. Suppose that the statement holds for all vectors of weight less than $\operatorname{wt}(\boldsymbol{u})$, $\operatorname{wt}(\boldsymbol{u}) = t \ge m+1$. Corollary 2.41 implies that

$$\sum_{\substack{\boldsymbol{v} \in V_n \\ \boldsymbol{v} \preccurlyeq \boldsymbol{u}}} W_f(\boldsymbol{v}) = 2^n - 2^{\operatorname{wt}(\boldsymbol{u})+1} \operatorname{wt}(f^{0,\dots,0}_{i_1,\dots,i_t}),$$

where $i_1, i_2, \dots, i_t$ are the numbers of zeros in the vector $\boldsymbol{u}$. It is obvious that the right-hand side of the equation is divisible by 2^{m+1}. According to the induction hypothesis, all summands on the left-hand side of the equation are divisible by 2^{m+1}, possibly excluding $W_f(\boldsymbol{u})$. But then $W_f(\boldsymbol{u}) \equiv 0 \pmod{2^{m+1}}$. Thus, the first statement of the theorem is proved.

Suppose now that f is balanced, $\operatorname{cor} f = m \le n-2$. The proof will also be done by induction on $\operatorname{wt}(\boldsymbol{u})$. If $\boldsymbol{u} = \mathbf{0}$, then $W_f(\mathbf{0}) = 0$ because the function is balanced. For $1 \le \operatorname{wt}(\boldsymbol{u}) \le m$ all Walsh–Hadamard coefficients $W_f(\boldsymbol{u})$ are equal to zero by virtue of Theorem 7.10. Suppose that the statement holds for all vectors of weights less than $\operatorname{wt}(\boldsymbol{u})$, $\operatorname{wt}(\boldsymbol{u}) = t \ge m+1$. Corollary 2.41 shows that

$$\sum_{\substack{\boldsymbol{v} \in V_n \\ \boldsymbol{v} \preccurlyeq \boldsymbol{u}}} W_f(\boldsymbol{v}) = 2^n - 2^{\operatorname{wt}(\boldsymbol{u})+1} \operatorname{wt}(f^{0,\dots,0}_{i_1,\dots,i_t}),$$

where $i_1, i_2, \dots, i_t$ are the coordinates of zeros in the vector $\boldsymbol{u}$. Obviously, the right-hand side of the equality is divisible by 2^{m+2}. According to the induction hypothesis, all summands in the left-hand side of the equality are divisible by 2^{m+2}, possibly excluding $W_f(\boldsymbol{u})$. But then $W_f(\boldsymbol{u}) \equiv 0 \pmod{2^{m+2}}$. □

COROLLARY 7.23. 1. *If* $f \in \mathcal{F}_n$, $\operatorname{cor} f \le n-1$ *then* $NW_f \le 2^{2(n-1-\operatorname{cor} f)}$.

2. *If* $f \in \mathcal{F}_n$ *is balanced,* $\operatorname{cor} f \le n-2$, *then* $NW_f \le 2^{2(n-2-\operatorname{cor} f)}$.

PROBLEM 7.24. Prove that:

(1) If $f \in \mathcal{F}_n$, $\operatorname{cor} f = n-1$, then f is an affine function.
(2) If $f \in \mathcal{F}_n$ is balanced, $\operatorname{cor} f = n-2$, then f is an affine function.

Using Theorem 7.22 and Problem 2.47, we obtain the following important statement:

THEOREM 7.25 (Siegenthaler inequality). *For any* $f \in \mathcal{F}_n$ *the following inequality holds:*

$$\deg f + \operatorname{cor} f \le n.$$

If $f \in \mathcal{F}_n$ *is balanced,* $\operatorname{cor} f \le n-2$, *then*

$$\deg f + \operatorname{cor} f \le n-1.$$

The concept of resilient mappings was introduced in [**21, 87**]. These mappings were connected with research areas such as distributed error-resistant computations, and development of common keys for quantum-cryptographic communication channels. An essential property of resilient mappings is their ability to stay balanced after some restrictions of their domain.

DEFINITION 7.26. Suppose that $\Phi = (f_1, \dots, f_k) \in \mathcal{F}_{n,k}$, $n \geq k \geq 1$.

1. A mapping Φ is called *resilient* relative to a subset $M = \{j_1, \dots, j_d\}$ of the set $\{1, 2, \dots, n\}$ if for any set $(a^{(1)}, a^{(2)}, \dots, a^{(d)}) \in V_d$ the mapping $\Phi_{j_1,\dots,j_d}^{a^{(1)},\dots,a^{(d)}}$ is balanced.

2. A mapping Φ is called (n, k, d)*-resilient* if it is resilient relative to any subset M of the set $\{1, 2, \dots, n\}$ of cardinality d.

It is obvious that if a mapping $\Phi \in \mathcal{F}_{n,k}$ is (n, k, d)-resilient, then it is (n, k, d_1)-resilient whenever $d_1 < d$. Therefore, it is natural to introduce the following notation:

$$\operatorname{sut} \Phi = \max\{d \in \mathbb{N} \mid \Phi \text{ is } (n, k, d)\text{-resilient}\}.$$

It is clear that for any mapping $\Phi \in \mathcal{F}_{n,k}$ one has

$$n - k \geq \operatorname{sut} \Phi.$$

PROBLEM 7.27. Suppose that $f \in \mathcal{F}_n$. Prove that

$$\operatorname{sut} f = \begin{cases} 0 & \text{if } f \text{ is not balanced;} \\ \operatorname{cor} f & \text{if } f \text{ is balanced.} \end{cases}$$

Observe that, according to Theorem 7.10, one has $\operatorname{sut} f = d$ if and and only if $W_f(\boldsymbol{u}) = 0$ for all vectors $\boldsymbol{u} \in V_n$ such that $\operatorname{wt}(\boldsymbol{u}) \leq d$.

PROBLEM 7.28. Suppose that a function $f \in \mathcal{F}_n$ depends linearly on certain s variables. Prove that

$$\operatorname{sut} f \geq s - 1.$$

DEFINITION 7.29. Any balanced correlation-immune function $f \in \mathcal{F}_n$ of order d is called *d-resilient*.

DEFINITION 7.30. Suppose that $d \leq n - 2$. Then a d-resilient function $f \in \mathcal{F}_n$ is called *d-optimal* if $\deg f = n - d - 1$.

If f is a d-optimal function, then $d = \operatorname{cor} f$ by Siegenthaler's inequality.

REMARK 7.31. From the formal point of view, which is accepted in many works, we can regard any balanced Boolean function as 0-resilient and an arbitrary Boolean function as (-1)-resilient (for a function of n variables there are no subfunctions of $n + 1$ variables, so any condition is true for such a subfunction).

It is obvious that the properties of mappings we introduced in Definition 7.26 are secondary (see §2.5), and Theorem 2.143 can be applied to them. Thus, the following statements are true:

LEMMA 7.32. *A mapping $\Phi = (f_1, f_2, \dots, f_k) \in \mathcal{F}_{n,k}$, $n \geq k \geq 1$, is resilient relative to the subset $M = \{j_1, \dots, j_d\}$ of the set $\{1, 2, \dots, n\}$ if and only if for any nonzero vector $\boldsymbol{b} = (b^{(1)}, b^{(2)}, \dots, b^{(k)})^\top \in V_k$ the Boolean function $f_{\boldsymbol{b}} = \bigoplus_{i=1}^k b^{(i)} f_i$ is resilient relative to the subset M.*

LEMMA 7.33. *A mapping* $\Phi = (f_1, f_2, \ldots, f_k) \in \mathcal{F}_{n,k}$, $n \geq k \geq 1$, *is* (n, k, d)*-resilient if and only if for any nonzero vector* $\boldsymbol{b} = (b^{(1)}, b^{(2)}, \ldots, b^{(k)})^\top \in V_k$ *the Boolean function* $f = \bigoplus_{i=1}^k b^{(i)} f_i$ *is* $(n, 1, d)$*-resilient.*

If a mapping $\Phi = (f_1, f_2, \ldots, f_k) \in \mathcal{F}_{n,k}$, $n \geq k \geq 1$ is (n, k, d)-resilient, then from Lemma 7.33 it follows that $\Phi' = (f_1, f_2, \ldots, f_s) \in \mathcal{F}_{n,s}$ is (n, s, d)-resilient for any s such that $1 \leq s \leq k$. Lemma 7.33 shows that any (n, k, d)-resilient mapping yields $2^k - 1$ balanced correlation-immune Boolean functions of order d.

PROBLEM 7.34. Suppose that $\Phi_i \in \mathcal{F}_{n_i,k}$, $i = 1, \ldots, s$, are mappings of nonintersecting variables. Prove that

$$\operatorname{sut}\left(\bigoplus_{i=1}^{s} \Phi_i\right) = \sum_{i=1}^{s} \operatorname{sut} \Phi_i + s - 1.$$

Now we give examples of resilient linear mappings.

EXAMPLE 7.35. 1. $k = 1$, $d = n - 1$, $f(x^{(1)}, x^{(2)}, \ldots, x^{(n)}) = x^{(1)} \oplus x^{(2)} \oplus \cdots \oplus x^{(n)}$.

2. $k = n - 1$, $d = 1$, $\Phi(\boldsymbol{x}) = (f_1(\boldsymbol{x}), f_2(\boldsymbol{x}), \ldots, f_{n-1}(\boldsymbol{x}))$, $f_i(\boldsymbol{x}) = x^{(i)} \oplus x^{(i+1)}$, $i = 1, 2, \ldots, n - 1$.

3. $n = 3h$, $k = 2$, $d = 2h - 1$, $\Phi(\boldsymbol{x}) = (f_1(\boldsymbol{x}), f_2(\boldsymbol{x}))$, $f_1(\boldsymbol{x}) = x^{(1)} \oplus \cdots \oplus x^{(2h)}$, $f_2(\boldsymbol{x}) = x^{(h+1)} \oplus \cdots \oplus x^{(3h)}$.

It is easy to derive the following statement from Lemma 7.33 and Problem 7.20.

LEMMA 7.36. *For any* $\Phi \in \mathcal{F}_{n,k}$ *and for any* $\mathfrak{g} \in \mathfrak{D}_n$,

$$\operatorname{sut} \Phi^{\mathfrak{g}} = \operatorname{sut} \Phi.$$

The problem of the existence of resilient mappings with various values of parameters n, k, and d is very important and interesting. As will be shown later, this problem is equivalent to the problem of the existence of other well-studied mathematical objects. Let us start with a question of the existence of linear resilient mappings.

In order to use methods of coding theory for the description of resilient linear mappings, we will need the following simple fact.

LEMMA 7.37. *A matrix* G *of size* $k \times n$ *and* $\operatorname{rank} G = k$ *is a generator matrix of a linear* $[n, k, d]$*-code if and only if* d *is the smallest number such that after removing any* d *columns from the matrix* G *one gets a matrix* G' *of size* $k \times (n - d)$ *and* $\operatorname{rank} G' < k$.

PROOF. Denote by d the minimum distance of the code generated by the matrix G. There exists a nonzero vector $\boldsymbol{x} \in V_k$ such that $\operatorname{wt}(\boldsymbol{x}^\top G) = d$. If we remove d nonzero coordinates from $\boldsymbol{x}^\top G$, we get $\boldsymbol{x}^\top G' = (\mathbf{0})$, where the matrix G' of size $k \times (n - d)$ is obtained from G by removing the columns numbered the same way as nonzeros in $\boldsymbol{x}^\top G$. Therefore, $\operatorname{rank} G' < k$. Suppose now that having removed $t < d$ columns of the matrix G we obtained a matrix G' with $\operatorname{rank} G' < k$. Then there exists a nonzero $\boldsymbol{x} \in V_k$ such that $\boldsymbol{x}^\top G' = \mathbf{0}$. From this it follows that for a nonzero code word $\boldsymbol{x}^\top G$ one has $\operatorname{wt}(\boldsymbol{x}^\top G) \leq t < d$, which is impossible.

Conversely, suppose that G is a $(k \times n)$ matrix, $\operatorname{rank} G = k$, and for a nonnegative d the hypothesis of the lemma holds. Suppose now that for some nonzero $\boldsymbol{x} \in V_k$ one has $\operatorname{wt}(\boldsymbol{x}^\top G) = t < d$. Then if we remove $t < d$ appropriate columns of the matrix G, we will get a $(k \times (n-d))$ matrix G' for which $\boldsymbol{x}^\top G' = (\mathbf{0})$, i.e., $\operatorname{rank} G' < k$. This contradicts the statement of the theorem, so $\operatorname{wt}(\boldsymbol{x}^\top G) \geq d$. □

DEFINITION 7.38. An (n, k, d)-resilient mapping $\Phi \in \mathcal{F}_{n,k}$ is called *linear* if it can be represented as

$$\Phi(\boldsymbol{x}) = \boldsymbol{x}^\top M$$

for every $\boldsymbol{x} \in V_n$, where M is an $(n \times k)$ matrix.

The following theorem shows that a linear resilient mapping is equivalent to a linear code.

THEOREM 7.39. *The existence of a linear $[n, k, d]$-code is equivalent to the existence of a linear $(n, k, d-1)$-resilient mapping.*

PROOF. Let G be a generator matrix of a linear $[n, k, d]$-code. We define the mapping $\Phi \in \mathcal{F}_{n,k}$ as follows:

$$\Phi(\boldsymbol{x}) = \boldsymbol{x}^\top G^\top$$

for any $\boldsymbol{x} \in V_n$. We will show that Φ is an $(n, k, d-1)$-resilient mapping. Suppose that $1 \leq j_1 < \cdots < j_t \leq n$, $t < d$, and $\boldsymbol{a} = (a^{(1)}, \ldots, a^{(t)})^\top \in V_t$ are chosen arbitrarily; denote by G' the $(k \times (n-t))$ matrix obtained from G by removing the columns with indexes $j_1, \ldots, j_t$. Then for any $\boldsymbol{x}' \in V_{n-t}$ we have

$$\Phi_{j_1,\ldots,j_t}^{a^{(1)},\ldots,a^{(t)}}(\boldsymbol{x}') = (\boldsymbol{x}')^\top (G')^\top \oplus \boldsymbol{b},$$

where $\boldsymbol{b}$ is a fixed vector. According to Lemma 7.37, $\operatorname{rank} G' = k$. Therefore, the mapping $\Psi(\boldsymbol{x}') = (\boldsymbol{x}')^\top (G')^\top \in \mathcal{F}_{n-t,k}$ is linear and surjective. Thus, for any $\boldsymbol{c} \in V_k$, the cardinality $\#\big(\Phi_{j_1,\ldots,j_t}^{a^{(1)},\ldots,a^{(t)}}\big)^{-1}(\boldsymbol{c})$ equals the cardinality of the kernel of the mapping Ψ, whence

$$\#\big(\Phi_{j_1,\ldots,j_t}^{a^{(1)},\ldots,a^{(t)}}\big)^{-1}(\boldsymbol{c}) = 2^{n-t-k}.$$

This shows that the mapping Φ is $(n, k, d-1)$-resilient.

Conversely, let $\Phi(\boldsymbol{x}) = \boldsymbol{x}^\top G^\top$ be a linear $(n, k, d-1)$-resilient mapping, where G is a $(k \times n)$ matrix. Suppose that $\mathcal{C}$ is a linear code of length n with generator matrix G. Since $\Phi(\boldsymbol{x}) = \boldsymbol{x}^\top G^\top$ is a linear resilient mapping, one has $\operatorname{rank} G = k$. It remains to show that the code $\boldsymbol{C}$ has minimum distance d. Assume the converse, i.e., that the minimum distance of the code $\boldsymbol{C}$ is equal to $t < d$. Then, according to Lemma 7.37, we can remove at most $d-1$ columns from the matrix G for the resulting matrix G' to have rank less than k. If we fix the corresponding $d-1$ zero coordinates of $\boldsymbol{x}$, then the function $\Phi_{j_1,\ldots,j_{d-1}}^{0,\ldots,0} \in \mathcal{F}_{n-d+1,k}$ is not surjective, which is impossible. Therefore, the minimum distance of the code $\boldsymbol{C}$ is at least d. As Φ is not (n, k, d)-resilient, there exist $1 \leq j_1 < \cdots < j_d \leq n$ and $\boldsymbol{a} = (a^{(1)}, \ldots, a^{(d)})^\top \in V_d$ such that $\Phi_{j_1,\ldots,j_d}^{a^{(1)},\ldots,a^{(d)}}$ is not a resilient mapping from V_{n-d} to V_k. It follows that $\operatorname{rank} G' < k$ and there exists a nonzero vector $\boldsymbol{y} \in V_k$ such that $\boldsymbol{y}^\top G' = \mathbf{0}$, whence $\operatorname{wt}(G) = d$. □

Now we proceed with correlation-immune functions and resilient mappings of a general type. We will need the following definition for a well-known combinatorial object which is of a great importance for the algebraic approach to the scheme of relations in coding theory.

DEFINITION 7.40. An *orthogonal array* of size $m \times n$ with constraints of level 2 (in general, if the alphabet has cardinality q, the level of constraints will be q), strength t, and index ν is an $(m \times n)$ matrix M over the field $\mathbb{F}_2$ having the following property: in any subset of t columns of the matrix M, any 2^t vectors of the space V_t occur as a row exactly ν times.

Such an array is denoted by $OA_\nu(m, n, 2, t)$. It is clear from a counting argument that $m = \nu \cdot 2^t$, i.e., the parameter m is uniquely determined by ν and t. Obviously, $OA_\nu(m, n, 2, t)$ is $OA_\nu(m, n, 2, i)$ for any $i \in \{1, 2, \dots, t\}$.

For a function $f \in \mathcal{F}_n$, we will denote by M_f a $(\mathrm{wt}(f) \times n)$ matrix the rows of which are exactly the vectors in V_n at which the value of the function f is equal to 1. Even though in this case the ordering of the rows of the matrix does not matter, we will assume that they are ordered lexicographically (top-down). The matrix M_f is also called the truth table of the function f.

THEOREM 7.41. *The inequality* $\operatorname{cor} f \geq t$ *is satisfied for a function* $f \in \mathcal{F}_n$ *if and only if its truth table* M_f *is an orthogonal array* $OA_\nu(\mathrm{wt}(f), n, 2, t)$.

PROOF. Suppose that $f \in \mathcal{F}_n$ and M_f is the corresponding truth table. Assume that M_f is an orthogonal array $OA_\nu(\mathrm{wt}(f), n, 2, t)$, $\mathrm{wt}(f) = 2^t\nu$. From the definition of orthogonal array it follows that for any $\boldsymbol{u} \in V_n$ of weight $\mathrm{wt}(u) = k$, $1 \leq k \leq t$, the following chain of equations holds:

$$\begin{aligned}\overline{W}_f(\boldsymbol{u}) &= \sum_{\boldsymbol{x} \in V_n} f(\boldsymbol{x})(-1)^{\langle \boldsymbol{x}, \boldsymbol{u} \rangle} = \nu \cdot \sum_{\boldsymbol{y} \in V_k} (-1)^{\mathrm{wt}(\boldsymbol{y})} \\ &= \nu \cdot (\#\{\boldsymbol{y} \in V_k \mid \mathrm{wt}(\boldsymbol{y}) \text{ even}\} - \#\{\boldsymbol{y} \in V_k \mid \mathrm{wt}(\boldsymbol{y}) \text{ odd}\}) = 0.\end{aligned}$$

Therefore, $\overline{W}_f(\boldsymbol{u}) = 0$ for all $\boldsymbol{u}$ such that $1 \leq \mathrm{wt}(\boldsymbol{u}) \leq t$. In view of Theorem 2.36, one has $W_f(\boldsymbol{u}) = 0$ for all $\boldsymbol{u}$ such that $1 \leq \mathrm{wt}(\boldsymbol{u}) \leq t$, which is equivalent to the inequality $\operatorname{cor} f \geq t$.

We prove the converse. Suppose that $\operatorname{cor} f \geq t$ for some $f \in \mathcal{F}_n$. According to Theorem 7.10, for all $\boldsymbol{u}$ such that $1 \leq \mathrm{wt}(\boldsymbol{u}) \leq t$, the equation $W_f(\boldsymbol{u}) = 0$ holds. We define the support of an element $\boldsymbol{u} \in V_n$ as the set $\mathrm{supp}(\boldsymbol{u}) = \{i \in \{1, 2, \dots, n\} \mid u^{(i)} = 1\}$; the projection $\pi_{\boldsymbol{u}} \colon V_n \to V_t$ of a vector $\boldsymbol{u} \in V_n$, $\mathrm{wt}(\boldsymbol{u}) = t$, is, by definition,

$$\pi_{\boldsymbol{u}}(\boldsymbol{x}) = (x^{(i)}), \quad i \in \mathrm{supp}(\boldsymbol{u}).$$

Moreover, define the mapping $\psi_f \colon V_t \to \mathbb{Z}$ by the equality

$$\psi_f(\boldsymbol{y}) = \#Q_{\boldsymbol{y}}$$

for each $\boldsymbol{y} \in V_t$, where $Q_{\boldsymbol{y}} = \{\boldsymbol{x} \in M_f \mid \pi_{\boldsymbol{u}}(\boldsymbol{x}) = \boldsymbol{y}\}$. For any nonzero $\boldsymbol{u}' \in V_n$, $\mathrm{supp}(\boldsymbol{u}') \subset \mathrm{supp}(\boldsymbol{u})$, one has

$$\begin{aligned}\overline{W}_f(\boldsymbol{u}') &= \sum_{\boldsymbol{x} \in M_f} (-1)^{\langle \boldsymbol{u}', \boldsymbol{x} \rangle} = \sum_{\boldsymbol{y} \in V_t} \sum_{\boldsymbol{x} \in Q_{\boldsymbol{y}}} (-1)^{\langle \boldsymbol{u}', \boldsymbol{x} \rangle} \\ &= \sum_{\boldsymbol{y} \in V_t} \psi_f(\boldsymbol{y})(-1)^{\langle \pi_{\boldsymbol{u}}(\boldsymbol{u}'), \boldsymbol{y} \rangle} = \overline{W}_{\psi_f}(\pi_{\boldsymbol{u}}(\boldsymbol{u}')) = 0.\end{aligned}$$

If the vector $\boldsymbol{u}'$ runs over all vectors satisfying the condition $\mathrm{supp}(\boldsymbol{u}') \subset \mathrm{supp}(\boldsymbol{u})$, then $\pi_{\boldsymbol{u}}(\boldsymbol{u}')$ runs over the whole space V_t. Hence, for all nonzero $\boldsymbol{v} \in V_t$ the equation $\overline{W}_{\psi_f}(\boldsymbol{v}) = 0$ holds and $\overline{W}_f(\boldsymbol{0}) = \sum_{\boldsymbol{y} \in V_t} \psi_f(\boldsymbol{y}) = \mathrm{wt}(f)$. Write down the Fourier inversion formula for the function ψ_f:

$$\psi_f(\boldsymbol{y}) = \frac{1}{2^t} \sum_{\boldsymbol{v} \in V_t} \overline{W}_{\psi_f}(\boldsymbol{v})(-1)^{\langle \boldsymbol{v}, \boldsymbol{y} \rangle} = \frac{\mathrm{wt}(f)}{2^t} = \nu.$$

This equality holds for any vector $\boldsymbol{u} \in V_n$, $\mathrm{wt}(\boldsymbol{u}) = t$. Therefore, in any set of t columns from the array M_f, each vector $\boldsymbol{y} \in V_t$ occurs as a row exactly $\frac{\mathrm{wt}(f)}{2^t} = \nu$ times and M_f is an orthogonal array $OA_\nu(\mathrm{wt}(f), n, 2, t)$. □

It is clear that if a function $f \in \mathcal{F}_n$ is correlation-immune of order t, then the function $f \oplus 1$ is also correlation-immune of order t. Moreover, if f is a balanced function, then M_f and $M_{f\oplus 1}$ represent orthogonal arrays with the same parameters $OA_\nu(2^{n-1}, n, 2, t)$, $\nu = 2^{n-t-1}$. However, a correlation-immune function of order t may be nonbalanced. In this case M_f and $M_{f\oplus 1}$ are orthogonal arrays with different parameters $OA_\nu(\mathrm{wt}(f), n, 2, t)$ and $OA_{\nu_1}(2^n - \mathrm{wt}(f), n, 2, t)$, are connected by the equation

$$\nu + \nu_1 = 2^{n-t}.$$

EXAMPLE 7.42. Define a correlation-immune function of order 1 on $\mathcal{F}_4$ with the algebraic normal form

$$\begin{aligned} f(x^{(1)}, x^{(2)}, x^{(3)}, x^{(4)}) &= x^{(1)}x^{(2)} \oplus x^{(1)}x^{(4)} \oplus x^{(2)}x^{(3)} \oplus x^{(2)}x^{(4)} \\ &\quad \oplus x^{(1)} \oplus x^{(2)} \oplus x^{(3)} \oplus x^{(4)}. \end{aligned}$$

Then

$$M_f = \begin{pmatrix} 0 & 0 & 0 & 1 \\ 0 & 0 & 1 & 0 \\ 0 & 1 & 0 & 0 \\ 0 & 1 & 0 & 1 \\ 0 & 1 & 1 & 0 \\ 0 & 1 & 1 & 1 \\ 1 & 0 & 0 & 0 \\ 1 & 0 & 0 & 1 \\ 1 & 0 & 1 & 0 \\ 1 & 0 & 1 & 1 \\ 1 & 1 & 0 & 1 \\ 1 & 1 & 1 & 0 \end{pmatrix}, \quad M_{f\oplus 1} = \begin{pmatrix} 0 & 0 & 0 & 0 \\ 0 & 0 & 1 & 1 \\ 1 & 1 & 0 & 0 \\ 1 & 1 & 1 & 1 \end{pmatrix}$$

are the orthogonal arrays $OA_6(12, 4, 2, 1)$ and $OA_2(4, 4, 2, 1)$, respectively.

DEFINITION 7.43. A *large set of orthogonal arrays* over the field $\mathbb{F}_2$ is a set of $\frac{2^{n-t}}{\nu}$ orthogonal arrays $OA_\nu(m, n, 2, t)$, $m = \nu 2^t$ over $\mathbb{F}_2$ such that each element of V_n occurs in exactly one table.

It is clear from the definition that the parameter ν is a power of 2. Besides, the union of tables from the large set of orthogonal arrays is the trivial orthogonal array $OA_1(2^n, n, 2, n)$ corresponding to the array of all elements in V_n.

THEOREM 7.44. *The existence of an (n, k, t)-resilient mapping is equivalent to the existence of a large set of 2^k orthogonal arrays $OA_{2^{n-k-t}}(2^{n-k}, n, 2, t)$.*

PROOF. Suppose that Φ is an (n, k, t)-resilient mapping. We will construct a large set of orthogonal arrays as follows. For any $\boldsymbol{y} \in V_k$, put $\Phi^{-1}(\boldsymbol{y}) = \{\boldsymbol{x} \in V_n \mid \Phi(\boldsymbol{x}) = \boldsymbol{y}\}$. We regard $\Phi^{-1}(\boldsymbol{y})$ as not just a set of vectors of V_n but as an array of vectors written in lexicographic order (top-down). It is clear that there are 2^k such arrays and that any element of V_n will occur in exactly one array. We now show that these are orthogonal arrays $OA_{2^{n-k-t}}(2^{n-k}, n, 2, t)$. Since the mapping Φ is balanced, the number of rows in any array is 2^{n-k}. Suppose that $1 \leq j_1 < \cdots < j_t \leq n$ are arbitrary vectors and $\boldsymbol{a} = (a^{(1)}, \ldots, a^{(t)})^\top \in V_t$. Since the mapping $\Phi^{a^{(1)},\ldots,a^{(t)}}_{j_1,\ldots,j_t}$ is balanced, there are exactly 2^{n-k-t} rows $\boldsymbol{x}$ in any array for which $x^{(j_1)} = a^{(1)}, \ldots, x^{(j_t)} = a^{(t)}$. This shows that the constructed arrays form a large set of orthogonal arrays $OA_{2^{n-k-t}}(2^{n-k}, n, 2, t)$.

Conversely, let a large set of 2^k orthogonal arrays $OA_{2^{n-k-t}}(2^{n-k}, n, 2, t)$ be given. We will construct the mapping Φ as follows. Establish an arbitrary 1-1 correspondence between the 2^k orthogonal arrays and the elements of V_k. Now for each vector $v \in V_n$ we choose the array from the large set to which it belongs; $\Phi(v)$ is the element of V_k corresponding to this array. The mapping Φ lies in $\mathcal{F}_{n,k}$. Using the properties of orthogonal arrays, it is easy to show that Φ is (n, k, t)-resilient. □

7.2. The inheritance of properties under restrictions of Boolean functions

The development of the notion of correlation immunity of Boolean functions has led to the following general method for introducing their properties:

– choose a property $\mathcal{PROP}$ of Boolean functions;

– for a given Boolean function $f(\boldsymbol{x}) \in \mathcal{F}_n$ and a given subset $H \subseteq V_n$, consider all restrictions of $f(\boldsymbol{x})$ to cosets of H in V_n (restriction is denoted by $f|_{\boldsymbol{u}\oplus H}$);

– if a Boolean function $f(\boldsymbol{x})$ itself and all its $2^{n-\dim H}$ restrictions have the property $\mathcal{PROP}$, then it is natural to say that the property $\mathcal{PROP}$ is inherited by the given restrictions of the Boolean function $f(\boldsymbol{x})$, and to take it to a new "derivative" property H-$\mathcal{PROP}$.

We will apply this approach to the following property of Boolean functions: the value $\frac{W_f(\boldsymbol{\alpha})}{2^n}$ is equal to a fixed number.

DEFINITION 7.45. We will say that a Boolean function $f(\boldsymbol{x}) \in \mathcal{F}_n$ is $(H, \boldsymbol{\alpha})$-*stable*, $H \subseteq V_n$, $\boldsymbol{\alpha} \in V_n$, if

$$\frac{W_{f|_{\boldsymbol{u}\oplus H}}(\boldsymbol{\alpha})}{\#H} = \frac{W_f(\boldsymbol{\alpha})}{2^n}$$

for every $\boldsymbol{u} \in V_n$. For a given Boolean function $f(\boldsymbol{x}) \in \mathcal{F}_n$, we define the following family of subsets:

$$\mathcal{H}_{f,\boldsymbol{\alpha}} = \{H \subseteq V_n \mid f(\boldsymbol{x}) \text{ is } (H, \boldsymbol{\alpha})\text{-stable}\}.$$

Observe that without loss of generality we can restrict ourselves to the case $\boldsymbol{\alpha} = \mathbf{0}$, as the following lemma is obvious.

LEMMA 7.46. *A function $f(\boldsymbol{x})$ is $(H, \boldsymbol{\alpha})$-stable if and only if the function $f(\boldsymbol{x}) \oplus \langle \boldsymbol{\alpha}, \boldsymbol{x} \rangle$ is $(H, \mathbf{0})$-stable.*

The definition of $(H, \mathbf{0})$-stable functions immediately yields the following.

LEMMA 7.47. *If a Boolean function $f(\boldsymbol{x}) \in \mathcal{F}_n$ is $(H, \mathbf{0})$-stable, then*

$$W_f(\mathbf{0}) \equiv 0 \pmod{2^{n-t+1}},$$

where $t = \dim H$.

PROOF. If a function $f(\boldsymbol{x})$ is $(H,\mathbf{0})$-stable, then for any $\boldsymbol{u} \in V_n$,

$$\sum_{\boldsymbol{x}\in\boldsymbol{u}+H} (-1)^{f(\boldsymbol{x})} = \frac{W_f(\mathbf{0})}{2^{n-t}}.$$

Since

$$\sum_{\boldsymbol{x}\in\boldsymbol{u}+H} (-1)^{f(\boldsymbol{x})} \equiv 0 \pmod 2,$$

one has

$$W_f(\mathbf{0}) \equiv 0 \pmod{2^{n-t+1}}.$$ □

EXAMPLE 7.48. Suppose that a function $f(\boldsymbol{x})$ is $(H,\mathbf{0})$-stable and $\dim H = 1$, i.e., $H = \{\mathbf{0},\boldsymbol{h}\}$, $\boldsymbol{h} \in V_n\backslash\{\mathbf{0}\}$. By Definition 7.45, this is equivalent to the condition that, for any vector $\boldsymbol{u} \in V_n$,

$$(-1)^{f(\boldsymbol{u})} + (-1)^{f(\boldsymbol{u}\oplus\boldsymbol{h})} = \text{const}\,.$$

It is easy to see that this is possible only in the following two cases:

1) the function $f(\boldsymbol{x})$ is constant: $f(\boldsymbol{x}) = \text{const}$;
2) the vector $\boldsymbol{h}$ satisfies the condition

$$f(\boldsymbol{u}\oplus\boldsymbol{h}) = f(\boldsymbol{u})\oplus 1 \text{ for each vector } \boldsymbol{u}\in V_n.$$

Note that if the vector $\boldsymbol{h}$ satisfies the condition $f(\boldsymbol{u}\oplus\boldsymbol{h}) = f(\boldsymbol{u})\oplus\varepsilon$ for each vector $\boldsymbol{u}\in V_n$, where $\varepsilon\in\{0,1\}$, then a weaker condition is necessary:

$$|(-1)^{f(\boldsymbol{u})} + (-1)^{f(\boldsymbol{u}\oplus\boldsymbol{h})}| = \text{const}\,.$$

In order to obtain criteria for $(H,\mathbf{0})$-resiliency, we need two identities which connect certain sums of values of the autocorrelation function, certain sums of squares of Walsh–Hadamard coefficients, and certain sums of values of the Boolean function itself. These identities are interesting in their own right; they form the content of the following lemma.

LEMMA 7.49. *For any Boolean function $f(\boldsymbol{x}) \in \mathcal{F}_n$, any subspace $H \subseteq V_n$, $\dim H = t$, and any $\boldsymbol{z}\in V_n$, the following equalities hold:*

$$\sum_{\boldsymbol{h}\in H} \Delta_f(\boldsymbol{h})(-1)^{\langle\boldsymbol{h},\boldsymbol{z}\rangle} = \sum_{\boldsymbol{u}\in V_n/H}\left(\sum_{\boldsymbol{x}\in\boldsymbol{u}\oplus H} (-1)^{f(\boldsymbol{x})\oplus\langle\boldsymbol{z},\boldsymbol{x}\rangle}\right)^2 = \frac{1}{2^{n-t}} \sum_{\boldsymbol{\alpha}\in\boldsymbol{z}\oplus H^\perp} W_f^2(\boldsymbol{\alpha}).$$

PROOF. It is known that

$$\Delta_f(\boldsymbol{h}) = \sum_{\boldsymbol{x}\in V_n} (-1)^{f(\boldsymbol{x})\oplus f(\boldsymbol{x}\oplus\boldsymbol{h})} = \frac{1}{2^n}\sum_{\boldsymbol{\alpha}\in V_n} W_f^2(\boldsymbol{\alpha})(-1)^{\langle\boldsymbol{\alpha},\boldsymbol{h}\rangle}. \tag{7.8}$$

Multiply both sides of the first equality by $(-1)^{\langle\boldsymbol{h},\boldsymbol{z}\rangle}$ and sum over $\boldsymbol{h}\in H$:

$$\begin{aligned}\sum_{\boldsymbol{h}\in H} \Delta_f(\boldsymbol{h})(-1)^{\langle\boldsymbol{h},\boldsymbol{z}\rangle} &= \sum_{\boldsymbol{x}\in V_n} (-1)^{f(\boldsymbol{x})} \sum_{\boldsymbol{h}\in H} (-1)^{f(\boldsymbol{x}\oplus\boldsymbol{h})}(-1)^{\langle\boldsymbol{h},\boldsymbol{z}\rangle} \\ &= \sum_{\boldsymbol{u}\in V_n/H}\ \sum_{\boldsymbol{x}\in\boldsymbol{u}\oplus H} (-1)^{f(\boldsymbol{x})}(-1)^{\langle\boldsymbol{x},\boldsymbol{z}\rangle} \sum_{\boldsymbol{y}\in\boldsymbol{x}\oplus H} (-1)^{f(\boldsymbol{y})}(-1)^{\langle\boldsymbol{y},\boldsymbol{z}\rangle}.\end{aligned}$$

Note that if $\boldsymbol{x}\in\boldsymbol{u}\oplus H$, then

$$\sum_{\boldsymbol{y}\in\boldsymbol{x}\oplus H} (-1)^{f(\boldsymbol{y})}(-1)^{\langle\boldsymbol{y},\boldsymbol{z}\rangle} = \sum_{\boldsymbol{y}\in\boldsymbol{u}\oplus H} (-1)^{f(\boldsymbol{y})}(-1)^{\langle\boldsymbol{y},\boldsymbol{z}\rangle}.$$

Hence, the equality becomes

$$\sum_{\boldsymbol{h}\in H} \Delta_f(\boldsymbol{h})(-1)^{\langle \boldsymbol{h},\boldsymbol{z}\rangle} = \sum_{\boldsymbol{u}\in V_n/H} \Big(\sum_{\boldsymbol{x}\in \boldsymbol{u}\oplus H} (-1)^{f(\boldsymbol{x})}(-1)^{\langle \boldsymbol{x},\boldsymbol{z}\rangle} \Big)^2.$$

Let us now multiply both sides of the second equality (7.8) by $(-1)^{\langle \boldsymbol{h},\boldsymbol{z}\rangle}$ and sum over $\boldsymbol{h} \in H$:

$$\begin{aligned}\sum_{\boldsymbol{h}\in H} \Delta_f(\boldsymbol{h})(-1)^{\langle \boldsymbol{h},\boldsymbol{z}\rangle} &= \frac{1}{2^n} \sum_{\boldsymbol{\alpha}\in V_n} W_f^2(\boldsymbol{\alpha}) \sum_{\boldsymbol{h}\in H} (-1)^{\langle \boldsymbol{\alpha},\boldsymbol{h}\rangle}(-1)^{\langle \boldsymbol{h},\boldsymbol{z}\rangle} \\ &= \frac{1}{2^{n-t}} \sum_{\boldsymbol{\alpha}\in \boldsymbol{z}\oplus H^\perp} W_f^2(\boldsymbol{\alpha})\end{aligned}$$

since it is known that

$$\sum_{\boldsymbol{h}\in H} (-1)^{\langle \boldsymbol{\alpha},\boldsymbol{h}\rangle} = \begin{cases} 2^t & \text{if } \boldsymbol{\alpha} \in H^\perp, \\ 0 & \text{if } \boldsymbol{\alpha} \notin H^\perp. \end{cases}$$

□

THEOREM 7.50. *A Boolean function $f(\boldsymbol{x}) \in \mathcal{F}_n$ is $(H, \mathbf{0})$-stable, $\dim H = t$, if and only if one of the following two equivalent conditions is satisfied:*

1) $\sum_{\boldsymbol{h}\in H} \Delta_f(\boldsymbol{h}) = \frac{W_f^2(\mathbf{0})}{2^{n-t}}$;
2) $W_f(\boldsymbol{\alpha}) = 0$ *for any* $\boldsymbol{\alpha} \in H^\perp \backslash \{\mathbf{0}\}$.

PROOF. If a function $f(\boldsymbol{x}) \in \mathcal{F}_n$ is $(H, \mathbf{0})$-stable, then for any $\boldsymbol{u} \in V_n$ one has

$$\sum_{\boldsymbol{x}\in \boldsymbol{u}\oplus H} (-1)^{f(\boldsymbol{x})} = \frac{W_f(\mathbf{0})}{2^{n-t}},$$

whence

$$\sum_{\boldsymbol{u}\in V_n/H} \Big(\sum_{\boldsymbol{x}\in \boldsymbol{u}\oplus H} (-1)^{f(\boldsymbol{x})} \Big)^2 = \frac{W_f^2(\mathbf{0})}{2^{n-t}}.$$

Now using Lemma 7.49 (for $\boldsymbol{z} = \mathbf{0}$), it is easy to obtain both hypotheses of Theorem 7.50.

If one of the hypotheses of Theorem 7.50 holds, then by Lemma 7.49, the other holds too, as well as the following equality:

$$\sum_{\boldsymbol{u}\in V_n/H} \Big(\sum_{\boldsymbol{x}\in \boldsymbol{u}\oplus H} (-1)^{f(\boldsymbol{x})} \Big)^2 = \frac{W_f^2(\mathbf{0})}{2^{n-t}} = \frac{1}{2^{n-t}} \Big(\sum_{\boldsymbol{u}\in V_n/H} \Big(\sum_{\boldsymbol{x}\in \boldsymbol{u}\oplus H} (-1)^{f(\boldsymbol{x})} \Big) \Big)^2.$$

From this it is easy to infer (see Remark 7.51 below) that for any $\boldsymbol{u} \in V_n$ we have

$$\sum_{\boldsymbol{x}\in \boldsymbol{u}\oplus H} (-1)^{f(\boldsymbol{x})} = \frac{1}{2^{n-t}} \sum_{\boldsymbol{u}\in V_n/H} \Big(\sum_{\boldsymbol{x}\in \boldsymbol{u}\oplus H} (-1)^{f(\boldsymbol{x})} \Big) = \frac{W_f(\mathbf{0})}{2^{n-t}},$$

i.e., the function $f(\boldsymbol{x})$ is $(H, \mathbf{0})$-stable. □

REMARK 7.51. If $\sum_{i=1}^N a_i^2 = \frac{1}{N}\big(\sum_{i=1}^N a_i\big)^2$, then $a_i = \frac{1}{N}\sum_{i=1}^N a_i$. Put

$$b_i = a_i - \frac{1}{N}\sum_{i=1}^N a_i.$$

Then

$$b_i^2 = a_i^2 - 2a_i \frac{1}{N} \sum_{i=1}^{N} a_i + \frac{1}{N^2} \left(\sum_{i=1}^{N} a_i \right)^2 .$$

Summing over i, we see that

$$\sum_{i=1}^{N} b_i^2 = \sum_{i=1}^{N} a_i^2 - 2 \sum_{i=1}^{N} a_i \frac{\sum_{i=1}^{N} a_i}{N} + N \frac{\left(\sum_{i=1}^{N} a_i\right)^2}{N^2} = \sum_{i=1}^{N} a_i^2 - \frac{\left(\sum_{i=1}^{N} a_i\right)^2}{N} = 0.$$

Therefore, $b_i = 0$ for all i.

We give another proof of the second part of Theorem 7.50, based on an identity, which is important in its own right.

LEMMA 7.52. *For any Boolean function $f(\boldsymbol{x}) \in \mathcal{F}_n$, any subset $H \subseteq V_n$, and any vector $\boldsymbol{\alpha} \in H^{\perp}$, the following equality holds:*

$$W_f(\boldsymbol{\alpha}) = \sum_{\boldsymbol{u} \in V_n/H} \left(\sum_{\boldsymbol{x} \in \boldsymbol{u} \oplus H} (-1)^{f(\boldsymbol{x})} \right) (-1)^{\langle \boldsymbol{\alpha}, \boldsymbol{u} \rangle}.$$

PROOF. Since $\boldsymbol{\alpha} \in H^{\perp}$, we can transform the expression for $W_f(\boldsymbol{\alpha})$ as follows:

$$\begin{aligned} W_f(\boldsymbol{\alpha}) &= \sum_{\boldsymbol{x} \in V_n} (-1)^{f(\boldsymbol{x})} (-1)^{\langle \boldsymbol{\alpha}, \boldsymbol{x} \rangle} = \sum_{\boldsymbol{u} \in V_n/H} \sum_{\boldsymbol{x} \in \boldsymbol{u} \oplus H} (-1)^{f(\boldsymbol{x})} (-1)^{\langle \boldsymbol{\alpha}, \boldsymbol{x} \rangle} \\ &= \sum_{\boldsymbol{u} \in V_n/H} \sum_{\boldsymbol{h} \in H} (-1)^{f(\boldsymbol{u} \oplus \boldsymbol{h})} (-1)^{\langle \boldsymbol{\alpha}, \boldsymbol{u} \oplus \boldsymbol{h} \rangle} \\ &= \sum_{\boldsymbol{u} \in V_n/H} \left(\sum_{\boldsymbol{x} \in \boldsymbol{u} \oplus H} (-1)^{f(\boldsymbol{x})} \right) (-1)^{\langle \boldsymbol{\alpha}, \boldsymbol{u} \rangle}. \end{aligned}$$

□

Part 2) of Theorem 7.50 follows immediately from Lemma 7.52 since $\sum_{\boldsymbol{x} \in \boldsymbol{u} \oplus H} (-1)^{f(\boldsymbol{x})}$ does not depend on $\boldsymbol{u} \in V_n$ if and only if $W_f(\boldsymbol{\alpha}) = 0$ for any $\boldsymbol{\alpha} \in H^{\perp} \backslash \{\mathbf{0}\}$.

Part 2) of Theorem 7.50 allows us to obtain more properties of Walsh–Hadamard coefficients for $(H, \mathbf{0})$-stable functions. First we prove the following lemma.

LEMMA 7.53. *For any Boolean function $f(\boldsymbol{x}) \in \mathcal{F}_n$, any subspace $L \subseteq V_n$, and any $\boldsymbol{\beta} \in V_n \backslash L$ one has*

$$\sum_{\boldsymbol{\alpha} \in L} W_f(\boldsymbol{\alpha}) + \sum_{\boldsymbol{\alpha} \in \boldsymbol{\beta} \oplus L} W_f(\boldsymbol{\alpha}) \equiv 0 \pmod{2^T},$$

where

$$T = \begin{cases} n & \text{if } \dim L = n - 1, \\ \dim L + 2 & \text{if } \dim L < n - 1. \end{cases}$$

PROOF. Let M be a subspace spanned by L and $\boldsymbol{\beta}$, $\dim L = s$. We have

$$\sum_{\boldsymbol{\alpha} \in M} W_f(\boldsymbol{\alpha}) = \sum_{\boldsymbol{x} \in V_n} (-1)^{f(\boldsymbol{x})} \sum_{\boldsymbol{\alpha} \in M} (-1)^{\langle \boldsymbol{\alpha}, \boldsymbol{x} \rangle} = 2^{s+1} \sum_{\boldsymbol{x} \in M^{\perp}} (-1)^{f(\boldsymbol{x})}.$$

On the other hand, M is such that

$$\sum_{\boldsymbol{\alpha} \in M} W_f(\boldsymbol{\alpha}) = \sum_{\boldsymbol{\alpha} \in L} W_f(\boldsymbol{\alpha}) + \sum_{\boldsymbol{\alpha} \in \boldsymbol{\beta} \oplus L} W_f(\boldsymbol{\alpha}).$$

To complete the proof of Lemma 7.53, observe that

$$\sum_{\boldsymbol{x}\in M^{\perp}} (-1)^{f(\boldsymbol{x})} \equiv \begin{cases} 1 & \bmod 2 \quad \text{if } \dim L = n-1, \\ 0 & \bmod 2 \quad \text{if } \dim L < n-1. \end{cases} \qquad \square$$

Repeating the proof of Lemma 7.53 for sums of the form

$$\sum_{\boldsymbol{\alpha}\in M} W_f(\boldsymbol{\alpha})(-1)^{\langle\boldsymbol{\alpha},\boldsymbol{z}\rangle},$$

where $\boldsymbol{z} \in V_n$, it is easy to obtain the following statement (the proof is straightforward).

COROLLARY 7.54. *For any Boolean function $f(\boldsymbol{x}) \in \mathcal{F}_n$, any subspace $L \subseteq V_n$, and any $\boldsymbol{z}, \boldsymbol{\beta} \in V_n \backslash L$, one has*

$$\sum_{\boldsymbol{\alpha}\in L} W_f(\boldsymbol{\alpha})(-1)^{\langle\boldsymbol{\alpha},\boldsymbol{z}\rangle} + \sum_{\boldsymbol{\alpha}\in\boldsymbol{\beta}\oplus L} W_f(\boldsymbol{\alpha})(-1)^{\langle\boldsymbol{\alpha},\boldsymbol{z}\rangle} \equiv 0 \pmod{2^{T}},$$

where

$$T = \begin{cases} n & \text{if } \dim L = n-1, \\ \dim L + 2 & \text{if } \dim L < n-1. \end{cases}$$

COROLLARY 7.55. *Let $f(\boldsymbol{x}) \in \mathcal{F}_n$ be an $(H, \mathbf{0})$-stable function. Then for any $\boldsymbol{\beta} \in V_n$ one has*

$$\sum_{\boldsymbol{\alpha}\in\boldsymbol{\beta}\oplus H^{\perp}} W_f(\boldsymbol{\alpha})(-1)^{\langle\boldsymbol{\alpha},\boldsymbol{z}\rangle} \equiv 0 \pmod{2^{n-\dim H+1}}.$$

The proof can be easily obtained from Lemmas 7.53, 7.47 and statement 2) of Theorem 7.50.

In addition to a criterion for $(H, \mathbf{0})$-resiliency, Theorem 7.50 also gives a method of description of the family $\mathcal{H}_{f,\mathbf{0}}$. Observe that if $H \in \mathcal{H}_{f,\mathbf{0}}$, then $H' \in \mathcal{H}_{f,\mathbf{0}}$ as well for any $H' \supseteq H$. Hence, the family $\mathcal{H}_{f,\mathbf{0}}$ can be represented by a set of minimal subspaces $H_1, H_2, \ldots, H_l$ such that

1) $f(\boldsymbol{x})$ is $(H_i, \mathbf{0})$-stable for all $i = 1, \ldots, l$;
2) for any $H \in \mathcal{H}_{f,\mathbf{0}}$ there exists an index i, $i = 1, \ldots, l$, such that $H \supseteq H_i$;
3) there is no pair $1 \leq i \neq j \leq l$ such that $H_i \subseteq H_j$.

Applying Theorem 7.50 we obtain:

$$W_f(\boldsymbol{\alpha}) = 0 \text{ for all } \boldsymbol{\alpha} \in H_i^{\perp} \backslash \{\mathbf{0}\}, \quad i = 1, 2, \ldots, l.$$

This condition allows us to state the following algorithm for construction of a family $\mathcal{H}_{f,\mathbf{0}}$ for a given Boolean function $f(\boldsymbol{x})$.

Algorithm for construction of $\mathcal{H}_{f,\mathbf{0}}$.

1. Find all coefficients $W_f(\boldsymbol{\alpha})$ for $\boldsymbol{\alpha} \in V_n$.
2. Determine the set of zeros of $\mathrm{CI}_f = \{\boldsymbol{\alpha} \in V_n \mid W_f(\boldsymbol{\alpha}) = 0\}$.
3. Find all maximal subsets L which are in $\mathrm{CI}_f \cup \{\mathbf{0}\}$.
4. The set of all maximal subsets found in step 3 gives the family $\mathcal{H}_{f,\mathbf{0}}$.

It is obvious from Lemma 7.46 that the algorithm of constructing $\mathcal{H}_{f,\boldsymbol{\gamma}}$ for any $\boldsymbol{\gamma} \in V_n$ is different from the algorithm that constructs $\mathcal{H}_{f,\mathbf{0}}$ by replacing the set CI_f with the set $\mathrm{CI}_f \oplus \boldsymbol{\gamma}$.

REMARK 7.56. The set CI_f cannot be arbitrary. For example, it is easy to see that the set CI_f cannot be a union of the form

$$\mathrm{CI}_f = H_{\boldsymbol{\alpha}_1} \cup \cdots \cup H_{\boldsymbol{\alpha}_t},$$

where $H_{\boldsymbol{\alpha}_i} = \{\boldsymbol{x} \in V_n \mid \langle \boldsymbol{x}, \boldsymbol{\alpha}_i \rangle = \varepsilon\}$, $\varepsilon \in \{0,1\}$, $i \in 1, \ldots, t$, is a hyperplane, if the vectors $\boldsymbol{\alpha}_1, \ldots, \boldsymbol{\alpha}_t$ do not form a subspace in V_n. This follows immediately because all vectors $\boldsymbol{\alpha} \in V_n$ are such that $H_{\boldsymbol{\alpha}} \subseteq \mathrm{CI}_f$. They (and only they) are linear shifts of the function f, and it is known that all linear shifts form a subspace in V_n.

Consider an example of an $(H, \mathbf{0})$-stable function as an illustration of the notions introduced and results obtained.

EXAMPLE 7.57. Suppose that a basis in V_n is chosen in such a way that $H = \{(\mathbf{0}, \boldsymbol{y}) \mid \boldsymbol{y} \in V_t\}$. Define a Boolean function

$$f(\boldsymbol{u}, \boldsymbol{y}) = \langle \Phi(\boldsymbol{u}), \boldsymbol{y} \rangle = \varphi_1(\boldsymbol{u}) y_1 \oplus \cdots \oplus \varphi_t(\boldsymbol{u}) y_t,$$

where $\Phi \in \mathcal{F}_{n-t,t}$ is a certain mapping with $\Phi(\boldsymbol{u}) \neq \mathbf{0}$ for any $\boldsymbol{u} \in V_{n-t}$. Then for any choice of the variable $\boldsymbol{u} = \boldsymbol{u}_0 \in V_{n-t}$, the function $f(\boldsymbol{u}_0, \boldsymbol{y})$ is not linearly stable. Thus,

$$W_f(\mathbf{0})\big|_{\boldsymbol{u} \oplus H} = W_f(\mathbf{0}) = 0 \text{ for all } \boldsymbol{u} \in V_{n-t},$$

i.e., the function $f(\boldsymbol{x})$ is $(H, \mathbf{0})$-stable.

EXAMPLE 7.58. Suppose now that $n = 2t$ and that a mapping $\Phi \in \mathcal{F}_{t,t}$ satisfies the following conditions:

1) $\Phi(\boldsymbol{u}_1) = \Phi(\boldsymbol{u}_2) = \boldsymbol{y}_1 \neq \mathbf{0}$ for some $\boldsymbol{u}_1, \boldsymbol{u}_2, \boldsymbol{y}_1 \in V_t$;
2) $\Phi(\boldsymbol{u}) \neq \mathbf{0}$ for any $\boldsymbol{u} \in V_t$;
3) the mapping $\Phi' \colon V_t \backslash \{\boldsymbol{u}_1, \boldsymbol{u}_2\} \to V_t \backslash \{\mathbf{0}, \boldsymbol{y}_1\}$ is one-to-one.

We apply the algorithm for constructing $\mathcal{H}_{f,\mathbf{0}}$ to the function $f(\boldsymbol{u}, \boldsymbol{y})$. We find all $W_f(\boldsymbol{\alpha}, \boldsymbol{\beta})$ for $\boldsymbol{\alpha}, \boldsymbol{\beta} \in V_t$:

$$\begin{aligned} W_f(\boldsymbol{\alpha}, \boldsymbol{\beta}) &= \sum_{\boldsymbol{u}, \boldsymbol{y} \in V_t} (-1)^{f(\boldsymbol{u}, \boldsymbol{y})} (-1)^{\langle \boldsymbol{\alpha}, \boldsymbol{u} \rangle} (-1)^{\langle \boldsymbol{\beta}, \boldsymbol{y} \rangle} \\ &= \sum_{\boldsymbol{u} \in V_t} (-1)^{\langle \boldsymbol{\alpha}, \boldsymbol{u} \rangle} \sum_{\boldsymbol{y} \in V_t} (-1)^{\langle \Phi(\boldsymbol{u}) \oplus \boldsymbol{\beta}, \boldsymbol{y} \rangle} = 2^t \sum_{\boldsymbol{u} \in \Phi^{-1}(\boldsymbol{\beta})} (-1)^{\langle \boldsymbol{\alpha}, \boldsymbol{u} \rangle}. \end{aligned}$$

Hence,

$$W_f(\boldsymbol{\alpha}, \boldsymbol{\beta}) = \begin{cases} 2^t \cdot (-1)^{\langle \boldsymbol{\alpha}, \Phi^{-1}(\boldsymbol{\beta}) \rangle} & \text{if } \boldsymbol{\beta} \neq \mathbf{0}, \boldsymbol{y}_1; \\ 0 & \text{if } \boldsymbol{\beta} = \mathbf{0}; \\ 2^t \cdot \left((-1)^{\langle \boldsymbol{\alpha}, \boldsymbol{u}_1 \rangle} + (-1)^{\langle \boldsymbol{\alpha}, \boldsymbol{u}_2 \rangle} \right) & \text{if } \boldsymbol{\beta} = \boldsymbol{y}_1. \end{cases}$$

Thus, the set of zeros CI_f has the following form:

$$\mathrm{CI}_f = \{(\boldsymbol{\alpha}, \mathbf{0}) \mid \boldsymbol{\alpha} \in V_t\} \cup \{(\boldsymbol{\alpha}, \boldsymbol{y}_1) \mid \boldsymbol{\alpha} \in V_t \colon \langle \boldsymbol{\alpha}, \boldsymbol{u}_1 \rangle \oplus \langle \boldsymbol{\alpha}, \boldsymbol{u}_2 \rangle = 1\}.$$

The cardinality of the set CI_f is equal to $2^t + 2^{t-1}$. It is obvious that the maximal subspace which is not contained in CI_f is the subspace $\{(\boldsymbol{\alpha}, \mathbf{0}) \mid \boldsymbol{\alpha} \in V_t\}$. Therefore the family $\mathcal{H}_{f,\mathbf{0}}$ contains all subspaces in which $H = \{(\mathbf{0}, \boldsymbol{y}) \mid \boldsymbol{y} \in V_t\}$ is contained.

The connection between the properties of $(H, \mathbf{0})$-resiliency and correlation immunity is obvious. For the reader's convenience we state this connection in a separate lemma.

LEMMA 7.59. *A Boolean function $f(\boldsymbol{x}) \in \mathcal{F}_n$ is correlation-immune of order m if and only if it is $(H, \mathbf{0})$-stable for all $\binom{n}{m}$ subspaces spanned by $n-m$ basis vectors of the the space V_n.*

Some additional advantages of the notion of $(H, \mathbf{0})$-resiliency over correlation immunity are the following:

– the property of being $(H, \mathbf{0})$-resilient is "weaker", since it is one of the components of the correlation immunity;

– the property of being $(H, \mathbf{0})$-resilient is more general because its definition does not depend on the choice of a basis for V_n and is applicable to any subspace H.

On the other hand, for correlation immunity, which is a "stronger" property, more results can be obtained. One such result is connected with the power of the nonlinearity of correlation-immune functions. A similar statement for $(H, \mathbf{0})$-stable functions has a more complicated statement.

THEOREM 7.60. *Suppose that a Boolean function $f(\boldsymbol{x}) \in \mathcal{F}_n$ is $(H, \mathbf{0})$-stable and that the basis in V_n is chosen so that H is spanned by the last t basis vectors, i.e., $H = \{(\mathbf{0}, \boldsymbol{y}) \mid \boldsymbol{y} \in V_t\}$. Then the Zhegalkin polynomial of the function $f(\boldsymbol{u}, \boldsymbol{y})$ has the following properties:*

1) *it contains no monomials of the form $y_1 y_2 \cdots y_t u_{i_1} u_{i_2} \cdots u_{i_l}$ for $l \geq 1$ (i.e., only monomials of the form $y_1 y_2 \cdots y_t$ can occur);*

2) *if the number $\frac{\mathrm{wt}(f)}{2^{n-t}}$ is even, then there are no monomials of the form $y_1 y_2 \cdots y_t$.*

In other words, if the number $\frac{\mathrm{wt}(f)}{2^{n-t}}$ is even, then the degree of the function $f(\boldsymbol{x})$ in the variables $y_1, y_2, \ldots, y_t$ is at most $t-1$.

PROOF. Decompose the Zhegalkin polynomial of the function $f(\boldsymbol{x})$ with respect to the variables $y_1, y_2, \ldots, y_t$:

$$f(\boldsymbol{u}, \boldsymbol{y}) = y_1 y_2 \cdots y_t \varphi(\boldsymbol{u}) \oplus \cdots .$$

Suppose that the function $\varphi(\boldsymbol{u})$ is not a constant. Then there exist $\boldsymbol{u}_1 \neq \boldsymbol{u}_2 \in V_{n-t}$ such that $\varphi(\boldsymbol{u}_1) = 0$, $\varphi(\boldsymbol{u}_2) = 1$. Thus the restriction of the function $f(\boldsymbol{u}, \boldsymbol{y})$ to the coset $\boldsymbol{u}_1 \oplus H$ has even weight, and its restriction to the coset $\boldsymbol{u}_2 \oplus H$ has odd weight, which contradicts $(H, \mathbf{0})$-resiliency. This proves the first statement.

Suppose there exists a monomial $y_1 y_2 \cdots y_t$ in the Zhegalkin polynomial of the function $f(\boldsymbol{u}, \boldsymbol{y})$. Then for all cosets $\boldsymbol{u} \oplus H$ the span of the function $f(\boldsymbol{u}, \boldsymbol{y})$ has odd weight. On the other hand, it follows from the $(H, \mathbf{0})$-resiliency that this weight is equal to $\frac{\mathrm{wt}(f)}{2^{n-t}}$, so it is even. This contradiction concludes the proof of statement 2). □

Theorem 7.60 allows us to state the relationship between $(H, \mathbf{0})$-resiliency and correlation immunity as follows: $(H, \mathbf{0})$-resiliency is correlation immunity of order $n-t$ with respect to the variables $u_1, u_2, \ldots, u_{n-t}$.

7.3. General methods for constructing correlation-immune functions and resilient mappings

In this section we discuss how to construct correlation-immune functions and resilient mappings independently of other cryptographic properties and how to calculate the order of correlation immunity and resiliency.

LEMMA 7.61. *Suppose that $f \in \mathcal{F}_n$, $\deg f \leq n-1$. Then there exists a variable $x^{(i)}$, $i = 1, \dots, n$, such that the function f can be represented as follows:*

$$f = (x^{(i)} \oplus 1) f_1 \oplus x^{(i)} f_2, \tag{7.9}$$

where

$$f_1 \in \mathcal{F}_{n-1},\ f_2 \in \mathcal{F}_{n-1}; \quad \deg f_1 = \deg f_2 = \deg f; \quad \deg f_1 \oplus f_2 < \deg f.$$

PROOF. The decomposition of the algebraic normal form with respect to one of the variables is well known:

$$f = x^{(i)} h_1 \oplus h_2.$$

Furthermore, $h_1, h_2 \in \mathcal{F}_{n-1}$, $\deg h_1 \leq \deg f - 1$, and $\deg h_2 \leq \deg f$. It is easy to observe that $\deg h_2 = \deg f$ if and only if there exists a monomial of higher degree in the algebraic normal form in which the variable $x^{(i)}$ does not occur. The last condition is fulfilled for some i if

$$\deg f \leq n-1.$$

Choose i such that $\deg h_2 = \deg f$; put $f_2 = h_1 \oplus h_2$ and $f_1 = h_2$. Then

$$(x^{(i)} \oplus 1) f_1 \oplus x^{(i)} f_2 = x^{(i)}[f_1 \oplus f_2] \oplus f_1 = x^{(i)} h_1 \oplus h_2,$$

and the functions f_1 and f_2 satisfy the hypotheses of the lemma. □

This lemma allows us to relate correlation immunity of a function to correlation immunity of its subfunctions.

THEOREM 7.62. *The equation* $\operatorname{sut} f = d$ *is satisfied for a balanced Boolean function $f \in \mathcal{F}_n$ if and only if the following two conditions hold:*

1) $\operatorname{sut} f_1 = \operatorname{sut} f_2 = d - 1$, *where $f_1, f_2 \in \mathcal{F}_{n-1}$ are the functions occurring in the decomposition* (7.9);

2) *for any $\boldsymbol{v} \in V_{n-1}$, where* $\operatorname{wt}(\boldsymbol{v}) = d$, *we have*

$$W_{f_1}(\boldsymbol{v}) + W_{f_2}(\boldsymbol{v}) = 0.$$

PROOF. Suppose $\boldsymbol{v} \in V_n$, $\boldsymbol{v}^\top = (\boldsymbol{v}'^\top, v^{(n)})$, $v^{(n)} \in \mathbb{F}_2$, $\boldsymbol{v}' = (v^{(1)}, \dots, v^{(n-1)})^\top \in V_{n-1}$, $\boldsymbol{x} = (x^{(1)}, \dots, x^{(n-1)}, x^{(n)})^\top \in V_n$, and $\boldsymbol{x}' = (x^{(1)}, \dots, x^{(n-1)})^\top \in V_{n-1}$. Without loss of generality (up to renumbering the variables) we assume that the variable that satisfies (7.9) is $x^{(n)}$. Then

$$\begin{aligned} W_f(\boldsymbol{v}) &= \sum_{\boldsymbol{x} \in V_n} (-1)^{f(\boldsymbol{x})} (-1)^{\langle \boldsymbol{x}', \boldsymbol{v}' \rangle \oplus x^{(n)} v^{(n)}} \\ &= \sum_{\substack{\boldsymbol{x}' \in V_{n-1}; \\ x^{(n)}=0}} (-1)^{f_1(\boldsymbol{x}')} (-1)^{\langle \boldsymbol{x}', \boldsymbol{v}' \rangle} + \sum_{\substack{\boldsymbol{x}' \in V_{n-1}; \\ x^{(n)}=1}} (-1)^{f_2(\boldsymbol{x}')} (-1)^{\langle \boldsymbol{x}', \boldsymbol{v}' \rangle \oplus v^{(n)}}, \end{aligned} \tag{7.10}$$

$$W_f(\boldsymbol{v}) = W_{f_1}(\boldsymbol{v}') + (-1)^{v^{(n)}} W_{f_2}(\boldsymbol{v}').$$

Therefore, for any $\boldsymbol{v}' \in V_{n-1}$ one has

$$W_f(\boldsymbol{v}, 0) = W_{f_1}(\boldsymbol{v}') + W_{f_2}(\boldsymbol{v}'), \quad W_f(\boldsymbol{v}, 1) = W_{f_1}(\boldsymbol{v}') - W_{f_2}(\boldsymbol{v}').$$

It is easy to infer from these equations that the condition "$W_f(\boldsymbol{v}) = 0$ for all $\boldsymbol{v} \in V_n$, $\operatorname{wt}(\boldsymbol{v}) \leq d$" is equivalent to the following two conditions:

1) $W_{f_1}(\boldsymbol{v}') = W_{f_2}(\boldsymbol{v}') = 0$ for all $\boldsymbol{v}' \in V_{n-1}$, $\operatorname{wt}(\boldsymbol{v}') \leq d-1$;

2) $W_{f_1}(\boldsymbol{v}') + W_{f_2}(\boldsymbol{v}') = 0$ for all $\boldsymbol{v}' \in V_{n-1}$, $\operatorname{wt}(\boldsymbol{v}') = d$.

In view of Theorem 7.10 this is equivalent to the statement of the theorem. □

COROLLARY 7.63. *Let $f \in \mathcal{F}_n$, $f_1, f_2 \in \mathcal{F}_{n-1}$ satisfy the conditions of Lemma 7.61. The function f is d-optimal if and only if the functions f_1 and f_2 are $(d-1)$-optimal.*

PROBLEM 7.64 (Open problem). Find necessary and sufficient conditions for a pair of functions $f_1, f_2 \in \mathcal{F}_{n-1}$ to satisfy condition 2) of Theorem 7.62.

PROBLEM 7.65. Let $f \in \mathcal{F}_n$, $l \in \mathcal{F}_k$ be functions with disjoint variables and l an affine function. Prove that:

1) $\operatorname{sut}(f \oplus l) = k + \operatorname{sut} f$;
2) the function $f \oplus l$ is d-optimal if and only if f is $(d-k)$-optimal.

PROBLEM 7.66. Suppose that

$$g(\boldsymbol{z}) = g(\boldsymbol{x}, \boldsymbol{y}) = \bigoplus_{\boldsymbol{\sigma} \in \boldsymbol{V_l}} (y^{(1)} \oplus \sigma^{(1)} \oplus 1) \cdots (y^{(l)} \oplus \sigma^{(l)} \oplus 1) f_{\boldsymbol{\sigma}}(\boldsymbol{x}),$$

$\boldsymbol{x} \in V_n$, $\boldsymbol{y} \in V_l$, $f_{\boldsymbol{\sigma}} \in \mathcal{F}_n$. Prove that $\operatorname{sut} g = \min_{\boldsymbol{\sigma} \in V_l} \operatorname{sut} f_{\boldsymbol{\sigma}}$.

PROBLEM 7.67. Suppose that $\Phi = (f_1, \dots, f_{k_1})$ is an (n_1, k_1, d_1)-resilient mapping and $\Psi = (g_1, \dots, g_{k_2})$ is an (n_2, k_2, d_2)-resilient mapping. Prove that the mapping

$$\Xi(\boldsymbol{z}) = \Xi(\boldsymbol{x}, \boldsymbol{y}) = (f_1(\boldsymbol{x}), \dots, f_{k_1}(\boldsymbol{x}), g_1(\boldsymbol{y}), \dots, g_{k_2}(\boldsymbol{y})),$$

where $\boldsymbol{x} \in V_{n_1}$, $\boldsymbol{y} \in V_{n_2}$, is $(n_1 + n_2, k_1 + k_2, d)$-resilient, where $d = \min\{d_1, d_2\}$.

LEMMA 7.68. *Suppose that $\Phi = (f_1, \dots, f_k) \in \mathcal{F}_{n,k}$ and*

$$\Psi(\boldsymbol{x}, \boldsymbol{y}, \boldsymbol{z}) = (\Phi(\boldsymbol{x}) \oplus \Phi(\boldsymbol{y}), \Phi(\boldsymbol{y}) \oplus \Phi(\boldsymbol{z})) \in \mathcal{F}_{3n,2k}.$$

Then

$$\operatorname{sut} \Psi = 2 \operatorname{sut} \Phi + 1.$$

PROOF. By virtue of Lemma 7.33, the value of $\operatorname{sut} \Psi$ is equal to the minimum value of sut for nonzero linear combinations of coordinate functions of the mapping Ψ, i.e., of the following Boolean functions:

$$f_1(\boldsymbol{x}) \oplus f_1(\boldsymbol{y}), \dots, f_k(\boldsymbol{x}) \oplus f_k(\boldsymbol{y}), f_1(\boldsymbol{y}) \oplus f_1(\boldsymbol{z}), \dots, f_k(\boldsymbol{y}) \oplus f_k(\boldsymbol{z}).$$

Put

$$\begin{aligned} f(\boldsymbol{x}, \boldsymbol{y}, \boldsymbol{z}) &= \bigoplus_{j=1}^{k} a^{(j)}(f_j(\boldsymbol{x}) \oplus f_j(\boldsymbol{y})) \oplus \bigoplus_{j=1}^{k} b^{(j)}(f_j(\boldsymbol{y}) \oplus f_j(\boldsymbol{z})) \\ &= \bigoplus_{j=1}^{k} a^{(j)} f_j(\boldsymbol{x}) \oplus \bigoplus_{j=1}^{k} (a^{(j)} \oplus b^{(j)}) f_j(\boldsymbol{y}) \oplus \bigoplus_{j=1}^{k} b^{(j)} f_j(\boldsymbol{z}), \end{aligned}$$

where either $\boldsymbol{a} = \left(a^{(1)}, \dots, a^{(k)}\right)^{\top} \neq \boldsymbol{0}$ or $\boldsymbol{b} = \left(b^{(1)}, \dots, b^{(k)}\right)^{\top} \neq \boldsymbol{0}$.

The function $f(\boldsymbol{x}, \boldsymbol{y}, \boldsymbol{z})$ is a function of disjoint variables; thus, according to Problem 7.34, it is easy to see that $\boldsymbol{a}, \boldsymbol{b} \in V_n$ can be chosen so that $\operatorname{sut} f(\boldsymbol{x}, \boldsymbol{y}, \boldsymbol{z})$ has the values $3 \operatorname{sut} \Phi + 2$ and $2 \operatorname{sut} \Phi + 1$. In other cases $\operatorname{sut} f(\boldsymbol{x}, \boldsymbol{y}, \boldsymbol{z})$ is greater. Therefore,

$$\operatorname{sut} \Psi = 2 \operatorname{sut} \Phi + 1.$$

□

The construction from Lemma 7.68 allows us to obtain a $(3n, 2k, 2d+1)$-resilient mapping from an (n, k, d)-resilient mapping. This technique can be generalized in the opposite direction. If for an (n, k, d)-resilient mapping $\Phi = (f_1, \dots, f_k)$ we put

$$\Psi(\boldsymbol{x}, \boldsymbol{y}, \boldsymbol{z}, \boldsymbol{u}) = (\Phi(\boldsymbol{x}) \oplus \Phi(\boldsymbol{y}), \Phi(\boldsymbol{y}) \oplus \Phi(\boldsymbol{z}), \Phi(\boldsymbol{z}) \oplus \Phi(\boldsymbol{u})), \tag{7.11}$$

where $\boldsymbol{x}, \boldsymbol{y}, \boldsymbol{z}, \boldsymbol{u} \in V_n$, then the mapping $\Psi\colon V_{4n} \to V_{3n}$ will be $(4n, 3k, 2d+1)$-resilient. Using this construction t times, we obtain a $(4^t n, 3^t k, 2^t(d+1)-1)$-resilient mapping.

PROBLEM 7.69. Suppose that Φ is an (n, k, d)-resilient mapping. Suggest a method for constructing $((l+1)^t n, l^t k, 2^t(1+d)-1)$-resilient mappings for all $l = 2, 3, \dots$ and $k = 2, 3, \dots$.

PROBLEM 7.70. Let $\Phi \in \mathcal{F}_{n,k}$, g be an arbitrary permutation of V_k. Prove that $\operatorname{sut}(g \circ \Phi) = \operatorname{sut} \Phi$.

Now we turn to the problem of resiliency of Boolean function given as linear branchings (see §2.5). Recall the construction. The Boolean function $f \in \mathcal{F}_n$ is defined by the following equation:

$$f(\boldsymbol{z}) = f(\boldsymbol{x}, \boldsymbol{y}) = \langle \boldsymbol{x}, \Phi(\boldsymbol{y}) \rangle \oplus h(\boldsymbol{y}), \tag{7.12}$$

where $\boldsymbol{z}^\top = (\boldsymbol{x}^\top, \boldsymbol{y}^\top) \in V_n$, $\boldsymbol{x} \in V_t$, $\boldsymbol{y} \in V_s$, $n = t + s$, $\Phi \in \mathcal{F}_{s,t}$, $h \in \mathcal{F}_s$.

Suppose that $\boldsymbol{u}^\top = (\boldsymbol{v}^\top, \boldsymbol{w}^\top) \in V_n$, $\boldsymbol{v} \in V_t$, $\boldsymbol{w} \in V_s$. According to Problem 2.135,

$$W_f(\boldsymbol{u}) = W_f(\boldsymbol{v}, \boldsymbol{w}) = 2^t \sum_{\boldsymbol{y} \in \Phi^{-1}(\boldsymbol{v})} (-1)^{h(\boldsymbol{y}) \oplus \langle \boldsymbol{y}, \boldsymbol{w} \rangle}. \tag{7.13}$$

To continue the study of resilient functions, we need a new definition. We will use the notion of partially defined Boolean function from §6.5.

DEFINITION 7.71. We will say that a partially defined Boolean function $f_S \in \mathcal{F}_n(S)$ is a *partially defined d-resilient function* (p.d. d-resilient function), if $W_{f_S}(\boldsymbol{u}) = 0$ for any $\boldsymbol{u} \in V_n$ such that $\operatorname{wt}(\boldsymbol{u}) \le d$.

This definition together with formula (7.13) allows us to prove the following methodologically useful theorem.

THEOREM 7.72. *A function $f \in \mathcal{F}_n$ defined as in (7.12) is d-resilient if and only if, for each $\boldsymbol{v} \in V_t$ such that $\operatorname{wt}(\boldsymbol{v}) \le d$, the restriction of the function $h \in \mathcal{F}_s$ to the set $\Phi^{-1}(\boldsymbol{v})$ is a p.d. $(d - \operatorname{wt}(\boldsymbol{v}))$-resilient function.*

PROOF. Denote the restriction of the function $h \in \mathcal{F}_s$ to the set $\Phi^{-1}(\boldsymbol{v})$ by $h_{\boldsymbol{v}}$. Now (7.13) can be rewritten as follows:

$$W_f(\boldsymbol{v}, \boldsymbol{w}) = 2^t W_{h_{\boldsymbol{v}}}(\boldsymbol{w}).$$

Thus, $W_f(\boldsymbol{v}, \boldsymbol{w}) = 0$ for any $\boldsymbol{v} \in V_t$, $\boldsymbol{w} \in V_s$ such that $\operatorname{wt}(\boldsymbol{v}) + \operatorname{wt}(\boldsymbol{w}) \le d$ if and only if $W_{h_{\boldsymbol{v}}}(\boldsymbol{w}) = 0$ for any $\boldsymbol{v} \in V_t$, $\boldsymbol{w} \in V_s$ such that $\operatorname{wt}(\boldsymbol{v}) + \operatorname{wt}(\boldsymbol{w}) \le d$. □

This theorem means that in a sense any resilient function can be obtained by "gluing" p.d. resilient functions. Note that for a given set $S \subseteq V_n$, p.d. resilient functions in $\mathcal{F}_n(S)$ do not necessarily exist. In particular, if $S = \varnothing$, then any function in $\mathcal{F}_n(S)$ is resilient. Obviously, if the cardinality of S is odd, then there are no resilient functions in $\mathcal{F}_n(S)$.

COROLLARY 7.73. *Suppose that the mapping* Φ *in* (7.12) *satisfies the following condition:*

$$\min\{\mathrm{wt}(\Phi(\boldsymbol{y})) \mid \boldsymbol{y} \in V_s\} \geq d+1$$

for some d, $0 \leq d < t$. *Then any function* f *as in* (7.12) *is* d*-resilient for any* $h \in \mathcal{F}_s$.

PROOF. It is obvious that if $\mathrm{wt}(\boldsymbol{v}) \leq d$, then $\Phi^{-1}(\boldsymbol{v}) = \varnothing$, whence all functions in $\Phi^{-1}(\boldsymbol{v})$ are resilient. Thus, the corollary follows from the theorem we just proved. □

PROBLEM 7.74. Prove the following inequalities assuming the hypotheses of Corollary 7.73:

$$2^{n-1} - 2^{t-1} \max_{\boldsymbol{v} \in V_t} \#\Phi^{-1}(\boldsymbol{v}) \leq N_f \leq 2^{n-1} - 2^{t-1} \left\lceil \frac{2^{s/2}}{\sqrt{\sum_{i=k+1}^{t} \binom{t}{i}}} \right\rceil .$$

(*Hint*: Use the results of Problem 2.136.)

COROLLARY 7.75. *Suppose that mapping* Φ *from* (7.12) *satisfies the following condition: For a certain* $\boldsymbol{v}_0 \in V_t$ *such that* $\mathrm{wt}(\boldsymbol{v}_0) \leq d$, *the cardinality of the set* $\Phi^{-1}(\boldsymbol{v})$ *is odd. Then none of the functions of the form described in* (7.12) *is* d*-resilient for any* $h \in \mathcal{F}_s$.

COROLLARY 7.76. *Among the functions defined as in* (7.12) *there are* $(t-2)$*-optimal functions.*

PROOF. Suppose that $2 < t \leq 2^s$. We define a mapping Φ as follows:

$$\Phi(\boldsymbol{y}) = (f_1(\boldsymbol{y}) \oplus 1, \ldots, f_t(\boldsymbol{y}) \oplus 1),$$

$\boldsymbol{y} \in V_{n-t}$, where for any $1 \leq i < j \leq t$,

$$\{\boldsymbol{y} \in V_{n-t} \mid f_i(\boldsymbol{y}) = 1\} \cap \{\boldsymbol{y} \in V_{n-t} \mid f_j(\boldsymbol{y}) = 1\} = \varnothing,$$

and there exists a number i for which $\deg f_i = n-t$. Then for any $\boldsymbol{y} \in V_{n-t}$ the following equation holds: $\mathrm{wt}(\Phi(\boldsymbol{y})) \geq t-1$. Thus, according to Corollary 7.73, the function f defined as in (7.12) is $(t-2)$-resilient. Furthermore, $\deg f = n-t+1$ and therefore the function f is optimal. □

7.4. Nonlinearity of correlation-immune and resilient functions

The possibility of using resilient functions for constructing various cryptographic objects explains the interest in studying the nonlinearity of these functions. In this section we study bounds for nonlinearity of correlation-immune functions as well as conditions for these bounds to be attained.

THEOREM 7.77. 1. *If* $f \in \mathcal{F}_n$, $\mathrm{cor}\, f = m \leq n-1$, *then*

$$N_f \leq 2^{n-1} - 2^m. \tag{7.14}$$

2. *If* $f \in \mathcal{F}_n$, $\mathrm{sut}\, f = m \leq n-2$, *then*

$$N_f \leq 2^{n-1} - 2^{m+1}. \tag{7.15}$$

PROOF. It is enough to use Theorem 7.22 and the fact that, by definition, $N_f = 2^{n-1} - \frac{1}{2}\max_{\boldsymbol{u} \in V_n} |W_f(\boldsymbol{u})|$. □

EXAMPLE 7.78. The bound in (7.14) is sharp.

1. $m = n-1$. The Siegenthaler inequality implies that $\deg f \leq 1$. If $\deg f = 1$, then the function f is balanced. Otherwise $f_1 = \text{const}$ and $N_{f_1} = 0$, i.e., the bound (7.14) is attained.

2. $n = 2$, $m = 0$. Define a function $f_2(x^{(1)}, x^{(2)}) = x^{(1)} \cdot x^{(2)}$, which is a nonbalanced correlation-immune function of order 0. Its nonlinearity is equal to 1: $N_{f_2} = 1$, i.e., the bound (7.14) is attained.

3. $n = 3$, $m = 1$. Define a function

$$f_3(x^{(1)}, x^{(2)}, x^{(3)}) = \bigoplus_{1\leq i<j\leq 3} x^{(i)}x^{(j)} \oplus \bigoplus_{i=1}^{3} x^{(i)} \oplus 1,$$

which is a nonbalanced correlation-immune function of order 1. Its nonlinearity is equal to 2: $N_{f_3} = 2$, i.e., the bound (7.14) is attained.

4. $n = 6$, $m = 3$. Define a function

$$f_6(x^{(1)}, x^{(2)}, \ldots, x^{(6)}) = \bigoplus_{1\leq i<j<k\leq 6} x^{(i)}x^{(j)}x^{(k)} \oplus \bigoplus_{i=1}^{4} x^{(i)}x^{(i+1)}$$

$$\oplus\, x^{(1)}x^{(5)} \oplus \bigoplus_{i=1}^{5} x^{(i)} \oplus 1,$$

which is nonbalanced, correlation-immune of order 3. Its nonlinearity is equal to 24: $N_{f_6} = 24$, i.e., the bound (7.14) is attained.

5. $n = 5$, $m = 2$. Define a function $f_5(x^{(1)}, x^{(2)}, \ldots, x^{(5)}) = (f_6)_6^0$, which is obviously nonbalanced, correlation-immune of order 2. Its nonlinearity is equal to 12: $N_{f_5} = 12$, i.e., the bound (7.14) is attained.

PROBLEM 7.79. Give examples of Boolean functions from $\mathcal{F}_n$, $n \geq 7$, for which the bounds (7.14) are attained.

Imposing extra conditions on a Boolean function f allows us, on one hand, to strengthen inequality (7.15) and on the other hand, to derive an important conclusion.

THEOREM 7.80. *Suppose that* $f \in \mathcal{F}_n$, $\operatorname{sut} f = m \leq n-3$, $\deg f + \operatorname{sut} f \leq n-2$. *Then*

$$N_f \leq 2^{n-1} - 2^{m+2}. \tag{7.16}$$

PROOF. Since $\operatorname{sut} f = m$, we can find a vector $\boldsymbol{u} \in V_n$ with $\operatorname{wt}(\boldsymbol{u}) = m+1$ such that

$$\operatorname{wt}\bigl(f_{i_1,\ldots,i_{m+1}}^{a^{(1)},\ldots,a^{(m+1)}}\bigr) \neq 2^{n-m-2}$$

for some $a^{(1)}, \ldots, a^{(m+1)} \in \mathbb{F}_2$. Without loss of generality we may assume that $a^{(1)} = \cdots = a^{(m+1)} = 0$. Using Problem 2.41, we see that

$$W_f(\boldsymbol{u}) = 2^n - 2^{\operatorname{wt}(\boldsymbol{u})+1} \operatorname{wt}\bigl(f_{i_1,\ldots,i_{m+1}}^{0,\ldots,0}\bigr).$$

Since $\deg f_{i_1,\ldots,i_{m+1}}^{0,\ldots,0} \leq \deg f \leq n-m-2$ and $f_{i_1,\ldots,i_{m+1}}^{0,\ldots,0} \in \mathcal{F}_{n-m-1}$, the weight $\operatorname{wt}\bigl(f_{i_1,\ldots,i_{m+1}}^{0,\ldots,0}\bigr)$ is even and we can assume that $\operatorname{wt}\bigl(f_{i_1,\ldots,i_{m+1}}^{0,\ldots,0}\bigr) \leq 2^{n-m-2} - 2$. Plugging this in the expression for $W_f(\boldsymbol{u})$, we get

$$W_f(\boldsymbol{u}) \geq 2^{m+3},$$

whence $N_f \leq 2^{n-1} - 2^{m+2}$. □

COROLLARY 7.81. *Suppose that* $f \in \mathcal{F}_n$, $f \neq \text{const}$, $\operatorname{sut} f = m \leq n-2$, *and* $N_f = 2^{n-1} - 2^{m+1}$. *Then* $\deg f + \operatorname{sut} f = n-1$.

PROOF. If $\operatorname{sut} f = n-2$, then $\deg f \leq 1$ and $N_f = 0$. Since $f \neq \text{const}$, we conclude that $\deg f = 1 = (n-1) - \operatorname{sut} f$.

If $\operatorname{sut} f \leq n-3$ but $\deg f + \operatorname{sut} f \leq n-2$, then according to Theorem 7.80,

$$N_f \leq 2^{n-1} - 2^{m+2},$$

which contradicts the hypothesis of the theorem. □

Let us now derive a bound for the nonlinearity of a Boolean function in terms of its algebraic degree. We will also need a relation between Fourier coefficients and the coefficients of the numerical normal form of a Boolean function f (see §2.2):

$$\overline{W}_f(\boldsymbol{a}) = (-1)^{\operatorname{wt}(\boldsymbol{a})} \sum_{\boldsymbol{u} \in V_n \mid \boldsymbol{a} \preccurlyeq \boldsymbol{u}} 2^{n-\operatorname{wt}(\boldsymbol{u})} \lambda_f(\boldsymbol{u}), \tag{7.17}$$

$$\lambda_f(\boldsymbol{u}) = 2^{-n}(-2)^{\operatorname{wt}(\boldsymbol{u})} \sum_{\boldsymbol{x} \in V_n \mid \boldsymbol{u} \preccurlyeq \boldsymbol{x}} \overline{W}_f(\boldsymbol{x}). \tag{7.18}$$

LEMMA 7.82. *Suppose that* $f \in \mathcal{F}_n$ *and* $0 \leq m < n$. *The equality* $\operatorname{sut} f = m$ *holds if and only if the function* $g(\boldsymbol{x}) = f(\boldsymbol{x}) \oplus x^{(1)} \oplus \cdots \oplus x^{(n)}$ *has numerical degree not greater than* $n-m-1$.

PROOF. For any $\boldsymbol{a} \in V_n$ we have $W_g(\boldsymbol{a}) = W_f(\boldsymbol{a} \oplus \mathbf{1})$. The equation $\operatorname{sut} f = m$ holds if and only if $W_g(\boldsymbol{a}) = 0$ for any vector $\boldsymbol{a}$ such that $n-m \leq \operatorname{wt}(\boldsymbol{a}) \leq n$. The equation $W_g(\boldsymbol{a}) = 2^n \delta(\boldsymbol{a}) - 2\overline{W}_g(\boldsymbol{a})$ implies a similar statement for the Fourier coefficients $\overline{W}_g(\boldsymbol{a})$. Applying (7.18) to the function g, we obtain the lemma. □

THEOREM 7.83. *Suppose that* $f \in \mathcal{F}_n$, $\operatorname{sut} f = m$, $0 \leq m \leq n-2$, *and* $\deg f = d \neq 0$. *Then* $\overline{W}_f(\boldsymbol{a})$ *is divisible by* $2^{m+1+\left\lfloor \frac{n-m-2}{d} \right\rfloor}$ *for any* $\boldsymbol{a} \in V_n$.

PROOF. Put $g(\boldsymbol{x}) = f(\boldsymbol{x}) \oplus x^{(1)} \oplus \cdots \oplus x^{(n)}$. Since $\deg g = \deg f = d$, Corollary 2.57 implies that the coefficients $\lambda_f(\boldsymbol{u})$ of the numerical normal form of the function g are divisible by $2^{\left\lfloor \frac{\operatorname{wt}(\boldsymbol{u})-1}{d} \right\rfloor}$. Hence, for any $s = 1, \ldots, n$ and for all $\boldsymbol{u} \in V_n$ such that $\operatorname{wt}(\boldsymbol{u}) \leq s$, the numbers $2^{n-\operatorname{wt}(\boldsymbol{u})}\lambda_g(\boldsymbol{u})$ are divisible by $2^{n-s+\left\lfloor \frac{s-1}{d} \right\rfloor}$. Since $\operatorname{sut} f = m$, Lemma 7.82 implies that $\lambda_g(\boldsymbol{u}) = 0$ for $\operatorname{wt}(\boldsymbol{u}) \geq n-m$, and that all the numbers $2^{n-\operatorname{wt}(\boldsymbol{u})}\lambda_g(\boldsymbol{u})$, $\boldsymbol{u} \in V_n$, are divisible by $2^{m+1+\left\lfloor \frac{n-m-2}{d} \right\rfloor}$. Now relation (7.17) implies that $\overline{W}_g(\boldsymbol{a})$ is divisible by $2^{m+1+\left\lfloor \frac{n-m-2}{d} \right\rfloor}$ for any $\boldsymbol{a} \in V_n$. Obviously, this is also true for $\overline{W}_f(\boldsymbol{a})$. □

COROLLARY 7.84. *If f satisfies the hypothesis of Theorem* 7.83, *then the nonlinearity of the function f satisfies the inequality*

$$N_f \leq 2^{n-1} - 2^{m+\left\lfloor \frac{n-m-2}{d} \right\rfloor}. \tag{7.19}$$

It is clear that Theorem 7.77 is a particular case of this corollary for $\deg f \leq n-m-2$.

From Corollary 7.81 it follows that equality in the bound (7.15) can be attained only for m-optimal functions. Now, the natural question is how to construct such functions. The answer is contained in the following statement:

THEOREM 7.85. *Suppose that* $f \in \mathcal{F}_n$, $f \neq \text{const}$, $\operatorname{sut} f = m \leq n-2$, *and* $N_f = 2^{n-1} - 2^{m+1}$. *Then the function f is plateaued of order* $2(n-m-2)$.

PROOF. Since $\operatorname{sut} f = m$, according to Theorem 7.22 we have $W_f(\boldsymbol{u}) \equiv 0 \pmod{2^{m+2}}$ for any $\boldsymbol{u} \in V_n$. On the other hand, it follows from

$$N_f = 2^{n-1} - \frac{1}{2} \max_{\boldsymbol{u}\in V_n} |W_f(\boldsymbol{u})| = 2^{n-1} - 2^{m+1}$$

that $\max_{\boldsymbol{u}\in V_n} \left|W_f(\boldsymbol{u})\right| = 2^{m+2}$. Therefore, Walsh–Hadamard coefficients can take only the values 0 and $\pm 2^{m+2}$, so the function f is a plateaued function of order $2n - 2m - 4$. □

The bounds (7.14) and (7.15) can be strengthened in some cases.

THEOREM 7.86. *Suppose that* $f \in \mathcal{F}_n$.

(1) *If* $\operatorname{cor} f = m \le n/2 - 1$, *then*

$$N_f \le 2^{n-1} - 2^{n/2-1} - 2^m. \tag{7.20}$$

(2) *If* $\operatorname{sut} f = m \le n/2 - 2$, *then*

$$N_f \le 2^{n-1} - 2^{n/2-1} - 2^{m+1}. \tag{7.21}$$

PROOF. Since f is a correlation-immune (resilient) function, it is not a bent function, whence $\max_{\boldsymbol{u}\in V_n} \left|W_f(\boldsymbol{u})\right| \ne 2^{n/2}$. On the other hand, by Theorem 7.22, $\max_{\boldsymbol{u}\in V_n} \left|W_f(\boldsymbol{u})\right|$ is divisible by 2^{m+1} (by 2^{m+2}). Therefore, $\max_{\boldsymbol{u}\in V_n} \left|W_f(\boldsymbol{u})\right| \ge 2^{n/2} + 2^{m+1}$ ($\ge 2^{n/2} + 2^{m+2}$) and the theorem follows. □

THEOREM 7.87. *Suppose that* $f \in \mathcal{F}_n$, $\operatorname{sut} f = m \le n/2 - 2$. *Then*

$$N_f \le 2^{n-1} - 2^{m+1} \left\lceil \frac{2^{n-m-2}}{\sqrt{2^n - \sum_{i=1}^{m} \binom{n}{i}}} \right\rceil. \tag{7.22}$$

PROOF. Theorem 7.22 implies that for any vector $\boldsymbol{u} \in V_n$, the Walsh–Hadamard coefficients $W_f(\boldsymbol{u})$ can be represented as $W_f(\boldsymbol{u}) = \theta_f(\boldsymbol{u}) \cdot 2^{m+2}$, where $\theta_f(\boldsymbol{u})$ is a numerical function of V_n. It follows from Parseval's identity that

$$\sum_{\boldsymbol{u}\in V_n} W_f^2(\boldsymbol{u}) = \sum_{\boldsymbol{u}\in V_n} \theta_f^2(\boldsymbol{u}) \cdot 2^{2m+4} = 2^{2n}.$$

Since $W_f(\boldsymbol{u}) = 0$ for all vectors $\boldsymbol{u}$ such that $0 \le \operatorname{wt}(\boldsymbol{u}) \le m$, we have

$$\sum_{\substack{\boldsymbol{u}\in V_n \\ \operatorname{wt}(\boldsymbol{u})>m}} \theta_f^2(\boldsymbol{u}) = 2^{2n-2m-4}.$$

Thus,

$$\max_{\boldsymbol{u}\in V_n} \left|\theta_f(\boldsymbol{u})\right| \ge \sqrt{\frac{2^{2n-2m-4}}{2^n - \sum_{i=1}^{m} \binom{n}{i}}} = \frac{2^{n-m-2}}{\sqrt{2^n - \sum_{i=1}^{m} \binom{n}{i}}},$$

and since $\theta_f(\boldsymbol{u})$ is an integer-valued function,

$$\max_{\boldsymbol{u}\in V_n} \left|\theta_f(\boldsymbol{u})\right| \ge \left\lceil \frac{2^{n-m-2}}{\sqrt{2^n - \sum_{i=1}^{m} \binom{n}{i}}} \right\rceil.$$

Therefore,

$$\max_{\boldsymbol{u}\in V_n}\left|W_f(\boldsymbol{u})\right| \geq 2^{m+2}\left\lceil \frac{2^{n-m-2}}{\sqrt{2^n-\sum_{i=1}^m \binom{n}{i}}}\right\rceil,$$

and the theorem follows. □

7.5. Construction of resilient Boolean functions with good cryptographic properties

The upper bounds for nonlinearity of correlation-immune and resilient functions given in §7.4 suggest the following question: Can these bounds be attained in various ranges of values of n and m? As a rule, problems of this type are solved by constructing classes of Boolean functions parameterized by n and m whose nonlinearity is close to the bound or coincides with it.

Let n and m be integers, $-1 \leq m \leq n$.

DEFINITION 7.88. We denote by $N_{\max}(n,m)$ the maximum possible nonlinearity of an m-resilient Boolean function in $\mathcal{F}_n$.

It follows from Theorems 7.77, 7.80, and 7.86 that

1) $N_{\max}(n,m) \leq 2^{n-1}-2^{m+1}$ for $\operatorname{sut} f = m \leq n-2$;

2) $N_{\max}(n,m) \leq 2^{n-1}-2^{m+2}$ for $\operatorname{sut} f = m \leq n-3$, $\deg f + \operatorname{sut} f \leq n-2$;

3) if $N_{\max}(n,m) = 2^{n-1}-2^{m+1}$, then there exists a function $f \in \mathcal{F}_n$ such that $\operatorname{sut} f = m \leq n-2$, $\deg f + \operatorname{sut} f = n-1$;

4) $N_{\max}(n,m) \leq 2^{n-1}-2^{m+1}-2^{n/2-1}$ for $\operatorname{sut} f = m \leq n/2-2$.

Taking into account the comments about (-1)-resilient Boolean functions made in §7.1, we have

$$N_{\max}(n,-1) \leq 2^{n-1}-2^{n/2-1}.$$

This bound is attained for even n and maximum-nonlinear (bent) functions. Using certain results (see §5.3) on the covering radius of Reed–Muller codes of first order for odd $n \leq 7$, one can show that $N_{\max}(n,-1) = 2^{n-1}-2^{(n-1)/2}$ and that for even $n > 15$ the following inequality holds: $N_{\max}(n,-1) > 2^{n-1}-2^{(n-1)/2}$. For a balanced (0-resilient) function $f \in \mathcal{F}_n$ we have $N_f < 2^{n-1}-2^{n/2-1}$, and $N_{\max}(n,m) < 2^{n-1}-2^{n/2-1}$ for $m \geq 0$.

If f is an m-resilient function in $\mathcal{F}_n$, $m \geq n-2$, then according to Siegenthaler's inequality one has $\deg f \leq 1$ and $N_{\max}(n,m) = 0$. It is also known that $N_{\max}(n,n-3) = 2^{n-2}$ and $N_{\max}(n,n-4) = 2^{n-1}-2^{n-3}$ (see Theorem 7.94).

We give the exact values of the maximum nonlinearity for some small n and m: $N_{\max}(4,0) = 4$, $N_{\max}(5,-1) = N_{\max}(5,0) = N_{\max}(5,1) = 12$, $N_{\max}(6,0) = 26$, $N_{\max}(6,1) = N_{\max}(6,2) = 24$, $N_{\max}(7,-1) = N_{\max}(7,0) = N_{\max}(7,1) = 56$. All these values are obtained by concatenation of particular functions of maximum possible nonlinearity or by an exhaustive search.

If we characterize all existing methods for constructing Boolean functions with high (or maximal) nonlinearity, we must say that, as a rule, all of them are constructed by an iterative procedure that produces functions of many variables from functions of a smaller number of variables. Every iteration is a kind of operator acting on the truth tables of Boolean functions. In short, these methods can be characterized as "tabular-operator" methods.

We will give only a few methods for constructing resilient functions with high nonlinearity. It will be clear from the description of these methods why we have chosen them.

For the needs of the exposition that follows we recall some facts and ideas. A variable $x^{(i)}$ for a function $f(x^{(1)}, \dots, x^{(i-1)}, x^{(i)}, x^{(i+1)}, \dots, x^{(n)}) \in \mathcal{F}_n$ is called linear if $\deg(f, x^{(i)}) = 1$. If a variable $x^{(i)}$ is linear for a function f, then the function f can be represented as follows:

$$\begin{aligned} &f(x^{(1)}, \dots, x^{(i-1)}, x^{(i)}, x^{(i+1)}, \dots, x^{(n)}) \\ &\quad = g(x^{(1)}, \dots, x^{(i-1)}, x^{(i+1)}, \dots, x^{(n)}) \oplus x^{(i)}. \end{aligned}$$

A Boolean function $f(x^{(1)}, \dots, x^{(n)}) \in \mathcal{F}_n$ depends *quasi-linearly* on a pair of its variables $(x^{(i)}, x^{(j)})$, $1 \le i < j \le n$, if $f(\boldsymbol{x}') \ne f(\boldsymbol{x}'')$ for any two vectors $\boldsymbol{x}'$ and $\boldsymbol{x}''$ differing only in the ith and jth components. In this case one says that the pair of variables $(x^{(i)}, x^{(j)})$ is a *pair of quasi-linear variables* for the function f.

PROBLEM 7.89. Suppose that $f(x^{(1)}, \dots, x^{(n)}) \in \mathcal{F}_n$. Prove that the pair of variables $(x^{(i)}, x^{(j)})$, $1 \le i < j \le n$, is a pair of quasi-linear variables for the function f if and only if the function f can be written as

$$\begin{aligned} f(x^{(1)}, \dots, x^{(n)}) = g(&x^{(1)}, \dots, x^{(i-1)}, x^{(i+1)}, \dots, \\ &x^{(j-1)}, x^{(j+1)}, \dots, x^{(n)}, x^{(i)} \oplus x^{(j)}) \oplus x^{(i)}, \end{aligned} \tag{7.23}$$

where $N_f = 2N_g$.

LEMMA 7.90. *Suppose that $f_1, f_2 \in \mathcal{F}_n$ and there exist variables $x^{(i)}$ and $x^{(j)}$, $1 \le i < j \le n$, such that f_2 depends on the variables $x^{(i)}$ and $x^{(j)}$ linearly, and f_1 depends on the pairs of variables $(x^{(i)}, x^{(j)})$ quasi-linearly. Then for any affine function l, at least one of the functions $f_1 \oplus l$ and $f_2 \oplus l$ is balanced.*

PROOF. If l does not depend on the variable $x^{(i)}$ (or $x^{(j)}$), then $f_1 \oplus l$ is linearly dependent on the variable $x^{(i)}$ (or $x^{(j)}$), so it is balanced.

If l depends on both variables $x^{(i)}$ and $x^{(j)}$, i.e., $l = l' \oplus x^{(i)} \oplus x^{(j)}$, then, obviously, $f_2 \oplus l = f_2 \oplus l' \oplus x^{(i)} \oplus x^{(j)}$ is quasi-linearly dependent on the pairs of variables $(x^{(i)}, x^{(j)})$, and so it is balanced. □

LEMMA 7.91. *Suppose $f_1, f_2 \in \mathcal{F}_n$, $\operatorname{sut} f_1 = \operatorname{sut} f_2 = m$, $N_0 = \min(N_{f_1}, N_{f_2})$, and there exist variables $x^{(i)}$ and $x^{(j)}$ such that f_1 depends on the variables $x^{(i)}$, $x^{(j)}$ linearly and f_2 depends on the pairs of variables $(x^{(i)}, x^{(j)})$ quasi-linearly. Then the function*

$$\begin{aligned} &f_1'(x^{(1)}, \dots, x^{(n)}, x^{(n+1)}) \\ &\quad = (x^{(n+1)} \oplus 1) f_1(x^{(1)}, \dots, x^{(n)}) \oplus x^{(n+1)} f_2(x^{(1)}, \dots, x^{(n)}) \end{aligned} \tag{7.24}$$

is an m-resilient function in $\mathcal{F}_{n+1}$ with nonlinearity $N_{f_1'} \ge 2^{n-1} + N_0$, and the function

$$\begin{aligned} &f_2'(x^{(1)}, \dots, x^{(n)}, x^{(n+1)}, x^{(n+2)}) \\ &\quad = (x^{(n+1)} \oplus x^{(n+2)} \oplus 1) f_1(x^{(1)}, \dots, x^{(n)}) \\ &\quad\quad \oplus (x^{(n+1)} \oplus x^{(n+2)}) f_2(x^{(1)}, \dots, x^{(n)}) \oplus x^{(n+1)} \end{aligned} \tag{7.25}$$

is an $(m+1)$-resilient function in $\mathcal{F}_{n+2}$ with nonlinearity $N_{f_2'} \ge 2^n + 2N_0$. Besides, f_2' is quasi-linearly dependent on the pairs of variables $(x^{(n+1)}, x^{(n+2)})$.

PROOF. Since both functions f_1 and f_2 are m-resilient, Problem 7.66 implies that the function in (7.24) is also m-resilient. Put $l(\boldsymbol{x}) = l_{\boldsymbol{a},b}(\boldsymbol{x}) = \bigoplus_{s=1}^{n+1} a^{(t)}x^{(t)}$, $\boldsymbol{a} \in V_{n+1}$, $b \in \mathbb{F}_2$. Then

$$\operatorname{dist}(f_1', l) = \operatorname{dist}(f_1, l_{n+1}^0) + \operatorname{dist}(f_2, l_{n+1}^1) = \operatorname{wt}(f_1 \oplus l_{n+1}^0) + \operatorname{wt}(f_2 \oplus l_{n+1}^1).$$

It follows from Lemma 7.90 that at least one of the functions $f_1 \oplus l_{n+1}^0$ or $f_2 \oplus l_{n+1}^1$ is balanced. Taking this into account it is obvious that $N_{f_1'} \geq 2^{n-1} + N_0$.

Let us now turn our attention to the function in (7.25). From Problem 7.89 it follows that the function f_2' depends on the pairs of variables $(x^{(n+1)}, x^{(n+2)})$ quasi-linearly. Observe that

$$f_2'(x^{(1)}, \ldots, x^{(n+2)}) = \begin{cases} f_1(x^{(1)}, \ldots, x^{(n)}) \oplus x^{(n+1)} & \text{for } x^{(n+1)} \oplus x^{(n+2)} = 0, \\ f_2(x^{(1)}, \ldots, x^{(n)}) \oplus x^{(n+1)} & \text{for } x^{(n+1)} \oplus x^{(n+2)} = 1. \end{cases}$$

Problem 7.66 implies that $\operatorname{sut}(f_1 \oplus x^{(n+1)}) = \operatorname{sut}(f_2 \oplus x^{(n+1)}) = m+1$, and Problem 7.89 implies that $\min(N_{f_1 \oplus x^{(n+1)}}, N_{f_2 \oplus x^{(n+1)}}) = 2N_0$. Repeating the same argument for f_1' and for f_2', we find that f_2' is an $(m+1)$-resilient function and $N_{f_2'} \geq 2^n + 2N_0$. □

LEMMA 7.92. *Suppose that $f_{n,1}$ is an m-resilient Boolean function in $\mathcal{F}_n$, $N_{f_{n,1}} \geq N_0$, and $f_{n+1,2}$ is an $(m+1)$-resilient Boolean function in $\mathcal{F}_{n+1}$, $N_{f_{n+1,2}} \geq 2N_0$. If the function $f_{n+1,2}$ depends on some pairs of its variables $(x^{(i)}, x^{(j)})$ quasi-linearly, then there exists an $(m+2)$-resilient Boolean function $f_{n+3,1} \in \mathcal{F}_{n+3}$ with $N_{f_{n+3,1}} \geq 2^{n+1} + 4N_0$, and an $(m+3)$-resilient Boolean function $f_{n+4,2} \in \mathcal{F}_{n+4}$ with $N_{f_{n+4,2}} \geq 2^{n+2} + 8N_0$. Furthermore, the function $f_{n+4,2}$ depends on some of its pairs of variables quasi-linearly.*

PROOF. Suppose that $1 \leq i < j \leq n$. We put

$$\begin{aligned} f_1(x^{(1)}, \ldots, x^{(n+2)}) = f_{n,1}(x^{(1)}, \ldots, x^{(i-1)}, x^{(i+1)}, \ldots, x^{(j-1)}, \\ x^{(j+1)}, \ldots, x^{(n)}, x^{(n+1)}, x^{(n+2)}) \oplus x^{(i)} \oplus x^{(j)}, \end{aligned}$$

$$f_2(x^{(1)}, \ldots, x^{(n+2)}) = f_{n+1,2}(x^{(1)}, \ldots, x^{(n+1)}) \oplus x^{(n+2)}.$$

It is clear that the functions f_1 and f_2 are $(m+2)$-resilient in $\mathcal{F}_{n+2}$, so Problem 7.89 shows that $N_{f_1} \geq 4N_0$ and $N_{f_2} \geq 4N_0$. Besides, f_1 is linearly dependent on the variables $x^{(i)}$ and $x^{(j)}$ and f_2 depends on the pairs of variables $(x^{(i)}, x^{(j)})$ quasi-linearly. If we substitute f_1 and f_2 in (7.24) and (7.25) (replacing n by $n+2$), we get

$$\begin{aligned} f_1'(x^{(1)}, \ldots, x^{(n+2)}, x^{(n+3)}) &= (x^{(n+3)} \oplus 1) f_1(x^{(1)}, \ldots, x^{(n+2)}) \\ &\oplus x^{(n+3)} f_2(x^{(1)}, \ldots, x^{(n+2)}), \\ f_2'(x^{(1)}, \ldots, x^{(n+2)}, x^{(n+3)}, x^{(n+4)}) &= (x^{(n+3)} \oplus x^{(n+4)} \oplus 1) f_1(x^{(1)}, \ldots, x^{(n+2)}) \\ &\oplus (x^{(n+3)} \oplus x^{(n+4)}) f_2(x^{(1)}, \ldots, x^{(n+2)}). \end{aligned}$$

According to Lemma 7.91, we have constructed an $(m+2)$-resilient Boolean function $f_{n+3,1} = f_1' \in \mathcal{F}_{n+3}$ with $N_{f_{n+3,1}} \geq 2^{n+1} + 4N_0$ and an $(m+3)$-resilient Boolean function $f_{n+4,2} = f_2' \in \mathcal{F}_{n+4}$ with $N_{f_{n+4,2}} \geq 2^{n+2} + 8N_0$. Furthermore, the function $f_{n+4,2}$ depends on the pairs of variables $(x^{(n+3)}, x^{(n+4)})$ quasi-linearly. □

COROLLARY 7.93. *Let $m \leq n-2$, $f_{n,1}$ be an m-resilient Boolean function in $\mathcal{F}_n$, and let $N_{f_{n,1}} = 2^{n-1} - 2^{m+1}$, $f_{n+1,2}$ be an $(m+1)$-resilient Boolean function in $\mathcal{F}_{n+1}$, $N_{f_{n+1,2}} = 2^n - 2^{m+2}$. Suppose that the function $f_{n+1,2}$ depends on some pairs of variables $(x^{(i)}, x^{(j)})$, $1 \leq i < j \leq n$, quasi-linearly. Then there exists an $(m+2)$-resilient Boolean function $f_{n+3,1} \in \mathcal{F}_{n+3}$, $N_{f_{n+3,1}} = 2^{n+2} - 2^{m+3}$, and an $(m+3)$-resilient Boolean function $f_{n+4,2} \in \mathcal{F}_{n+4}$, with $N_{f_{n+4,2}} = 2^{n+3} - 2^{m+4}$. The function $f_{n+4,2}$ is quasi-linearly dependent on some pairs of variables.*

PROOF. Corollary 7.93 coincides with Lemma 7.92 for $N_0 = 2^{n-1} - 2^{m+1}$. According to Lemma 7.92 we can construct functions $f_{n+3,1}$ and $f_{n+4,2}$ with the necessary properties and nonlinearities $N_{f_{n+3,1}} \geq 2^{n+1} + 4N_0 = 2^{n+2} - 2^{m+3}$, $N_{f_{n+4,2}} \geq 2^{n+2} + 8N_0 = 2^{n+3} - 2^{m+4}$, respectively. Theorem 7.77 shows that the right-hand sides of the last two inequalities are also upper bounds for the nonlinearities of the corresponding functions. Therefore, we have the equations $N_{f_{n+3,1}} = 2^{n+2} - 2^{m+3}$, and $N_{f_{n+4,2}} = 2^{n+3} - 2^{m+4}$. □

In the following theorem the exact value of $N_{\max}(n, m)$ for a large range of values of $(m; n)$ is found.

THEOREM 7.94. *If $\frac{2n-7}{3} \leq m \leq n-2$, then $N_{\max}(n, m) = 2^{n-1} - 2^{m+1}$.*

PROOF. By virtue of Siegenthaler's inequality, any $(n-2)$-resilient function in $\mathcal{F}_n$ is affine. Thus, $N_{\max}(n, n-2) = 0$.

Put $f_{2,1} = x^{(1)}x^{(2)}$ and $f_{3,2} = x^{(1)}(x^{(2)} \oplus x^{(3)}) \oplus x^{(2)}$. These functions satisfy the hypothesis of Corollary 7.93 for $n = 2$, $m = -1$. According to this corollary we can construct functions $f_{5,1}$ and $f_{6,2}$ such that $f_{5,1}$ is a 1-resilient Boolean function in $\mathcal{F}_5$, with $N_{f_{5,1}} = 2^4 - 2^2 = 12$, and $f_{6,2}$ is a 2-resilient Boolean function in $\mathcal{F}_6$, with $N_{f_{6,2}} = 2^5 - 2^3 = 24$. Then we use the functions $f_{5,1}$ and $f_{6,2}$ in Corollary 7.93, and so on. This method allows one to construct, for any natural number k, $k \geq 3$, an m-resilient Boolean function $f_{l,1} \in \mathcal{F}_l$ with nonlinearity $2^{l-1} - 2^{m+1}$, where $l = 3k - 7$, $m = 2k - 7$, i.e., $k = l - m$.

Now let n and m be such that $\frac{2n-7}{3} \leq m \leq n-3$. It is clear that in this case $3(n-m)-7 \leq n$. Let us use the above method for $k = n-m$ to construct a function $f_{3(n-m)-7,1}(x^{(1)}, \ldots, x^{(3(n-m)-7)})$ which is $2(n-m)-7$-resilient with nonlinearity $N_{f_{3(n-m)-7,1}} = 2^{3(n-m)-8} - 2^{2(n-m)-6}$. We define the function as follows:

$$f(x^{(1)}, \ldots, x^{(n)}) = f_{3(n-m)-7,1}(x^{(1)}, \ldots, x^{(3(n-m)-7)}) \oplus \bigoplus_{s=3(n-m)-6}^{n} x^{(s)}.$$

According to Problem 7.21, the resiliency of this function equals $(2(n-m)-7) + (n - (3(n-m)-7)) = m$. The nonlinearity of the function f is, by virtue of Problem 2.120, equal to $2^{n-(3(n-m)-7)}(2^{3(n-m)-8} - 2^{2(n-m)-6}) = 2^{n-1} - 2^{m+1}$. Thus, for $\frac{2n-7}{3} \leq m \leq n-2$ we have constructed an m-resilient Boolean function with nonlinearity $2^{n-1} - 2^{m+1}$. □

REMARK 7.95. From Theorem 7.94 it follows that $N_{\max}(n, n-4) = 2^{n-1} - 2^{n-3}$ for $n \geq 5$.

PROBLEM 7.96. Prove that for any $g \in \mathcal{F}_n$ the function $f(x^{(1)}, \ldots, x^{(2n)}) = g(x^{(1)} \oplus x^{(n+1)}, x^{(2)} \oplus x^{(n+2)}, \ldots, x^{(n)} \oplus x^{(2n)}) \oplus x^{(1)} \oplus x^{(2)} \oplus \cdots \oplus x^{(n)} \in \mathcal{F}_{2n}$ is $(n-1)$-resilient.

Unfortunately, the construction of resilient functions with maximum possible nonlinearity does not solve the problem of constructing of Boolean functions with good cryptographic characteristics. In particular, one of the undesirable characteristics is that the constructed Boolean functions may depend linearly on many variables, and as a consequence, the corresponding cryptographic problems can be linearized. As the results show, resilient functions with maximum possible nonlinearity obtained by the known methods, as a rule, have variables on which they depend linearly. For example, the method described in Theorem 7.94, in the case $\frac{2n-7}{3} < m$ leads to the function that depends linearly on some variables. Only in a few cases one obtains functions that do not depend linearly on any of their variables.

We will describe without proofs another method for constructing resilient functions with maximum possible nonlinearity, which at the same time optimizes these functions relative to Siegenthaler's inequality in all variables. A more detailed exposition of these results can be found in [**386, 389**].

DEFINITION 7.97. A variable $x^{(i)}$ of a Boolean function $f \in \mathcal{F}_n$ is called *covering for the function* if any other variable of the function f is contained together with $x^{(i)}$ in a summand of maximum length in the algebraic normal form of the function f.

DEFINITION 7.98. A pair of variables $(x^{(i)}, x^{(j)})$, $1 \leq i < j \leq n$, of a Boolean function $f \in \mathcal{F}_n$ is called a *quasi-linear pair of covering variables for the function* if any other variable of the function f is contained together with $x^{(i)}$ in a summand of maximum length in the algebraic normal form of the function f (respectively, is contained together with $x^{(j)}$ in a summand of maximum length in the algebraic normal form of f).

LEMMA 7.99. *For integers k and n satisfying the inequalities $k \geq 3$ and $3k-7 \leq n < 3 \cdot 2^{k-2} - 2$, there exists a Boolean function $f_{n,1}^k \in \mathcal{F}_n$ with the following properties:*

1a) *$f_{n,1}^k$ is an $(n-k)$-resilient function;*
1b) *$N_{f_{n,1}^k} = 2^{n-1} - 2^{n-k+1}$;*
1c) *$\deg(f_{n,1}^k, x^{(i)}) = k-1$ for any variable $x^{(i)}$, $1 \leq i \leq n$;*
1d) *$f_{n,1}^k$ has a covering variable.*

For integers k and n that satisfy the inequalities $k \geq 3$ and $3k-7 < n \leq 3 \cdot 2^{k-2} - 2$ there exists a Boolean function $f_{n,2}^k \in \mathcal{F}_n$ with the following properties:

2a) *$f_{n,2}^k$ is an $(n-k)$-resilient function;*
2b) *$N_{f_{n,2}^k} = 2^{n-1} - 2^{n-k+1}$;*
2c) *$\deg(f_{n,2}^k, x^{(i)}) = k-1$ for any variable $x^{(i)}$, $1 \leq i \leq n$;*
2d) *$f_{n,2}^k$ has a quasi-linear pair of covering variables.*

THEOREM 7.100. *For integers m and n satisfying the inequalities $\frac{2n-7}{3} \leq m \leq n - \log_2 \frac{n+2}{3} - 2$ there exists an m-resilient Boolean function in $\mathcal{F}_n$ with nonlinearity $2^{n-1} - 2^{m+1}$ which satisfies Siegenthaler's inequality for any particular variable.*

7.6. Covering sequences of correlation-immune and resilient functions

We turn to the characterization of correlation-immune and resilient functions in terms of covering sequences. Before we present the basic concepts, we recall the definition of the Kravchuk polynomials in a form suitable for our purposes. Suppose that n is a fixed natural number, $i, l \leq n$, and let the vector $\boldsymbol{u} \in V_n$ have weight l. We obtain from Definition 4.41, replacing k by i and z by l:

$$P_i(l) = \sum_{j=0}^{i} \binom{l}{j} \binom{n-l}{i-j} = \sum_{\substack{\boldsymbol{x} \in V_n \\ \mathrm{wt}(\boldsymbol{x})=i}} (-1)^{\langle \boldsymbol{u}, \boldsymbol{x} \rangle}. \tag{7.26}$$

THEOREM 7.101. *For $f \in \mathcal{F}_n$ the equality* $\operatorname{cor} f = m$ *(respectively,* $\operatorname{sut} f = m$*) holds if and only if there exists a trivial (nontrivial) covering sequence $\boldsymbol{\theta} = (\theta_{\boldsymbol{a}})$, $\boldsymbol{a} \in V_n$, with level ρ such that*

$$m = \min \{\mathrm{wt}(\boldsymbol{b}) \mid \boldsymbol{b} \in V_n,\ \boldsymbol{b} \neq \boldsymbol{0},\ \overline{W}_{\boldsymbol{\theta}}(\boldsymbol{b}) = \overline{W}_{\boldsymbol{\theta}}(\boldsymbol{0}) - 2\rho\} - 1. \tag{7.27}$$

PROOF. Suppose there exists a covering sequence $\boldsymbol{\theta}$ for f, with level ρ, for which (7.27) is satisfied. According to Theorem 2.149,

$$V_n \setminus \mathrm{CI}_f \subseteq \{\boldsymbol{b} \in V_n \mid \overline{W}_{\boldsymbol{\theta}}(\boldsymbol{b}) = \overline{W}_{\boldsymbol{\theta}}(\boldsymbol{0}) - 2\rho\};$$

thus, (7.27) leads to the following conclusions:

1) $W_f(\boldsymbol{b}) = 0$ for all $\boldsymbol{b} \in V_n$, $1 \leq \mathrm{wt}(\boldsymbol{b}) \leq m$;
2) there exists a $\boldsymbol{b} \in V_n$, where $\mathrm{wt}(\boldsymbol{b}) = m+1$, such that $W_f(\boldsymbol{b}) \neq 0$.

This means that $\operatorname{cor} f = m$, and if $\rho \neq 0$, then $\operatorname{sut} f = m$.

We will now prove the converse. Suppose that $\operatorname{cor} f = m$. Then $V_n \setminus \mathrm{CI}_f \subseteq M = \{\boldsymbol{b} \in V_n \mid \mathrm{wt}(\boldsymbol{b}) \geq m+1\} \cup \{\boldsymbol{0}\}$. According to Theorem 2.149, the sequence $\boldsymbol{\theta}^M$ is covering for the function f, where

$$\overline{W}_{\boldsymbol{\theta}^M}(\boldsymbol{b}) = \begin{cases} 2^n & \text{if } \boldsymbol{b} \in M, \\ 0 & \text{if } \boldsymbol{b} \notin M, \end{cases}$$

whence (7.27) holds.

The argument for the case $\operatorname{sut} f = m$ is similar. □

It follows from the proof of Theorem 7.101 that the same covering sequence can be taken for any correlation-immune function of order m (respectively, any m-resilient function). Since such a sequence may be quite complex, it makes sense to try to find simpler covering sequences for individual correlation-immune functions of order m (resp., m-resilient functions). The covering sequence $\boldsymbol{\theta}$, whose Fourier transform assumes the constant value r on the set of vectors of weight greater than m and value $r_i \neq r$ on the set of vectors of weight $i \leq m$, can be expressed in terms of Kravchuk polynomials as follows:

$$\theta_{\boldsymbol{a}} = \sum_{i \leq m} \sum_{\substack{\boldsymbol{c} \in V_n \\ \mathrm{wt}(\boldsymbol{c})=i}} r_i (-1)^{\langle \boldsymbol{a}, \boldsymbol{c} \rangle} + \sum_{\substack{\boldsymbol{c} \in V_n \\ \mathrm{wt}(\boldsymbol{c})>m}} r(-1)^{\langle \boldsymbol{a}, \boldsymbol{c} \rangle} = \sum_{i \leq m} r_i P_i(l) + r \sum_{i>m} P_i(l),$$

where $l = \mathrm{wt}(\boldsymbol{b})$.

Now we try to use the characterization obtained in Theorem 7.101 for distinguishing certain subclasses of resilient functions. These subclasses allow simple covering sequences, which makes it possible to find the order of resilience as well as upper bounds for algebraic degrees of nonlinearity. We recall that the support of a vector $\boldsymbol{x} \in V_n$ is

$$\mathrm{supp}(\boldsymbol{x}) = \{i \in 1, 2, \ldots, n \mid x^{(i)} = 1\}.$$

LEMMA 7.102. *Suppose that* $f \in \mathcal{F}_n$ *and* E *is a set of vectors in* V_n *satisfying the following condition: For any pair of different vectors* $\boldsymbol{x}$ *and* $\boldsymbol{y}$ *in* E, $\operatorname{supp}(\boldsymbol{x}) \cap \operatorname{supp}(\boldsymbol{y}) = \varnothing$.

If the indicator of the set E *is a nontrivial covering sequence of the function* f *with level* ρ, *then* $\operatorname{sut} f \geq \rho - 1$.

PROOF. We denote by $\boldsymbol{\theta}$ the indicator of the set E ($\theta_{\boldsymbol{a}} = 1$ if $\boldsymbol{a} \in E$, and $\theta_{\boldsymbol{a}} = 0$ otherwise). Since f admits $\boldsymbol{\theta}$ as a nontrivial covering sequence of level ρ, Corollary 2.151 implies that the Boolean function $f \oplus l_{\boldsymbol{b},0}$ is balanced for any vector $\boldsymbol{b} \in V_n$ such that $\sum_{\boldsymbol{a} \in H_{\boldsymbol{b}}} \theta_{\boldsymbol{a}} \neq \rho$ (where $H_{\boldsymbol{b}}$ is the affine hyperplane defined by the equation $\langle \boldsymbol{a}, \boldsymbol{b} \rangle = 1$). We have

$$\sum_{\boldsymbol{a} \in H_{\boldsymbol{b}}} \theta_{\boldsymbol{a}} = \#(E \cap H_{\boldsymbol{b}}) \leq \#\{\boldsymbol{a} \in E \mid \operatorname{supp}(\boldsymbol{a}) \cap \operatorname{supp}(\boldsymbol{b}) \neq \varnothing\} \leq \operatorname{wt}(\boldsymbol{b}).$$

The last inequality follows from the hypothesis that $\operatorname{supp}(\boldsymbol{x}) \cap \operatorname{supp}(\boldsymbol{y}) = \varnothing$ for any $\boldsymbol{x}, \boldsymbol{y} \in E$, $\boldsymbol{x} \neq \boldsymbol{y}$. Thus, if $\operatorname{wt}(\boldsymbol{b}) \leq \rho - 1$, then $\sum_{\boldsymbol{a} \in H_{\boldsymbol{b}}} \theta_{\boldsymbol{a}} \neq \rho$. Therefore, if $\operatorname{wt}(\boldsymbol{b}) \leq \rho - 1$, then $f \oplus l_{\boldsymbol{b},0}$ is a balanced function and $W_f(\boldsymbol{b}) = 0$, i.e., f is a correlation-immune function of order $\rho - 1$. As $\rho \neq 0$, f is a $(\rho - 1)$-resilient function according to Corollary 2.152. □

DEFINITION 7.103. A covering sequence $\boldsymbol{\theta} = (\theta_{\boldsymbol{a}})$, $\boldsymbol{a} \in V_n$ of a function f is called *simple* if $\theta_{\boldsymbol{a}} \in \{0, 1\}$ for any $\boldsymbol{a} \in V_n$ and $\{\boldsymbol{a} \in V_n \mid \theta_{\boldsymbol{a}} = 1\} \subseteq \{\boldsymbol{e}_1, \ldots, \boldsymbol{e}_n\}$.

LEMMA 7.104. *A Boolean function* $f \in \mathcal{F}_n$ *allows a simple covering sequence* $\boldsymbol{\theta} = I_E$ *of level* ρ, *where* $E = \{\boldsymbol{e}_1, \boldsymbol{e}_2, \ldots, \boldsymbol{e}_n\}$, *if and only if the function* f *is* ρ*-regular (see Definition* 2.21*).*

PROOF. Let $\boldsymbol{\theta} = I_E$ be a simple covering sequence of level ρ for the function f. Then $\sum_{\boldsymbol{a} \in V_n} \theta_{\boldsymbol{a}} D_{\boldsymbol{a}} f = \sum_{i=1}^{n} D_{\boldsymbol{e}_i} f = \rho$. Therefore, for any vector $\boldsymbol{x} \in V_n$ we have

$$\sum_{i=1}^{n} [f(\boldsymbol{x}) \oplus f(\boldsymbol{x} \oplus \boldsymbol{e}_i)] = \rho. \tag{7.28}$$

According to (7.28), for any vector $\boldsymbol{x} \in V_n$ such that $f(\boldsymbol{x}) = 0$, we see that the value of $f(\boldsymbol{y})$ is equal to 1 only for ρ vectors $\boldsymbol{y}$ at distance 1 from $\boldsymbol{x}$. Similarly, for any vector $\boldsymbol{x} \in V_n$ such that $f(\boldsymbol{x}) = 1$, (7.28) implies that only for ρ vectors $\boldsymbol{y} \in V_n$ at distance 1 from $\boldsymbol{x}$ the value of $f(\boldsymbol{y})$ is equal to 0. Therefore, f is a ρ-regular function.

Now we prove the converse. Suppose that f is a ρ-regular function. We put $\boldsymbol{\theta} = I_E$. Then it is easy to see from (7.28) that the sequence $\boldsymbol{\theta}$ defined in this way is a simple covering sequence of level ρ for the function f. □

Now look at the construction of codes in Lemma 7.104. Suppose that $\boldsymbol{C}_1 = \{\boldsymbol{x} \in V_n \mid f(\boldsymbol{x}) = 1\}$ and $\boldsymbol{C}_2 = \{\boldsymbol{x} \in V_n \mid f(\boldsymbol{x}) = 0\}$ are nonlinear codes (see [**241**]) in V_n. A Boolean function f is ρ-regular if and only if for any vector $\boldsymbol{u} \in \boldsymbol{C}_i$, $i = 1, 2$, we have

$$\#\{\boldsymbol{x} \in \boldsymbol{C}_1 \mid \operatorname{dist}(\boldsymbol{u}, \boldsymbol{x}) = 1\} = n - \rho, \quad \#\{\boldsymbol{y} \in \boldsymbol{C}_2 \mid \operatorname{dist}(\boldsymbol{u}, \boldsymbol{y}) = 1\} = n - \rho.$$

Further on, for convenience, we will use the notation $\widetilde{E} = \bigcup_{\boldsymbol{x} \in E} \operatorname{supp}(\boldsymbol{x})$ for the set of indices of nonzero components of vectors in the set $E = \{\boldsymbol{e}_i \in V_n \mid \theta_{\boldsymbol{e}_i} = 1\}$ for the simple covering sequence $\boldsymbol{\theta} = I_E$.

PROBLEM 7.105. Prove that a Boolean function $f \in \mathcal{F}_n$ allows a simple covering sequence $\boldsymbol{\theta} = I_E$ ($E = \{\boldsymbol{e}_i \in V_n \mid \theta_{\boldsymbol{e}_i} = 1\}$) of level ρ if and only if any subfunction f obtained by fixing the variables $x^{(j)}$, $j \in \{1, 2, \dots, n\} \setminus \widetilde{E}$, is ρ-regular.

EXAMPLE 7.106. Suppose that a Boolean function f is given as a linear branching (see §2.5), i.e., that

$$f(\boldsymbol{z}) = f(\boldsymbol{x}, \boldsymbol{y}) = \langle \boldsymbol{x}, \Phi(\boldsymbol{y}) \rangle \oplus h(\boldsymbol{y}),$$

where $\boldsymbol{z} = (\boldsymbol{x}, \boldsymbol{y}) \in V_n$, $\boldsymbol{x} \in V_t$, $\boldsymbol{y} \in V_s$, $n = t + s$, $\Phi = (\varphi_1, \varphi_2, \dots, \varphi_t) \in \mathcal{F}_{s,t}$, $h \in \mathcal{F}_s$. Assume that for the mapping Φ the condition $\mathrm{wt}(\Phi(\boldsymbol{y})) = \rho$ holds for any $\boldsymbol{y} \in V_s$. Then the function f allows a simple covering sequence $\boldsymbol{\theta} = I_E$, where $E = \{(\boldsymbol{e}_1, \mathbf{0}), (\boldsymbol{e}_2, \mathbf{0}), \dots, (\boldsymbol{e}_t, \mathbf{0})\}$. It is not hard to show that

$$\sum_{j=1}^{t} D_{(\boldsymbol{e}_j, \mathbf{0})} f(\boldsymbol{x}) = \sum_{j=1}^{t} \varphi_j(\boldsymbol{y}) = \rho.$$

We will now use the characterization of regular functions to make statements that enable us to evaluate the algebraic degree of a function having simple covering sequences.

LEMMA 7.107. *A Boolean function $f \in \mathcal{F}_n$ is ρ-regular, $1 \le \rho \le n - 1$, if and only if*

$$V_n \setminus \mathrm{CI}_f \subseteq \{\boldsymbol{a} \in V_n \mid \mathrm{wt}(\boldsymbol{a}) = \rho\}. \tag{7.29}$$

This condition is equivalent to the following two conditions:

1) *$f(\boldsymbol{x})$ is a $(\rho - 1)$-resilient function;*
2) *$f(\boldsymbol{x}) \oplus x^{(1)} \oplus \cdots \oplus x^{(n)}$ is an $(n - \rho - 1)$-resilient function.*

PROOF. Let $f \in \mathcal{F}_n$ be a ρ-regular function. Then according to Lemma 7.104 the function f allows a simple covering sequence $\boldsymbol{\theta} = I_E$ of level ρ, where $E = \{\boldsymbol{e}_1, \boldsymbol{e}_2, \dots, \boldsymbol{e}_n\}$. Then for any vector $\boldsymbol{b} \in V_n$ we have

$$\overline{W}_{\boldsymbol{\theta}}(\boldsymbol{b}) = \sum_{\boldsymbol{a} \in V_n} \theta_{\boldsymbol{a}} (-1)^{\langle \boldsymbol{a}, \boldsymbol{b} \rangle} = \sum_{i=1}^{n} (-1)^{b^{(i)}} = n - 2\,\mathrm{wt}(\boldsymbol{b}).$$

By Theorem 2.149, $\overline{W}_{\boldsymbol{\theta}}(\boldsymbol{u})$ is constant on the set $V_n \setminus \mathrm{CI}_f$, with the value $\overline{W}_{\boldsymbol{\theta}}(\mathbf{0}) - 2\rho = n - 2\rho$. It is clear that this is possible only if $\mathrm{wt}(\boldsymbol{b}) = \rho$. Therefore, (7.29) holds.

We will now prove the converse. Suppose that (7.29) holds and we are given a sequence $\boldsymbol{\theta} = I_E$, where $E = \{\boldsymbol{e}_1, \boldsymbol{e}_2, \dots, \boldsymbol{e}_n\}$. Then for any $\boldsymbol{b} \in V_n \setminus \mathrm{CI}_f$ we have

$$\overline{W}_{\boldsymbol{\theta}}(\boldsymbol{b}) = \sum_{\boldsymbol{a} \in V_n} \theta_{\boldsymbol{a}} (-1)^{\langle \boldsymbol{a}, \boldsymbol{b} \rangle} = \sum_{i=1}^{n} (-1)^{b^{(i)}} = n - 2\,\mathrm{wt}(\boldsymbol{b}) = n - 2\rho = \overline{W}_{\boldsymbol{\theta}}(\mathbf{0}) - 2\rho.$$

Therefore, in view of Theorem 2.149, the function f allows $\boldsymbol{\theta}$ as a simple covering sequence of level ρ, and f is a ρ-regular function.

It is clear that (7.29) is equivalent to the equality $W_f(\boldsymbol{b}) = 0$ for any vector $\boldsymbol{b}$ of weight at most $\rho - 1$ (i.e., to f being $(\rho - 1)$-resilient) and for any vector $\boldsymbol{b}$ of weight at least $\rho + 1$. We put $g(\boldsymbol{x}) = f(\boldsymbol{x}) \oplus x^{(1)} \oplus \cdots \oplus x^{(n)}$. Then $W_g(\boldsymbol{b}) = W_f(\boldsymbol{b} \oplus \mathbf{1})$, and the vector $\boldsymbol{b}$ has weight greater than ρ if and only if the vector $\boldsymbol{b} \oplus \mathbf{1}$ has weight less than $n - \rho$. □

LEMMA 7.108. *The algebraic degree of any ρ-regular function f ($1 \le \rho \le n-1$) in $\mathcal{F}_n$ satisfies the inequality*

$$\deg f \le \min(\rho, n-\rho).$$

If $E \subseteq \{\boldsymbol{e}_1, \dots, \boldsymbol{e}_n\}$ and ρ is a natural number $1 \le \rho \le \#E - 1$, then the algebraic degree of any function f allowing a simple covering sequence $\boldsymbol{\theta} = I_E$ of level ρ satisfies the inequality

$$\deg f \le \min(n - \#E + \rho, n - \rho).$$

PROOF. For a ρ-regular function the lemma follows from Lemma 7.107 and the Siegenthaler inequality.

In the general case put by $l = n - \#E$. Without loss of generality we may assume that $E = \{\boldsymbol{e}_{l+1}, \dots, \boldsymbol{e}_n\}$. Then for a vector $\boldsymbol{x} \in V_n$ we have $\boldsymbol{x}^\top = (\boldsymbol{x}'^\top, \boldsymbol{x}''^\top)$, where $\boldsymbol{x}' = (x^{(1)}, \dots, x^{(l)})^\top$, $\boldsymbol{x}'' = (x^{(l+1)}, \dots, x^{(n)})^\top$. For an arbitrary $\boldsymbol{u} = (u^{(1)}, \dots, u^{(l)})^\top \in V_l$ we have

$$f_{1,\dots,l}^{u^{(1)},\dots,u^{(l)}}(\boldsymbol{x}'') = f(u^{(1)}, \dots, u^{(l)}, x^{(l+1)}, \dots, x^{(n)}) = f_{1,\dots,l}^{\boldsymbol{u}}(\boldsymbol{x}'').$$

Now we can represent the function f as follows:

$$f(\boldsymbol{x}) = \bigoplus_{\boldsymbol{u} \in V_l} f_{1,\dots,l}^{\boldsymbol{u}}(\boldsymbol{x}'') \prod_{i=1}^{l} (x^{(i)} \oplus u^{(i)} \oplus 1).$$

According to Problem 7.105, for any $\boldsymbol{u} \in V_l$ the subfunction $f_{1,\dots,l}^{\boldsymbol{u}'}(\boldsymbol{x}'')$ is ρ-regular. The first part of this lemma shows that $\deg f_{1,\dots,l}^{\boldsymbol{u}'}(\boldsymbol{x}'') \le \min(\rho, n - l - \rho)$ for any $\boldsymbol{u} \in V_l$. Therefore, for the function f,

$$\deg f \le l + \min(\rho, n - l - \rho) = \min(l + \rho, n - \rho) = \min(n - \#E + \rho, n - \rho). \quad \square$$

We proceed by stating results which make it possible to evaluate the nonlinearity of regular Boolean functions and functions that allow simple covering sequences.

THEOREM 7.109. *The nonlinearity of any Boolean function $f \in \mathcal{F}_n$ that allows for a simple covering sequence $\boldsymbol{\theta} = I_E$ of level ρ satisfies the inequality*

$$N_f \le 2^{n-1} - \frac{2^{\frac{n+\#E}{2}-1}}{\sqrt{\binom{\#E}{\rho}}}. \tag{7.30}$$

The nonlinearity of any ρ-regular function $f \in \mathcal{F}_n$ satisfies the inequality

$$N_f \le 2^{n-1} - \max\left\{\frac{2^{n-1}}{\sqrt{\binom{n}{\rho}}}, 2^{\rho}, 2^{n-\rho}\right\}. \tag{7.31}$$

PROOF. Let $f \in \mathcal{F}_n$ be a Boolean function allowing a simple covering sequence $\boldsymbol{\theta} = I_E$ of level ρ. Then according to Corollary 2.151, for any $\boldsymbol{b} \in V_n$ such that $\sum_{\boldsymbol{a} \in H_{\boldsymbol{b}}} \theta_{\boldsymbol{a}} \ne \rho$, one has $W_f(\boldsymbol{b}) = 0$. We put $B = \{\boldsymbol{b} \in V_n \mid \sum_{\boldsymbol{a} \in H_{\boldsymbol{b}}} \theta_{\boldsymbol{a}} = \rho\}$. Parseval's identity implies that

$$\sum_{\boldsymbol{b} \in B} W_f^2(\boldsymbol{b}) = 2^{2n}.$$

Hence,

$$\max_{\boldsymbol{b} \in B} W_f^2(\boldsymbol{b}) \ge \frac{2^{2n}}{\#B}.$$

The sum $\sum_{\boldsymbol{a}\in H_{\boldsymbol{b}}} \theta_{\boldsymbol{a}}$ is equal to the cardinality of the set $E \cap H_{\boldsymbol{b}} = \{\boldsymbol{e}_i \in E \mid b^{(i)} = 1\}$. As $\sum_{\boldsymbol{a}\in H_{\boldsymbol{b}}} \theta_{\boldsymbol{a}} = \rho$, from $\#E$ components $b^{(i)}$ of the vector $\boldsymbol{b}$ such that $i \in E$, exactly ρ must equal 1 and the rest must equal 0. The remaining $n - \#E$ components of the vector $\boldsymbol{b}$ can be defined arbitrarily. Thus, the number of such vectors $\boldsymbol{b}$ is equal to $\binom{\#E}{\rho} \cdot 2^{n-\#E}$, whence

$$\frac{2^{2n}}{\#B} = \frac{2^{2n}}{\binom{\#E}{\rho} \cdot 2^{n-\#E}}$$

and the inequality (7.30) holds.

Suppose now that $f \in \mathcal{F}_n$ is a ρ-regular Boolean function. According to Lemma 7.104, the function f allows a simple covering sequence $\boldsymbol{\theta} = I_E$ of level ρ, where $E = \{\boldsymbol{e}_1, \boldsymbol{e}_2, \dots, \boldsymbol{e}_n\}$, i.e., $\#E = n$. Then the inequality (7.30) becomes

$$N_f \le 2^{n-1} - \frac{2^{n-1}}{\sqrt{\binom{n}{\rho}}}.$$

On the other hand, Lemma 7.107 for the ρ-regular function f implies that

$$N_f \le 2^{n-1} - \max(2^{\rho}, 2^{n-\rho}).$$

The last two inequalities together give the inequality (7.31). □

REMARK 7.110. Using the Stirling formula (see, for example, [**123**]), it is easy to show that $2^{n-1}/\sqrt{\binom{n}{\rho}}$ is greater than $2^{\max(\rho,n-\rho)}$ when ρ is close to $n/2$ (the minimum value for this case is $n = 10$, $\rho = 5$), and that in the other cases $2^{n-1}/\sqrt{\binom{n}{\rho}}$ is less than $2^{\max(\rho,n-\rho)}$.

Let g and h be Boolean functions in $\mathcal{F}_n$ and $\mathcal{F}_l$ respectively. Suppose that both allow simple covering sequences. For convenience we label the vectors in the canonical bases of these spaces as follows: $\{\boldsymbol{e}_1, \dots, \boldsymbol{e}_n\}$ and $\{\boldsymbol{e}_{n+1}, \dots, \boldsymbol{e}_{n+l}\}$. Then the covering sequences of the functions g and h are $\boldsymbol{\theta} = I_E$ ($E \subseteq \{\boldsymbol{e}_1, \dots, \boldsymbol{e}_n\}$) and $\boldsymbol{\theta}' = I_{E'}$ ($E' \subseteq \{\boldsymbol{e}_{n+1}, \dots, \boldsymbol{e}_{n+l}\}$). A function $f \in \mathcal{F}_{n+l}$ of the form

$$f(\boldsymbol{x}, \boldsymbol{y}) = g(\boldsymbol{x}) \oplus h(\boldsymbol{y}), \quad (\boldsymbol{x}, \boldsymbol{y}) \in V_{n+l}, \quad \boldsymbol{x} \in V_n, \quad \boldsymbol{y} \in V_l,$$

allows a simple covering sequence $\boldsymbol{S} = I_M$, where, for example, $S_1 = E$, $S_2 = E'$, and $S_3 = E \cup E'$. This can be verified by a simple check.

LEMMA 7.111. *Suppose that $g, h \in \mathcal{F}_n$ allow the same simple covering sequence $\boldsymbol{\theta} = I_E$ of level ρ. Define a Boolean function $f \in \mathcal{F}_{n+1}$ by the formula*

$$f(x^{(1)}, \dots, x^{(n)}, x^{(n+1)}) = x^{(n+1)} g(x^{(1)}, \dots, x^{(n)}) \oplus (x^{(n+1)} \oplus 1) h(x^{(1)}, \dots, x^{(n)}).$$

Then the function f allows the same simple covering sequence $\boldsymbol{\theta} = I_E$ of level ρ. Furthermore,

– *If g and h are such that for any vector $\boldsymbol{u} \in V_n$ at least one of the Walsh–Hadamard coefficients $W_g(\boldsymbol{u})$, $W_h(\boldsymbol{u})$ is zero, then $N_f = 2^{n-1} + \min(N_g, N_h)$. A sufficient condition for this is that the functions g and h have another common covering sequence with different levels (for example, this second covering sequence may be trivial for g and nontrivial for h).*

– *If the algebraic normal forms of the functions g and h do not have equal monomials of maximum length, then $\deg f = 1 + \max(\deg g, \deg h)$.*

PROOF. For any $i \leq n$ we have

$$\begin{aligned} &D_{\boldsymbol{e}_i} f(x^{(1)}, \ldots, x^{(n)}, x^{(n+1)}) \\ &\quad = x^{(n+1)} D_{\boldsymbol{e}_i} g(x^{(1)}, \ldots, x^{(n)}) \oplus (x^{(n+1)} \oplus 1) D_{\boldsymbol{e}_i} h(x^{(1)}, \ldots, x^{(n)}). \end{aligned} \tag{7.32}$$

Since g and h allow the same simple covering sequence $\boldsymbol{\theta} = I_E$ of level ρ, i.e., $\sum_{\boldsymbol{e}_i \in E} D_{\boldsymbol{e}_i} g = \sum_{\boldsymbol{e}_i \in E} D_{\boldsymbol{e}_i} h = \rho$, we obtain from (7.32) that $\sum_{\boldsymbol{e}_i \in E} D_{\boldsymbol{e}_i} f = \rho$. For any $\boldsymbol{u} \in V_{n+1}$ we have

$$W_f(u^{(1)}, \ldots, u^{(n)}, u^{(n+1)}) = (-1)^{u^{(n+1)}} W_g(u^{(1)}, \ldots, u^{(n)}) + W_h(u^{(1)}, \ldots, u^{(n)}).$$

If for any vector $\boldsymbol{u}$ at least one of the coefficients $W_g(u^{(1)}, \ldots, u^{(n)})$ or $W_h(u^{(1)}, \ldots, u^{(n)})$ is zero, then

$$\begin{aligned} &\max_{\boldsymbol{u} \in V_{n+1}} |W_f(u^{(1)}, \ldots, u^{(n)}, u^{(n+1)})| \\ &\quad = \max_{(u^{(1)}, \ldots, u^{(n)})^\top \in V_n} \max(|W_g(u^{(1)}, \ldots, u^{(n)})|, |W_h(u^{(1)}, \ldots, u^{(n)})|). \end{aligned}$$

Using the expression for nonlinearity of a Boolean function, we see that

$$2^{n+1} - 2N_f = \max(2^n - 2N_g, 2^n - 2N_h),$$

whence

$$N_f = 2^{n-1} + \min(N_g, N_h).$$

If the functions g and h allow the same covering sequence with different levels, then by Theorem 2.149 the sets $V_n \setminus \mathrm{CI}_g$ and $V_n \setminus \mathrm{CI}_h$ do not intersect. Therefore, for any $\boldsymbol{u} \in V_n$ at least one of the coefficients $W_g(\boldsymbol{u})$ or $W_h(\boldsymbol{u})$ is zero.

The last statement of the lemma is easy. □

LEMMA 7.112. *Suppose that $g \in \mathcal{F}_n$ and that f allows a simple covering sequence $\boldsymbol{\theta} = I_E$, where $E \subseteq \{\boldsymbol{e}_1, \ldots, \boldsymbol{e}_{n-1}\}$, of level ρ. Then the Boolean function*

$$f(x^{(1)}, \ldots, x^{(n)}, x^{(n+1)}) = x^{(n+1)} \oplus g(x^{(1)}, \ldots, x^{(n)} \oplus x^{(n+1)}) \in \mathcal{F}_{n+1}$$

allows a simple covering sequence $\boldsymbol{\theta}' = I_{E \cup \{\boldsymbol{e}_n, \boldsymbol{e}_{n+1}\}}$. The level of $\boldsymbol{\theta}'$ is equal to $\rho + 1$. Furthermore, $N_f = 2N_g$ and $\deg f = \deg g$.

PROOF. We have $D_{\boldsymbol{e}_{n+1}} f = 1 \oplus D_{\boldsymbol{e}_n} f = 1 - D_{\boldsymbol{e}_n} f$, whence $D_{\boldsymbol{e}_n} f + D_{\boldsymbol{e}_{n+1}} f = 1$. It follows from the hypothesis of the lemma that $\sum_{\boldsymbol{e}_j \in E} D_{\boldsymbol{e}_j} f = \rho$ and $E \subseteq \{\boldsymbol{e}_1, \ldots, \boldsymbol{e}_{n-1}\}$; hence,

$$\sum_{\boldsymbol{e}_j \in E} D_{\boldsymbol{e}_j} f + D_{\boldsymbol{e}_n} f + D_{\boldsymbol{e}_{n+1}} f = \sum_{\boldsymbol{e}_j \in E \cup \{\boldsymbol{e}_n, \boldsymbol{e}_{n+1}\}} D_{\boldsymbol{e}_j} f = \rho + 1.$$

The equality $N_f = 2N_g$ holds because f can be obtained from g by an affine substitution of variables and addition of an affine function. The last statement of the lemma is obvious. □

DEFINITION 7.113. A Boolean function $f \in \mathcal{F}_n$ is called *weakly nondegenerate* if $D_{\boldsymbol{e}_i} f \not\equiv 0$ for any $i = 1, 2, \ldots, n$.

THEOREM 7.114. *For any $\rho \geq 2$, put $n(\rho) = 3 \cdot 2^{\rho-1} - 2$. There exists a ρ-regular weakly nondegenerate Boolean function $f_\rho \in \mathcal{F}_{n(\rho)}$ such that $N_{f_\rho} = 2^{n(\rho)-1} - 2^{n(\rho)-\rho}$ and $\deg f_\rho = \rho$.*

PROOF. We will use induction on ρ. For $\rho = 2$, $n(2) = 4$ we can define a function

$$f_2(x^{(1)}, \dots, x^{(4)}) = (x^{(1)} \oplus x^{(2)})(x^{(3)} \oplus x^{(4)}) \oplus x^{(1)} \oplus x^{(3)}.$$

This is a weakly nondegenerate 2-regular function, because $D_{\mathbf{e}_1} f_2 + D_{\mathbf{e}_2} f_2 = D_{\mathbf{e}_3} f_2 + D_{\mathbf{e}_4} f_2 = 1$, and its nonlinearity coincides with the nonlinearity of the function $x^{(1)}x^{(3)}$, i.e., is equal to 4. Suppose that two (not necessarily different) functions $g_\rho(x^{(1)}, \dots, x^{(n(\rho))})$ and $h_\rho(x^{(n(\rho)+1)}, \dots, x^{(2n(\rho))})$ satisfying the conditions of the theorem are constructed, and let their simple covering sequences be $\boldsymbol{\theta} = I_E$ and $\boldsymbol{\theta}' = I_{E'}$, respectively. Substitute

$$\begin{aligned} f_{\rho+1}&(x^{(1)}, \dots, x^{(2n(\rho)+2)}) \\ &= x^{(2n(\rho)+1)} \oplus (x^{(2n(\rho)+1)} \oplus x^{(2n(\rho)+2)}) g_\rho(x^{(1)}, \dots, x^{(n(\rho))}) \\ &\quad \oplus (x^{(2n(\rho)+1)} \oplus x^{(2n(\rho)+2)} \oplus 1) h_\rho(x^{(n(\rho)+1)}, \dots, x^{(2n(\rho))}). \end{aligned}$$

The functions $g_\rho(x^{(1)}, \dots, x^{(n(\rho))})$ and $h_\rho(x^{(n(\rho)+1)}, \dots, x^{(2n(\rho))})$ regarded as functions of the variables $x^{(1)}, \dots, x^{(2n(\rho)+2)}$ allow a simple covering sequence $\boldsymbol{\Lambda} = I_{E \cup E'}$ of level ρ, and their nonlinearities are equal to $2^{n(\rho)}$ times the nonlinearity of the functions g_ρ and h_ρ. By virtue of Lemmas 7.111 and 7.112 we see that $f_{\rho+1}$ is a $(\rho+1)$-regular weakly nondegenerate function of $2n(\rho) + 2 = 2 \cdot (3 \cdot 2^{\rho-1} - 2) + 2 = 3 \cdot 2^\rho - 2 = n(\rho+1)$ variables. Besides, $g_\rho(x^{(1)}, \dots, x^{(n(\rho))})$ and $h_\rho(x^{(n(\rho)+1)}, \dots, x^{(2n\rho)})$ allow the same simple covering sequence I_E with different levels. Lemma 7.111 implies that the function $f_{\rho+1}$ has nonlinearity $2[2^{2n(\rho)-1} + 2^{n(\rho)}(2n(\rho) - 1 - 2^{n(\rho)-\rho})] = 2^{2n(\rho)+1} - 2^{2n(\rho)-\rho+1} = 2^{n(\rho+1)-1} - 2^{n(\rho)-\rho-1}$ and algebraic degree $\rho + 1$. □

The expression for $n(\rho)$ shows that ρ is much smaller than $n(\rho)$. The difference between these parameters can be decreased as shown in the following theorem.

THEOREM 7.115. *For any $\rho \geq 2$, put $n'(\rho) = 4\rho - 4$. There exists a ρ-regular weakly nondegenerate Boolean function $f'_\rho \in \mathcal{F}_{n'(\rho)}$ such that $N_{f'_\rho} = 2^{n'(\rho)-1} - 2^{n'(\rho)-\rho}$ and $\deg f'_\rho = \rho$.*

PROOF. We will use induction on ρ to construct a sequence of functions such that $D_{\mathbf{e}_{n'(\rho)-1}} f' + D_{\mathbf{e}_{n'(\rho)}} f' = 1$. For $\rho = 2$ we can construct the function

$$f'_2(x^{(1)}, \dots, x^{(4)}) = (x^{(1)} \oplus x^{(2)})(x^{(3)} \oplus x^{(4)}) \oplus x^{(1)} \oplus x^{(3)}.$$

Suppose that the function $f'_2(x^{(1)}, \dots, x^{(n'(\rho))})$ is constructed and satisfies the conditions of the theorem. We set

$$\begin{aligned} &f'_{\rho+1}(x^{(1)}, \dots, x^{(n'(\rho)+4)}) \\ &= x^{(n'(\rho)+3)} \oplus (x^{(n'(\rho)+3)} \oplus x^{(n'(\rho)+4)}) f'_\rho(x^{(1)}, \dots, x^{(n'(\rho)-2)}, x^{(n'(\rho)+1)}, x^{(n'(\rho)+2)}) \\ &\quad \oplus (x^{(n'(\rho)+3)} \oplus x^{(n'(\rho)+4)} \oplus 1) f'_\rho(x^{(1)}, \dots, x^{(n'(\rho))}). \end{aligned}$$

Then

$$\begin{aligned} D_{\mathbf{e}_{n'(\rho)-1}} &f'_\rho(x^{(1)}, \dots, x^{(n'(\rho)-2)}, x^{(n'(\rho)+1)}, x^{(n'(\rho)+2)}) \\ &= D_{\mathbf{e}_{n'(\rho)}} f'_\rho(x^{(1)}, \dots, x^{(n'(\rho)-2)}, x^{(n'(\rho)+1)}, x^{(n'(\rho)+2)}) \equiv 0, \end{aligned}$$

$$D_{\mathbf{e}_{n'(\rho)-1}} f'_\rho(x^{(1)}, \dots, x^{(n'(\rho))}) + D_{\mathbf{e}_{n'(\rho)}} f'_\rho(x^{(1)}, \dots, x^{(n'(\rho))}) = 1.$$

Applying Lemma 7.111 to the functions $f'_\rho(x^{(1)},\dots,x^{(n'(\rho)-2)},x^{(n'(\rho)+1)},x^{(n'(\rho)+2)})$ and $f'_\rho(x^{(1)},\dots,x^{(n'(\rho))})$, and using in addition Lemma 7.112, it is easy to see that the function $f'_{\rho+1}$ is a $(\rho+1)$-regular function of $4(\rho+1)-4$ variables that has nonlinearity equal to

$$2[4\cdot 2^{n'(\rho)-1}+4(2^{n'(\rho)-1}-2^{n'(\rho)-\rho})]=2^{n'(\rho+1)-1}-2^{n'(\rho+1)-(\rho+1)},$$

and algebraic degree equal to $\deg f'_{\rho+1}=\rho+1$. From the construction of the function $f'_{\rho+1}$ it is also clear that $D_{e_{n'(\rho+1)-1}}f'_{\rho+1}+D_{e_{n'(\rho+1)}}f'_{\rho+1}=1$. $\square$

Note that the function $g(x^{(1)},\dots,x^{(n'(\rho))})=f'_\rho(x^{(1)},\dots,x^{(n'(\rho))})\oplus x^{(1)}\oplus\cdots\oplus x^{(n'(\rho))}$ is $(n'(\rho)\rho-1)$-resilient. Therefore, the bound obtained from Siegenthaler's inequality is attained for functions constructed in Theorem 7.115. For these functions the maximum possible nonlinearity for $(\rho-1)$-resilient functions is attained.

THEOREM 7.116. *The maximum nonlinearity of a ρ-regular function in $\mathcal{F}_n$ is equal to $2^{n-1}-2^{\max(\rho,n-\rho)}$ for $1\le\min(\rho,n-\rho)\le\frac{n}{4}+1$.*

PROOF. According to Theorem 7.109, the maximum possible nonlinearity of any ρ-regular function in $\mathcal{F}_n$ is bounded above by $2^{n-1}-2^{\max(\rho,n-\rho)}$. We put $n'=4\rho-r\le n$. Then, by Theorem 7.115, there exists a ρ-regular function $f'(x^{(1)},\dots,x^{(n')})\in\mathcal{F}_{n'}$ with nonlinearity $N_{f'}=2^{n'-1}-2^{n'-\rho}$. Adding $n-n'$ new dummy variables, we get a ρ-regular function $f(x^{(1)},\dots,x^{(n)})\in\mathcal{F}_n$ with nonlinearity $N_{f'}=2^{n-1}-2^{n-\rho}$.

If $n-\rho\le\frac{n}{4}+1$, we construct in the same way an $(n-\rho)$-regular function $f(x^{(1)},\dots,x^{(n)})\in\mathcal{F}_n$ of nonlinearity $N_{f'}=2^{n-1}-2^{\rho}$. Then the function

$$g(x^{(1)},\dots,x^{(n)})=f(x^{(1)},\dots,x^{(n)})\oplus x^{(1)}\oplus\cdots\oplus x^{(n)}$$

is obviously a ρ-regular function from $\mathcal{F}_n$ with nonlinearity $N_g=2^{n-1}-2^{\rho}$. We have $N_f=N_g$. $\square$

Let $\boldsymbol{\theta}=(\theta_{\boldsymbol{a}})$, $\boldsymbol{a}\in V_n$, be a covering sequence of the function $f\in\mathcal{F}_n$. It is clear that the elements $\theta_{\boldsymbol{a}}$ for vectors $\boldsymbol{a}\in V_n$ with $D_{\boldsymbol{a}}f(\boldsymbol{x})\equiv 0$ do not contribute to the sum $\sum_{\boldsymbol{a}\in V_n}\theta_{\boldsymbol{a}}D_{\boldsymbol{a}}f$. They can be replaced by zero and the resulting sequence $\boldsymbol{\theta}'$ will also be covering for the function f with the same level.

DEFINITION 7.117. A covering sequence $\boldsymbol{\theta}=(\theta_{\boldsymbol{a}})$, $\boldsymbol{a}\in V_n$, of a Boolean function f is called *perfect* if for any vector $\boldsymbol{a}\in V_n$ such that $\theta_{\boldsymbol{a}}\ne 0$, one has $D_{\boldsymbol{a}}f(\boldsymbol{x})\not\equiv 0$.

THEOREM 7.118. *Suppose that $f\in\mathcal{F}_n$ allows a perfect nonnegative covering sequence $\boldsymbol{\theta}=(\theta_{\boldsymbol{a}})$ $\boldsymbol{a}\in V_n$ of level ρ. Then*

$$\overline{W}_{\boldsymbol{\theta}}(\mathbf{0})\le\rho\cdot 2^{\deg f-1}.$$

PROOF. By definition of a covering sequence we have, for any $\boldsymbol{x}\in V_n$,

$$\sum_{\boldsymbol{a}\in V_n}\theta_{\boldsymbol{a}}D_{\boldsymbol{a}}f(\boldsymbol{x})=\rho.$$

Summing these equations for all $\boldsymbol{x}\in V_n$ and using the inequality $\deg D_{\boldsymbol{a}}f(\boldsymbol{x})\le\deg f-1$, we see that

$$\rho\cdot 2^n=\sum_{\boldsymbol{a}\in V_n}\theta_{\boldsymbol{a}}\sum_{\boldsymbol{x}\in V_n}D_{\boldsymbol{a}}f(\boldsymbol{x})=\sum_{\boldsymbol{a}\in V_n}\theta_{\boldsymbol{a}}\,\mathrm{wt}(D_{\boldsymbol{a}}f(\boldsymbol{x}))\ge 2^{n-\deg f+1}\sum_{\boldsymbol{a}\in V_n}\theta_{\boldsymbol{a}}.$$

Since $\sum_{\boldsymbol{a}\in V_n} \theta_{\boldsymbol{a}} = \overline{W}_{\boldsymbol{\theta}}(\mathbf{0})$, the last inequality yields the theorem. $\square$

COROLLARY 7.119. *Suppose that f is an arbitrary ρ-regular weakly nondegenerate Boolean function from $\mathcal{F}_n$. Then $n \leq \rho \cdot 2^{\rho-1}$.*

Now suppose that f is a ρ-regular Boolean function and $D_{\boldsymbol{e}_i} f(\boldsymbol{x}) \not\equiv 1$ for any $i = 1, 2, \ldots, n$. Then $n \leq (n-\rho) \cdot 2^{n-\rho-1}$.

PROOF. For f a ρ-regular function in $\mathcal{F}_n$ we have $\overline{W}_{\boldsymbol{\theta}}(\mathbf{0}) = n$. According to Lemma 7.108, the inequality $\deg f \leq \rho$ holds, and this gives the first statement of the corollary.

Define a function $g(x^{(1)}, \ldots, x^{(n)}) = f(x^{(1)}, \ldots, x^{(n)}) \oplus x^{(1)} \oplus \cdots \oplus x^{(n)}$. According to Lemma 7.107, the function g is $(n-\rho)$-regular. Besides, g is a weakly nondegenerate function. Using the first part of the corollary, we obtain the inequality $n \leq (n-\rho) \cdot 2^{n-\rho-1}$. $\square$

COROLLARY 7.120. *For a given $\rho \geq 2$, the maximum n for which there exists a ρ-regular weakly nondegenerate Boolean function satisfies the inequalities*

$$3 \cdot 2^{\rho-1} - 2 \leq n \leq \rho \cdot 2^{\rho-1}.$$

PROOF. The lower bound follows from Theorem 7.114. The upper bound follows from Corollary 7.119. $\square$

COROLLARY 7.121. *The minimum ρ for which there exists a ρ-regular weakly nondegenerate Boolean function in $\mathcal{F}_n$ satisfies the following asymptotic relation:*

$$\min \rho = \log_2 n \pm \mathcal{O}(\log_2 \log_2 n)$$

as $n \to \infty$.

PROOF. Corollary 7.120 implies that $3 \cdot 2^{\min \rho - 1} - 2 \leq n$, from which it follows that $\rho \leq \log_2 n$. From the inequality $n \leq \min \rho \cdot 2^{\min \rho - 1}$ we get $\rho \geq \log_2 n - \log_2 \log_2 n + 1$. $\square$

7.7. Quadratic resilient Boolean functions of maximum order

Quadratic Boolean functions are very useful mathematical objects since the corresponding mathematical problems can be completely solved. For example, we mention the complete classification of quadratic functions based on group equivalence (§3.3), complete description of the weight spectrum of the Reed–Muller code of second order (§5.4), and so on. In this section we give a complete description of resilient quadratic functions of minimum order. Results and ideas from Chapters 2 and 3 will be used.

Suppose that $f \in \mathcal{F}_n$, $\deg f \leq 2$. As before (§3.3), we denote by $q_f(\boldsymbol{u}, \boldsymbol{v}) = f(\boldsymbol{u} \oplus \boldsymbol{v}) \oplus f(\boldsymbol{u}) \oplus f(\boldsymbol{v}) \oplus f(\mathbf{0})$ the symplectic bilinear form associated with the function f. If $f(\boldsymbol{x}) = f(x^{(1)}, \ldots, x^{(n)}) = \boldsymbol{x}^\top Q_f \boldsymbol{x} \oplus \boldsymbol{b}_f^\top \boldsymbol{x} \oplus f(\mathbf{0})$ (Q_f is the upper triangular matrix with zeros on the main diagonal, see §3.4, and $\boldsymbol{b}_f \in V_n$), then the symplectic matrix corresponding to q_f is $\widetilde{Q}_f = Q_f \oplus Q_f^\top$. Furthermore, $q_f(\boldsymbol{u}, \boldsymbol{v}) = \boldsymbol{u}^\top \widetilde{Q}_f \boldsymbol{v}$. The rank of the bilinear form q_f coincides with the rank of the matrix $\widetilde{Q}_f$, and it is always even (see Lemma 3.30). It is clear that if $g(\boldsymbol{x}) = f(\boldsymbol{x}) \oplus l_{\boldsymbol{a},\varepsilon}(\boldsymbol{x})$, then the functions g and f have the same associated symplectic bilinear form (and the same symplectic matrix $\widetilde{Q}_g = \widetilde{Q}_f$).

According to Problem 2.126, $L_f = \ker q_f$ and for a quadratic Boolean function f, the kernel q_f can be easily found as

$$\ker q_f = \ker \widetilde{Q}_f = \{\boldsymbol{u} \in V_n \mid \boldsymbol{u}^\top \widetilde{Q}_f \boldsymbol{x} = 0 \text{ for any } \boldsymbol{x} \in V_n\}.$$

Since any quadratic function is a partially mn-bent function, Problems 6.56, 6.58 imply the following:

1) f is balanced if and only if $f|_{\ker q_f} \neq \text{const}$;

2) $V_n \setminus \mathrm{CI}_f = \boldsymbol{u}_f \oplus (\ker q_f)^\perp$, where $\boldsymbol{u}_f \in V_n$ is a certain fixed vector.

For an m-resilient function $f \in \mathcal{F}_n$ the Siegenthaler inequality $\deg f \leq n-m-1$ holds. Since $\deg f = 2$, the inequality $m \leq n-3$ is true, whence the maximum value of $\operatorname{sut} f$ for a quadratic function is equal to $n-3$. The $(n-3)$-resilient quadratic functions in $\mathcal{F}_n$ will be described in this section. It is known that for balanced Boolean functions one has

$$\operatorname{sut} f = \min\{\mathrm{wt}(\boldsymbol{u}) \mid \boldsymbol{u} \in V_m \setminus \mathrm{CI}_f\} - 1. \tag{7.33}$$

PROBLEM 7.122. Prove the following statement.

A quadratic Boolean function $f \in \mathcal{F}_n$ is $(n-3)$-resilient if and only if its algebraic normal form is one of (7.34)–(7.37) below:

$$f(\boldsymbol{x}) = x^{(i)}x^{(j)} \oplus \bigoplus_{t\in\{1,\dots,n\}\setminus\{i,j\}} x^{(t)} \oplus \varepsilon^{(i)}x^{(i)} \oplus \varepsilon^{(j)}x^{(j)} \oplus \varepsilon, \tag{7.34}$$

where $\varepsilon^{(i)}, \varepsilon^{(j)}, \varepsilon$ are arbitrary elements of $\mathbb{F}_2$, and i and j are distinct elements of the set $\{1,\dots,n\}$;

$$f(\boldsymbol{x}) = (x^{(i)} \oplus x^{(j)})(x^{(i)} \oplus x^{(k)}) \oplus \langle \boldsymbol{v}, \boldsymbol{x}\rangle \oplus \varepsilon, \tag{7.35}$$

where ε is an arbitrary element of $\mathbb{F}_2$, the linear function $\langle \boldsymbol{v}, \boldsymbol{x}\rangle$ is equal to either $\bigoplus_{t=1}^{n} x^{(t)}$ or $\bigoplus_{t\in\{1,\dots,n\}\setminus\{i,j\}} x^{(t)}$, or $\bigoplus_{t\in\{1,\dots,n\}\setminus\{i,k\}} x^{(t)}$, or $\bigoplus_{t\in\{1,\dots,n\}\setminus\{j,k\}} x^{(t)}$, and i, j, k are arbitrary distinct elements of the set $\{1,\dots,n\}$;

$$f(\boldsymbol{x}) = (x^{(i)} \oplus x^{(j)})x^{(k)} \oplus \langle \boldsymbol{v}, \boldsymbol{x}\rangle \oplus \varepsilon, \tag{7.36}$$

where ε is an arbitrary element of the field $\mathbb{F}_2$, the linear function $\langle \boldsymbol{v}, \boldsymbol{x}\rangle$ is equal to either $\bigoplus_{t\in\{1,\dots,n\}\setminus\{i\}} x^{(t)}$ or $\bigoplus_{t\in\{1,\dots,n\}\setminus\{j\}} x^{(t)}$, or $\bigoplus_{t\in\{1,\dots,n\}\setminus\{i,k\}} x^{(t)}$, or $\bigoplus_{t\in\{1,\dots,n\}\setminus\{j,k\}} x^{(t)}$, and i, j, k are arbitrary distinct elements of the set $\{1,\dots,n\}$;

$$f(\boldsymbol{x}) = (x^{(j)} \oplus x^{(k)})(x^{(i)} \oplus x^{(l)}) \oplus \langle \boldsymbol{v}, \boldsymbol{x}\rangle \oplus \varepsilon, \tag{7.37}$$

where ε is an arbitrary element of $\mathbb{F}_2$, the linear function $\langle \boldsymbol{v}, \boldsymbol{x}\rangle$ is equivalent to either $\bigoplus_{t\in\{1,\dots,n\}\setminus\{i,j\}} x^{(t)}$ or $\bigoplus_{t\in\{1,\dots,n\}\setminus\{i,k\}} x^{(t)}$, or $\bigoplus_{t\in\{1,\dots,n\}\setminus\{l,j\}} x^{(t)}$, or $\bigoplus_{t\in\{1,\dots,n\}\setminus\{l,k\}} x^{(t)}$, and i, j, l, k are any distinct elements of the set $\{1,\dots,n\}$.

COROLLARY 7.123. *Let Q_n be the number of $(n-3)$-resilient quadratic functions in $\mathcal{F}_n$. Then*

$$Q_n = \frac{1}{3}n(n-1)(3n-2)(n+1).$$

PROOF. The corollary follows immediately from (7.34)–(7.37). □

PROBLEM 7.124. Let f be a quadratic k-resilient function from $\mathcal{F}_n$, $n-k \geq 2$, $\deg(f, x^{(i)}) \geq 2$ for all $i = 1, 2, \dots, n$. Prove that

$$n \leq (n-k-1)2^{\deg f - 1}. \tag{7.38}$$

THEOREM 7.125. *Suppose that f is a quadratic k-resilient function from $\mathcal{F}_n$, $n \geq 3$, $\deg(f, x^{(i)}) = 2$ for all $i = 1, 2, \ldots, n$. Then*

$$k \leq \left\lfloor \frac{n}{2} \right\rfloor - 1.$$

PROOF. Using Problem 7.124, we substitute $\deg f = 2$ in the inequality (7.38). As a result we get $k \leq \frac{n}{2} - 1$, which for integer k is equivalent to the inequality $k \leq \left\lfloor \frac{n}{2} \right\rfloor - 1$. □

REMARK 7.126. In the inequality (7.38), equality can be attained. According to Problem 7.96, there exist $(k-1)$-resilient functions of $2k$ variables. In order to construct a quadratic function which satisfies the conditions of Theorem 7.125, it is enough to define g as the function

$$g(x^{(1)}, \ldots, x^{(n)}) = \bigoplus_{i<j} x^{(i)} x^{(j)}.$$

To conclude this section, we give a table of cardinalities of the sets of all quadratic functions (NQ_n), balanced quadratic functions (BQ_n) and $(n-3)$-resilient quadratic functions (Q_n) for $n = 2, \ldots, 8$.

TABLE 7.1

n	NQ_n	BQ_n	Q_n
3	112	56	56
4	2 016	840	200
5	65 472	36 456	520
6	4 194 176	1 828 008	1 120
7	536 870 656	300 503 336	2 128
8	137 438 952 960	60 273 666 600	3 696

Comments to Chapter 7

A slightly different proof of Theorem 3.34 can be found in [**22, 225, 264**]. This theorem is basic for solving the problem of complete description of weight enumerators of codes $RM(2, m)$ (see Chapter 5).

In 1989, an approach for evaluation of the nonlinearity of Boolean functions was suggested in the work [**267**]. It is important to have criteria for measuring the nonlinearity of a Boolean function. The authors made an attempt to classify the existing criteria and to suggest new criteria. As a base of such classification they used the algebraic idea that any useful criterion must be invariant under a certain group of transformations of the base space. It is also clear that nonlinearity criterion must remain invariant under transformation groups as big as possible. In view of this approach, a function is regarded as weak if it can be reduced to a cryptographically weak function by a simple (for example, linear or affine) transformation. The most promising group from this point of view is the affine group $\mathfrak{GA}(V_n)$. The criterion for distance to an affine group, the Rothaus criterion, and some others are invariant under this group. Unfortunately, the enormous quantity of equivalence classes with respect to $\mathfrak{GA}(V_n)$ for $n > 6$ prevents any efficient use of these criteria.

Attempts for classification of cubic Boolean functions with a small number of variables using RM-equivalence were made in [**86**]. The results of §3.5 concerning RM-equivalence can serve as a theoretic base for further attempts to build classifications of Boolean functions in the general case.

Even more complicated is the problem of classification for Boolean mappings $\Phi \in \mathcal{F}_{m,n}$. By definition, Boolean mappings $\Phi, \Psi \in \mathcal{F}_{m,n}$ are equivalent with respect to a pair of groups $\mathfrak{G}$ and $\mathfrak{H}$ acting on V_m and V_n, respectively, if there exist elements $\mathfrak{g} \in \mathfrak{G}$ and $\mathfrak{h} \in \mathfrak{H}$ such that

$$\mathfrak{h} \circ \Phi \circ \mathfrak{g} = \Psi.$$

In contrast to the case of Boolean functions, the group $\mathfrak{H}$ can be large if $n > 1$, which seriously complicates the classification.

In [**32**], the problem of linear and affine classification of Boolean permutations $\Phi \in \mathcal{F}_{n,n}$ (i.e., the case where $\mathfrak{G} = \mathfrak{H}$ and either $\mathfrak{G} = \mathfrak{GL}(V_n)$ or $\mathfrak{G} = \mathfrak{GA}(V_n)$) was attacked. An algorithm was suggested which for a given pair of permutations $\Phi, \Psi \in \mathcal{F}_{n,n}$, produces a pair $\mathfrak{g}_1, \mathfrak{g}_2 \in \mathfrak{G}$ such that

$$\mathfrak{g}_1 \circ \Phi \circ \mathfrak{g}_2 = \Psi$$

if the permutations in question are equivalent or states that the permutations Φ and Ψ are nonequivalent. The asymptotical complexity of this algorithm is equal to $\mathcal{O}(n^3 2^n)$ for the linear classification problem and to $\mathcal{O}(n^3 2^{2n})$ for the affine classification problem (it is easy to see that the complexity of the trivial complete search algorithm is $\mathcal{O}(2^{n^2})$). In the same work [**32**], the authors generalize their algorithm to mappings $\Phi, \Psi \in \mathcal{F}_{m,n}$, which are not permutations if $|n-m|$ is small. Besides, the authors introduce an important (from the cryptographic point of view) notion of "almost equivalent mappings" and adapt their algorithm to the case of linear and affine almost equivalence. By definition, Boolean mappings $\Phi, \Psi \in \mathcal{F}_{m,n}$ are *almost equivalent*with respect to a pair of groups $\mathfrak{G}$ and $\mathfrak{H}$ if there exist elements $\mathfrak{g} \in \mathfrak{G}$ and $\mathfrak{h} \in \mathfrak{H}$ such that for almost all $\boldsymbol{x} \in V_m$,

$$\mathfrak{h} \circ \Phi \circ \mathfrak{g}(\boldsymbol{x}) = \Psi(\boldsymbol{x}).$$

CHAPTER 8

Codes, Boolean Mappings, and Their Cryptographic Properties

8.1. Almost perfect nonlinear and almost bent mappings

Suppose that $\Phi = (f_1, f_2, \ldots, f_l) \in \mathcal{F}_{m,l}$. The *characteristic function of the mapping* Φ, denoted by I_Φ, is the following Boolean function in $\mathcal{F}_{m+l}$:

$$I_\Phi(\boldsymbol{x}, \boldsymbol{y}) = \begin{cases} 1 & \text{if } \boldsymbol{y} = \Phi(\boldsymbol{x}), \\ 0 & \text{otherwise.} \end{cases}$$

Let f and g be Boolean functions in $\mathcal{F}_m$ or arbitrary integer-valued functions in V_m. Then the *convolution* $f \otimes g$ of these functions is defined as the integer-valued function in V_m defined by the formula

$$(f \otimes g)(\boldsymbol{a}) = \sum_{\boldsymbol{x} \in V_m} f(\boldsymbol{x}) g(\boldsymbol{x} \oplus \boldsymbol{a}),$$

for any $\boldsymbol{a} \in V_m$ (addition is over the real numbers).

If we regard the mapping Φ from the cryptographic point of view, we have to note the following. In the differential method, an important role is played by the sets

$$S_\Phi(\boldsymbol{a}, \boldsymbol{b}) = \{\boldsymbol{x} \in V_m \mid \Phi(\boldsymbol{x} \oplus \boldsymbol{a}) \oplus \Phi(\boldsymbol{x}) = \boldsymbol{b}\} \tag{8.1}$$

for $\boldsymbol{a} \in V_m$, $\boldsymbol{b} \in V_l$.

Let us form a $((2^m - 1) \times 2^l)$ matrix s_Φ as follows:

$$s_\Phi = (s_\Phi(\boldsymbol{a}, \boldsymbol{b})), \quad \boldsymbol{a} \in V_m \setminus \{\boldsymbol{0}\}, \quad \boldsymbol{b} \in V_l, \tag{8.2}$$

where $s_\Phi(\boldsymbol{a}, \boldsymbol{b}) = \# S_\Phi(\boldsymbol{a}, \boldsymbol{b})$. The matrix s_Φ is called a *difference table*, and its entries are subject to some relations. In particular, for any $\boldsymbol{a} \in V_m \setminus \{\boldsymbol{0}\}$, the following equality holds:

$$\sum_{\boldsymbol{b} \in V_l} s_\Phi(\boldsymbol{a}, \boldsymbol{b}) = 2^m. \tag{8.3}$$

PROBLEM 8.1. Find out how the matrix s_Φ, where $\Phi \in \mathcal{F}_{m,l}$, is transformed by the following operations:

1) multiplication on the left (right) by a nondegenerate affine mapping in $\mathcal{F}_{l,l}$ ($\mathcal{F}_{m,m}$);

2) addition of an arbitrary affine mapping in $\mathcal{F}_{m,l}$;

3) inverting Φ (if Φ is a one-to-one mapping in $\mathcal{F}_{m,m}$).

If the mapping Φ is a part of a cryptographic transformation, the difference table s_Φ has various applications in differential cryptanalysis (§9.5). The maximum element of the difference table

$$\mu l(\Phi) = \max_{\boldsymbol{a}\in V_m\setminus\{\boldsymbol{0}\}} \max_{\boldsymbol{b}\in V_l} s_\Phi(\boldsymbol{a},\boldsymbol{b}) \tag{8.4}$$

is used for efficient evaluation of the differential method of cryptanalysis.

The quantity $\mu l(\Phi)$ characterizes the deviation of the mapping Φ from the set of linear mappings from V_m to V_l. For an affine mapping $A\colon V_m \to V_l$, exactly one element in any row of the difference table is nonzero, and this element equals 2^m. Thus, the equality $\mu l(A) = 2^m$ holds. In terms of μl, the mappings Φ which minimize $\mu l(\Phi)$ are at maximum distance from the linear mappings and are most resistant to differential cryptanalysis. From equality (8.3) it follows that for any mapping $\Phi\colon V_m \to V_l$ the inequality $\mu l(\Phi) \geq 2^{m-l}$ holds (this inequality is nontrivial only if $m \geq l$). Besides, in order to achieve this estimate, it is necessary that

$$s_\Phi(\boldsymbol{a},\boldsymbol{b}) = 2^{m-l} \tag{8.5}$$

for all $\boldsymbol{a} \in V_m \setminus \{\boldsymbol{0}\}$ and $\boldsymbol{b} \in V_l$. This means that the derivatives of mappings Φ in nonzero directions are balanced mappings. If $l = 1$ and Φ is a Boolean function in $\mathcal{F}_m$, the last property is equivalent to the condition that the function Φ is an mn-bent function (see the Rothaus criterion, §6.2).

For the linear method we define the sets

$$T_\Phi(\boldsymbol{a},\boldsymbol{b}) = \{\boldsymbol{x} \in V_m \mid \langle \boldsymbol{a},\boldsymbol{x}\rangle \oplus \langle \boldsymbol{b},\Phi(\boldsymbol{x})\rangle = 0\}$$

for $\boldsymbol{b} \in V_l$, $\boldsymbol{a} \in V_m$ such that $\#T_\Phi(\boldsymbol{a},\boldsymbol{b}) \neq 2^{m-1}$. The efficiency of the linear method based on such sets depends on the absolute value of

$$t_\Phi(\boldsymbol{a},\boldsymbol{b}) = \#T_\Phi(\boldsymbol{a},\boldsymbol{b}) - 2^{m-1}.$$

PROBLEM 8.2. Find out how the matrix $\big(t_\Phi\big)$ is transformed under the following operations where $\Phi \in \mathcal{F}_{m,l}$:

1) left (right) multiplication by a nondegenerate affine mapping in $\mathcal{F}_{l,l}$ ($\mathcal{F}_{m,m}$);
2) addition of an arbitrary affine mapping in $\mathcal{F}_{m,l}$;
3) inverting the mapping if Φ is a one-to-one mapping in $\mathcal{F}_{m,m}$.

Cryptographic properties of a mapping Φ can be characterized by the quantities

$$\mu l(\Phi) = \max_{\substack{\boldsymbol{a}\in V_m \\ \boldsymbol{a}\neq\boldsymbol{0}}} \max_{\boldsymbol{b}\in V_l} s_\Phi(\boldsymbol{a},\boldsymbol{b}), \tag{8.6}$$

$$\mu t(\Phi) = \max_{\boldsymbol{a}\in V_m} \max_{\substack{\boldsymbol{b}\in V_l \\ \boldsymbol{b}\neq\boldsymbol{0}}} |t_\Phi(\boldsymbol{a},\boldsymbol{b})|. \tag{8.7}$$

If the quantities in (8.6) and (8.7) are small, we can deduce that the mapping Φ is in a sense stable with respect to the methods above.

The following notions will be useful later.

LEMMA 8.3. *For any* $(\boldsymbol{a},\boldsymbol{b}) \in V_m \times V_l$ *one has*

$$s_\Phi(\boldsymbol{a},\boldsymbol{b}) = (I_\Phi \otimes I_\Phi)(\boldsymbol{a},\boldsymbol{b}).$$

PROOF. Suppose that $\boldsymbol{a} \in V_m$, $\boldsymbol{b} \in V_l$. Then

$$\begin{aligned}(I_\Phi \otimes I_\Phi)(\boldsymbol{a},\boldsymbol{b}) &= \sum_{\boldsymbol{x}\in V_m}\sum_{\boldsymbol{y}\in V_l} I_\Phi(\boldsymbol{x},\boldsymbol{y}) I_\Phi(\boldsymbol{x}\oplus\boldsymbol{a},\boldsymbol{y}\oplus\boldsymbol{b}) = \sum_{\boldsymbol{x}\in V_m} I_\Phi(\boldsymbol{x}\oplus\boldsymbol{a},\Phi(\boldsymbol{x})\oplus\boldsymbol{b}) \\ &= \#\{\boldsymbol{x} \in V_m \mid \Phi(\boldsymbol{x}\oplus\boldsymbol{a}) = \Phi(\boldsymbol{x})\oplus\boldsymbol{b}\} = s_\Phi(\boldsymbol{a},\boldsymbol{b}).\end{aligned}$$

□

LEMMA 8.4. *If* $(\boldsymbol{a}, \boldsymbol{b}) \in V_m \times V_l$, *then*

$$t_\Phi(\boldsymbol{a}, \boldsymbol{b}) = \frac{1}{2}\overline{W}_{I_\Phi}(\boldsymbol{a}, \boldsymbol{b}) = \frac{1}{2} W_{\langle \boldsymbol{b}, \Phi(\boldsymbol{x})\rangle}(\boldsymbol{a}).$$

PROOF. If $\boldsymbol{a} \in V_m$, $\boldsymbol{b} \in V_l$, then

$$\begin{aligned}\overline{W}_{I_\Phi}(\boldsymbol{a}, \boldsymbol{b}) &= \sum_{\boldsymbol{x}\in V_m}\sum_{\boldsymbol{y}\in V_l} I_\Phi(\boldsymbol{x}, \boldsymbol{y})(-1)^{\langle \boldsymbol{a}, \boldsymbol{x}\rangle \oplus \langle \boldsymbol{b}, \boldsymbol{y}\rangle} = \sum_{\boldsymbol{x}\in V_m} (-1)^{\langle \boldsymbol{a}, \boldsymbol{x}\rangle \oplus \langle \boldsymbol{b}, \Phi(\boldsymbol{x})\rangle} \\ &= W_{\langle \boldsymbol{b}, \Phi(\boldsymbol{x})\rangle}(\boldsymbol{a}) = T_\Phi(\boldsymbol{a}, \boldsymbol{b}) - (2^m - \# T_\Phi(\boldsymbol{a}, \boldsymbol{b})) = 2t_\Phi(\boldsymbol{a}, \boldsymbol{b}).\end{aligned}$$

□

COROLLARY 8.5. *For any* $\Phi \in \mathcal{F}_{m,l}$, *the inequality* $\mu t(\Phi) \geq 2^{m/2-1}$ *holds.*

THEOREM 8.6. *For any* $\Phi \in \mathcal{F}_{m,l}$, *the inequality* $\mu l(\Phi) \geq 2^{m-l}$ *holds.*

PROOF. Let $\boldsymbol{a} \in V_m \setminus \{\boldsymbol{0}\}$ be a fixed vector. Summing $s_\Phi(\boldsymbol{a}, \boldsymbol{b})$ for all vectors $\boldsymbol{b} \in V_l$, we get $\sum_{\boldsymbol{b}\in V_l} s_\Phi(\boldsymbol{a}, \boldsymbol{b}) = 2^m$. Since in the latter sum we have 2^l terms and all of them are nonnegative, we may conclude that $\max_{\boldsymbol{b}\in V_l} s_\Phi(\boldsymbol{a}, \boldsymbol{b}) \geq 2^{m-l}$ for any vector $\boldsymbol{a} \in V_m \setminus \{\boldsymbol{0}\}$. Hence, $\mu l(\Phi) \geq 2^{m-l}$. □

Let us discuss the relations between $\mathrm{ill}(\Phi)$ (§2.5) and $\mu l(\Phi)$.

THEOREM 8.7. *For any* $\Phi \in \mathcal{F}_{m,l}$, *the following inequality holds:*

$$\mu l(\Phi) \geq 2^{m-\mathrm{ill}(\Phi)}.$$

PROOF. Suppose that the image of Φ is written as in (2.28) and put $k = \mathrm{ill}(\Phi)$. We will represent any of the branching mappings $A_{\boldsymbol{u}}$, $\boldsymbol{u} \in V_k$, as

$$A_{\boldsymbol{u}}(\boldsymbol{x}) = L_{\boldsymbol{u}}\boldsymbol{x} \oplus \boldsymbol{c}_{\boldsymbol{u}},$$

where $L_{\boldsymbol{u}}$ is an $(l \times m)$ matrix over the field $\mathbb{F}_2$, $\boldsymbol{c}_{\boldsymbol{u}} \in V_l$. Using this representation, one can write the following expressions for the derivative of the mapping Φ in the direction of $\boldsymbol{a} \in V_m$:

$$\begin{aligned}D_{\boldsymbol{a}}\Phi(\boldsymbol{x}) &= \Phi(\boldsymbol{x} \oplus \boldsymbol{a}) \oplus \Phi(\boldsymbol{x}) = A_{h(\boldsymbol{x}\oplus\boldsymbol{a})}(\boldsymbol{x} \oplus \boldsymbol{a}) \oplus A_{h(\boldsymbol{x})}(\boldsymbol{x}) \\ &= L_{h(\boldsymbol{x})\oplus h(\boldsymbol{a})}(\boldsymbol{x}) \oplus L_{h(\boldsymbol{x})\oplus h(\boldsymbol{a})}(\boldsymbol{a}) \oplus \boldsymbol{c}_{h(\boldsymbol{x})\oplus h(\boldsymbol{a})} \oplus L_{h(\boldsymbol{x})}(\boldsymbol{x}) \oplus \boldsymbol{c}_{h(\boldsymbol{x})}.\end{aligned} \tag{8.8}$$

First we consider the case where the kernel of a linear surjective mapping $h \in \mathcal{F}_{m,k}$ is nontrivial, i.e., $\ker h \neq \{\boldsymbol{0}\}$, and in particular $m > k$. Suppose that a nonzero vector $\boldsymbol{a} \in V_m \setminus \{\boldsymbol{0}\}$ belongs to the kernel $\ker h$, i.e., that $h(\boldsymbol{a}) = \boldsymbol{0}$. Then it follows from (8.8) that

$$D_{\boldsymbol{a}}\Phi(\boldsymbol{x}) = L_{h(\boldsymbol{x})}(\boldsymbol{a}).$$

Thus, for such a vector $\boldsymbol{a} \in V_m \setminus \{\boldsymbol{0}\}$, one has

$$\ker(h) \subseteq \{\boldsymbol{x} \in V_m \mid D_{\boldsymbol{a}}\Phi(\boldsymbol{x}) = L_{\boldsymbol{0}}(\boldsymbol{a})\}.$$

Therefore, for the element $s_\Phi(\boldsymbol{a}, L_{\boldsymbol{0}}(\boldsymbol{a}))$ in the difference table, the following inequality holds:

$$s_\Phi(\boldsymbol{a}, L_{\boldsymbol{0}}(\boldsymbol{a})) \geq \#\ker h. \tag{8.9}$$

Since $2^m = 2^k \cdot \#\ker h = \mathrm{ill}(\Phi) \cdot \#\ker h$, inequality (8.9) is equivalent to the inequality

$$s_\Phi(\boldsymbol{a}, L_{\boldsymbol{0}}(\boldsymbol{a})) \geq 2^{m-\mathrm{ill}(\Phi)}.$$

Theorem 8.7 follows from this inequality if one takes into account the inequality $\mu l(\Phi) \geq s_\Phi(\boldsymbol{a}, L_{\boldsymbol{0}}(\boldsymbol{a}))$.

If the kernel of the linear mapping h is trivial, i.e., if $\ker h = \{\boldsymbol{0}\}$ and $m = k = \mathrm{ill}(\Phi)$, Theorem 8.7 is trivial. □

The result of Theorem 8.7 cannot be improved since the bound is attained, for example, for affine mappings Φ. From Theorem 8.7 it follows that mappings with small values of $\mu l(\Phi)$ must be searched for among the mappings with large values of $\mathrm{ill}(\Phi)$.

THEOREM 8.8. *If the space of linear structures of the mapping* $\Phi \in \mathcal{F}_{m,l}$ *is trivial, then* $s_\Phi(\boldsymbol{a}, \boldsymbol{b}) \leq 2^m - 2^{m-\mathrm{ill}(\Phi)}$ *for any* $\boldsymbol{a} \in V_m \setminus \{\boldsymbol{0}\}$, $\boldsymbol{b} \in V_l$.

PROOF. An element $s_\Phi(\boldsymbol{a}, \boldsymbol{b})$ of the difference table for which $\boldsymbol{a} \in V_m \setminus \{\boldsymbol{0}\}$, $\boldsymbol{b} \in V_l$, equals the number of solutions for the unknown $\boldsymbol{x} \in V_m$ of the equation

$$D_{\boldsymbol{a}}\Phi(\boldsymbol{x}) = \boldsymbol{b}.$$

Suppose that the image of Φ is represented as in (2.28) and put $k = \mathrm{ill}(\Phi)$. We will express any of the branching mappings $A_{\boldsymbol{u}}$, $\boldsymbol{u} \in V_k$, as

$$A_{\boldsymbol{u}}(\boldsymbol{x}) = L_{\boldsymbol{u}}\boldsymbol{x} \oplus \boldsymbol{c}_{\boldsymbol{u}},$$

where $L_{\boldsymbol{u}}$ is an $(l \times m)$ matrix over the field $\mathbb{F}_2$, $\boldsymbol{c}_{\boldsymbol{u}} \in V_l$. With this representation, the equality (8.8) becomes

$$L_{h(\boldsymbol{x})\oplus h(\boldsymbol{a})}(\boldsymbol{x}) \oplus L_{h(\boldsymbol{x})\oplus h(\boldsymbol{a})}(\boldsymbol{a}) \oplus \boldsymbol{c}_{h(\boldsymbol{x})\oplus h(\boldsymbol{a})} \oplus L_{h(\boldsymbol{x})}(\boldsymbol{x}) \oplus \boldsymbol{c}_{h(\boldsymbol{x})} = \boldsymbol{b}. \tag{8.10}$$

We set $h(\boldsymbol{a}) = \boldsymbol{r} \in V_k$ and introduce a new parameter, $h(\boldsymbol{x}) = \boldsymbol{y} \in V_k$. Then equality (8.8) is equivalent to the union (in terms of the union of the sets of solutions) of the following 2^k systems of linear equations:

$$\begin{cases} (L_{\boldsymbol{y}\oplus\boldsymbol{r}} \oplus L_{\boldsymbol{y}})(\boldsymbol{x}) = \boldsymbol{b} \oplus \boldsymbol{c}_{\boldsymbol{y}\oplus\boldsymbol{r}} \oplus \boldsymbol{c}_{\boldsymbol{y}} \oplus L_{\boldsymbol{y}\oplus\boldsymbol{r}}(\boldsymbol{a}), \\ h(\boldsymbol{x}) = \boldsymbol{y}. \end{cases} \tag{8.11}$$

It is known from linear algebra that in the case of compatibility, the set of solutions of the nonhomogeneous system of linear equations (8.11) is a coset of the subset $\ker(h) \cap \ker(L_{\boldsymbol{y}\oplus\boldsymbol{r}} \oplus L_{\boldsymbol{y}})$.

We will denote by $Y(\boldsymbol{a}, \boldsymbol{b})$ the set of values of the parameter $\boldsymbol{y} \in V_k$ for which the system (8.11) is compatible; the number of solutions of the system (8.11) will be denoted by $t_{\boldsymbol{y},\boldsymbol{a},\boldsymbol{b}}$. Since for different $\boldsymbol{y} \in V_k$ the sets of solutions of the system (8.11) are disjoint, one has

$$s_\Phi(\boldsymbol{a}, \boldsymbol{b}) = \sum_{\boldsymbol{y} \in V_k} t_{\boldsymbol{y},\boldsymbol{a},\boldsymbol{b}}. \tag{8.12}$$

For the numbers $t_{\boldsymbol{y},\boldsymbol{a},\boldsymbol{b}}$ the equality

$$t_{\boldsymbol{y},\boldsymbol{a},\boldsymbol{b}} = \begin{cases} \#(\ker(h) \cap \ker(L_{\boldsymbol{y}\oplus\boldsymbol{r}} \oplus L_{\boldsymbol{y}})) & \text{if } \boldsymbol{y} \in Y(\boldsymbol{a}, \boldsymbol{b}), \\ 0 & \text{otherwise} \end{cases}$$

holds. If $\boldsymbol{y} \in Y(\boldsymbol{a}, \boldsymbol{b})$, then either $\ker(h) \subseteq \ker(L_{\boldsymbol{y}\oplus\boldsymbol{r}} \oplus L_{\boldsymbol{y}})$ and then $t_{\boldsymbol{y},\boldsymbol{a},\boldsymbol{b}} = \#\ker(h) = 2^{m-\mathrm{ill}(\Phi)}$ or $\ker(h) \not\subseteq \ker(L_{\boldsymbol{y}\oplus\boldsymbol{r}} \oplus L_{\boldsymbol{y}})$ and then $t_{\boldsymbol{y},\boldsymbol{a},\boldsymbol{b}}$ is not greater than the order of a maximal subspace of the space $\ker(h)$, i.e., $t_{\boldsymbol{y},\boldsymbol{a},\boldsymbol{b}} \leq 2^{m-\mathrm{ill}(\Phi)-1}$.

Let us fix $\boldsymbol{a} \in V_m \setminus \{\boldsymbol{0}\}$, $\boldsymbol{b} \in V_l$ and consider two cases.

1. $\ker(h) \subseteq \ker(L_{\boldsymbol{y}\oplus\boldsymbol{r}} \oplus L_{\boldsymbol{y}})$ for any $\boldsymbol{y} \in Y(\boldsymbol{a}, \boldsymbol{b})$. It follows from (8.12) that

$$s_\Phi(\boldsymbol{a}, \boldsymbol{b}) = \#Y(\boldsymbol{a}, \boldsymbol{b}) \cdot 2^{m-\mathrm{ill}(\Phi)}.$$

If $Y(\boldsymbol{a}, \boldsymbol{b}) = V_k$, then $\#Y(\boldsymbol{a}, \boldsymbol{b}) = 2^{\mathrm{ill}(\Phi)}$, whence $s_\Phi(\boldsymbol{a}, \boldsymbol{b}) = 2^m$, i.e., $\boldsymbol{a}$ is a linear structure of the mapping Φ.

If $Y(\boldsymbol{a},\boldsymbol{b}) \neq V_k$ as well, then $\#Y(\boldsymbol{a},\boldsymbol{b}) \leq 2^{\mathrm{ill}(\Phi)} - 1$, whence

$$s_\Phi(\boldsymbol{a},\boldsymbol{b}) \leq (2^{\mathrm{ill}(\Phi)} - 1) \cdot 2^{m-\mathrm{ill}(\Phi)} = 2^m - 2^{m-\mathrm{ill}(\Phi)}.$$

2. $\ker(h) \not\subseteq \ker(L_{\boldsymbol{y}'\oplus \boldsymbol{r}} \oplus L_{\boldsymbol{y}'})$ for some $\boldsymbol{y}' \in Y(\boldsymbol{a},\boldsymbol{b})$. Then

$$t_{\boldsymbol{y}',\boldsymbol{a},\boldsymbol{b}} \leq 2^{m-\mathrm{ill}(\Phi)-1}.$$

First, we will consider the case $\boldsymbol{r} = h(\boldsymbol{a}) \neq \boldsymbol{0}$. From the system (8.11) we choose $\boldsymbol{y} = \boldsymbol{y}'$ and $\boldsymbol{y} = \boldsymbol{y}' \oplus \boldsymbol{r}$. For $\boldsymbol{y} = \boldsymbol{y}' \oplus \boldsymbol{r}$ two cases are possible: either $\boldsymbol{y}' \oplus \boldsymbol{r} \notin Y(\boldsymbol{a},\boldsymbol{b})$ and then $t_{\boldsymbol{y}'\oplus\boldsymbol{r},\boldsymbol{a},\boldsymbol{b}} = 0$, or $\boldsymbol{y}' \oplus \boldsymbol{r} \in Y(\boldsymbol{a},\boldsymbol{b})$ and then $t_{\boldsymbol{y}'\oplus\boldsymbol{r},\boldsymbol{a},\boldsymbol{b}} \leq 2^{m-\mathrm{ill}(\Phi)-1}$ as $\ker(h) \not\subseteq \ker(L_{(\boldsymbol{y}'\oplus\boldsymbol{r})\oplus\boldsymbol{r}} \oplus L_{\boldsymbol{y}'\oplus\boldsymbol{r}}) = \ker(L_{\boldsymbol{y}'} \oplus L_{\boldsymbol{y}'\oplus\boldsymbol{r}})$. Thus, $t_{\boldsymbol{y},\boldsymbol{a},\boldsymbol{b}} \neq 2^{m-\mathrm{ill}(\Phi)}$ for $\boldsymbol{y} = \boldsymbol{y}'$ and for $\boldsymbol{y} = \boldsymbol{y}' \oplus \boldsymbol{r}$. Therefore, the number of $\boldsymbol{y} \in Y(\boldsymbol{a},\boldsymbol{b})$ such that $t_{\boldsymbol{y},\boldsymbol{a},\boldsymbol{b}} = 2^{m-\mathrm{ill}(\Phi)}$ is at most $2^{\mathrm{ill}(\Phi)} - 2$. Thus,

$$s_\Phi(\boldsymbol{a},\boldsymbol{b}) \leq (2^{\mathrm{ill}(\Phi)} - 2) \cdot 2^{m-\mathrm{ill}(\Phi)} + 2 \cdot 2^{m-\mathrm{ill}(\Phi)-1} = 2^m - 2^{m-\mathrm{ill}(\Phi)}.$$

In the case $\boldsymbol{r} = h(\boldsymbol{a}) = \boldsymbol{0}$ the system of linear equations (8.11) in the unknown $\boldsymbol{x} \in V_m$ becomes

$$\begin{cases} \boldsymbol{b} = L_{\boldsymbol{y}}(\boldsymbol{a}), \\ h(\boldsymbol{x}) = \boldsymbol{y}. \end{cases} \tag{8.13}$$

If $\boldsymbol{b} = L_{\boldsymbol{y}}(\boldsymbol{a})$, then the system (8.13) has $\#\ker(h) = 2^{m-\mathrm{ill}(\Phi)}$ solutions. If $\boldsymbol{b} \neq L_{\boldsymbol{y}}(\boldsymbol{a})$, then the system (8.13) has no solutions. If for any $\boldsymbol{y} \in V_k$ all vectors $L_{\boldsymbol{y}}(\boldsymbol{a})$ coincide, we have $s_\Phi(\boldsymbol{a},\boldsymbol{b}) = 2^m$, i.e., $\boldsymbol{a}$ is a linear structure of the mapping Φ. If not all vectors $L_{\boldsymbol{y}}(\boldsymbol{a})$ coincide, we have

$$s_\Phi(\boldsymbol{a},\boldsymbol{b}) \leq (2^{\mathrm{ill}(\Phi)} - 1) \cdot 2^{m-\mathrm{ill}(\Phi)} = 2^m - 2^{m-\mathrm{ill}(\Phi)}. \qquad \square$$

COROLLARY 8.9. *If the space of linear structures of a mapping $\Phi \in \mathcal{F}_{m,l}$ is trivial, then*

$$\mu l(\Phi) \leq 2^m - 2^{m-\mathrm{ill}(\Phi)}. \tag{8.14}$$

PROBLEM 8.10. Prove that if the space of linear structures of a mapping $\Phi \in \mathcal{F}_{m,l}$ is trivial and $\mathrm{ill}(\Phi) = 1$, then $\mu l(\Phi) = 2^{m-1}$.

DEFINITION 8.11. A mapping $\Phi \in \mathcal{F}_{m,l}$ is called a *bent mapping* if for any $\boldsymbol{b} \in V_l$, $\boldsymbol{b} \neq \boldsymbol{0}$ the Boolean function $\langle \boldsymbol{b}, \Phi(\boldsymbol{x})\rangle$ is a bent function.

The property of "being a bent mapping" is a reducible property of mappings (§2.5).

PROBLEM 8.12. Let $\Phi \in \mathcal{F}_{m,l}$ be a bent mapping. Prove that the following mappings are also bent:

1) $g_1 \circ \Phi \circ g_2$, where $g_1 \in \mathcal{F}_{m,m}$, $g_2 \in \mathcal{F}_{l,l}$ are arbitrary nondegenerate affine mappings;

2) $\Phi \oplus h$, where $h \in \mathcal{F}_{m,l}$ is an arbitrary affine mapping.

PROBLEM 8.13. Prove that $\Phi \in \mathcal{F}_{m,l}$ is a bent mapping if and only if for any $\boldsymbol{a} \in V_m$ and $\boldsymbol{b} \in V_l$, $\boldsymbol{b} \neq \boldsymbol{0}$, one has $\overline{W}_{I_\Phi}(\boldsymbol{a},\boldsymbol{b}) = \pm 2^{m/2}$ or $t_\Phi(\boldsymbol{a},\boldsymbol{b}) = \frac{1}{2}\overline{W}_{I_\Phi}(\boldsymbol{a},\boldsymbol{b}) = \pm 2^{m/2-1}$.

DEFINITION 8.14. A mapping $\Phi \in \mathcal{F}_{m,l}$ is called *perfect nonlinear* if $\mu l(\Phi) = 2^{m-l}$.

THEOREM 8.15. *A mapping $\Phi \in \mathcal{F}_{m,l}$ is perfect nonlinear if and only if Φ is a bent mapping.*

PROOF. Let Φ be a perfect nonlinear mapping. Then $\mu l(\Phi) = 2^{m-l}$, and for any $\boldsymbol{a} \in V_m$, $\boldsymbol{a} \neq \boldsymbol{0}$, we have $s_\Phi(\boldsymbol{a}, \boldsymbol{b}) = (I_\Phi \otimes I_\Phi)(\boldsymbol{a}, \boldsymbol{b}) = 2^{m-l}$. Furthermore, $s_\Phi(\boldsymbol{0}, \boldsymbol{0}) = 2^m$ and for $\boldsymbol{b} \neq \boldsymbol{0}$ the equality $s_\Phi(\boldsymbol{0}, \boldsymbol{b}) = 0$ holds. Therefore,

$$\begin{aligned}\overline{W}^{\,2}_{I_\Phi}(\boldsymbol{a}, \boldsymbol{b}) &= \overline{W}_{I_\Phi \otimes I_\Phi}(\boldsymbol{a}, \boldsymbol{b}) \\ &= \sum_{\boldsymbol{x} \in V_m} \sum_{\boldsymbol{y} \in V_m} (I_\Phi \otimes I_\Phi)(\boldsymbol{x}, \boldsymbol{y})(-1)^{\langle \boldsymbol{a}, \boldsymbol{x}\rangle \oplus \langle \boldsymbol{b}, \boldsymbol{y}\rangle} = \begin{cases} 2^m & \text{if } \boldsymbol{b} \neq \boldsymbol{0}, \\ 0 & \text{if } \boldsymbol{b} = \boldsymbol{0},\ \boldsymbol{a} \neq \boldsymbol{0}, \\ 2^{2m} & \text{if } \boldsymbol{b} = \boldsymbol{0},\ \boldsymbol{a} = \boldsymbol{0}, \end{cases}\end{aligned}$$

i.e., Φ is a bent mapping.

We will now prove the converse. If Φ is a bent mapping, then $\overline{W}_{I_\Phi}(\boldsymbol{a}, \boldsymbol{b}) = \pm 2^{m/2}$ for any $(\boldsymbol{a}, \boldsymbol{b})$, $\boldsymbol{b} \neq \boldsymbol{0}$. It is easy to obtain the desired statement from

$$s_\Phi(\boldsymbol{a}, \boldsymbol{b}) = (I_\Phi \otimes I_\Phi)(\boldsymbol{a}, \boldsymbol{b}) = \sum_{\boldsymbol{x} \in V_m} \sum_{\boldsymbol{y} \in V_m} I_\Phi(\boldsymbol{x}, \boldsymbol{y}) I_\Phi(\boldsymbol{x} \oplus \boldsymbol{a}, \boldsymbol{y} \oplus \boldsymbol{b}),$$

using the Fourier transform for the functions $I_\Phi(\boldsymbol{x}, \boldsymbol{y})$ and $I_\Phi(\boldsymbol{x} \oplus \boldsymbol{a}, \boldsymbol{y} \oplus \boldsymbol{b})$. □

THEOREM 8.16. *Bent mappings in $\mathcal{F}_{m,l}$ exist only for even $m \geq 2l$.*

PROOF. Let Φ be a bent mapping. According to the result of Problem 8.13, for any $\boldsymbol{b} \in V_l$, $\boldsymbol{b} \neq \boldsymbol{0}$, we have $\overline{W}_{I_\Phi}(\boldsymbol{a}, \boldsymbol{b}) = \pm 2^{m/2}$. Therefore, m is even. We introduce the following objects:

$$U = 2^{-n/2} \sum_{\substack{\boldsymbol{b} \in V_l \\ \boldsymbol{b} \neq \boldsymbol{0}}} \overline{W}_{I_\Phi}(\boldsymbol{0}, \boldsymbol{b}), \quad r_0 = \#\{\boldsymbol{b} \in V_l \setminus \{\boldsymbol{0}\} \mid \overline{W}_{I_\Phi}(\boldsymbol{0}, \boldsymbol{b}) = 2^{m/2}\}.$$

One has $U = 2r_0 - 2^l + 1$, so U is odd. In addition we have

$$\begin{aligned}\sum_{\substack{\boldsymbol{b} \in V_l \\ \boldsymbol{b} \neq \boldsymbol{0}}} \overline{W}_{I_\Phi}(\boldsymbol{0}, \boldsymbol{b}) &= \sum_{\boldsymbol{b} \in V_l} \overline{W}_{I_\Phi}(\boldsymbol{0}, \boldsymbol{b}) - \overline{W}_{I_\Phi}(\boldsymbol{0}, \boldsymbol{0}) \\ &= \sum_{\boldsymbol{b} \in V_l} \sum_{\boldsymbol{x} \in V_m} (-1)^{\langle \boldsymbol{b}, \Phi(\boldsymbol{x})\rangle} - 2^m = \sum_{\boldsymbol{x} \in V_m} \sum_{\boldsymbol{b} \in V_l} (-1)^{\langle \boldsymbol{b}, \Phi(\boldsymbol{x})\rangle} - 2^m = 2^l r_1 - 2^m,\end{aligned}$$

where $r_1 = \#\{\boldsymbol{x} \in V_m \mid \Phi(\boldsymbol{x}) = \boldsymbol{0}\}$. Then we get $U = 2^{-m/2}(2^l r_1 - 2^m) = 2^{l-m/2} r_1 - 2^{m/2}$ and $r_1 = (U + 2^{m/2})2^{m/2-l}$. As r_1 is an integer and U is odd, the number $2^{m/2-l}$ must be an integer. Therefore, $m \geq 2l$. □

EXAMPLE 8.17. Similarly to the class $\mathcal{M}$ (§6.3) of bent functions, a class of bent mappings can be constructed. For any one-to-one mapping $\pi \in \mathcal{F}_{m,m}$ and for any mapping $\psi \in \mathcal{F}_{m,m}$, we define the mapping Φ from $V_m \times V_m$ to V_m as follows:

$$\Phi(\boldsymbol{x}, \boldsymbol{y}) = \boldsymbol{x} \cdot \pi(\boldsymbol{y}) \oplus \psi(\boldsymbol{y}),$$

where $\boldsymbol{x} \in V_m$, $\boldsymbol{y} \in V_m$, and $\boldsymbol{x} \cdot \pi(\boldsymbol{y})$ are products of elements of the field $\mathbb{F}_{2^m}$ (we assume that there is a natural identification between the elements of V_m and those of $\mathbb{F}_{2^m}$). It is easy to show by a simple check that the mapping $\Phi(\boldsymbol{x}, \boldsymbol{y})$ defined in this way is a bent mapping.

The results above describe rather detailed constructions of bent functions. But, as was noted before, perfect nonlinear bent mappings (and bent functions) are extremal mathematical examples of mappings resistant against certain cryptanalytic methods, and so they cannot be used as primitives for cryptographic schemes and algorithms since they lack the fundamental property of being balanced. For this reason, methods for constructing balanced mappings with properties close to those of perfect nonlinear bent mappings have been developed.

PROPOSITION 8.18. *The number $s_\Phi(\boldsymbol{a},\boldsymbol{b})$ is even for any $\Phi \in \mathcal{F}_{m,l}$, and the inequality $\mu l(\Phi) \geq 2$ holds.*

PROOF. Put

$$S_\Phi(\boldsymbol{a},\boldsymbol{b}) = \{\boldsymbol{x} \in V_m \mid \Phi(\boldsymbol{x} \oplus \boldsymbol{a}) \oplus \Phi(\boldsymbol{x}) = \boldsymbol{b}\}$$

for $\boldsymbol{a} \in V_m$, $\boldsymbol{a} \neq \boldsymbol{0}$. If $S_\Phi(\boldsymbol{a},\boldsymbol{b}) \neq \varnothing$, then it contains a vector $\boldsymbol{x} \in V_m$. But it is clear that the vector $\boldsymbol{x}' = \boldsymbol{x} \oplus \boldsymbol{a}$ also belongs to the set $S_\Phi(\boldsymbol{a},\boldsymbol{b})$. Therefore, $s_\Phi(\boldsymbol{a},\boldsymbol{b})$ is an even number, $s_\Phi(\boldsymbol{a},\boldsymbol{b}) \geq 2$ and $\mu l(\Phi) \geq 2$. □

DEFINITION 8.19. A mapping $\Phi \in \mathcal{F}_{m,l}$ is called *almost perfect nonlinear, or APN*, if $\mu l(\Phi) = 2$ and $s_\Phi(\boldsymbol{a},\boldsymbol{b}) = 2$ for all $\boldsymbol{a} \in V_m$, $\boldsymbol{a} \neq \boldsymbol{0}$, $\boldsymbol{b} \in V_l$.

According to Theorem 8.18, the inequality $\mu l(\Phi) \geq 2^{m-l}$ holds, so APN mappings can exist only for $l \geq m-1$.

APN mappings are the most resistant against differential cryptanalysis amongst all resilient mappings.

LEMMA 8.20. *For a mapping $\Phi \in \mathcal{F}_{m,l}$ the following inequality holds:*

$$\sum_{\boldsymbol{a}\in V_m} \sum_{\substack{\boldsymbol{b}\in V_l \\ \boldsymbol{b}\neq\boldsymbol{0}}} \overline{W}^{\,4}_{I_\Phi}(\boldsymbol{a},\boldsymbol{b}) \geq 2^{2m}(3\cdot 2^{m+l} - 2^{l+1} - 2^{2m});$$

this inequality becomes equality if and only if Φ is an APN mapping.

PROOF. From the definition of $t_\Phi(\boldsymbol{a},\boldsymbol{b})$ we have

$$t_\Phi(\boldsymbol{a},\boldsymbol{0}) = \begin{cases} 2^{n-1} & \text{if } \boldsymbol{a} = \boldsymbol{0}, \\ 0 & \text{otherwise}, \end{cases}$$

and from the definition of $s_\Phi(\boldsymbol{a},\boldsymbol{b})$ we have $s_\Phi(\boldsymbol{0},\boldsymbol{0}) = 2^n$. Using the equality $\overline{W}^{\,2}_f(\boldsymbol{a},\boldsymbol{b}) = \overline{W}_{f\otimes f}(\boldsymbol{a},\boldsymbol{b})$, which holds for any integer-valued function, and Lemmas 8.3 and 8.4, we get

$$\begin{aligned}
\sum_{\boldsymbol{a}\in V_m} \sum_{\substack{\boldsymbol{b}\in V_l \\ \boldsymbol{b}\neq\boldsymbol{0}}} \overline{W}^{\,4}_{I_\Phi}(\boldsymbol{a},\boldsymbol{b}) &= \sum_{\boldsymbol{a}\in V_m} \sum_{\boldsymbol{b}\in V_l} \overline{W}^{\,4}_{I_\Phi}(\boldsymbol{a},\boldsymbol{b}) - \sum_{\boldsymbol{a}\in V_m} \overline{W}^{\,4}_{I_\Phi}(\boldsymbol{a},\boldsymbol{0}) \\
&= \sum_{\boldsymbol{a}\in V_m} \sum_{\boldsymbol{b}\in V_l} \overline{W}^{\,2}_{I_\Phi\otimes I_\Phi}(\boldsymbol{a},\boldsymbol{b}) - \sum_{\boldsymbol{a}\in V_m} (2t_\Phi(\boldsymbol{a},\boldsymbol{0}))^4 \\
&= \sum_{\boldsymbol{a}\in V_m} \sum_{\boldsymbol{b}\in V_l} \overline{W}^{\,2}_{s_\Phi}(\boldsymbol{a},\boldsymbol{b}) - 2^4 \sum_{\boldsymbol{a}\in V_m} t^4_\Phi(\boldsymbol{a},\boldsymbol{0}) \qquad (8.15) \\
&= \sum_{\boldsymbol{a}\in V_m} \sum_{\boldsymbol{b}\in V_l} \overline{W}_{s_\Phi\otimes s_\Phi}(\boldsymbol{a},\boldsymbol{b}) - 2^{4m} \\
&= 2^{m+l}(s_\Phi \otimes s_\Phi)(\boldsymbol{0},\boldsymbol{0}) - 2^{4m}.
\end{aligned}$$

Using the definition of convolution, we obtain

$$(8.16)\qquad \begin{aligned}(s_\Phi \otimes s_\Phi)(\mathbf{0},\mathbf{0}) &= \sum_{\boldsymbol{a}\in V_m}\sum_{\boldsymbol{b}\in V_l} s_\Phi(\boldsymbol{a},\boldsymbol{b})s_\Phi(\boldsymbol{a},\boldsymbol{b})\\ &= \sum_{\substack{\boldsymbol{a}\in V_m\\ \boldsymbol{a}\neq\mathbf{0}}}\sum_{\boldsymbol{b}\in V_l} s_\Phi^2(\boldsymbol{a},\boldsymbol{b}) + \sum_{\boldsymbol{b}\in V_l} s_\Phi^2(\mathbf{0},\boldsymbol{b})\\ &= \sum_{\substack{\boldsymbol{a}\in V_m\\ \boldsymbol{a}\neq\mathbf{0}}}\sum_{\boldsymbol{b}\in V_l} s_\Phi^2(\boldsymbol{a},\boldsymbol{b}) + s_\Phi^2(\mathbf{0},\mathbf{0}),\end{aligned}$$

as $s_\Phi(\mathbf{0},\boldsymbol{b}) = 0$ for $\boldsymbol{b}\neq\mathbf{0}$. Besides, $s_\Phi(\mathbf{0},\mathbf{0}) = 2^n$. In view of the above and by combining equalities (8.15) and (8.16), we get

$$(8.17)\qquad \sum_{\boldsymbol{a}\in V_m}\sum_{\substack{\boldsymbol{b}\in V_l\\ \boldsymbol{b}\neq\mathbf{0}}} \overline{W}^4_{I_\Phi}(\boldsymbol{a},\boldsymbol{b}) = 2^{m+l}\sum_{\substack{\boldsymbol{a}\in V_m\\ \boldsymbol{a}\neq\mathbf{0}}}\sum_{\boldsymbol{b}\in V_l} s_\Phi^2(\boldsymbol{a},\boldsymbol{b}) + 2^{3m+l} - 2^{4m}.$$

For any even number $k\geq 0$ we have $k^2\geq 2k$, and the equality $k^2 = 2k$ holds only if $k=2$ or $k=0$. Since $s_\Phi(\boldsymbol{a},\boldsymbol{b})$ is even (§8.18), for any $\boldsymbol{a}\neq\mathbf{0}$ and for any $\boldsymbol{b}\in V_l$ we have $s_\Phi^2(\boldsymbol{a},\boldsymbol{b})\geq 2s_\Phi(\boldsymbol{a},\boldsymbol{b})$. This inequality becomes equality if and only if the mapping Φ is an APN mapping. Using the above-mentioned inequalities, we obtain:

$$(8.18)\qquad \sum_{\substack{\boldsymbol{a}\in V_m\\ \boldsymbol{a}\neq\mathbf{0}}}\sum_{\boldsymbol{b}\in V_l} s_\Phi^2(\boldsymbol{a},\boldsymbol{b}) \geq 2\sum_{\substack{\boldsymbol{a}\in V_m\\ \boldsymbol{a}\neq\mathbf{0}}}\sum_{\boldsymbol{b}\in V_l} s_\Phi(\boldsymbol{a},\boldsymbol{b}) = 2\sum_{\substack{\boldsymbol{a}\in V_m\\ \boldsymbol{a}\neq\mathbf{0}}} 2^m = 2\cdot(2^m-1)\cdot 2^m.$$

If we substitute (8.18) in (8.17), we get

$$\begin{aligned}\sum_{\boldsymbol{a}\in V_m}\sum_{\substack{\boldsymbol{b}\in V_l\\ \boldsymbol{b}\neq\mathbf{0}}} \overline{W}^4_{I_\Phi}(\boldsymbol{a},\boldsymbol{b}) &\geq 2^{m+l}[2\cdot(2^m-1)2^m] + 2^{3m+l} - 2^{4m}\\ &= 2^{2m}(3\cdot 2^{m+l} - 2^{l+1} - 2^{2m}),\end{aligned}$$

and the equality holds if and only if the mapping Φ is an APN mapping. □

THEOREM 8.21. *For any mapping $\Phi\in\mathcal{F}_{m,l}$, the following inequality holds:*

$$(8.19)\qquad \mu t(\Phi) \geq \frac{1}{2}\Big(3\cdot 2^m - 2 - 2\frac{(2^m-1)(2^{m-1}-1)}{2^l-1}\Big)^{1/2}.$$

PROOF. Observe that according to Lemma 8.4 the following equalities hold:

$$\mu t^2(\Phi) = \max_{\boldsymbol{a}\in V_m}\max_{\substack{\boldsymbol{b}\in V_l\\ \boldsymbol{b}\neq\mathbf{0}}} t_\Phi^2(\boldsymbol{a},\boldsymbol{b}) = \max_{\boldsymbol{a}\in V_m}\max_{\substack{\boldsymbol{b}\in V_l\\ \boldsymbol{b}\neq\mathbf{0}}}\Big(\frac{1}{4}\overline{W}^2_{I_\Phi}(\boldsymbol{a},\boldsymbol{b})\Big).$$

For any function $\theta(\boldsymbol{a},\boldsymbol{b})$, $\boldsymbol{a}\in V_m$, $\boldsymbol{b}\in V_l$, one has

$$M = \max_{\boldsymbol{a}\in V_m}\max_{\substack{\boldsymbol{b}\in V_l\\ \boldsymbol{b}\neq\mathbf{0}}}\theta^2(\boldsymbol{a},\boldsymbol{b}) \geq \frac{\sum_{\boldsymbol{a}\in V_m}\sum_{\substack{\boldsymbol{b}\in V_l\\ \boldsymbol{b}\neq\mathbf{0}}}\theta^4(\boldsymbol{a},\boldsymbol{b})}{\sum_{\boldsymbol{a}\in V_m}\sum_{\substack{\boldsymbol{b}\in V_l\\ \boldsymbol{b}\neq\mathbf{0}}}\theta^2(\boldsymbol{a},\boldsymbol{b})},$$

and this becomes equality if and only if all nonzero values of $\theta^2(\boldsymbol{a},\boldsymbol{b})$ are equal to M. For an arbitrary mapping Φ, we have

$$\begin{aligned}\sum_{\boldsymbol{a}\in V_m}\sum_{\substack{\boldsymbol{b}\in V_l\\ \boldsymbol{b}\neq\boldsymbol{0}}}\overline{W}^2_{I_\Phi}(\boldsymbol{a},\boldsymbol{b}) &= \sum_{\boldsymbol{a}\in V_m}\sum_{\boldsymbol{b}\in V_l}\overline{W}^2_{I_\Phi}(\boldsymbol{a},\boldsymbol{b}) - \sum_{\boldsymbol{a}\in V_m}\overline{W}^2_{I_\Phi}(\boldsymbol{a},\boldsymbol{0})\\ &= \sum_{\boldsymbol{a}\in V_m}\sum_{\boldsymbol{b}\in V_l}\overline{W}_{I_\Phi\otimes I_\Phi}(\boldsymbol{a},\boldsymbol{b}) - 4\sum_{\boldsymbol{a}\in V_m} t^2_\Phi(\boldsymbol{a},\boldsymbol{0})\\ &= \sum_{\boldsymbol{a}\in V_m}\sum_{\boldsymbol{b}\in V_l}\overline{W}_{s_\Phi}(\boldsymbol{a},\boldsymbol{b}) - 4(2^{m-1})^2 = 2^{m+l}s_\Phi(\boldsymbol{0},\boldsymbol{0}) - 2^{2m}\\ &= 2^{2m+l} - 2^{2m} = 2^{2m}(2^l-1).\end{aligned}$$

Now using the above inequalities and Lemma 8.20, we get

$$\begin{aligned}4\mu t^2(\Phi) &= \max_{\boldsymbol{a}\in V_m}\max_{\substack{\boldsymbol{b}\in V_l\\ \boldsymbol{b}\neq\boldsymbol{0}}}(\overline{W}^2_{I_\Phi}(\boldsymbol{a},\boldsymbol{b})) \geq \frac{2^{2m}(3\cdot 2^{m+l} - 2^{l+1} - 2^{2m})}{2^{2m}(2^l-1)}\\ &= \frac{3\cdot 2^{m+l} - 2^{l+1} - 2^{2m}}{2^l-1} = 3\cdot 2^m - 2 - 2\frac{(2^m-1)(2^{m-1}-1)}{2^l-1}. \qquad \square\end{aligned}$$

DEFINITION 8.22. A mapping $\Phi\in\mathcal{F}_{m,l}$ is called an *almost bent (AB) mapping* or a *plateaued mapping* if $\mu t(\Phi) = \frac{1}{2}\big(3\cdot 2^m - 2 - 2\frac{(2^m-1)(2^{m-1}-1)}{2^l-1}\big)^{1/2}$ and thus all nonzero values of $t^2_\Phi(\boldsymbol{a},\boldsymbol{b})$ equal $\frac{1}{4}\big(3\cdot 2^m - 2 - 2\frac{(2^m-1)(2^{m-1}-1)}{2^l-1}\big)$, i.e., for any $\boldsymbol{b}\in V_l$, $\boldsymbol{b}\neq\boldsymbol{0}$, the Boolean functions $\langle\boldsymbol{b},\Phi(\boldsymbol{x})\rangle$ are plateaued with the same order.

As can be seen from the definition, the property of being a plateaued mapping is a reducible property of mappings.

PROPOSITION 8.23. *Any AB mapping is an APN mapping.*

PROOF. If Φ is an AB mapping, then

$$\sum_{\boldsymbol{a}\in V_m}\sum_{\substack{\boldsymbol{b}\in V_l\\ \boldsymbol{b}\neq\boldsymbol{0}}}\overline{W}^4_{I_\Phi}(\boldsymbol{a},\boldsymbol{b}) = 2^{2m}(3\cdot 2^{m+l} - 2^{l+1} - 2^{2m}).$$

Then according to Lemma 8.20, Φ is an APN mapping. $\square$

EXAMPLE 8.24. Let $\Phi(\boldsymbol{x}) = \boldsymbol{x}^{2^k+1}$ be a polynomial over the field $\mathbb{F}_{2^m}$. If m is odd, $1<k<m$, and $\gcd(m,k)=1$, then Φ is an almost bent mapping (and at the same time an almost perfect nonlinear mapping).

EXAMPLE 8.25. Put $\Phi(\boldsymbol{x}) = \boldsymbol{x}^{-1}$ for all $\boldsymbol{x}\in\mathbb{F}_{2^m}\setminus\{\boldsymbol{0}\}$ and $\Phi(\boldsymbol{0})=\boldsymbol{0}$. If m is odd, then the mapping Φ is an almost perfect nonlinear mapping but not an almost bent mapping.

COROLLARY 8.26. *Let $\Phi\in\mathcal{F}_{m,l}$ be a bent mapping as well as an almost bent mapping. Then $m=2$, $l\leq 1$.*

PROOF. Since Φ is a bent mapping, it follows from Theorem 8.16 that $m\geq 2l$ and m is even. Since Φ is an almost bent mapping, Proposition 8.23 shows that Φ is an APN mapping and $l\geq m-1$. The statement follows from these inequalities. $\square$

LEMMA 8.27. *Suppose that $m>2$ and a mapping $\Phi\in\mathcal{F}_{m,l}$ is an almost bent mapping. Then $m\leq l$.*

PROOF. For any mapping $\Phi \in \mathcal{F}_{m,l}$ the inequality $t^*_\Phi \geq \frac{1}{2}2^{m/2}$ holds. It becomes equality only for bent mappings. Since $m > 2$, one derives from Corollary 8.26 and the hypothesis of the lemma that Φ is an AB mapping but not a bent mapping. Moreover, the following strict inequality holds:

$$\frac{1}{2}\sqrt{\frac{3 \cdot 2^{m+l} - 2^{l+1} - 2^{2m}}{2^l - 1}} > \frac{1}{2}\sqrt{2^m}.$$

It is easy to obtain the latter inequality in the form

$$2^{l+1}(2^m - 1) - 2^m(2^m - 1) > 0,$$

which gives $l + 1 > m$. □

PROBLEM 8.28. Prove that the number

$$\frac{(2^m - 1)(2^{m-1} - 1)}{2^l - 1}$$

is not an integer for any $m < l$.

THEOREM 8.29. *Let $m > 2$ and suppose that a mapping $\Phi \in \mathcal{F}_{m,l}$ is almost bent. Then $m = l$ and*

$$\mu t(\Phi) = \frac{1}{2} 2^{\frac{m+1}{2}}.$$

PROOF. Suppose that the hypotheses of the theorem are satisfied and (8.19) is an equality. Then the number

$$\frac{(2^m - 1)(2^{m-1} - 1)}{2^l - 1}$$

is an integer and by Problem 8.28, $m \geq l$. On the other hand, according to Lemma 8.27, $m \leq l$. Therefore, $m = l$ and $\mu t(\Phi) = \frac{1}{2}\, 2^{(m+1)/2}$. □

COROLLARY 8.30. *For $m > 2$ a mapping $\Phi \in \mathcal{F}_{m,l}$ is an almost bent mapping if and only if $m = l$ and for any $\boldsymbol{b} \in V_l$, $\boldsymbol{b} \neq \boldsymbol{0}$, the Boolean function $\langle \boldsymbol{b}, \Phi(\boldsymbol{x})\rangle$ is plateaued of order $m - 1$.*

PROBLEM 8.31. Suppose that $m > 2$ and that $\Phi \in \mathcal{F}_{m,l}$ is an almost bent mapping. Prove that $\deg\langle \boldsymbol{b}, \Phi(\boldsymbol{x})\rangle \leq \frac{m-1}{2}$ for any $\boldsymbol{b} \in V_l$, $\boldsymbol{b} \neq \boldsymbol{0}$.

PROBLEM 8.32. Introduce for mappings a notion similar to "a pair of mutually complementary plateaued mappings of order $m-1$" (see Definition 6.95) and prove that a mapping $\Phi \in \mathcal{F}_{m,m}$ is a bent mapping if and only if the restrictions of Φ to an arbitrary hyperplane and to its coset form a pair of mutually complementary plateaued mappings of order $m - 2$.

Suppose that $\Phi = (f_1, f_2, \ldots, f_m) \in \mathcal{F}_{m,m}$. We use the algebraic normal form (§2.1) of the coordinate components of this mapping:

$$\begin{aligned} f_i(\boldsymbol{x}) &= f_i(x^{(1)}, x^{(2)}, \ldots, x^{(m)}) \\ &= \bigoplus_{(u^{(1)},\ldots,u^{(m)})^\top \in V_m} g_i(u^{(1)}, u^{(2)}, \ldots, u^{(m)})(x^{(1)})^{u^{(1)}} \cdot \cdots \cdot (x^{(m)})^{u^{(m)}} \\ &= \bigoplus_{\boldsymbol{u} \in V_m} g_i(\boldsymbol{u})\boldsymbol{x}^{\boldsymbol{u}}, \end{aligned}$$

where $\boldsymbol{x} \in V_m$, $i = 1, 2, \dots, m$. Then the mapping Φ can be written as follows:

$$\begin{aligned}\Phi(\boldsymbol{x}) &= (f_1(\boldsymbol{x}), \dots, f_m(\boldsymbol{x})) \\ &= \bigoplus_{(u^{(1)},\dots,u^{(m)})\in V_m} (g_1(u^{(1)}, \dots, u^{(m)}), \dots, g_n(u^{(1)}, \dots, u^{(m)}))(x^{(1)})^{u^{(1)}} \cdots (x^{(m)})^{u^{(m)}} \\ &= \bigoplus_{\boldsymbol{u}\in V_m} \Psi(\boldsymbol{u})\boldsymbol{x}^{\boldsymbol{u}},\end{aligned}$$

where $\Psi = (g_1, g_2, \dots, g_m) \in \mathcal{F}_{m,m}$ and g_i $(i = 1, \dots, m)$ are its coordinate functions.

DEFINITION 8.33. The *algebraic degree of a mapping* Φ is the maximum algebraic degree of its coordinate functions: $\deg \Phi = \max\{\mathrm{wt}(\boldsymbol{u}) \mid \boldsymbol{u} \in V_m, \Psi(\boldsymbol{u}) \neq \boldsymbol{0}\}$.

From Problem 8.31 it follows that if $m > 2$ and $\Phi \in \mathcal{F}_{m,m}$ is an almost bent mapping, then $\deg \Phi \leq \frac{m-1}{2}$.

We note that $\deg \Phi = m - 1$ for the self-mapping of the field $\mathbb{F}_{2^m}$ defined by the formula $\Phi(\boldsymbol{x}) = \boldsymbol{x}^{-1} = \boldsymbol{x}^{2^m-2}$ (Example 8.25). Therefore, Φ is not an AB mapping.

We consider some transformations which preserve the properties of "being an APN mapping" or "being an AB mapping". From the definitions of APN and AB mappings and taking into consideration problems in §§8.1, 8.2, it is easy to obtain the following statements.

LEMMA 8.34. *Let Φ be an APN mapping from V_m to V_m, $\mathfrak{g} \in \mathfrak{GA}(V_m)$, and $\mathfrak{h}$ an arbitrary affine mapping from V_m to V_m. Then the following statements are true:*

1) *the mappings $\Phi' = \mathfrak{g} \circ \Phi$ and $\Phi'' = \Phi \circ \mathfrak{g}$ are APN mappings;*
2) *if Φ is a one-to-one mapping, then Φ^{-1} is also an APN mapping;*
3) *the mapping $\Psi(\boldsymbol{x}) = \Phi(\boldsymbol{x}) \oplus \mathfrak{h}$ is an APN mapping.*

LEMMA 8.35. *Let Φ be an AB mapping from V_m to V_m, $\mathfrak{g} \in \mathfrak{GA}(V_m)$, and $\mathfrak{h}$ an arbitrary affine mapping from V_m to V_m. Then the following statements are true:*

1) *the mappings $\Phi' = \mathfrak{g} \circ \Phi$ and $\Phi'' = \Phi \circ \mathfrak{g}$ are AB mappings;*
2) *if Φ is a one-to-one mapping, then Φ^{-1} is also an AB mapping;*
3) *the mapping $\Psi(\boldsymbol{x}) = \Phi(\boldsymbol{x}) \oplus \mathfrak{h}$ is an AB mapping.*

8.2. Coding-theoretic approach to the study of APN and AB mappings

In this section we will consider an interpretation of APN and AB mappings in terms of coding theory. As usual, we will denote by n the length of a code. By m we will denote the degree of a given extension over the field $\mathbb{F}_2$; we will assume that $m > 1$. Codes, as used in this section, will be of primitive length, i.e., $n = 2^m - 1$. We will interpret an arbitrary element $\boldsymbol{a}$ of the field $\mathbb{F}_{2^m}$ as a column vector in the space V_m with coordinates in a suitable basis, which will make it possible to go from the finite field $\mathbb{F}_{2^m}$ to the vector space V_m and vice versa. We will denote a primitive element of the field $\mathbb{F}_{2^m}$ by $\boldsymbol{\alpha} \in \mathbb{F}_{2^m}$.

It is known that an arbitrary mapping $\Phi \in \mathcal{F}_{m,m}$ can be uniquely represented as a polynomial in the ring $\mathbb{F}_{2^m}[\boldsymbol{x}]$ whose degree is at most $2^m - 1$, i.e.,

$$\Phi(\boldsymbol{x}) = \bigoplus_{i=0}^{2^m-1} \boldsymbol{a}_i \boldsymbol{x}^i \tag{8.20}$$

for some $\boldsymbol{a}_i \in \mathbb{F}_{2^m}$. The properties of mappings studied in this section are either satisfied or not, simultaneously for both mappings $\Phi(\boldsymbol{x})$ and $\Phi(\boldsymbol{x}) \oplus \boldsymbol{c}$, $\boldsymbol{c} \in \mathbb{F}_{2^n}$. Thus without loss of generality we can assume that $\Phi(\boldsymbol{0}) = \boldsymbol{0}$, i.e., $\boldsymbol{a}_0 = \boldsymbol{0}$. We will say that the mapping $\Phi \in \mathcal{F}_{m,m}$ defined by (8.20) with $\boldsymbol{a}_0 = \boldsymbol{0}$ is a *polynomial mapping*.

Let $\boldsymbol{C}$ be a linear code of length n, $\boldsymbol{C}^\perp$ its dual code, and $(A'_0, A'_1, \ldots, A'_n)$ the weight spectrum of the code $\boldsymbol{C}^\perp$. We denote by $0 \le j_1, \ldots, j_s \le n$ $(0 < j_1 < \cdots < j_s \le n)$ those indices i for which $A'_i \ne 0$.

DEFINITION 8.36. For a linear code $\boldsymbol{C}$ we call $d_{\boldsymbol{C}}^\perp = j_1$ the *dual distance* and $d'_{\boldsymbol{C}} = s$ the *external distance* of this code. The set $M_{\boldsymbol{C}} = \{j_1, \ldots, j_s\}$ will be called the *characteristic set* of the code $\boldsymbol{C}$.

Later we will need a technical statement which describes the properties of the weight spectra of some classes of codes.

THEOREM 8.37. *Let $\boldsymbol{C}$ be a linear $[n, n-k, d]$-code, where $d \ge 3$, $n = 2^m - 1$, m is even, and the code $\boldsymbol{C}^\perp$ does not contain the code word $(1, \ldots, 1)$ (i.e., the code $\boldsymbol{C}$ contains at least one code word of odd weight). Let $(A_0, A_1, \ldots, A_n)$ (resp. $(A'_0, A'_1, \ldots, A'_n)$) be the weight spectrum of the code $\boldsymbol{C}$ (resp. $\boldsymbol{C}^\perp$). Let i_0 be the minimum integer i, $0 < i < 2^{m-1}$, such that*

$$A'_i + A'_{2^m - i} \ne 0.$$

Then $k \ge m$ and the following properties hold:

1. If $k \ge 2m$, then $i_0 \le 2^{m-1} - 2^{(m-1)/2}$. In addition, if $i_0 = 2^{m-1} - 2^{(m-1)/2}$, then $A_3 = A_4 = 0$, $k = 2m$, $M_{\boldsymbol{C}} = \{2^{m-1}, 2^{m-1} \pm 2^{(m-1)/2}\}$ and the weight spectrum of the code $\boldsymbol{C}^\perp$ is the same as the weight spectrum of the dual of the two-error correcting BCH code (see Table 8.1).

TABLE 8.1

i	A'_i
0	1
$2^{m-1} - 2^{(m-1)/2}$	$(2^m - 1)(2^{m-2} + 2^{(m-3)/2})$
2^{m-1}	$(2^m - 1)(2^{m-1} + 1)$
$2^{m-1} + 2^{(m-1)/2}$	$(2^m - 1)(2^{m-2} - 2^{(m-3)/2})$

2. If $m \le k < 2m$, then the minimum distance of $\boldsymbol{C}$ is at most 4. Furthermore, if $i_0 \ge 2^{m-1} - 2^{(m-1)/2}$, then

$$A_3 + A_4 \le \frac{(2^{m-1} - 1)(2^{3m-3} - 2^{k+m-3})}{3 \cdot 2^{k-1}}. \tag{8.21}$$

PROOF. We need to use the Pless identities [**314**] for the fourth power moments of the weight spectrum of the code $\boldsymbol{C}^{\perp}$. They are

$$\sum_{i=0}^{n} iA'_i = 2^{k-1}n, \quad \sum_{i=0}^{n} i^2 A'_i = 2^{k-2}n(n+1),$$
$$\sum_{i=0}^{n} i^3 A'_i = 2^{k-3}(n^2(n+3) - 3!A_3), \tag{8.22}$$
$$\sum_{i=0}^{n} i^4 A'_i = 2^{k-4}(n(n+1)(n^2+5n-2) + 4!(A_4 - nA_3)).$$

For a positive number l we consider the quantity

$$J_l = \sum_{i=1}^{n} (i - 2^{m-1})^l A'_i.$$

Obviously $J_l \geq 0$ for even l. Since, for even l, $(i - 2^{m-1})^l = ((2^m - i) - 2^{m-1})^l$, one has

$$J_l = \sum_{i=1}^{n} (i - 2^{m-1})^l A'_i = \sum_{i=i_0}^{2^{m-1}-1} (i - 2^{m-1})^l (A'_i + A'_{2^m - i}). \tag{8.23}$$

We simplify the expressions for J_2 and J_4 using (8.22) and get

$$J_2 = 2^{k+m-2} - 2^{2m-2}, \tag{8.24}$$

$$J_4 = 2^{k+m-4}(3 \cdot 2^m - 2) - 2^{4m-4} + 3 \cdot 2^{k-1}(A_3 + A_4). \tag{8.25}$$

Since $J_2 \geq 0$, it follows from (8.24) that $k \geq m$. From (8.24) and (8.25) we get

$$J_4 - 2^{m-1} J_2 = (2^{m-1} - 1)(2^{k+m-3} - 2^{3m-3}) + 3 \cdot 2^{k-1}(A_3 + A_4). \tag{8.26}$$

From (8.23) we obtain another expression for this difference:

$$J_4 - 2^{m-1} J_2 = \sum_{i=i_0}^{2^{m-1}-1} (i - 2^{m-1})^2 \big((i - 2^{m-1})^2 - 2^{m-1}\big)(A'_i + A'_{2^m - i}). \tag{8.27}$$

Note that for all i such that $2^{m-1} - 2^{(m-1)/2} < i \leq 2^{m-1} - 1$ and $A'_i + A'_{2^m - i} \neq 0$, the ith terms in the sum (8.27) are negative.

1. Suppose that $k \geq 2m$. Then according to (8.26), $J_4 - 2^{m-1} J_2 \geq 0$, so the sum (8.27) must contain at least one nonnegative term. Therefore, $i_0 \leq 2^{m-1} - 2^{(m-1)/2}$.

Suppose that $i_0 = 2^{m-1} - 2^{(m-1)/2}$. If we substitute this value for i_0 in (8.27), we get $J_4 - 2^{m-1} J_2 \leq 0$. But as we concluded above, $J_4 - 2^{m-1} J_2 \geq 0$, so $J_4 - 2^{m-1} J_2 = 0$. Then from (8.26) it follows that $k = 2m$ and $A_3 + A_4 = 0$. We obtain from (8.27) that for all i such that $2^{m-1} - 2^{(m-1)/2} < i \leq 2^{m-1} - 1$, the equality $A'_i + A'_{2^m - i} = 0$ holds. Therefore, the code $\boldsymbol{C}^{\perp}$ has dimension $2m$, the code $\boldsymbol{C}$ has minimum distance less than 5, and only A'_i can be nonzero for $i = 2^{m-1} \pm 2^{(m-1)/2}$ and $i = 2^{m-1}$. Obviously the following relation holds for these quantities:

$$A'_{2^{m-1} - 2^{(m-1)/2}} + A'_{2^{m-1}} + A'_{2^{m-1} + 2^{(m-1)/2}} = 2^{2m} - 1.$$

This relation together with the first two Pless identities for power moments gives the weight spectrum of the code $\boldsymbol{C}^{\perp}$.

2. If $k = m$, then $J_2 = 0$. Therefore the weights of the nonzero code words of the code $\boldsymbol{C}^\perp$ equal 2^{m-1}, i.e., the code $\boldsymbol{C}^\perp$ has the same weight spectrum as the simplex code. Thus, the minimum distance of the code $\boldsymbol{C}$ is not greater than the minimum distance of the Hamming code (which is equal to 3).

Suppose now that $m < k < 2m$. If $d_{\boldsymbol{C}} \geq 5$, then $A_3 + A_4 = 0$ and from (8.25) we get

$$J_4 = 2^{m-4}(3 \cdot 2^{m+k} - 2^{k+1} - 2^{3m}).$$

Since $J_4 \geq 0$, we have $k \geq 2m-1$ and $k = 2m-1$. Thus, the code $\boldsymbol{C}$ has parameters $n = 2^m - 1$, $2^m - 2m$ and the minimum distance that is not less than 5. Therefore, there exists a linear code with parameters $[2^m - 1, 2^m - 2m, 5]$, which is impossible according to the results of [**40, 118**]. Hence the minimum distance of the code $\boldsymbol{C}$ is not greater than 4. If $i_0 \geq 2^{m-1} - 2^{(m-1)/2}$, then by (8.27), $J_4 - 2^{m-1}J_2 \leq 0$ and we get inequality (8.21) from (8.26). □

DEFINITION 8.38. Let $\Phi \in \mathcal{F}_{m,m}$ be a polynomial mapping. A *linear code* $\boldsymbol{C}_\Phi$ *determined by the mapping* Φ is an $[n = 2^m - 1, k, d]$-code with the parity-check matrix

$$H_\Phi = \begin{pmatrix} \mathbf{1} & \boldsymbol{\alpha} & \boldsymbol{\alpha}^2 & \dots & \boldsymbol{\alpha}^{n-1} \\ \Phi(\mathbf{1}) & \Phi(\boldsymbol{\alpha}) & \Phi(\boldsymbol{\alpha}^2) & \dots & \Phi(\boldsymbol{\alpha}^{n-1}) \end{pmatrix}. \tag{8.28}$$

(Recall that any element of the field $\mathbb{F}_{2^m}$ in the matrix H_Φ can be regarded as a column vector in the space V_m with an appropriate basis, for example, with the basis $\mathbf{1}, \boldsymbol{\alpha}, \dots, \boldsymbol{\alpha}^{m-1}$. The size of the binary matrix H_Φ will be $2m \times n$.)

By Definition 8.38, the code $\boldsymbol{C}_\Phi$ consists of binary rows $\boldsymbol{c} = (c_0, c_1, \dots, c_{n-1})$, which are solutions of the following system of equations:

$$\begin{cases} \bigoplus_{i=0}^{n-1} c_i \boldsymbol{\alpha}^i = 0, \\ \bigoplus_{i=0}^{n-1} c_i \Phi(\boldsymbol{\alpha}^i) = 0. \end{cases} \tag{8.29}$$

PROBLEM 8.39. Show that the system (8.29) is equivalent to the system

$$\begin{cases} \bigoplus_{i=0}^{n-1} c_i \boldsymbol{\alpha}^i = 0, \\ \bigoplus_{j=3}^{n} \boldsymbol{a}_j \left(\bigoplus_{i=0}^{n-1} c_i \boldsymbol{\alpha}^{ij} \right) = 0, \end{cases} \tag{8.30}$$

where $\boldsymbol{a}_j \in \mathbb{F}_{2^n}$ are the coefficients of the representation of Φ in (8.20).

Consider now the dual code $\boldsymbol{C}_\Phi^\perp$. By definition, its code words are linear combinations of rows of the matrix H_Φ.

PROBLEM 8.40. Prove that any code word of $\boldsymbol{C}_\Phi^\perp$ is a row of $2^m - 1$ values (excluding the value at zero) of the Boolean function $\langle \boldsymbol{a}, \boldsymbol{x} \rangle \oplus \langle \boldsymbol{b}, \Phi(\boldsymbol{x}) \rangle$ for some $\boldsymbol{a}, \boldsymbol{b} \in \mathbb{F}_{2^m}$.

THEOREM 8.41. *For any polynomial mapping* $\Phi \in \mathcal{F}_{m,m}$, *the following statements are true:*

1) *for the code* $\boldsymbol{C}_\Phi$ *the inequalities* $k_{\boldsymbol{C}_\Phi} \geq 2^m - 1 - 2m$, $3 \leq d_{\boldsymbol{C}_\Phi} \leq 5$ *hold;*

2) Φ *is an APN mapping if and only if* $d_{\boldsymbol{C}_\Phi} = 5$;

3) Φ *is an AB mapping if and only if the characteristic set* $M_{\boldsymbol{C}_\Phi}$ *is equal to* $\{2^{m-1}, 2^{m-1} \pm 2^{(m-1)/2}\}$ *and the weight spectrum of the code* $\boldsymbol{C}_\Phi^\perp$ *is given by Table* 8.1.

PROOF. 1. The inequality $k \geq 2^m - 1 - 2m$ is obvious. As any two columns of the matrix H_Φ are different, $d_{\boldsymbol{C}_\Phi} \geq 3$.

We recall that if a linear $[n, k, d]$-code exists, then there also exists a linear $[n-1, k, d-1]$-code (it is enough to remove from all code words of the first code a coordinate which takes the value of 1 in some code words of minimum weight). Then from the existence of the code $\boldsymbol{C}_\Phi$ with parameters $[2^m - 1, k, 6]$, $k \geq 2^m - 1 - 2m$, follows the existence of a $[2^m - 2, k, 5]$-code. But according to the results of [**40, 118**], such a code does not exist. Therefore $d_{\boldsymbol{C}_\Phi} \leq 5$.

2. By (8.29), $d_{\boldsymbol{C}_\Phi} \leq 4$ if and only if there exist four elements $\boldsymbol{x}$, $\boldsymbol{y}$, $\boldsymbol{x}'$, $\boldsymbol{y}'$ in the field $\mathbb{F}_{2^m}$ such that

$$\begin{cases} \boldsymbol{x} \oplus \boldsymbol{y} \oplus \boldsymbol{x}' \oplus \boldsymbol{y}' = \boldsymbol{0}, \\ \Phi(\boldsymbol{x}) \oplus \Phi(\boldsymbol{y}) \oplus \Phi(\boldsymbol{x}') \oplus \Phi(\boldsymbol{y}') = \boldsymbol{0}. \end{cases} \tag{8.31}$$

The minimum distance of the code $\boldsymbol{C}_\Phi$ is equal to 3 if one of these elements is 0 and equal to 4 if none of the elements is 0.

Let us consider the equation used above:

$$\Phi(\boldsymbol{x} \oplus \boldsymbol{a}) \oplus \Phi(\boldsymbol{x}) = \boldsymbol{b}, \tag{8.32}$$

$\boldsymbol{a}, \boldsymbol{b} \in \mathbb{F}_{2^m}$, $\boldsymbol{a} \neq \boldsymbol{0}$. It can be written in an equivalent form with the help of the system of two equations in two unknowns $\boldsymbol{x}, \boldsymbol{y} \in \mathbb{F}_{2^m}$,

$$\begin{cases} \boldsymbol{x} \oplus \boldsymbol{y} = \boldsymbol{a}, \\ \Phi(\boldsymbol{x}) \oplus \Phi(\boldsymbol{y}) = \boldsymbol{b}. \end{cases} \tag{8.33}$$

By Definition 8.19, Φ is an APN mapping if and only if the system (8.33) has a unique solution $(\boldsymbol{x}, \boldsymbol{y})$. On the other hand, it is obvious that the system (8.33) has at least the solutions $(\boldsymbol{x}, \boldsymbol{y})$, $(\boldsymbol{x}', \boldsymbol{y}')$ if and only if $\boldsymbol{x}, \boldsymbol{y}, \boldsymbol{x}', \boldsymbol{y}'$ satisfy (8.31) and $d_{\boldsymbol{C}_\Phi} \leq 4$. Thus, according to part 1) of the theorem, Φ is an APN mapping if and only if $d_{\boldsymbol{C}_\Phi} = 5$.

3. According to Problem 8.40, the characteristic set $M_{\boldsymbol{C}_\Phi}$ of a polynomial mapping Φ coincides with the set of weights of the Boolean functions $\langle \boldsymbol{a}, \boldsymbol{x} \rangle \oplus \langle \boldsymbol{b}, \Phi(\boldsymbol{x}) \rangle$ for all $\boldsymbol{a}, \boldsymbol{b} \in \mathbb{F}_{2^m}$. According to Corollary 8.30, the mapping Φ is an AB mapping if and only if the set of weights of the Boolean functions $\langle \boldsymbol{a}, \boldsymbol{x} \rangle \oplus \langle \boldsymbol{b}, \Phi(\boldsymbol{x}) \rangle$ for all $\boldsymbol{a}, \boldsymbol{b} \in \mathbb{F}_{2^m}$ coincides with $\{2^{m-1}, 2^{m-1} \pm 2^{(m-1)/2}\}$. □

COROLLARY 8.42. *Let $\Phi \in \mathcal{F}_{m,m}$ be a polynomial APN mapping. Then*
1) $k_{\boldsymbol{C}_\Phi} = 2^m - 2m - 1$;
2) the code $\boldsymbol{C}_\Phi$ does not contain the code word $(1, \ldots, 1)$.

PROOF. 1) By Theorem 8.41, the linear code $\boldsymbol{C}_\Phi$ is an $[n, k, d]$-code with parameters $n = 2^m - 1$, $d = 5$, and $k \geq n - 2m$. If $k = n + 1 - 2m$, then we have a linear $[2^m - 1, 2^m - 2m, 5]$-code. But the results of [**40, 118**] show that such a code does not exist. If $k > n + 1 - 2m$, we have a linear $[n, k, 5]$-code, from which it is easy to obtain an $[n, n + 1 - 2m, 5]$-code and again get a contradiction. Therefore $k = 2^m - 2m - 1$.

2) The statement is obvious since $d_{\boldsymbol{C}_\Phi} = 5$, so the code $\boldsymbol{C}_\Phi$ contains at least one code word of weight 5. □

Let us introduce a partial ordering $\prec$ for the set of polynomial mappings in the following way: $\Phi_1 \prec \Phi_2$ if $\boldsymbol{C}_{\Phi_1} \subseteq \boldsymbol{C}_{\Phi_2}$.

COROLLARY 8.43. *Let $\Phi \in \mathcal{F}_{m,m}$ be a polynomial APN mapping. Then any polynomial mapping Φ' such that $\Phi' \prec \Phi$ is an APN mapping.*

PROOF. If we suppose that Φ' is not an APN mapping, then according to Theorem 8.41 we get that $d_{\boldsymbol{C}_{\Phi'}} \leq 4$ and therefore $d_{\boldsymbol{C}_\Phi} \leq 4$. By Theorem 8.41, the last equality contradicts the fact that Φ is an APN mapping. □

We recall that the number

$$r_{\boldsymbol{C}} = \max_{\boldsymbol{u} \in V_n} \min_{\boldsymbol{c} \in \boldsymbol{C}} \{\mathrm{dist}(\boldsymbol{u}, \boldsymbol{c})\}.$$

is called the covering radius of a code $\boldsymbol{C} \subseteq V_n$ and is denoted by $r_{\boldsymbol{C}}$.

LEMMA 8.44. *For any polynomial APN mapping $\Phi \in \mathcal{F}_{m,m}$, the following inequalities hold:*

$$3 \leq r_{\boldsymbol{C}_\Phi} \leq 4.$$

PROOF. As Φ is an APN mapping, $d_{\boldsymbol{C}_\Phi} = 5$ and the code $\boldsymbol{C}_\Phi$ corrects up to two errors. A t-error correcting code is perfect if its covering radius is t. It is known that binary perfect codes of lengths $n \geq 7$ and minimum distance 5 do not exist (see [**241**]). Obviously, the code $\boldsymbol{C}_\Phi$ is not a trivial perfect code of length 5. Therefore $r_{\boldsymbol{C}_\Phi} \geq 3$.

Let us now suppose that $r_{\boldsymbol{C}_\Phi} \geq 5$. We consider a coset $\boldsymbol{C}_\Phi \oplus f$, where $f \in V_{2^m-1}$, of minimum weight $r_{\boldsymbol{C}_\Phi}$. The dimension of the code $\boldsymbol{C}_\Phi$ is equal to $k = 2^m - 2m - 1$ according to Corollary 8.42. Consider now the linear code $\boldsymbol{C}' = \boldsymbol{C}_\Phi \cup \boldsymbol{C}_\Phi \oplus f$. It is clear that this is a linear code with parameters $[2^m - 1, 2^m - 2m, 5]$. But according to the results of [**40, 118**], linear codes with these parameters do not exist. □

DEFINITION 8.45. Let $\boldsymbol{C} \subseteq V_n$ be a binary code (not necessarily linear) of length n, minimum distance $d_{\boldsymbol{C}} = 2t + 1$, and covering radius $r_{\boldsymbol{C}}$. For any $f \in V_n$ we denote by $C_j(f)$ the number of code words of the code $\boldsymbol{C}$ which are at distance j from f $(0 \leq j \leq r_{\boldsymbol{C}})$. A code is called *uniformly packed* if there exist rational numbers $h_0, h_1, \ldots, h_{r_{\boldsymbol{C}}}$ such that for any $f \in V_n$ the following holds:

$$\sum_{j=0}^{r_{\boldsymbol{C}}} h_j C_j(f) = 1.$$

Later we will need a special type of uniformly packed codes (introduced in [**351**]) for which

$$r_{\boldsymbol{C}} = t + 1 \quad \text{and} \quad h_0 = h_1 = \cdots = h_{t-1} = 1, \quad h_t = h_{t+1} = 1/l, \tag{8.34}$$

where l is a positive integer.

DEFINITION 8.46. A binary code $\boldsymbol{C} \subseteq V_n$ (not necessarily linear) is called a *completely regular code* if for each $f \in V_n$ the weight spectrum of the shift of this code $\boldsymbol{C}' = \boldsymbol{C} \oplus f$ is uniquely determined by the minimum weight of vectors in $\boldsymbol{C}'$.

In other words, for a completely regular code $\boldsymbol{C}$, its two shifts, for some f_1 and f_2, of the same minimum weight, have identical weight spectra.

THEOREM 8.47. *A polynomial mapping $\Phi \in \mathcal{F}_{m,m}$, for m odd, is an AB mapping if and only if the linear code $\boldsymbol{C}_\Phi$ is uniformly packed with minimum distance $d_{\boldsymbol{C}_\Phi} = 2t + 1 = 5$ and covering radius $r_{\boldsymbol{C}_\Phi} = t + 1 = 3$.*

PROOF. Let Φ be an AB mapping. According to Proposition 8.23, Φ is an APN mapping. By Theorem 8.41, the code $\boldsymbol{C}_\Phi$ has minimum distance $d_{\boldsymbol{C}_\Phi} = 5 = 2t+1$ and its characteristic set is

$$M_{\boldsymbol{C}_\Phi} = \{(n+1)/2, (n+1)/2 \pm \sqrt{(n+1)/2}\}, \quad \text{where } n = 2^m - 1.$$

Thus the internal distance $d'_{\boldsymbol{C}_\Phi}$ of the code $\boldsymbol{C}_\Phi$ is equal to 3. Using *Delsarte's inequality* ($r_{\boldsymbol{C}} \leq d'_{\boldsymbol{C}}$ for any code $\boldsymbol{C}$; see [**103**]), we get $r_{\boldsymbol{C}_\Phi} \leq 3$. But according to Lemma 8.44, $r_{\boldsymbol{C}_\Phi} \geq 3$. Hence, $r_{\boldsymbol{C}_\Phi} = d'_{\boldsymbol{C}_\Phi} = 3$. It is known (see [**18, 135**]) that the code $\boldsymbol{C}$ is uniformly packed if and only if $r_{\boldsymbol{C}} = d'_{\boldsymbol{C}}$. Thus, the code $\boldsymbol{C}_\Phi$ is uniformly packed, $d_{\boldsymbol{C}_\Phi} = 2t+1 = 5$, and $r_{\boldsymbol{C}_\Phi} = t+1 = 3$.

We will prove the converse. It is proved in [**17**] that a 2-error correcting BCH code of length $n = 2^m - 1$, where m is odd, is uniformly packed with parameters (8.34), where $l = (n-1)/6$, and the Lloyd polynomial (see [**17, 315**])

$$L_3(z) = h_0 P_0(z) + h_1 P_1(z) + h_2 P_2(z) + h_3 P_3(z)$$

of this code has roots

$$(8.35) \qquad z_1 = \frac{n+1}{2} - \sqrt{\frac{n+1}{2}}, \qquad z_2 = \frac{n+1}{2}, \qquad z_3 = \frac{n+1}{2} + \sqrt{\frac{n+1}{2}}.$$

It was shown in [**135**] that these codes are uniformly packed of length $n = 2^m - 1$, where $n \geq 31$ (and m is odd), with minimum distance $2t+1 = 5$ and covering radius $t+1 = 3$. Since $\boldsymbol{C}_\Phi$ is a linear code, the roots of its Lloyd polynomial determine the characteristic set $M_{\boldsymbol{C}_\Phi}$, i.e., the weight spectrum of the code $\boldsymbol{C}_\Phi^\perp$. Therefore, according to statement 3) of Theorem 8.41, the polynomial Φ defines an AB mapping. □

REMARK 8.48. By Theorem 8.41, if Φ is an AB mapping, then the weight spectrum of the code $\boldsymbol{C}_\Phi$ is uniquely determined and coincides with the weight spectrum of the 2-error correcting BCH code, and when m is even, it is uniformly packed. It is shown in [**351, 135**] that uniformly packed codes with parameters (8.34) are completely regular.

COROLLARY 8.49. *If Φ is a polynomial AB mapping and m is odd, then the code $\boldsymbol{C}_\Phi$ is completely regular.*

PROBLEM 8.50. Let $\Phi \in \mathcal{F}_{m,m}$ be an APN mapping and let m be odd. Prove that Φ is an AB mapping if and only if $d'_{\boldsymbol{C}_\Phi} = 3$.

8.3. Cyclic codes and Boolean mappings

We will consider primitive codes of length $n = 2^m - 1$. We associate with any vector $\boldsymbol{c} = (c_{(0)}, c_{(1)}, \ldots, c_{(n-1)})$ a polynomial $c(\lambda) = c_{(0)} \oplus c_{(1)}\lambda \oplus \cdots \oplus c_{(n-1)}\lambda^{n-1} \in \mathbb{F}_2[\lambda]$.

Let $\mathcal{N}$ be a union of cyclotomic classes (§1.3) modulo n. Let $\boldsymbol{C}$ be a cyclic code whose zeros (Definition 4.60) are elements of the set $\{\alpha^j \mid j \in \mathcal{N}\}$. The set $\mathcal{N}$ is called a *generating* set for the code $\boldsymbol{C}$. It is known (see Lemma 4.61) that a vector $\boldsymbol{c} = (c_{(0)}, c_{(1)}, \ldots, c_{(n-1)})$ belongs to the code $\boldsymbol{C}$ if and only if $c(\alpha^j) = 0$ for any $j \in \mathcal{N}$. The elements α^i for which $c(\alpha^i) \neq 0$ are called *locators of the vector* $\boldsymbol{c}$. We will denote by $\mathrm{loc}(\boldsymbol{c})$ the set of locators of the vector $\boldsymbol{c}$.

Consider polynomial mappings $\Phi \in \mathcal{F}_{m,m}$. Let $J_\Phi = \{j \in \{1, \ldots, n\} \mid \boldsymbol{a}_j \neq \boldsymbol{0}\}$, $\Gamma_\Phi = \{\mathcal{C}_{m,1}\} \cup \{\mathcal{C}_{m,\widehat{j}} \mid j \in J_\Phi\}$, where $\widehat{j}$ is the minimal representative from the cyclotomic class to which j belongs. The code $\boldsymbol{C}_\Phi$ is described in Definition 8.38.

LEMMA 8.51. *For any polynomial mapping $\Phi \in \mathcal{F}_{m,m}$, the code $\boldsymbol{C}_\Phi$ contains the cyclic code $\boldsymbol{B}_\Phi$ of length n with defining set Γ_Φ. If $\boldsymbol{a}_n = \boldsymbol{0}$, then the code $\boldsymbol{C}_\Phi^\perp$ does not contain the all-1 code word.*

PROOF. According to Problem 8.39 and the definition of the set Γ_Φ, we can conclude that if a vector $\boldsymbol{c} \in V_n$ is such that $c(\alpha^j) = 0$ for all $j \in \Gamma_\Phi$, then $\boldsymbol{c}$ is a code word of the code $\boldsymbol{C}_\Phi$. These code words form a cyclic code with defining set Γ_Φ. We denote it by $\boldsymbol{B}_\Phi$. If $\boldsymbol{a}_n = 0$, i.e., $\alpha^n = 1$ is not a zero of the code $\boldsymbol{B}_\Phi$, then $\boldsymbol{B}_\Phi$ contains code words of odd weight. Thus the vector consisting of only ones is not a code word of the code $\boldsymbol{B}_\Phi^\perp$. Therefore $\boldsymbol{C}_\Phi^\perp$ does not contain the all-1 code word either. □

Lemma 8.51 allows us to construct a class of polynomials in $\mathbb{F}_{2^m}[\boldsymbol{x}]$ that cannot represent APN mappings.

THEOREM 8.52. *Let $\Phi \in \mathcal{F}_{m,m}$ be a polynomial mapping. Let s be a divisor of the number m and Λ_s a set of numbers t, $1 \leq t \leq 2^m - 2$, such that $t \equiv 2^l \pmod{2^s - 1}$ for some l, $0 \leq l \leq s-1$.*

If $\boldsymbol{\Gamma}_\Phi \subset \Lambda_s$, then Φ is not an APN mapping.

PROOF. Let $\boldsymbol{C}$ be a cyclic code of length n with defining set Λ_s. It was shown in [**78**] that the minimum distance of this code is 3. Consider the vector $\boldsymbol{c} = (c_{(0)}, c_{(1)}, \ldots, c_{(n-1)})$ whose locators are the elements $\{1, \beta, \beta \oplus 1\}$, where $\beta \in \mathbb{F}_{2^s} \setminus \{0,1\}$. Then for any $t \in \Lambda_s$, $t \equiv 2^l \pmod{2^s - 1}$, we have

$$c(\alpha^t) = 1 \oplus \beta^t \oplus (1 \oplus \beta)^t = 1 \oplus \beta^{2^l} \oplus (1 \oplus \beta)^{2^l} = 0,$$

i.e., $\boldsymbol{c}$ is a code word of the code $\boldsymbol{C}$. Since $\boldsymbol{\Gamma}_\Phi \subset \Lambda_s$, one has $\boldsymbol{C} \subset \boldsymbol{B}_\Phi \subset \boldsymbol{C}_\Phi$, which gives $d_{\boldsymbol{C}_\Phi} = 3$. According to part 2) of Theorem 8.41, the mapping Φ is not an APN mapping. □

EXAMPLE 8.53. Suppose that $m = 9$ and $s = 3$. Let Λ_3 be the set of all integers t, $1 \leq t \leq 510$, which are powers of 2 modulo 7. The set Λ_3 is the union of the cyclotomic classes modulo 511 whose representatives form the following set:

$$R = (1, 9, 11, 15, 23, 25, 29, 37, 39, 43, 51, 53, 57, 79, \\ 85, 93, 95, 107, 109, 123, 127, 183, 191, 219, 239).$$

Let us take a polynomial mapping Φ. If for any $j \in J_\Phi$ some element from the cyclotomic class $\mathcal{C}_{m,\widehat{j}}$ belongs to R, the mapping Φ defined by the function x^t is not an APN mapping for any $t \in R$.

Denote by J_2 the set

$$J_2 = \{j \in \{1, 2, \ldots, 2^m - 1\} \mid w_2(j) \leq 2\},$$

where $w_2(j)$ is the number of ones in the binary representation of the number j. In other words, $J_2 = \{2^k + 2^l \mid 0 \leq k < l \leq m - 1\}$.

DEFINITION 8.54. A polynomial mapping $\Phi \in \mathcal{F}_{m,m}$ is called *quadratic* if $\Gamma_\Phi \subseteq J_2$, i.e., if

$$\Phi(x) = \bigoplus_{j \in J_2} \boldsymbol{a}_j \boldsymbol{x}^j, \quad \boldsymbol{a}_j \in \mathbb{F}_{2^m}. \tag{8.36}$$

Note that for a quadratic polynomial Φ, the Boolean function

$$f(\boldsymbol{x}) = \langle \boldsymbol{a}, \boldsymbol{x} \rangle \oplus \langle \boldsymbol{b}, \Phi(\boldsymbol{x}) \rangle$$

has an algebraic normal form whose degree is at most 2, for any $\boldsymbol{a}, \boldsymbol{b} \in \mathbb{F}_{2^m}$.

LEMMA 8.55. *The code $\boldsymbol{C}_\Phi^\perp$ is contained in the code $RM^*(2,m)$ for any quadratic polynomial Φ.*

PROOF. Recall (see Corollary 5.13) that the code $RM^*(r,m)$, where $r < m-1$, is a cyclic code of length $n = 2^m - 1$ whose zeros are the elements α^s with $1 \le s \le 2^m - 2$ and $1 \le w_2(s) \le m - r - 1$. It is easy to see that the set J_2 is exactly the generating set of the code $RM^*(m-3,m)$.

Consider now the cyclic code $\boldsymbol{B}_\Phi$ with defining set $\boldsymbol{\Gamma}_\Phi$. This means that the cyclic code $\boldsymbol{B}_\Phi$ contains a code $RM^*(m-3,m)$ and $\boldsymbol{B}_\Phi^\perp \subseteq RM^*(m-3,m)^\perp$. Then the nonzeros of the Reed–Muller code $RM^*(m-3,m)^\perp$ (see Theorem 4.64) will be the elements α^{n-s} with $1 \le s \le 2^m - 2$ and $1 \le w_2(s) \le 2$ (in other words, the nonzeros of $RM^*(m-3,m)^\perp$ will be the elements $\alpha^{s'}$ with $1 \le s' \le 2^m - 2$ and $w_2(s') = m-2$ or $m-1$). Therefore, the nonzeros of $RM^*(m-3,m)^\perp$ will be the elements α^t with $0 \le t \le 2^m - 2$ and $1 \le \mathrm{wt}(t) \le m-3$. It is clear that the defining set of the code $RM^*(m-3,m)^\perp$ contains the defining set of $RM^*(2,m)$, which gives

$$\boldsymbol{B}_\Phi^\perp \subseteq RM^*(m-3,m)^\perp \subseteq RM^*(2,m).$$

Since $\boldsymbol{C}_\Phi^\perp \subseteq \boldsymbol{B}_\Phi^\perp$, it follows that $\boldsymbol{C}_\Phi^\perp \subseteq RM^*(2,m)$. □

REMARK 8.56. The lemma we proved demonstrates the ability of the theory of cyclic codes to describe various properties of the mapping Φ. Together with this, Lemma 8.55 can be obtained by observing that for any quadratic polynomial $P(x) \in \mathbb{F}_{2^s}[x]$ and for any $b \in \mathbb{F}_{2^s}$, we have $\langle b, \Phi(x) \rangle = \mathrm{Tr}_{\mathbb{F}_{2^m}/\mathbb{F}_2}(b'x)$, where b' is a suitable element in the field $\mathbb{F}_{2^m}$.

LEMMA 8.57. *A quadratic polynomial Φ for m odd is an AB mapping if and only if Φ is an APN mapping.*

PROOF. If Φ is an AB mapping, then Φ is an APN mapping (see Proposition 8.23).

We now prove the converse. Let Φ be an APN mapping. According to Lemma 8.55, we have $\boldsymbol{C}_\Phi^\perp \subseteq RM^*(2,m)$. Any code word of the code $\boldsymbol{C}_\Phi^\perp$ is a vector consisting of values of the function $f(\boldsymbol{x}) = \langle \boldsymbol{a}, \boldsymbol{x} \rangle \oplus \langle \boldsymbol{b}, \Phi(\boldsymbol{x}) \rangle$ for some $\boldsymbol{a}, \boldsymbol{b} \in \mathbb{F}_{2^m}$, excluding the value of $f(\boldsymbol{0})$. Since $f(\boldsymbol{0}) = 0$ for any $\boldsymbol{a}$ and $\boldsymbol{b}$, there are no code words of weight i, $2^{m-1} - 2^{(m-1)/2} < i < 2^{m-1}$ in the code $\boldsymbol{C}_\Phi^\perp$. Besides, $\dim \boldsymbol{C}_\Phi^\perp = 2m$ (see statement 1) of Corollary 8.42), and $d_{\boldsymbol{C}_\Phi} = 5$ (see statement 2) of Theorem 8.41). Then using (8.27), (8.26), and the fact that $A_3 = A_4 = 0$, we get

$$(8.37)\ J_4 - 2^{m-1}J_2 = \sum_{i=i_0}^{2^{m-1}-2^{(m-1)/2}} (i - 2^{m-1})^2\big((i-2^{m-1})^2 - 2^{m-1}\big)(A'_i + A'_{2^m - i}) = 0.$$

As there are no code words of weight i in $\boldsymbol{C}_\Phi^\perp$ such that $2^{m-1} - 2^{(m-1)/2} < i < 2^{m-1}$, there are no negative terms in the sum (8.37). Therefore, all summands are equal to zero, i.e., $A'_i + A'_{2^m-i} = 0$ for all i which do not belong to the set $\{2^{m-1} \pm 2^{(m-1)/2}, 2^{m-1}\}$. Using Theorem 8.41, 3), we get that Φ is an APN mapping. □

PROBLEM 8.58. Put $\Phi(x) = x^k$. Suppose that there exists $j = 2^l + 1$ such that $jk \equiv 2^s(2^r + 1) \pmod{2^m - 1}$ for some s and r. Prove that Φ is an APN mapping if and only if Φ is an AB mapping.

LEMMA 8.59. *A quadratic polynomial (mapping) Φ is an AB mapping if and only if the code $\boldsymbol{C}_\Phi$ does not have code words of weight* 3.

PROOF. If Φ is an AB mapping, then Φ is an APN mapping. By Theorem 8.41, we have $d_{\boldsymbol{C}_\Phi} = 5$, whence $\boldsymbol{C}_\Phi$ does not contain code words of weight 3.

We will now prove the converse. Let Φ be a quadratic polynomial and suppose the code $\boldsymbol{C}_\Phi$ does not contain code words of weight 3. We assume that $\boldsymbol{C}' \in \boldsymbol{C}_\Phi$ and $\mathrm{wt}(\boldsymbol{c}') = 4$. A quadratic polynomial has the form $\Phi(x) = \bigoplus_{i\in I} a_i x^i$, where $I \subseteq J_2$ and $a_i \neq 0$. We recall that $\boldsymbol{c}' \in \boldsymbol{C}_\Phi$ if and only if

$$\boldsymbol{c}'(\alpha) = 0 \quad \text{and} \quad \bigoplus_{i\in I} a_i \boldsymbol{c}'(\alpha^i) = 0.$$

Put $\mathrm{loc}(\boldsymbol{c}') = \{X_1 = \alpha^{j_1}, X_2 = \alpha^{j_2}, X_3 = \alpha^{j_3}, X_4 = \alpha^{j_4}\}$, which gives $\boldsymbol{c}'(\alpha) = X_1 \oplus X_2 \oplus X_3 \oplus X_4 = 0$. Consider now the elements of the field $\mathbb{F}_{2^m}$ of the form $Y_1 = X_1 \oplus X_4$, $Y_2 = X_2 \oplus X_4$, $Y_3 = X_3 \oplus X_4$. We denote by $\boldsymbol{c}$ a binary vector of weight 3 such that $\mathrm{loc}(\boldsymbol{c}) = \{Y_1, Y_2, Y_3\}$. Then the following relation holds:

$$\boldsymbol{c}(\alpha) = Y_1 \oplus Y_2 \oplus Y_3 = X_1 \oplus X_2 \oplus X_3 \oplus X_4 = \boldsymbol{c}'(\alpha) = 0.$$

Since $i \in I \subseteq J_2$, one has $i = 2^k + 2^l$. Therefore,

$$\begin{aligned}\boldsymbol{c}'(\alpha^i) &= X_1^{2^k+2^l} \oplus X_2^{2^k+2^l} \oplus X_3^{2^k+2^l} \oplus X_4^{2^k+2^l} \\ &= (X_1 \oplus X_4)^{2^k+2^l} \oplus (X_2 \oplus X_4)^{2^k+2^l} \oplus (X_3 \oplus X_4)^{2^k+2^l} = \boldsymbol{c}(\alpha^j).\end{aligned}$$

From this we get

$$\bigoplus_{i\in I} a_i \boldsymbol{c}(\alpha^i) = \bigoplus_{i\in I} a_i \boldsymbol{c}'(\alpha^i) = 0.$$

Hence, $\boldsymbol{c} \in \boldsymbol{C}_\Phi$ and we arrive at a contradiction. Thus, the code $\boldsymbol{C}_\Phi$ does not contain code words of weight 4. According to parts 1) and 2) of Theorem 8.41, we have $d_{\boldsymbol{C}_\Phi} = 5$ and Φ is an APN mapping. Now the theorem follows from Lemma 8.57. □

THEOREM 8.60. *A quadratic polynomial (mapping) Φ is an AB mapping if and only if for each k, $k \in \{0, 1, \ldots, n-1\}$, and for each $\boldsymbol{b} \in \mathbb{F}_{2^m} \setminus \{\boldsymbol{0}, \boldsymbol{1}\}$ the following condition is satisfied:*

$$\bigoplus_{\substack{j\in J_2,\ w_2(j)=2 \\ j=2^s+2^l,\ s>l}} a_j \alpha^{jk} (\boldsymbol{b}^{2^s} \oplus \boldsymbol{b}^{2^l}) \neq 0. \tag{8.38}$$

PROOF. According to Lemma 8.59, a quadratic polynomial Φ is an AB mapping if and only if the code $\boldsymbol{C}_\Phi$ does not contain code words $\boldsymbol{c}$ for which $\mathrm{loc}(\boldsymbol{c}) \cup \{0\}$ is a 2-dimensional vector subspace of $\mathbb{F}_{2^m}$. Let K be the set of such subspaces. It is known that $\#K = (2^m - 1)(2^{m-1} - 1)/3$. There exist exactly $2^{m-1} - 1$ subspaces of K of the form $\{\boldsymbol{0}, \boldsymbol{1}, \boldsymbol{b}, \boldsymbol{b} \oplus \boldsymbol{1}\}$, $\boldsymbol{b} \in \mathbb{F}_{2^m} \setminus \{\boldsymbol{0}, \boldsymbol{1}\}$. Shifting the element $\boldsymbol{b}$, we obtain $2^m - 1$ subspaces of type

$$\{\boldsymbol{0}, \alpha^k, \alpha^k \boldsymbol{b}, \alpha^k(\boldsymbol{b} \oplus 1)\}, \quad k \in \{0, 1, \ldots, 2^m - 2\}. \tag{8.39}$$

In this way we get $(2^m - 1)(2^{m-1} - 1)$ subspaces, and any subspace of K of dimension 2 occurs three times. Therefore, any subspace of K has the form (8.39) for some $\boldsymbol{b}$ and k. We will denote by $\boldsymbol{c}_{\boldsymbol{b},k}$ the binary vectors with the set of locators $\{\alpha^k, \alpha^k\boldsymbol{b}, \alpha^k(\boldsymbol{b}\oplus 1)\}$.

Suppose that $j \in J_2$. Then for $j = 2^s$ we have

$$\boldsymbol{c}_{\boldsymbol{b},k}(\alpha^j) = \boldsymbol{c}_{\boldsymbol{b},k}(\alpha^{2^s}) = \boldsymbol{c}_{\boldsymbol{b},k}(\alpha)^{2^s} = 0,$$

as $\boldsymbol{c}_{\boldsymbol{b},k}(\alpha) = 0$ by the definition. Then if $j = 2^s + 2^l$, $s > l$, we get

$$\begin{aligned}\boldsymbol{c}_{\boldsymbol{b},k}(\alpha^j) &= \alpha^{kj} \oplus (\alpha^k\boldsymbol{b})^j \oplus (\alpha^k(\boldsymbol{b}\oplus 1))^j = \alpha^{kj}[1 \oplus \boldsymbol{b}^j \oplus (\boldsymbol{b}\oplus 1)^j] \\ &= \alpha^{kj}[1 \oplus \boldsymbol{b}^{2^s+2^l} \oplus (\boldsymbol{b}^{2^s} \oplus 1)(\boldsymbol{b}^{2^l} \oplus 1)] = \alpha^{kj}[\boldsymbol{b}^{2^s} \oplus \boldsymbol{b}^{2^l}].\end{aligned}$$

Suppose now that Φ is an AB mapping. Then (see Lemma 8.59) the code $\boldsymbol{C}_\Phi$ does not contain code words of weight 3, whence for any $\boldsymbol{c}_{\boldsymbol{b},k}$ the following holds:

$$\bigoplus a_j \boldsymbol{c}_{\boldsymbol{b},k} \neq 0, \tag{8.40}$$

$\boldsymbol{b} \in \mathbb{F}_{2^m} \setminus \{\boldsymbol{0}, \boldsymbol{1}\}$, $k \in \{0, 1, \ldots, n-1\}$. As $\boldsymbol{c}_{\boldsymbol{b},k} = 0$ for $j = 2^s$, the condition (8.40) is equivalent to the condition (8.38).

We will now prove the converse. It is easy to see that if (8.38) holds, then no binary vector of weight 3 can be a code word of the code $\boldsymbol{C}_\Phi$. Then, according to Lemma 8.59, the mapping Φ is an AB mapping. □

EXAMPLE 8.61. Suppose that $\Phi(x) = ax^j$, $j = 2^s + 2^l$, $s > l$. Then condition (8.38) is satisfied if and only if $\gcd(s-l, m) = 1$. In this case the following relation must be satisfied:

$$a\alpha^{kj}(\boldsymbol{b}^{2^s} \oplus \boldsymbol{b}^{2^l}) \neq 0$$

for any $\boldsymbol{b} \in \mathbb{F}_{2^m} \setminus \{\boldsymbol{0}, \boldsymbol{1}\}$ and $k \in \{0, 1, \ldots, n-1\}$. As $\boldsymbol{b}^{2^s} \oplus \boldsymbol{b}^{2^l} = (\boldsymbol{b}^{2^{s-l}} \oplus \boldsymbol{b})^{2^l}$, this is possible only if the polynomial $x^{2^{s-l}} \oplus x$ has only $\boldsymbol{0}$ and $\boldsymbol{1}$ as roots in the field $\mathbb{F}_{2^m}$.

Consider now the mapping of the field $\mathbb{F}_{2^m}$ into itself given by a power function. In this case, $\Phi(x) = x^s$ and the cardinality of the cyclotomic class $\mathcal{C}_{m,s}$ is equal to m (otherwise Φ is not an APN mapping; see Corollary 8.42). The code $\boldsymbol{C}_\Phi$ is a cyclic code of length $n = 2^m - 1$ whose zeros are α, α^s, and its conjugates. It is clear that this code is conjugate with the cyclic code with zeros α^j, α^{js} (multiplication $j \cdot s$ is performed modulo (mod $2^m - 1$)) and their conjugates.

DEFINITION 8.62. A binary cyclic code of length $n = 2^m - 1$ whose defining set coincides with $\mathcal{C}_{m,1} \cup \mathcal{C}_{m,s}$, $1 < s \leq 2^m - 2$, is called a *cyclic code with two zeros* and is denoted by $\boldsymbol{C}_{1,s}$.

The parity-check matrix of a cyclic code with two zeros is

$$H_{\boldsymbol{C}_{1,s}} = H_{1,s} = \begin{pmatrix} 1 & \alpha & \alpha^2 & \ldots & \alpha^{2^m-1} \\ 1 & \alpha^s & \alpha^{2s} & \ldots & \alpha^{(2^m-1)s} \end{pmatrix}.$$

No wonder the study of cryptographic properties and characteristics of the power function $\Phi(x) = x^s$ is connected with the study of certain characteristics of the code $\boldsymbol{C}_{1,s}^{\perp}$. Many authors investigate the power functions and the classes of powers s; we think those investigations have a rather technical character. In this connection, we give a few tables (Tables 8.2–8.5) containing different classes of exponents s such that the power function x^s leads to AB and APN mappings with high nonlinearity. The tables include appropriate references.

TABLE 8.2. Known and conjectured APN mappings x^s of $\mathbb{F}_{2^m}$ with $m = 2t + 1$.

Mapping	Exponent s	Source
quadratic	$2^i + 1$, $\gcd(i, m) = 1$, $1 \le i \le t$	Proved in [**137, 293**]
Kasami	$2^{2i} - 2^i + 1$, $\gcd(i, m) = 1$, $2 \le i \le t$	Proved in [**185**]
inverse	$2^{2t} - 1$	Proved in [**293, 26**]
Welch	$2^t + 3$	Proved in [**117**]
Niho	$\begin{cases} 2^t + 2^{t/2} - 1, & t \text{ is even;} \\ 2^t + 2^{(3t+1)/2} - 1, & t \text{ is odd} \end{cases}$	Proved in [**116**]
Dobbertin	$2^{4i} + 2^{3i} + 2^{2i} + 2^i - 1$, $m = 5i$	Conjectured in [**116**]

TABLE 8.3. Know and conjectured APN mappings x^s of $\mathbb{F}_{2^m}$ with $m = 2t$.

Mapping	Exponent s	Source
quadratic	$2^i + 1$, $\gcd(i, m) = 1$, $1 \le i < t$	Proved in [**137, 293**]
Kasami	$2^{2i} - 2^i + 1$, $\gcd(i, m) = 1$, $2 \le i < t$	Proved in [**185, 168**]
Dobbertin	$2^{4i} + 2^{3i} + 2^{2i} + 2^i - 1$, $m = 5i$	Conjectured in [**116**]

TABLE 8.4. Known power AB mappings (permutations) x^s of $\mathbb{F}_{2^m}$ with $m = 2t + 1$.

Mapping	Exponent s	Source
quadratic	$2^i + 1$, $\gcd(i, m) = 1$, $1 \le i \le t$	Proved in [**137, 293**]
Kasami	$2^{2i} - 2^i + 1$, $\gcd(i, m) = 1$, $2 \le i \le t$	Proved in [**185**]
Welch	$2^t + 3$	Proved in [**51, 52**]
Niho	$\begin{cases} 2^t + 2^{t/2} - 1, & t \text{ is even;} \\ 2^t + 2^{(3t+1)/2} - 1, & t \text{ is odd} \end{cases}$	Proved in [**165**]

TABLE 8.5. Known power mappings (permutations) x^s of $\mathbb{F}_{2^m}$ with maximum known nonlinearity for $m = 2t$.

Exponent s	Source
$2^{2t-1} - 1$	Proved in [**215**]
$2^i + 1$, $\gcd(i, m) = 2$, $1 \le i < t$, $m \equiv 2 \pmod 4$	Proved in [**137**]
$2^{2i} - 2^i + 1$, $\gcd(i, m) = 2$, $1 \le i < t$, $m \equiv 2 \pmod 4$	Proved in [**185**]
$\sum_{i=0}^{t} 2^{ik}$, t is even, $0 < k < t$, $\gcd(k, m) = 1$	Proved in [**111**]
$2^t + 2^{(t+1)/2} + 1$, t is odd	Proved in [**97**]
$2^t + 2^{t-1} + 1$, t is odd	Proved in [**97**]
$2^t + 2^{t/2} + 1$, $t \equiv 2 \pmod 4$	Proved in [**111**]

8.4. Avalanche criteria and propagation criteria

Suppose that $\Phi = (f_1, f_2, \dots, f_l) \in \mathcal{F}_{m,l}$. It is inadmissible from the cryptographic point of view for a coordinate function f of the mapping $\Phi = (f_1, f_2, \dots, f_l)$ to depend nonessentially on some of the variables $x^{(1)}, x^{(2)}, \dots, x^{(m)}$.

DEFINITION 8.63. A mapping $\Phi \in \mathcal{F}_{m,l}$ is called *complete* if all components of the vector

$$\sum_{\boldsymbol{x}\in V_m} [\Phi(\boldsymbol{x}) \oplus \Phi(\boldsymbol{x} \oplus \boldsymbol{e}_i)]$$
$$= \left(\sum_{\boldsymbol{x}\in V_m} [f_1(\boldsymbol{x}) \oplus f_1(\boldsymbol{x} \oplus \boldsymbol{e}_i)], \dots, \sum_{\boldsymbol{x}\in V_m} [f_l(\boldsymbol{x}) \oplus f_l(\boldsymbol{x} \oplus \boldsymbol{e}_i)] \right)$$

are nonzero for any i, $1 \le i \le m$.

DEFINITION 8.64. A mapping $\Phi \in \mathcal{F}_{m,l}$ satisfies the *avalanche criterion* if

$$\sum_{\boldsymbol{x}\in V_m} \mathrm{wt}(\Phi(\boldsymbol{x}) \oplus \Phi(\boldsymbol{x} \oplus \boldsymbol{e}_i)) = l \cdot 2^{m-1} \tag{8.41}$$

for any i, $1 \le i \le m$.

In other words, the table of size $2^m \times l$ with values of the mapping $D_{\boldsymbol{e}_i}\Phi$ contains an equal number of zeros and ones.

DEFINITION 8.65. A mapping $\Phi \in \mathcal{F}_{m,l}$ satisfies the *strict avalanche criterion* (SAC) if

$$\sum_{\boldsymbol{x}\in V_m} [\Phi(\boldsymbol{x}) \oplus \Phi(\boldsymbol{x} \oplus \boldsymbol{e}_i)] = (\underbrace{2^{m-1}, \dots, 2^{m-1}}_{l})^\top \tag{8.42}$$

for any i, $1 \le i \le m$.

It is easy to see that in the case $l = 1$ the strict avalanche criterion is equivalent to the avalanche criterion. It is obvious, too, that a mapping Φ satisfies the strict avalanche criterion if and only if the strict avalanche criterion is satisfied for all coordinate functions $f_1, f_2, \dots, f_l$ of Φ. Thus, when studying the strict avalanche criterion, it suffices to concentrate on the case $l = 1$.

PROBLEM 8.66. Prove that a Boolean function $f \in \mathcal{F}_m$ satisfies SAC if and only if $\Delta_f(\boldsymbol{c}) = 0$ for any $\boldsymbol{c} \in V_m$, $\mathrm{wt}(\boldsymbol{c}) = 1$.

PROBLEM 8.67. Suppose a Boolean function $f \in \mathcal{F}_m$ satisfies SAC. Prove that for any $\mathfrak{g} \in \mathfrak{D}_m$ the function $f^{\mathfrak{g}}$ also satisfies SAC.

DEFINITION 8.68. Suppose that $f \in \mathcal{F}_m$ and $\boldsymbol{a} \in V_m$. A Boolean function f satisfies the *propagation criterion* in the direction $\boldsymbol{a}$ if $\Delta_f(\boldsymbol{a}) = 0$, or in other words, if the derivative $D_{\boldsymbol{a}}f$ is a balanced function (cf. Definition 7.16).

Recall that, by definition,

$$\mathrm{PC}_f = \{\boldsymbol{a} \in V_m \mid \Delta_f(\boldsymbol{a}) = 0\}. \tag{8.43}$$

DEFINITION 8.69. A Boolean function $f \in \mathcal{F}_m$ satisfies the *propagation criterion of degree* k(notation: PC(k)) if $\Delta_f(\boldsymbol{a}) = 0$ for any $\boldsymbol{a} \in V_m$ such that $1 \le \mathrm{wt}(\boldsymbol{a}) \le k$.

It is clear that PC(1) is equivalent to SAC if f is a maximum-nonlinear function (and m is an even number), $\mathrm{PC}_f = V_m \setminus \{\mathbf{0}\}$, or in other words, the function f satisfies PC(m). The set PC_f does not have any algebraic structure. A characterization of the vectors in this set in terms of the Walsh–Hadamard coefficients of the function f is given in the following theorem, which can be easily proved using Corollary 2.81.

THEOREM 8.70. *A vector $\boldsymbol{a} \in V_m$ belongs to the set PC_f if and only if*

$$\sum_{\boldsymbol{z}\in V_m} (-1)^{\langle \boldsymbol{z},\boldsymbol{a}\rangle} W_f^2(\boldsymbol{z}) = 0.$$

Theorem 8.70 provides conditions for the propagation criterion to be satisfied over all the nonzero vectors of some subspace. We will state the corresponding results of [**416**] without proof.

THEOREM 8.71. *Let $L \subseteq V_n$ be a linear subspace of dimension d. The inclusion $\mathrm{PC}_f \supseteq L \setminus \{\mathbf{0}\}$ holds for a Boolean function f if and only if the Walsh–Hadamard coefficients of f satisfy the following system of 2^d equations:*

$$\begin{cases} \sum\limits_{\boldsymbol{\beta}\in L^\perp} W_f^2(\boldsymbol{\alpha} \oplus \boldsymbol{\beta}) = 2^{2n-d}, \\ \boldsymbol{\alpha} \in V_n/L^\perp. \end{cases}$$

COROLLARY 8.72. *If $\mathrm{PC}_f \supseteq L \setminus \{\mathbf{0}\}$, then $\max_{\boldsymbol{z}\in V_m} |W_f(\boldsymbol{z})| \le 2^{m-\frac{\dim L}{2}}$.*

PROBLEM 8.73. Suppose that $d_f = \max\{\dim L \mid L \setminus \{\mathbf{0}\} \subseteq \mathrm{PC}_f\}$. Prove that

$$N_f \ge 2^{m-1} - 2^{m-\frac{d_f}{2}-1}.$$

THEOREM 8.74. *Let $L \subseteq V_n$ be a linear subspace of dimension d; suppose that $[\boldsymbol{a}_0] \in V_n/L$, $\boldsymbol{a}_0 \neq \mathbf{0}$. Then the inclusion $\mathrm{PC}_f \supseteq \boldsymbol{a}_0 \oplus L$ holds for a Boolean function f if and only if the Walsh–Hadamard coefficients of f satisfy the following system of 2^d equations:*

$$\begin{cases} \sum\limits_{\boldsymbol{\beta}\in L^\perp} (-1)^{\langle \boldsymbol{\beta},\boldsymbol{a}_0\rangle} W_f^2(\boldsymbol{\alpha} \oplus \boldsymbol{\beta}) = 0, \\ [\boldsymbol{\alpha}] \in V_n/L^\perp. \end{cases}$$

PROBLEM 8.75. Prove that the numbers $\#\mathrm{PC}_f$ and $\{\dim L \mid L \setminus \{\mathbf{0}\} \subseteq \mathrm{PC}_f\}$ are affine invariants.

Further development of the concept of SAC has led to the introduction of SAC of order t, $1 \le t \le m-2$.

DEFINITION 8.76. A Boolean function $f \in \mathcal{F}_m$ satisfies the *strong avalanche criterion of order t* (SAC(t)), $1 \le t \le m-2$, if for any $1 \le j_1 < \cdots < j_t \le m$, $\boldsymbol{a} = (a^{(1)}, \ldots, a^{(t)})^\top \in V_t$, the subfunction $f_{j_1,\ldots,j_t}^{a^{(1)},\ldots,a^{(t)}}$ satisfies SAC.

We assume that a function satisfies SAC(0) if and only if it satisfies SAC. It is clear that t is not greater than $m-2$ since for $t > m-2$ the corresponding subfunction is affine and cannot satisfy SAC.

PROBLEM 8.77. Prove that if a Boolean function $f \in \mathcal{F}_m$ satisfies SAC(t), $1 \le t \le m-2$, then f satisfies SAC($t-1$).

PROBLEM 8.78. Prove that if a Boolean function $f \in \mathcal{F}_m$ satisfies SAC(t), then it satisfies SAC(i), $0 \le i < t$.

THEOREM 8.79. *Suppose that* $f \in \mathcal{F}_m$, $m > 2$. *Then*
1. *If* f *satisfies* SAC$(m-2)$, *then* $\deg f = 2$.
2. *If* f *satisfies* SAC(t), $0 \le t < m-2$, *then* $\deg f \le m-t-1$.

PROOF. 1. Let f satisfy SAC$(m-2)$. Affine functions do not satisfy SAC of any order, whence $\deg f \ge 2$. Suppose that $\deg f = d > 2$. Then monomials of length d occur in the algebraic normal form of function f. Take an arbitrary monomial and consider some variables $x^{(i_1)}$ and $x^{(i_2)}$ $(i_1 < i_2)$ occurring in it. The algebraic normal form of the function f can be represented as

$$\begin{aligned} &f(x^{(1)}, x^{(2)}, \dots, x^{(m)}) \\ &\quad = x^{(i_1)} x^{(i_2)} g_{12}(x^{(1)}, \dots, x^{(i_1-1)}, x^{(i_1+1)}, \dots, x^{(i_2-1)}, x^{(i_2+1)}, \dots, x^{(m)}) \\ &\quad\quad \oplus x^{(i_1)} g_1(x^{(1)}, \dots, x^{(i_1-1)}, x^{(i_1+1)}, \dots, x^{(i_2-1)}, x^{(i_2+1)}, \dots, x^{(m)}) \\ &\quad\quad \oplus x^{(i_2)} g_2(x^{(1)}, \dots, x^{(i_1-1)}, x^{(i_1+1)}, \dots, x^{(i_2-1)}, x^{(i_2+1)}, \dots, x^{(m)}) \\ &\quad\quad \oplus g_0(x^{(1)}, \dots, x^{(i_1-1)}, x^{(i_1+1)}, \dots, x^{(i_2-1)}, x^{(i_2+1)}, \dots, x^{(m)}), \end{aligned}$$

where the function g_{12} is not a constant. Then there exists a set of values of the variables

$$\begin{aligned} x^{(1)} = a^{(1)}, \quad \dots, \quad x^{(i_1-1)} = a^{(i_1-1)}, \quad x^{(i_1+1)} = a^{(i_1+1)}, \quad \dots, \\ x^{(i_2-1)} = a^{(i_2-1)}, \quad x^{(i_2+1)} = a^{(i_2+1)}, \quad \dots, \quad x^{(m)} = a^{(m)}, \end{aligned}$$

such that $g_{12}(a^{(1)}, \dots, a^{(i_1-1)}, a^{(i_1+1)}, \dots, a^{(i_2-1)}, a^{(i_2+1)}, \dots, a^{(m)}) = 0$. Hence, the corresponding subfunction of the function f is affine and does not satisfy SAC. We arrive at a contradiction, so $\deg f = 2$.

2. Suppose now that f satisfies SAC(t), $0 \le t < m-2$. For $t = 0$ it is easy to check that if $\deg f = m$, then f cannot satisfy SAC. Therefore, $\deg f \le m-1$.

Suppose that $0 < t < m-2$ and $\deg f = d > m-t-1$. Take an arbitrary monomial of higher degree occurring in the algebraic normal form of the function f: $x^{(j_1)} x^{(j_2)} \cdots x^{(j_d)}$. As $t \ge m-d$, Problem 8.78 implies that the function f satisfies SAC$(m-d)$. Hence, there exist $\boldsymbol{a} = (a^{(1)}, \dots, a^{(m-d)})^\top \in V_{m-d}$ and variables $\{x^{(i_1)}, \dots, x^{(i_{m-d})}\} = \{x^{(1)}, \dots, x^{(m)}\} \setminus \{x^{(j_1)}, \dots, x^{(j_d)}\}$ such that the subfunction $f_{i_1,\dots,i_{m-d}}^{a^{(1)},\dots,a^{(m-d)}} \in \mathcal{F}_d$ satisfies SAC. But $f_{i_1,\dots,i_{m-d}}^{a^{(1)},\dots,a^{(m-d)}}$ has degree d and (as has already been noted) cannot satisfy SAC, which is a contradiction. Therefore, $d \le m-t-1$. □

PROBLEM 8.80. Suppose that $f \in \mathcal{F}_m$, $m > 2$, $\deg f = 2$. Prove that f satisfies SAC(t), $0 \le t \le m-2$, if and only if any variable $x^{(i)}$ occurs at least $m+1$ times in the monomials of length two in the algebraic normal form of the function f.

PROBLEM 8.81. Describe all functions in $\mathcal{F}_m$, $m \ge 4$, satisfying SAC$(m-2)$ and find the cardinality of this class.

PROBLEM 8.82. Describe all functions in $\mathcal{F}_m$, $m \ge 5$, satisfying SAC$(m-3)$ and find the cardinality of this class.

PROBLEM 8.83. Prove that for $0 \le t \le m-3$ the bound in the inequality from statement 2) of Theorem 8.79 is attainable. (*Hint*: Show that the function $f(x^{(1)}, \dots, x^{(m)}) = x^{(1)} \cdots x^{(n-m-1)} \oplus \bigoplus_{1 \le i < j \le m} x^{(i)} x^{(j)}$ satisfies SAC$(n-m-1)$.)

A further generalization of the propagation criterion leads to the introduction of a new parameter, which is called an order.

DEFINITION 8.84. A Boolean function $f \in \mathcal{F}_m$ satisfies the *propagation criterion of degree k and order t*, $\mathrm{PC}(k,t)$, if for any set $1 \leq j_1 < \cdots < j_t \leq m$, $\boldsymbol{a} = (a^{(1)}, \dots, a^{(t)})^\top \in V_t$, the subfunction $f_{j_1,\dots,j_t}^{a^{(1)},\dots,a^{(t)}}$ satisfies $\mathrm{PC}(k)$.

A *propagation matrix* PM_m is an $(m \times m)$ matrix with entries

$$PM_m(k,t) = \frac{1}{2^{m+1}} \#\{f \in \mathcal{F}_m \mid f \text{ satisfies } \mathrm{PC}(k,t)\}.$$

Division by 2^{m+1} in the definition of the matrix PM_m is present because the monomials of lengths 1 and 0 in the algebraic normal form of a function do not influence the propagation characteristics and thus give the number of cosets of the code $RM(1,m)$. The following inequalities hold: $PM_m(k,t_1) \leq PM_m(k,t_2)$ if $t_1 \geq t_2$; $PM_m(k_1,t) \leq PM_m(k_2,t)$ if $k_1 \geq k_2$; and $PM_m(1,m-1) = 0$. For example, the matrix PM_5 has the form given in Table 8.6.

TABLE 8.6

k \ t	0	1	2	3	4
1	430 040	813	26	1	0
2	3 568	28	1	1	–
3	168	28	0	–	–
4	28	28	–	–	–
5	0	–	–	–	–

DEFINITION 8.85. Suppose that $f \in \mathcal{F}_m$. Then:

1) f satisfies the *extended propagation criterion of degree k and order 0* ($\mathrm{EPC}(k,0)$) if the derivative $D_{\boldsymbol{a}}f$ is a balanced function for all $\boldsymbol{a} \in V_m$ such that $1 \leq \mathrm{wt}(\boldsymbol{a}) \leq k$;

2) f satisfies the *extended propagation criterion of degree k and order $t > 0$* ($\mathrm{EPC}(k,t)$) if $D_{\boldsymbol{a}}f$ is a t-resilient function for all $\boldsymbol{a} \in V_m$ such that $1 \leq \mathrm{wt}(\boldsymbol{a}) \leq k$.

PROBLEM 8.86. Let $f \in \mathcal{F}_m$. Prove that:

1) if f satisfies $\mathrm{EPC}(k,t)$, $k+t \leq m$, then f satisfies $\mathrm{PC}(k,t)$;
2) if f satisfies $\mathrm{PC}(k,0)$ or $\mathrm{PC}(k,1)$, then f satisfies $\mathrm{EPC}(k,0)$ or $\mathrm{EPC}(k,1)$;
3) if f satisfies $\mathrm{PC}(1,t)$, then f satisfies $\mathrm{EPC}(1,t)$.

PROBLEM 8.87. Let $f \in \mathcal{F}_m$ satisfy $\mathrm{SAC}(t)$ (resp. $\mathrm{PC}(k)$, $\mathrm{PC}(k,t)$, $\mathrm{EPC}(k,t)$). Prove that for any $\mathfrak{g} \in \mathfrak{D}_m$ the function $f^{\mathfrak{g}}$ satisfies $\mathrm{SAC}(t)$ (resp. $\mathrm{PC}(k)$, $\mathrm{PC}(k,t)$, $\mathrm{EPC}(k,t)$).

PROBLEM 8.88. Put

$$f(\boldsymbol{x}) = f(x^{(1)}, x^{(2)}, \dots, x^{(m)}) = \bigoplus_{1 \leq i < j \leq m} x^{(i)}x^{(j)} \oplus \bigoplus_{s=1}^{m} c^{(s)}x^{(s)} \oplus c^{(0)},$$

where $\boldsymbol{c} = (c^{(1)}, \dots, c^{(m)})^\top$ is an arbitrary vector in V_m, $c^{(0)} \in \mathbb{F}_2$. Prove that f satisfies $\mathrm{PC}(k,t)$ if $k+t \leq m-1$ or $k+t = m$ and k is an even number. Prove in addition that

1) f is a unique (up to $RM(1,m)$) function which satisfies $\mathrm{PC}(1,m-2)$;
2) f is a unique (up to $RM(1,m)$) function which satisfies $\mathrm{PC}(2,m-2)$;
3) f is balanced if m is an odd number.

8.5. Construction of Boolean functions satisfying the propagation criterion of degree k and order t

In this section we return to the techniques developed in coding theory which can be used in cryptographic research. We state without proof the following theorem from [**203**].

THEOREM 8.89. *Suppose that the natural numbers k and t are such that there exists a binary $(s \times r)$ matrix M satisfying the following conditions:*

1) $s \geq \max\{k, t+1\}$ *and* $r \geq \max\{k, t+1\}$;
2) *for any vector* $\boldsymbol{u} \in V_r$ *such that* $1 \leq \mathrm{wt}(\boldsymbol{u}) \leq k$, *the inequality* $\mathrm{wt}(M\boldsymbol{u}) \geq t+1$ *holds;*
3) *for any vector* $\boldsymbol{v} \in V_s$, $1 \leq \mathrm{wt}(\boldsymbol{v}) \leq k$, *the inequality* $\mathrm{wt}(\boldsymbol{v}^\top M) \geq t+1$ *holds.*

We define the function f as follows:

$$f(\boldsymbol{x},\boldsymbol{y}) = f(x^{(1)},\dots,x^{(s)},y^{(1)},\dots,y^{(r)}) = \boldsymbol{x}^\top M\boldsymbol{y} \oplus g(\boldsymbol{x}), \tag{8.44}$$

where $\boldsymbol{x} \in V_s$, $\boldsymbol{y} \in V_r$, $g \in \mathcal{F}_s$, *and* $m = s+r$.

Then the Boolean function $f \in \mathcal{F}_m$ *satisfies* $\mathrm{PC}(k,t)$.

The matrix M, which has certain specific properties, plays a central role in Theorem 8.89. Thus, the construction of functions satisfying $\mathrm{PC}(k,t)$ with the aid of Theorem 8.89 is essentially reduced to the construction of classes of such matrices.

THEOREM 8.90. *Let G_1 be a generator matrix of a linear $[r,h,d_1]$-code $\boldsymbol{C}_1$ with dual distance $d_1^\perp$, and G_2 a generator matrix of a linear $[s,h,d_2]$-code $\boldsymbol{C}_2$ with dual distance $d_2^\perp$. Let $M = G_2^\top G_1$.*

Then the matrix M satisfies conditions 1)–3) *of Theorem* 8.89 *for*

$$k = \min\{d_1^\perp, d_2^\perp\} - 1, \quad t = \min\{d_1, d_2\} - 1.$$

PROOF. Condition 1) is obviously satisfied. We will show now that the matrix M satisfies condition 2). Suppose that $\boldsymbol{u} \in V_r$, $\leq \mathrm{wt}(\boldsymbol{u}) \leq k$. This vector is not a code word of the code $\boldsymbol{C}_1^\perp$ since $\mathrm{wt}(\boldsymbol{u}) \leq k < d_1^\perp$. Then $G_1\boldsymbol{u} \neq \boldsymbol{0}$ since G_1 is a parity-check matrix for $G_1^\perp$. Therefore, $M\boldsymbol{u} = G_2^\top(G_1\boldsymbol{u})$ is the all-zero code word of the code $\boldsymbol{C}_2$ since G_2 is a generator matrix of the code $\boldsymbol{C}_2$. Hence, $\mathrm{wt}(M\boldsymbol{u}) \geq d_2 \geq t+1$.

The proof that the matrix M satisfies condition 3) is similar. □

Consider now some examples borrowed from the book [**241**].

EXAMPLE 8.91. If $\boldsymbol{C}$ is the Hamming $[2^r-1, 2^r-1-r, 3]$-code, then $\boldsymbol{C}^\perp$ is the symplex $[2^r-1, r, 2^{r-1}]$-code.

If we take $\boldsymbol{C}_1 = \boldsymbol{C}_2 = \boldsymbol{C}$ in Theorem 8.90, we get a $(2^r-1)\times(2^r-1)$ matrix M, which we use to construct a Boolean function of $m = 2^{r+1}-2$ variables that satisfies the propagation criterion of degree $k = \min\{2^{r-1}, 2^{r-1}\} - 1 = 2^{r-1}-1$ and order $t = \min\{3,3\} - 1 = 2$, i.e., satisfies $\mathrm{PC}(2^{r-1}-1, 2)$.

If we put $\boldsymbol{C}_1 = \boldsymbol{C}_2 = \boldsymbol{C}^\perp$, we obtain a Boolean function of $m = 2^{r+1}-2$ variables which satisfies the propagation criterion of degree 2 and order $2^{r-1}-1$, i.e., $\mathrm{PC}(2, 2^{r-1}-1)$.

EXAMPLE 8.92. Let $\boldsymbol{C}$ be the extended Hamming $[2^r, 2^r - 1 - r, 4]$-code. Then $\boldsymbol{C}^\perp$ is a Reed–Muller $[2^r, r+1, 2^{r-1}]$-code of first order, $RM(1, r)$.

If we take $\boldsymbol{C}_1 = \boldsymbol{C}_2 = \boldsymbol{C}$ in Theorem 8.90, we obtain a Boolean function of $m = 2^{r+1}$ variables which satisfies the propagation criterion PC$(2^{r-1} - 1, 3)$. If we put $\boldsymbol{C}_1 = \boldsymbol{C}_2 = \boldsymbol{C}^\perp$, we get a Boolean function of $m = 2^{r+1}$ variables which satisfies the propagation criterion PC$(3, 2^{r-1} - 1)$.

The construction of Theorem 8.89 makes it possible to obtain balanced functions.

DEFINITION 8.93. Suppose that $g \in \mathcal{F}_s$ and M is a binary $(s \times r)$ matrix. A function g is called *balanced with respect to the matrix* M if the following equality is satisfied:

$$\#\{\boldsymbol{x} \in V_s \mid g(\boldsymbol{x}) = 0, \boldsymbol{x}^\top M = \boldsymbol{0}\} = \#\{\boldsymbol{x} \in V_s \mid g(\boldsymbol{x}) = 1, \boldsymbol{x}^\top M = \boldsymbol{0}\}.$$

PROBLEM 8.94. Prove that if a function g is balanced with respect to a matrix M, then the function f defined in (8.44) is balanced.

Using the construction of Theorem 8.89 one can get Boolean functions of high nonlinearity and high algebraic degree that satisfy PC(k, t).

PROBLEM 8.95. Construct a Boolean function $f \in \mathcal{F}_m$, $m = s + r$, of the form (8.44) such that
1) $\deg f = s/2$ and $N_f \geq 2^{r+s-1} - 2^{r+s/2-1}$ for s even;
2) $\deg f = (s-1)/2$ and $N_f \geq 2^{r+s-1} - 2^{r+(s-1)/2-1}$ for s odd.

8.6. Global avalanche characteristics of Boolean functions

The strong avalanche criterion (SAC) and its generalizations (SAC(k), PC(t), PC(t, k)) are just indicators of local properties for cryptographic functions. The reader can easily construct examples of functions which satisfy one of these criteria and at the same time have cryptographically "bad" properties (for example, a nontrivial space of linear structures).

The approach suggested in [**427**], which uses the so-called global avalanche characteristics, does not seem to have these defects. Roughly speaking, a Boolean function f has good global avalanche characteristics if $|\Delta_f(\boldsymbol{u})|$ is equal to zero for the majority of vectors $\boldsymbol{u} \in V_m$.

DEFINITION 8.96. Suppose that $f \in \mathcal{F}_m$. Then the *sum of squares*:

$$\sigma_f = \sum_{\boldsymbol{u} \in V_m} \Delta_f^2(\boldsymbol{u})$$

and the *absolute index*:

$$\Delta_f = \max_{\substack{\boldsymbol{u} \in V_m \\ \boldsymbol{u} \neq \boldsymbol{0}}} |\Delta_f(\boldsymbol{u})|$$

are called the *global avalanche characteristics (GAC)* of the Boolean function f.

The smaller σ_f and Δ_f, the better the global avalanche characteristics.

PROBLEM 8.97. Prove that σ_f and Δ_f are affine invariants.

PROBLEM 8.98. Suppose that $f \in \mathcal{F}_m$. Prove that:
1) $2^{2m} \leq \sigma_f \leq 2^{3m}$;
2) $\sigma_f = 2^{2m}$ if and only if f is an mn-bent function;
3) $\sigma_f = 2^{3m}$ if and only if f is an affine function.

EXAMPLE 8.99. 1. If $N\Delta_f = 2$, then $m = 2k+1$ is odd and $N_f = 2^{m-1} - 2^{(m-1)/2}$. Besides, there exists an invertible $(m \times m)$ matrix A over the field $\mathbb{F}_2$ such that

$$g(\boldsymbol{x}) = f^A(\boldsymbol{x}) = f(A\boldsymbol{x}) = cx^{(m)} \oplus h\left(x^{(1)}, \dots, x^{(m-1)}\right),$$

where $c \in \mathbb{F}_2$, and h is an mn-bent function in $\mathcal{F}_{m-1}$. In this case $\sigma_f = 2^{2m+1}$.

2. If $N\Delta_f = 4$, then $m = 2k$ is even and $N_f = 2^{m-1} - 2^{m/2}$. Furthermore, there exists an invertible $(m \times m)$ matrix A over the field $\mathbb{F}_2$ such that

$$g(\boldsymbol{x}) = f^A(\boldsymbol{x}) = f(A\boldsymbol{x}) = c^{(m-1)}x^{(m-1)} \oplus c^{(m)}x^{(m)} \oplus h\left(x^{(1)}, \dots, x^{(m-2)}\right),$$

where $c^{(m-1)}, c^{(m)} \in \mathbb{F}_2$, and h is an mn-bent function in $\mathcal{F}_{m-2}$. In this case, $\sigma_f = 2^{2m+2}$.

Note that in both cases of Example 8.99, the vectors in the set $V_m \setminus \mathrm{PC}_f$ are linear structures.

PROBLEM 8.100. Suppose that $f \in \mathcal{F}_m$. Prove that:

1) $0 \le \Delta_f \le 2^m$;
2) $\Delta_f = 0$ if and only if f is an mn-bent function;
3) $\Delta_f = 2^m$ if and only if f has a linear structure.

In particular, for any quadratic function f which is not a maximum-nonlinear bent function, $\Delta_f = 2^m$.

PROBLEM 8.101. Prove that for any vector $\boldsymbol{u} \in V_m$ and any $f \in \mathcal{F}_m$, the function $D_{\boldsymbol{u}}f$ is not an mn-bent function.

The result of the latter problem demonstrates the important observation to the effect that no mn-bent function is "integrable", i.e., an mn-bent function cannot be a derivative of some Boolean function.

The parameter Δ_f of a function $f \in \mathcal{F}_m$ with $\deg f = r \ge 3$ is connected with the weight spectrum of a code $RM(r-1, m)$.

PROBLEM 8.102. Prove that if $f \in \mathcal{F}_m$ and $\deg f = r \ge 3$, then

$$\Delta_f \ge 2^m - 2d_{RM(r-1,m)} = 2^m - 2^{m-r+2}.$$

The bound in Problem 8.102 is not too sharp since the weights of "nonintegrable" functions are taken into account.

For a cubic function the following statement holds.

PROBLEM 8.103. Suppose that $f \in \mathcal{F}_m$ is a cubic Boolean function that is not an mn-bent function. Prove that $\Delta_f \ge 2^{(m+1)/2}$.

Now we describe a method of constructing balanced Boolean functions with good global avalanche characteristics.

EXAMPLE 8.104. I. The case $m = 2k$. Suppose that $\boldsymbol{z} \in V_m$, $\boldsymbol{z} = (\boldsymbol{y}, \boldsymbol{x})$, where $\boldsymbol{y} \in V_k$, $\boldsymbol{x} \in V_k$. We choose a permutation π of $V_k \setminus \{\boldsymbol{0}\}$ and a fixed nonzero vector $\boldsymbol{u} \in V_k$. Put

$$f(\boldsymbol{z}) = f(\boldsymbol{y}, \boldsymbol{x}) = \begin{cases} \langle \boldsymbol{u}, \boldsymbol{x} \rangle & \text{if } \boldsymbol{y} = \boldsymbol{0}, \\ \langle \pi(\boldsymbol{y}), \boldsymbol{x} \rangle & \text{if } \boldsymbol{y} \ne \boldsymbol{0}. \end{cases}$$

The Boolean function f satisfies the following conditions:

1) f is balanced;
2) $N_f \geq 2^{2k-1} - 2^k$;
3) f satisfies the propagation criterion with respect to $2^{2k} - 2^{k+1} + 2^{k-1} - 1$ nonzero vectors;
4) $\sigma_f = 2^{4k} + 2^{3k+3} - 2^{3k+1}$;
5) $\Delta_f \leq 2^{k+1}$.

II. The case $m = 2k+1$. Let $(s_0, s_1, \dots, s_{2^k-2})$ be a code word of the binary simplex code $\boldsymbol{S}_k$ of a maximal period. Then the sequence $(\boldsymbol{u}_0, \boldsymbol{u}_1, \dots, \boldsymbol{u}_{2^k-2})$, defined by

$$\boldsymbol{u}_t = \left(s_{t \bmod 2^k-1}, s_{(t+1) \bmod 2^k-1}, \dots, s_{(t+k-1) \bmod 2^k-1},\right)^\top,$$

where $t = 0, 1, \dots, 2^k - 2$, is a sequence of all nonzero vectors of the space V_k in a given order. We define a permutation π of vectors of the space V_k as follows:

$$\pi = \begin{pmatrix} \mathbf{0}, & \boldsymbol{u}_0, & \boldsymbol{u}_1, & \dots, & \boldsymbol{u}_{2^k-2} \\ \mathbf{0}, & \boldsymbol{u}_1, & \boldsymbol{u}_2, & \dots, & \boldsymbol{u}_0 \end{pmatrix}.$$

For $\boldsymbol{z} \in V_{2k+1}$ we have $\boldsymbol{z} = (z^{(1)}, \boldsymbol{y}, \boldsymbol{x})$, where $z^{(1)} \in \mathbb{F}_2$, $\boldsymbol{y}, \boldsymbol{x} \in V_k$. We set

$$f(\boldsymbol{z}) = f(z^{(1)}, \boldsymbol{y}, \boldsymbol{x}) = \begin{cases} 1 \oplus \langle \boldsymbol{y}, \boldsymbol{x} \rangle & \text{if } z^{(1)} = 0, \\ \langle \pi(\boldsymbol{y}), \boldsymbol{x} \rangle & \text{if } z^{(1)} = 1. \end{cases}$$

The Boolean function f satisfies the following conditions:

1) f is balanced;
2) $N_f \geq 2^{2k} - 2^k$;
3) f satisfies the propagation criterion with respect to $2^{2k} - 1$ nonzero vectors;
4) $\sigma_f = 2^{4k+3}$;
5) $\Delta_f \leq 2^{k+1}$.

We will now prove an upper bound for $N\Delta_f$.

LEMMA 8.105. *Let $f \in \mathcal{F}_m$, $\deg f = d$. Then*

$$N\Delta_f \leq 1 + 2^{m-4-2\left\lfloor \frac{m-2}{d-1} \right\rfloor} \left(\max_{\boldsymbol{u} \in V_m} W_f^2(\boldsymbol{u}) - 2^m \right).$$

PROOF. Since $\deg(D_{\boldsymbol{v}} f) \leq d-1$ for any $\boldsymbol{v} \in V_m$ and the function $D_{\boldsymbol{v}} f$ has a linear structure, according to Corollary 5.25 the integer $W_{D_{\boldsymbol{v}} f}(\mathbf{0}) = \Delta_f(\boldsymbol{v})$ is divisible by $2^{\left\lfloor \frac{m-2}{d-1} \right\rfloor + 2}$. Then we get

$$\sigma_f = \sum_{\boldsymbol{x} \in V_m} \Delta_f^2(\boldsymbol{x}) = 2^{2m} + \sum_{\substack{\boldsymbol{x} \in V_m \setminus \mathrm{PC}_f \\ \boldsymbol{x} \neq \mathbf{0}}} \Delta_f^2(\boldsymbol{x}) \geq 2^{2m} + (N\Delta_f - 1) \cdot 2^{2\left\lfloor \frac{m-2}{d-1} \right\rfloor + 4}.$$

Using the inequality $\sigma_f \leq 2^m \max_{\boldsymbol{u} \in V_n} W_f^2(\boldsymbol{u})$, which can be easily checked, we obtain the required result. □

PROBLEM 8.106. Suppose that $f \in \mathcal{F}_m$ and $\dim L_f = k \geq 1$. Prove that $\max_{\boldsymbol{u} \in V_m} |W_f(\boldsymbol{u})| \geq 2^{\frac{n+k}{2}}$ and this inequality becomes equality if and only if $\Delta_f(\boldsymbol{x}) = 0$ for any $\boldsymbol{x} \in V_m \setminus L_f$. In this case k and m have the same parity and the set $\{\pm W_f(\boldsymbol{u}) \mid \boldsymbol{u} \in V_m\}$ consists of the elements 0 and $\pm \max_{\boldsymbol{u} \in V_m} |W_f(\boldsymbol{u})|$.

To conclude, we will prove a lower bound for Δ_f in the case where f is a k-resilient function in $\mathcal{F}_m$, $k > \frac{m-3}{2}$.

THEOREM 8.107. *Suppose that* $f \in \mathcal{F}_m$, $\operatorname{sut} f = k > \frac{m-3}{2}$. *Then*

$$\Delta_f \geq \Big(\frac{2k-m+3}{m+1}\Big) \cdot 2^m.$$

PROOF. We form a matrix M with m columns by using as rows the components of the vector $\boldsymbol{u} \in V_m$ exactly $W_f^2(\boldsymbol{u})$ times. According to Parseval's identity the matrix M has exactly 2^{2m} rows. In view of the spectral characterization of correlation-immune functions (see Theorem 7.10), any row of the matrix M contains at most $m-k-1$ zeros. Suppose that for some natural number α, $0 \leq \alpha \leq 1$, the matrix M contains exactly $\alpha \cdot 2^{2m}$ rows with less than $n-m-1$ zeros. Then the number of zeros in M is at most $\alpha 2^{2m}(m-k-2) + (2^{2m} - \alpha 2^{2m})(m-k-1) = (m-k-1-\alpha)2^{2m}$, whence the matrix M contains a column with index i having no more than $\frac{1}{m}(m-k-1-\alpha)2^{2m}$ zeros. From the construction of the matrix M it follows that

$$\sum_{\boldsymbol{x} \in V_m,\, x^{(i)}=0} W_f^2(\boldsymbol{x}) \leq \frac{1}{m}(m-k-1-\alpha)2^{2m}.$$

Using Corollary 2.81 we get

$$\begin{aligned}\Delta_f(\boldsymbol{e}_i) &= -2^m + 2^{1-m} \sum_{\boldsymbol{x} \in V_m,\, x^{(i)}=0} W_f^2(\boldsymbol{x}) \leq -2^m + \frac{1}{m}(m-k-1-\alpha) \cdot 2^{m+1} \\ &= \frac{1}{m}(m-2k-2-2\alpha) \cdot 2^m.\end{aligned}$$

From this it follows that

$$\Delta_f \geq \frac{1}{m}(2k-m+2+2\alpha) \cdot 2^m. \tag{8.45}$$

On the other hand,

$$\Delta_f(\mathbf{1}) = -2^m + 2^{1-m} \sum_{\substack{\boldsymbol{x} \in V_m \\ \operatorname{wt}(\boldsymbol{x}) \equiv 0 \bmod 2}} W_f^2(\boldsymbol{x}),$$

and

$$\Delta_f \geq |\Delta_f(\mathbf{1})| \geq (1-2\alpha)2^m. \tag{8.46}$$

The left-hand side of (8.45) increases with α, while the rightmost expression of (8.46) decreases with α. The rightmost expressions of (8.45) and (8.46) are equal for $\alpha = \frac{m-k-1}{m+1}$. Therefore,

$$\Delta_f \geq \Big(\frac{2k-m+3}{m+1}\Big) \cdot 2^m.$$ □

Comments to Chapter 8

The properties and criteria for perfect nonlinear and bent mappings and the conditions for their existence are studied in [**267, 290, 294**]. Some generalizations of the notions of almost perfect (APN) and almost bent mappings (AB) are discussed in [**77, 292, 293, 296**].

A code-theoretic approach to the generation of AB and APN mappings is given in [**53, 64**]. You can also consult [**315**].

Some relations between various propagation characteristics and nonlinearity of Boolean functions and mappings are presented in [**48, 348**]. The global avalanche characteristics are discussed in [**427**].

As a rule, cryptographic properties can be expressed in terms of values of certain parameters. Thus, there exist many optimal values for parameters of Boolean mappings which have to be satisfied by a given mapping. If the set of these parameters is large, then the class of mappings satisfying our criterion can be pretty small. On the other hand, it is not always necessary for cryptography that an optimal value of a parameter be attained. When choosing a particular mapping, we are happy if the values of its cryptographic parameters are sufficiently close to the optimal values in appropriate metrics. These metrics induce new modified cryptographic properties. This direction of research nowadays seems to be very promising.

CHAPTER 9

Basics of Cryptanalysis

This chapter is intended to acquaint the reader with some basic ideas of cryptanalysis. It logically connects the cryptographic properties of Boolean functions studied in this book with the practical resistance of cryptographic primitives (schemes, protocols, hash-functions, etc.) to certain attacks.

9.1. The Berlekamp–Massey algorithm. Linear complexity

The Berlekamp–Massey algorithm is the best example of close connection between a number of problems of coding theory and cryptography. The Berlekamp–Massey algorithm in its present form was developed by Berlekamp [**22**] as a tool for solving a problem which arises when building the Peterson–Gorenstein–Zierler decoder for BCH codes. Later, Massey [**256**] gave a new formulation of the original problem into a problem for constructing a linear feedback shift register. With this formulation the "cryptographic flavor" of the problem became obvious.

We will briefly present the basic ideas of Massey's approach. Sequences of binary symbols produced by the generators of pseudo-random sequences of stream ciphers are among the basic objects of cryptanalysis for this class of ciphers. These sequences must pass an enormous number of tests to ensure that they have properties "close" to those of a random uniform sequence. It seems natural to ask how the elements of the sequences depend on their preceding elements. In this case, a linear recursive sequence is taken as standard, and the minimum length of the linear feedback shift register which can generate the given sequence will become clear during the analysis. The length of the register could be considered as a measure of the complexity of the sequence. Presenting the algorithm for the Boolean case, we will follow[**144**], but we will slightly change the notation and the style of the exposition.

Let

$$V_\infty = \{\boldsymbol{u} = (u^{(0)}, u^{(1)}, \dots) \mid u^{(i)} \in \mathbb{F}_2,\ i = 0, 1, \dots\}$$

be the set of infinite sequences over the field $\mathbb{F}_2$. We define the sum of $\boldsymbol{u}$ and $\boldsymbol{v}$ as follows:

$$\boldsymbol{u} \oplus \boldsymbol{v} = (u^{(0)} \oplus v^{(0)}, u^{(1)} \oplus v^{(1)}, \dots).$$

It is clear that the set V_∞ is a vector space over the field $\mathbb{F}_2$ with respect to this operation. We denote by $\mathbf{0} = (0, 0, \dots)$ the sequence of zeroes in the field $\mathbb{F}_2$.

DEFINITION 9.1. A sequence $\boldsymbol{u} = (u^{(0)}, u^{(1)}, \dots) \in V_\infty$ is called an *ultimately periodic sequence* if there exist integers $d > 0$ and $i_0 \geq 0$ such that $u^{(i+d)} = u^{(i)}$ for all $i \geq i_0$. The number d is called a *period*, and the number i_0 is called a *pre-period* of the sequence $\boldsymbol{u}$. The least among all periods of the sequence is called the *minimum period* and is denoted by $\mathrm{per}(\boldsymbol{u})$. If $i_0 = 0$, the sequence is called *periodic*.

LEMMA 9.2. *Every period of an ultimately periodic sequence is divisible by its minimum period.*

PROOF. Let d be an arbitrary period of the ultimately periodic sequence $\boldsymbol{u} = (u^{(0)}, u^{(1)}, \dots)$ and let d_1 be its minimum period, i.e., $u^{(i+d)} = u^{(i)}$ for all $i \geq i_0$ and $u^{(i+d_1)} = u^{(i)}$ for all $i \geq i_1$ for some i_0 and i_1. If d is not divisible by d_1, then dividing d by d_1 we get $d = md_1 + r$, where $m \geq 1$, $0 < r < d_1$. For all $i \geq \max(i_0, i_1)$,

$$u^{(i)} = u^{(i+d)} = u^{(i+md_1+r)} = u^{(i+(m-1)d_1+r)} = \dots = u^{(i+r)}.$$

Therefore, r is also a period of the sequence $\boldsymbol{u}$, which contradicts the fact that d_1 is the minimum period. □

DEFINITION 9.3. A sequence $\boldsymbol{u} = (u^{(0)}, u^{(1)}, \dots) \in V_\infty$ is called a *linear recursive sequence (LRS) of nth series* if there exist $c^{(0)}, c^{(1)}, \dots, c^{(n-1)} \in \mathbb{F}_2$ such that the elements of the sequence $\boldsymbol{u}$ satisfy the following recursive relation:

$$u^{(n+k)} = \bigoplus_{i=0}^{n-1} c^{(i)} u^{(i+k)}, \quad k = 0, 1, \dots. \tag{9.1}$$

For a sequence $\boldsymbol{u} = (u^{(0)}, u^{(1)}, \dots) \in V_\infty$, we define the *shift operator* $\mathcal{L}$ as follows:

$$\mathcal{L}\boldsymbol{u} = (u^{(1)}, u^{(2)}, \dots).$$

It is easy to check that the shift operator $\mathcal{L}$ is a linear mapping from V_∞ to V_∞. For any natural number i we have

$$\mathcal{L}^i\boldsymbol{u} = (u^{(i)}, u^{(i+1)}, \dots),$$

where $\mathcal{L}^0\boldsymbol{u} = \mathcal{E}\boldsymbol{u} = \boldsymbol{u}$ and $\mathcal{E}$ is the identical image of V_∞. With the help of the shift operator $\mathcal{L}$ the relation (9.1) can be written as

$$\mathcal{L}^n\boldsymbol{u} = \bigoplus_{i=0}^{n-1} c^{(i)}\mathcal{L}^i\boldsymbol{u}$$

or

$$\left(\mathcal{L}^n \oplus \bigoplus_{i=0}^{n-1} c^{(i)}\mathcal{L}^i\right)\boldsymbol{u} = \mathbf{0}. \tag{9.2}$$

The latter relation is equivalent to the equality

$$c(\mathcal{L})\boldsymbol{u} = \mathbf{0}$$

for the polynomial $c(\lambda) = c^{(0)} \oplus c^{(1)}\lambda \oplus \dots \oplus c^{(n-1)}\lambda^{n-1} \oplus \lambda^n \in \mathbb{F}_2[\lambda]$. This proves the following lemma.

LEMMA 9.4. *A sequence $\boldsymbol{u} = (u^{(0)}, u^{(1)}, \dots) \in V_\infty$ is a linear recursive sequence of nth series if and only if there exists a polynomial $c(\lambda) \in \mathbb{F}_2[\lambda]$, $\deg c(\lambda) = n$, such that*

$$c(\mathcal{L})\boldsymbol{u} = \mathbf{0}.$$

The polynomial $c(\lambda)$ is called a *characteristic polynomial of the LRS $\boldsymbol{u}$*: $\boldsymbol{u}(\lambda) = c(\lambda)$.

Linear recursive sequences can be generated by a *linear feedback shift register*. This is a discrete electronic device processing information given elements of a finite field ($\mathbb{F}_2$ in this case) represented in a suitable form.

Binary registers consist of two types of constructive elements. Elements of the first type are adders. They have two inputs and one output. When the inputs of an adder are fed with two elements of the field $\mathbb{F}_2$, the output is their sum in the field $\mathbb{F}_2$ (i.e., sum modulo 2). The second type of constructive elements is a delay device (trigger). It has one input and one output. When an element of the field $\mathbb{F}_2$ is fed to the input at a given step of the operation of the device, it appears on the output at the next step. The linear feedback shift register is a closed chain of a finite number of constructive elements of these two types, connected in such a way that no two inputs and outputs are connected to each other.

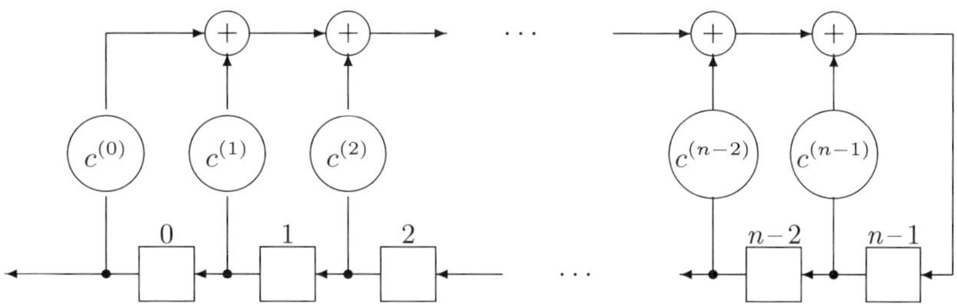

FIGURE 9.1

A linear feedback shift register is depicted in Figure 9.1. The output of the ith tap is connected with the existing adder if $c^{(i)} = 1$ and not connected if $c^{(i)} = 0$. The number of taps used in the device is called *the length of the register*, and the polynomial $\lambda^n \oplus \bigoplus_{i=0}^{n-1} \lambda^i$ is called a *characteristic polynomial of the register*.

PROPOSITION 9.5. *Let the word $u^{(0)}, u^{(1)}, \dots, u^{(n-1)}$ be written in the delay elements (see Figure 9.1) before the start of the operation of the device. Then the register generates an LRS $\boldsymbol{u}$ satisfying the recursive relation (9.1). The element $u^{(i)}$ appears as the output of the register at the ith ($i = 0, 1, 2, \dots$) step of the operation.*

PROOF. The structure of the shift register in Figure 9.1 is chosen to satisfy the proposition. Namely, the equation representing the operation of the register (i.e., dependence of an ith output symbol, for $i \geq n$, on the n preceding symbols) coincides with (9.1). □

For a nonzero polynomial $c(\lambda) \in \mathbb{F}_2[\lambda]$, denote by $B(c)$ the following set:

$$B(c) = \{\boldsymbol{u} \in V_\infty \mid c(\mathcal{L})\boldsymbol{u} = \boldsymbol{0}\}.$$

As $c(\mathcal{L})$ is a linear mapping from V_∞ to V_∞ and the set $B(c)$ is the kernel of this mapping, $B(c) = \ker c(\mathcal{L})$ is a linear subspace in V_∞.

For convenience, we will regard the polynomial $1 \in \mathbb{F}_2[\lambda]$: $\mathrm{char}_{\boldsymbol{0}}(\lambda) = 1$ as a characteristic polynomial of the zero sequence $\boldsymbol{0} = (0, 0, \dots)$.

THEOREM 9.6. *Let $c(\lambda) \in \mathbb{F}_2[\lambda]$ be a polynomial of degree n. Then*

$$\#B(c) = 2^n.$$

PROOF. The first n elements $(u^{(0)}, u^{(1)}, \dots, u^{(n-1)})$ of the sequence $\boldsymbol{u} = (u^{(0)}, u^{(1)}, \dots) \in B(c)$ completely determine the rest of the sequence, starting with $u^{(n)}$ and using the relation (9.1). There are exactly 2^n possibilities for choosing different n-tuples $(u^{(0)}, u^{(1)}, \dots, u^{(n-1)})$. Therefore, $\#B(c) = 2^n$. □

COROLLARY 9.7. *$B(c)$ is a linear space of dimension n over the field $\mathbb{F}_2$.*

Let $\boldsymbol{u}$ be a linear recursive sequence. According to Lemma 9.4, there exists a nonzero polynomial $\boldsymbol{u}(\lambda) = c(\lambda)$ of some degree n, $n \geq 0$, such that

$$c(\mathcal{L})\boldsymbol{u} = \mathbf{0}.$$

However, the polynomial $c(\lambda)$ may be not the only one satisfying this relation. For example, for the zero sequence $\mathbf{0} = (0, 0, \dots)$, every polynomial can be a characteristic polynomial. For the sequence $\boldsymbol{u}$ we define the set

$$A(\boldsymbol{u}) = \{c(\lambda) \in \mathbb{F}_2[\lambda] \mid c(\mathcal{L})\boldsymbol{u} = \mathbf{0}\}.$$

According to Lemma 9.4, the set $A(\boldsymbol{u})$ consists of all characteristic polynomials of the sequence $\boldsymbol{u}$.

PROBLEM 9.8. Prove that the set $A(\boldsymbol{u})$ is an ideal of the ring $\mathbb{F}_2[\lambda]$.

DEFINITION 9.9. The minimal normalized nonzero polynomial in the ideal $A(\boldsymbol{u})$ is called the *minimal polynomial of the sequence* $\boldsymbol{u}$ over the field $\mathbb{F}_2$. We will denote this polynomial by $\min_{\boldsymbol{u}}(\lambda)$.

It is easy to check that the minimal polynomial of the zero sequence $\mathbf{0} = (0, 0, \dots)$ equals one:

$$\min_{\mathbf{0}}(\lambda) = 1.$$

We can also easily calculate the minimal polynomial of the all-one sequence $\mathbf{1} = (1, 1, \dots)$:

$$\min_{\mathbf{1}}(\lambda) = \lambda \oplus 1.$$

LEMMA 9.10. *The minimal polynomial of an arbitrary sequence $\boldsymbol{u} \in V_\infty$ is unique.*

The polynomial $m(\lambda) \in \mathbb{F}_2[\lambda]$ is the minimal polynomial of the sequence $\boldsymbol{u}$, $\min_{\boldsymbol{u}}(\lambda) = m(\lambda)$, if and only if the following two conditions are satisfied:

1) $m(\mathcal{L})\boldsymbol{u} = \mathbf{0}$;

2) *for an arbitrary polynomial $c(\lambda) \in \mathbb{F}_2[\lambda]$ the equation $c(\mathcal{L})\boldsymbol{u} = \mathbf{0}$ is satisfied if and only if $m(\lambda)$ divides $c(\lambda)$.*

PROOF. Since $m(\lambda) \in A(\boldsymbol{u})$, condition 1) is satisfied. If $\boldsymbol{u} = \mathbf{0}$, then because $1 \mid c(\lambda)$ for every $c(\lambda) \in \mathbb{F}_2[\lambda]$, condition 2) is satisfied too. Now let $\boldsymbol{u}$ be a nonzero sequence. If $m(\lambda) \mid c(\lambda)$, then according to Problem 9.8 the inclusion $c(\lambda) \in A(\boldsymbol{u})$ holds and $c(\mathcal{L})\boldsymbol{u} = \mathbf{0}$.

Now suppose that $c(\mathcal{L})\boldsymbol{u} = \mathbf{0}$. If $m(\lambda)$ does not divide $c(\lambda)$, then we can divide $c(\lambda)$ by $m(\lambda)$ with a remainder:

$$c(\lambda) = q(\lambda)m(\lambda) \oplus r(\lambda),$$

where $q(\lambda), r(\lambda) \in \mathbb{F}_2[\lambda]$, $0 \leq \deg r(\lambda) < \deg m(\lambda)$. According to Problem 9.8, $q(\mathcal{L})m(\mathcal{L})\boldsymbol{u} = \mathbf{0}$. Therefore, the following chain of equalities is satisfied:

$$\mathbf{0} = c(\mathcal{L})\boldsymbol{u} = [q(\mathcal{L})m(\mathcal{L}) \oplus r(\mathcal{L})]\boldsymbol{u} = r(\mathcal{L})\boldsymbol{u}.$$

Thus, $r(\lambda) \in A(\boldsymbol{u})$; since $r(\lambda) \neq 0$ and $\deg r(\lambda) < \deg m(\lambda)$, we get a contradiction with the minimality of the polynomial $m(\lambda)$. Therefore $m(\lambda)$ divides $c(\lambda)$: $m(\lambda)|c(\lambda)$.

Let $m_1(\lambda) \in \mathbb{F}_2[\lambda]$ be another minimal polynomial of the sequence $\boldsymbol{u} \in A(\boldsymbol{u})$. According to condition 2) already proved and Definition 9.9 we have: $m(\lambda)|m_1(\lambda)$,

$m_1(\lambda)|m(\lambda)$, and $\deg m(\lambda) = \deg m_1(\lambda)$. Therefore, the polynomials $m(\lambda)$ and $m_1(\lambda)$ coincide. □

We note that for a sequence $\boldsymbol{u} \in B(c)$, the polynomial $c(\lambda)$ is not necessarily the minimal one. But the following statement is always true.

COROLLARY 9.11. *Suppose that* $\boldsymbol{u} \in B(c)$. *Then* $\min_{\boldsymbol{u}}(\lambda) \mid c(\lambda)$.

PROOF. By the definition of the set $B(c)$, the equation $c(\mathcal{L})\boldsymbol{u} = \mathbf{0}$ is always satisfied. Then, by Lemma 9.10, $\min_{\boldsymbol{u}}(\lambda)|c(\lambda)$. □

COROLLARY 9.12. *Suppose that* $c(\lambda) \neq 0$. *If* $c(\lambda)$ *is irreducible over the field* $\mathbb{F}_2$, *then for any sequence* $\boldsymbol{u} \in B(c)$, $\boldsymbol{u} \neq \mathbf{0}$ *the equation* $\min_{\boldsymbol{u}}(\lambda) = c(\lambda)$ *is satisfied.*

PROOF. Corollary 9.11 implies that we always have $\min_{\boldsymbol{u}}(\lambda) \mid c(\lambda)$. For $\boldsymbol{u} \neq \mathbf{0}$ we have $\min_{\boldsymbol{u}}(\lambda) \neq 1$, and since $c(\lambda)$ is irreducible, the equation $\min_{\boldsymbol{u}}(\lambda) = c(\lambda)$ is satisfied. □

By the definition of minimal polynomials, the degree of the minimal polynomial of a sequence $\boldsymbol{u}$ coincides with the minimal possible length of a linear feedback shift register generating the sequence $\boldsymbol{u}$. This quantity appears to be an important cryptographic characteristic of pseudo-random sequences used in stream ciphers.

LEMMA 9.13. *Every periodic sequence can be generated by a linear feedback shift register.*

PROOF. Let d be the period of the sequence $\boldsymbol{u}$. Then

$$u^{(k+d)} = u^{(k)}, \quad k = 0, 1, \ldots.$$

Thus, if $c(\lambda) = \lambda^d \oplus 1$, then $c(\mathcal{L})\boldsymbol{u} = \mathbf{0}$. Therefore, the sequence $\boldsymbol{u}$ can be generated by a linear feedback shift register of length d, defined by the polynomial $c(\lambda)$. □

According to Lemma 9.13 any periodic sequence can be regarded as a linear recursive sequence; thus, there exists a minimal polynomial of this sequence.

DEFINITION 9.14. Let $\boldsymbol{u}$ be a periodic sequence. Then the degree of the minimal polynomial $\min_{\boldsymbol{u}}(\lambda)$ of the sequence $\boldsymbol{u}$ is called the *linear span* or *linear complexity* of the sequence $\boldsymbol{u}$.

LEMMA 9.15. *Let* $\boldsymbol{u}$ *be a linear recurrent sequence with minimal polynomial* $\min_{\boldsymbol{u}}(\lambda) = m(\lambda)$. *Then the sequence* $\boldsymbol{u}$ *is periodic and* $\operatorname{per}(\boldsymbol{u}) = \operatorname{ord}(m(\lambda))$ *(i.e., the minimum period of a recurrent sequence coincides with the degree of its minimal polynomial).*

PROOF. Let $\operatorname{ord}(m(\lambda)) = d$. Then $m(\lambda) \mid \lambda^d \oplus 1$ and $\lambda^d \oplus 1 = m(\lambda)q(\lambda)$. In this case

$$(\mathcal{L}^d \oplus \mathcal{E})\boldsymbol{u} = q(\mathcal{L})m(\mathcal{L})\boldsymbol{u} = \mathbf{0},$$

so $\lambda^d \oplus 1 \in A(\boldsymbol{u})$ and d is the period of the sequence $\boldsymbol{u}$. Since d is the minimum number with this property, the number d coincides with the minimum period of the sequence $\boldsymbol{u}$. □

Various applications use sequences of finite length. Finding a linear feedback shift register of minimum length generating a given finite sequence is a rather hard problem. Before we start discussing this problem we must introduce some definitions and state some results.

Denote by $\boldsymbol{u}(N) = (u^{(0)}, u^{(1)}, \ldots, u^{(N-1)}) \in V_N$ a sequence of length N. Note that the sequence $\boldsymbol{u}(N)$ is not a periodic sequence with period N.

DEFINITION 9.16. We will say that a linear feedback shift register of length l generates a finite sequence $(u^{(0)}, u^{(1)}, \ldots, u^{(N-1)})$ if it can be obtained as the first N output symbols of a linear feedback shift register of length l with the starting state $(u^{(0)}, u^{(1)}, \ldots, u^{(l-1)})$.

We will use the notation $(c(\lambda), l)$, $c(\lambda) \in \mathbb{F}_2[\lambda]$, where l is a nonnegative integer, to refer to the linear feedback shift register of length l defined by the polynomial $c(\lambda)$, $l = \deg c(\lambda)$.

PROPOSITION 9.17. *If $l \geq N$, then the shift register $(c(\lambda), l)$ always generates the sequence $(u^{(0)}, u^{(1)}, \ldots, u^{(N-1)})$.*

If $l < N$, then the shift register $(c(\lambda), l)$ generates this sequence if and only if

$$u^{(l+k)} = \bigoplus_{i=0}^{l-1} c^{(i)} u^{(i+k)}, \quad k = 0, 1, \ldots, N-l-1,$$

where $c(\lambda) = \lambda^l \oplus \bigoplus_{i=0}^{l-1} c^{(i)} \lambda^i$.

DEFINITION 9.18. Suppose that $\boldsymbol{u}(N) \in V_N$. The *linear span* or *linear complexity* of this sequence is the minimum length of a linear feedback shift register which generates it:

$$LS(\boldsymbol{u}(N)) = \min\{0 \leq l \leq N \mid (c(\lambda), l) \text{ generates } \boldsymbol{u}(N)\}.$$

Later we will assume that the degenerate register $(1, 0)$ generates sequences of any finite length. Suppose that the shift register $(c_n(\lambda), l_n)$ generates the sequences $(u^{(0)}, u^{(1)}, \ldots, u^{(n)})$ for $n = 0, 1, \ldots, N-1$. Before we present the Berlekamp–Massey algorithm, we will discuss some properties of the shift operator in V_∞ and introduce the notion of discrepancy bit.

Let $\boldsymbol{u}(N) = (u^{(0)}, u^{(1)}, \ldots, u^{(N-1)}) \in V_N$. We will say that the infinite sequence $\boldsymbol{u} = (u^{(0)}, u^{(1)}, \ldots, u^{(N-1)}, 0, 0, \ldots) \in V_\infty$ is the *zero tail expansion* of the sequence $\boldsymbol{u}(N)$.

PROPOSITION 9.19. *If $l < N$, then the shift register $(c(\lambda), l)$, $c(\lambda) = \lambda^l \oplus \bigoplus_{i=0}^{l-1} c^{(i)} \lambda^i$, generates the sequence $\boldsymbol{u}(N) = (u^{(0)}, u^{(1)}, \ldots, u^{(N-1)})$ if and only if*

$$c(\mathcal{L})\boldsymbol{u} = (\underbrace{0, \ldots, 0}_{b} *, *, \ldots), \tag{9.3}$$

where $b = (N-1) - (l-1) = N - l$ (the symbol $*$ represents an arbitrary element of $\mathbb{F}_2$).

PROOF. Put

$$\theta^{(l+k)} = u^{(l+k)} \oplus \bigoplus_{i=0}^{l-1} c^{(i)} u^{(i+k)}, \quad k = 0, 1, \ldots, N-l-1.$$

Then

$$c(\mathcal{L})\boldsymbol{u} = (\theta^{(l)}, \theta^{(l+1)}, \ldots). \tag{9.4}$$

If the register $(c(\lambda), l)$ generates $\boldsymbol{u}(N)$, then according to Proposition 9.19 we have

$$\theta^{(l+k)} = 0, \quad k = 0, 1, \ldots, N-l-1. \tag{9.5}$$

Therefore, relation (9.3) holds.

Conversely, if relation (9.3) holds, then, using relations (9.4) and (9.5), it is easy to see that the register $(c(\lambda), l)$ generates $\boldsymbol{u}(N)$. □

We will say that the bits $\theta^{(j)}$ determined by relation (9.4) are the *discrepancy bits* of the relation depending on the polynomial $c(\lambda)$.

LEMMA 9.20. *Let $\boldsymbol{u}(N) = (u^{(0)}, u^{(1)}, \dots, u^{(N-1)})$ be a finite set and $\boldsymbol{u}$ its zero tail extension. Suppose also that $n < N$. Then the following statements are equivalent:*

1. *A shift register $(c(\lambda), l)$ generates $(u^{(0)}, u^{(1)}, \dots, u^{(n-1)})$ but does not generate $(u^{(0)}, u^{(1)}, \dots, u^{(n)})$.*
2. *The discrepancy bits which depend on $c(\lambda)$ satisfy the relation*

$$\theta^{(l+k)} = 0, \quad k = 0, 1, \dots, n-l-1, \quad \text{and} \quad \theta^{(n)} = 1.$$

3. $c(\mathcal{L})\boldsymbol{u} = (\underbrace{0, \dots, 0}_{n-l}, 1, *, *, \dots)$.

PROOF. The lemma follows from Proposition 9.19. □

For convenience we will denote by $\boldsymbol{v}_k$ any infinite sequence from V_∞ of the type $(0, \dots, 0, *, *, \dots)$ with k zeros, $k = 0, 1, \dots$.

LEMMA 9.21. *Put $c(\lambda) = \lambda^l \oplus \bigoplus_{i=0}^{l-1} c^{(i)} \lambda^i$. If $l < k$, then*

$$c(\mathcal{L})\boldsymbol{v}_k = (\underbrace{0, \dots, 0}_{k-l}, *, *, \dots) = \boldsymbol{v}_{k-l}.$$

PROOF. Note that

$$\boldsymbol{v}_k = (\underbrace{0, \dots, 0}_{k}, 1, *, *, \dots),$$

$$\mathcal{L}\boldsymbol{v}_k = (\underbrace{0, \dots, 0}_{k-1}, 1, *, *, \dots) = \boldsymbol{v}_{k-1},$$

$$\mathcal{L}^2\boldsymbol{v}_k = (\underbrace{0, \dots, 0}_{k-2}, 1, *, *, \dots) = \boldsymbol{v}_{k-2},$$

$$\dots\dots\dots\dots\dots\dots\dots\dots\dots\dots$$

$$\mathcal{L}^l\boldsymbol{v}_k = (\underbrace{0, \dots, 0}_{k-l}, 1, *, *, \dots) = \boldsymbol{v}_{k-l}.$$

The expression $c(\mathcal{L})\boldsymbol{v}_k$ is a linear combination of these sequences, i.e.,

$$c(\mathcal{L})\boldsymbol{v}_k = \mathcal{L}^l\boldsymbol{v}_k \oplus \bigoplus_{i=0}^{l-1} c^{(i)}\mathcal{L}^i\boldsymbol{v}_k = \mathcal{L}^l\boldsymbol{v}_k \oplus \bigoplus_{i=0}^{l-1} c^{(i)}\boldsymbol{v}_{k-i} = \boldsymbol{v}_{k-l}.$$ □

We now proceed with the presentation of the Berlekamp–Massey algorithm.

THEOREM 9.22. *Put $\boldsymbol{u}(N) = (u^{(0)}, u^{(1)}, \dots, u^{(N-1)}) \in V_N$, $N > 0$. The following algorithm constructs shift registers $(c_n(\lambda), l_n)$ which generate the sequence $(u^{(0)}, u^{(1)}, \dots, u^{(N-1)})$ for $n = 0, 1, \dots, N-1$.*

1. *Denote by $n_0 \geq 0$ the index of the first nonzero bit in the sequence $\boldsymbol{u}(N)$, i.e.,*

$$u^{(0)} = u^{(1)} = \dots = u^{(n_0-1)} = 0, \quad u^{(n_0)} = 1.$$

If $n_0 = 0$, *we put*

$$c_{-1}(\lambda) = 1, \quad l_{-1} = 0.$$

If $n_0 > 0$, *we put*

$$c_0(\lambda) = c_1(\lambda) = \cdots = c_{n_0-1}(\lambda) = 1, \quad l_0 = l_1 = \cdots = l_{n_0-1} = 0.$$

Then we put

$$c_{n_0}(\lambda) = \lambda^{n_0+1} \oplus 1, \quad l_{n_0} = n_0 + 1.$$

2. *Suppose that* $c_0(\lambda), c_1(\lambda), \dots, c_{n-1}(\lambda)$, $l_0, l_1 \dots, l_{n-1}$, $n > n_0$, *are already constructed and they are such that the shift register* $(c_{n-1}(\lambda), l_{n-1})$ *generates the sequence* $(u^{(0)}, u^{(1)}, \dots, u^{(n-1)})$. *Put*

$$c_{n-1}(\lambda) = \lambda^{l_{n-1}} \oplus \bigoplus_{i=0}^{l_{n-1}-1} c_{n-1}^{(i)} \lambda^i.$$

We count the discrepancy bit:

$$\theta^{(n)} = u^{(n)} \oplus \bigoplus_{i=0}^{l_{n-1}-1} c_{n-1}^{(i)} u^{n-l_{n-1}+i}.$$

If $\theta^{(n)} = 0$, *then the shift register* $(c_{n-1}(\lambda), l_{n-1})$ *also generates the sequence* $(u^{(0)}, u^{(1)}, \dots, u^{(n)})$. *In this case we put* $c_n(\lambda) = c_{n-1}(\lambda)$, $l_n = l_{n-1}$.

If $\theta^{(n)} = 1$, *then the shift register* $(c_{n-1}(\lambda), l_{n-1})$ *generates the sequence* $(u^{(0)}, u^{(1)}, \dots, u^{(n-1)})$ *but does not generate the sequence* $(u^{(0)}, u^{(1)}, \dots, u^{(n)})$. *Let* m $(0 \le m < n)$ *be the last number when the register has changed its length, i.e.,*

$$l_{m-1} < l_m = l_{m+1} = \cdots = l_{n-1}.$$

The shift register $(c_{m-1}(\lambda), l_{m-1})$ *generates the sequence* $(u^{(0)}, \dots, u^{(m-1)})$ *but does not generate the sequence* $(u^{(0)}, \dots, u^{(m)})$.

Put

$$a = (m-1) - l_{m-1} + 1, \quad b = (n-1) - l_{n-1} + 1.$$

If $a \ge b$, *we substitute*

$$c_n(\lambda) = c_{n-1}(\lambda) \oplus \lambda^{a-b} c_{m-1}(\lambda), \quad l_n = l_{n-1}.$$

If $a < b$, *we put*

$$c_n(\lambda) = \lambda^{b-a} c_{n-1}(\lambda) \oplus c_{m-1}(\lambda), \quad l_n = l_{n-1} + b - a.$$

3. *If* $n < N - 1$, *we go to step* 2.

PROOF. To prove the correctness of the algorithm it is obviously enough to prove that if $\theta^{(n)} = 1$ in step 2, then the suggested register $(c_n(\lambda), l_n)$ generates the sequence $(u^{(0)}, \dots, u^{(n)})$.

Using Lemma 9.20 for the two cases where the discrepancy bit is equal to 1, we obtain

$$c_{n-1}(\mathcal{L})\boldsymbol{u} = (\underbrace{0, \dots, 0}_{b}, 1, *, \dots), \quad b = n - l_{n-1}, \tag{9.6}$$

$$c_{m-1}(\mathcal{L})\boldsymbol{u} = (\underbrace{0, \dots, 0}_{a}, 1, *, \dots), \quad a = m - l_{m-1}. \tag{9.7}$$

Let us do the case $a \geq b$. Applying the operator $\mathcal{L}^{a-b}$ to equality (9.7), we get

$$\mathcal{L}^{a-b}c_{m-1}(\mathcal{L})\boldsymbol{u} = \mathcal{L}^{a-b}(\underbrace{0,\ldots,0}_{a},1,*,\ldots) = (\underbrace{0,\ldots,0}_{b},1,*,\ldots). \tag{9.8}$$

Adding (9.6) and (9.8), we obtain

$$[c_{n-1}(\mathcal{L}) \oplus \mathcal{L}^{a-b}c_{m-1}(\mathcal{L})]\boldsymbol{u} = (\underbrace{0,\ldots,0}_{b},0,*,\ldots). \tag{9.9}$$

Put $c_n(\lambda) = c_{n-1}(\lambda) \oplus \lambda^{a-b}c_{m-1}(\lambda)$ and $l_n = \deg c_n(\lambda)$. Then the relation (9.9) shows that the shift register $(c_n(\lambda), l_n)$ generates the sequence $(u^{(0)}, u^{(1)}, \ldots, u^{(n)})$. Since $n > m$, we have

$$\begin{aligned}\deg \lambda^{a-b}c_{m-1}(\lambda) &= a - b + l_{m-1}\\ &= (m-1) - l_{m-1} + 1 - ((n-1) - l_{n-1} + 1) + l_{m-1}\\ &= l_{n-1} - (n-m) < l_{n-1}.\end{aligned}$$

Therefore, $\deg c_n(\lambda) = l_{n-1}$. Thus, we set

$$(c_n(\lambda), l_n) = (c_n(\lambda), l_{n-1}).$$

It remains to consider the case $a < b$. In a similar manner as in the previous case we obtain

$$[\mathcal{L}^{b-a}c_{n-1}(\mathcal{L}) \oplus c_{m-1}(\mathcal{L})]\boldsymbol{u} = (\underbrace{0,\ldots,0}_{a},0,*,\ldots). \tag{9.10}$$

We set $c_n(\lambda) = \lambda^{b-a}c_{n-1}(\lambda) \oplus c_{m-1}(\lambda)$ and $l_n = \deg c_n(\lambda)$. It is easy to check that $l_n > l_{n-1}$. The relation (9.10) shows that the shift register $(c_n(\lambda), l_n)$ generates the sequence $(u^{(0)}, u^{(1)}, \ldots, u^{(n)})$. □

Now we show that any characteristic polynomial obtained by the Berlekamp–Massey algorithm has minimum possible degree (which is equivalent to the minimality of the length of the corresponding shift register).

LEMMA 9.23. *Suppose that the shift register $(c_{n-1}(\lambda), l_{n-1})$ generates the sequence $(u^{(0)}, u^{(1)}, \ldots, u^{(n-1)})$ but does not generate $(u^{(0)}, u^{(1)}, \ldots, u^{(n)})$. The shift register $(c_n(\lambda), l_n)$, which generates the sequence $(u^{(0)}, u^{(1)}, \ldots, u^{(n)})$, satisfies the following condition:*

$$l_n \geq n + 1 - l_{n-1}.$$

PROOF. Let $\boldsymbol{u} = (u^{(0)}, u^{(1)}, \ldots, u^{(n-1)}, 0, \ldots) \in V_\infty$ be an extension of the sequence $\boldsymbol{u}(n)$. Since the shift register $(c_{n-1}(\lambda), l_{n-1})$ generates the sequence $(u^{(0)}, u^{(1)}, \ldots, u^{(n-1)})$ but does not generate the sequence $(u^{(0)}, u^{(1)}, \ldots, u^{(n)})$, we have

$$c_{n-1}(\mathcal{L})\boldsymbol{u} = (\underbrace{0,0,\ldots,0}_{a},1,*,\ldots), \quad a = n - 1 - l_{n-1}. \tag{9.11}$$

Since the shift register $(c_n(\lambda), l_n)$ generates the sequence $(u^{(0)}, u^{(1)}, \ldots, u^{(n)})$, we have

$$c_N(\mathcal{L})\boldsymbol{u} = (\underbrace{0,0,\ldots,0}_{b},*,\ldots), \quad b = n - l_n. \tag{9.12}$$

Suppose that the statement of the lemma is false. Then $n + 1 - l_{n-1} > l_n$, whence $a + 1 \geq l_n$, $b \geq l_{n-1}$. This means that we can apply the operators $c_n(\mathcal{L})$

and $c_{n-1}(\mathcal{L})$ to (9.11) and (9.12) respectively and use Lemma 9.21. As a result we get

$$c_n(\mathcal{L})c_{n-1}(\mathcal{L})\boldsymbol{u} = (\underbrace{0,\dots,0}_{a-l_n},1,*,\dots), \tag{9.13}$$

$$c_{n-1}(\mathcal{L})c_n(\mathcal{L})\boldsymbol{u} = (\underbrace{0,\dots,0}_{b-l_{n-1}},*,\dots). \tag{9.14}$$

According to (9.11) and (9.12),

$$a - l_n = b - l_{n-1} - 1;$$

comparing (9.13) and (9.14) for the same sequence, we get a contradiction. □

LEMMA 9.24. *Suppose the shift register $(c_{n-1}(\lambda), l_{n-1})$ generates the sequence $(u^{(0)}, u^{(1)}, \dots, u^{(n-1)})$ but does not generate the sequence $(u^{(0)}, u^{(1)}, \dots, u^{(n)})$. Then the shift register $(c_n(\lambda), l_n)$, which generates the sequence $(u^{(0)}, u^{(1)}, \dots, u^{(n)})$, satisfies the following condition:*

$$l_n \geq \max\{l_{n-1}, n+1-l_{n-1}\}.$$

PROOF. Since l_n is monotonic, we have $l_n \geq l_{n-1}$. According to Lemma 9.23 the inequality $l_n \geq n+1-l_{n-1}$ holds. These two inequalities together prove the lemma. □

THEOREM 9.25. *Let $(c_n(\lambda), l_n)$, $n = 0, 1, \dots, N-1$, be shift registers obtained by the Berlekamp–Massey algorithm. The linear span of l_n is the least possible. In other words, if the shift register $(c_{n-1}(\lambda), l_{n-1})$ generates the sequence $(u^{(0)}, u^{(1)}, \dots, u^{(n-1)})$ but does not generate the sequence $(u^{(0)}, u^{(1)}, \dots, u^{(n)})$, then the shift register $(c_n(\lambda), l_n)$, which generates the sequence $(u^{(0)}, u^{(1)}, \dots, u^{(n)})$, satisfies the condition*

$$l_n = \max\{l_{n-1}, n+1-l_{n-1}\} \quad \textit{for } n \geq n_0, \tag{9.15}$$

where n_0 is the index of the first nonzero element of the sequence $(u^{(0)}, u^{(1)}, \dots, u^{(n-1)})$.

PROOF. According to Lemma 9.24, if (9.15) holds, then the linear span l_n has the minimum possible value for any n in the family of the shift registers generating the sequence $(u^{(0)}, u^{(1)}, \dots, u^{(n)})$. Now we use induction. First consider n_0. Since $l_0 = l_1 = \cdots = l_{n_0-1} = 0$ and

$$l_{n_0} = n_0 + 1 = \max\{l_{n_0-1}, n_0 + 1 - l_{n_0-1}\}$$

by Theorem 9.22, the statement holds for $n = n_0$.

Suppose that it holds for $c_{n_0}(\lambda), c_{n_0+1}(\lambda), \dots, c_{n-1}(\lambda)$. We will show that it holds also for $c_n(\lambda)$. Looking at the polynomial $c_n(\lambda)$, we can say (Theorem 9.22) that there exists m, $0 < m < n$, for which

$$l_{m-1} < l_m = l_{m+1} = \cdots = l_{n-1}.$$

We will also use the same a and b as in Theorem 9.22:

$$a = (m-1) - l_{m-1} + 1, \quad b = (n-1) - l_{n-1} + 1.$$

By the induction hypothesis, $l_m = m + 1 - l_{m-1}$, but since $l_m = l_{n-1}$, one has

$$l_{n-1} = m + 1 - l_{m-1}. \tag{9.16}$$

Now we must examine the following two cases.

Case $a \geq b$: according to the algorithm we have $l_n = l_{n-1}$. Using (9.16) and substituting the expressions for a and b, we get $l_{n-1} = m + 1 - l_{m-1} = a + 1 \geq b + 1 = n + 1 - l_{n-1}$. Thus, $l_n = \max\{l_{n-1}, n + 1 - l_{n-1}\}$.

Case $a < b$: in this case $l_n = b - a + l_{n-1} \geq l_{n-1}$. Using (9.16) and substituting the expressions for a and b, we get

$$b - a + l_{n-1} = n - m + l_{m-1} = (n + 1) - (m + 1 - l_{m-1}) = n + 1 - l_{n-1}.$$

Thus, $l_n = \max\{l_{n-1}, n + 1 - l_{n-1}\}$. □

COROLLARY 9.26. *Under the hypotheses of Theorem* 9.25 *we have:*

1. $0 \leq l_0 \leq l_1 \leq \cdots \leq l_n \leq \cdots$.
2. $l_n = l_{n-1}$ *or* $l_n = n + 1 - l_{n-1}$.
3. *If* $2l_{n-1} > n$, *then* $l_n = l_{n-1}$.
4. *If there exists* m, $0 < m < n$, *such that*

$$l_{m-1} < l_m = \cdots = l_{n-1} < l_n,$$

then $n - m = l_n - l_{m-1}$.

PROOF. We need to prove only statement 3. If $l_n > l_{n-1}$, then $l_{n-1} < l_n = n + 1 - l_{n-1}$, whence $2l_{n-1} \leq n$, which contradicts the hypotheses. □

9.2. Principles of the statistical method for cryptanalysis of block ciphers

In this section we consider the general approach of the statistical method for cryptanalysis as applied to block ciphers. Practical implementations of this method for a number of block ciphers are more efficient at determining the secret key than the exhaustive key search ("brute force attack").

Iterated block ciphers. Before we describe the method of cryptanalysis, we will define iterated block ciphers (see Figure 9.2).

From the cryptanalytical point of view, a *block cipher* is a mapping

$$F \colon V_m \times V_n \to V_m$$

of vector spaces over the field of two elements such that for any fixed value of the second argument $\boldsymbol{K} \in V_n$ the mapping $F(\,\cdot\,, \boldsymbol{K}) \colon V_m \to V_m$ is bijective, i.e., is a permutation $\pi_{\boldsymbol{K}} \in \mathfrak{S}_{V_m}$.

If $F(\boldsymbol{X}, \boldsymbol{K}) = \boldsymbol{Y}$, then the first argument $\boldsymbol{X} \in V_m$ of the mapping F is called the *plaintext block* (or simply *plaintext*), the second argument $\boldsymbol{K} \in V_n$ is called the *key*, and the output $\boldsymbol{Y} \in V_m$ is called the *ciphertext block* (or simply *ciphertext*).

REMARK 9.27. From the point of view of a cryptographer who is designing a cipher, the mapping F must satisfy a number of requirements, which are hard to formalize. In particular, it is always necessary that for some given values of the arguments $\boldsymbol{X} \in V_m$ and $\boldsymbol{K} \in V_n$, the value of the mapping $\boldsymbol{Y} = F(\boldsymbol{X}, \boldsymbol{K})$ could be computed "easily", i.e., that the mapping F could be effectively implemented on various platforms (an algorithm which computes F is sometimes called an *enciphering algorithm*). Similar requirements apply to a *deciphering* algorithm, i.e., it is necessary that, for given $\boldsymbol{Y} \in V_m$ and $\boldsymbol{K} \in V_n$, it should be "easy" to compute the value of $\boldsymbol{X} \in V_m$ such that $\boldsymbol{Y} = F(\boldsymbol{X}, \boldsymbol{K})$.

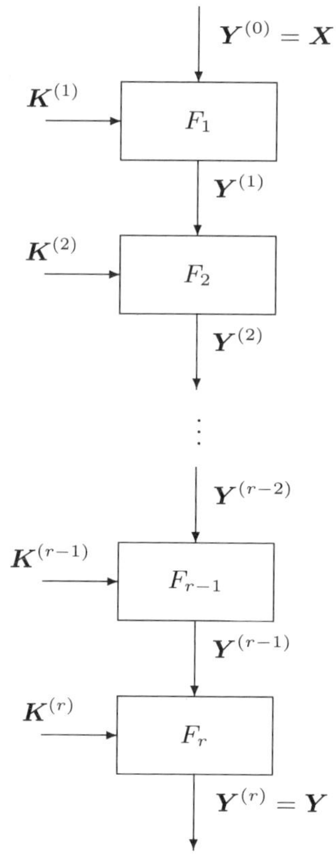

FIGURE 9.2

The most important requirement for the mapping F is that it should be cryptographically strong. In particular, this means that if we do not know the key $\boldsymbol{K}$, it should be practically impossible to find the value $\boldsymbol{X} \in V_m$ such that $\boldsymbol{Y} = F(\boldsymbol{X}, \boldsymbol{K})$ if we are given $\boldsymbol{Y} \in V_m$. Similarly, if we know a number of pairs of plaintexts and ciphertexts $(\boldsymbol{X}_i, \boldsymbol{Y}_i)$ obtained by applying the mapping F with some fixed key $\boldsymbol{K}$, $F(\boldsymbol{X}_i, \boldsymbol{K}) = \boldsymbol{Y}_i$, $i = 1, \dots, N$, it should be practically impossible to find the key $\boldsymbol{K}$.

Most block ciphers are examples of *iteration ciphers*. This means that the mapping F is defined as composition of a number of simpler transformations all of which are of the same form (see Figure 9.2):

$$\boldsymbol{Y}^{(0)} = \boldsymbol{X}; \quad \boldsymbol{Y}^{(i)} = F_i(\boldsymbol{Y}^{(i-1)}, \boldsymbol{K}^{(i)}) \text{ for } i = 1, \dots, r; \quad \boldsymbol{Y} = \boldsymbol{Y}^{(r)},$$

where $\boldsymbol{Y}^{(i)} \in V_m$, $\boldsymbol{K}^{(i)} \in V_{n'}$, $i = 1, \dots, r$. The mappings $F_i \colon V_m \times V_{n'} \to V_m$, $i = 1, \dots, r$, are called *round transformations* (or simply *rounds*). As the mapping F itself, any round mapping F_i, $i = 1, \dots, r$, with a fixed argument $\boldsymbol{K}^{(i)} \in V_{n'}$ is bijective, i.e., it is a permutation $F_i(\,\cdot\,, \boldsymbol{K}^{(i)}) \colon V_m \to V_m$ of the symmetric group $\mathfrak{S}_{V_m}$. The result $\boldsymbol{Y}^{(i)} \in V_m$ obtained after the ith round ($i = 1, \dots, r-1$) is called an *intermediate ciphertext*.

In each round, the round *subkeys* $\boldsymbol{K}^{(i)} \in V_{n'}$, $i = 1, \dots, r$, are used, which are derived from the secret key $\boldsymbol{K} \in V_n$, usually by simple transformations:

$$\boldsymbol{K}^{(i)} = L_i(\boldsymbol{K}),$$

where $L_i\colon V_n \to V_{n'}$, $i = 1, \dots, r$. The list of round subkeys $\boldsymbol{K}^{(1)}, \dots, \boldsymbol{K}^{(r)}$ obtained from the key $\boldsymbol{K}$ is called the *key schedule*. Often the mappings L_i simply pick a subvector of length $n' \leq n$ of a binary vector of length n.

EXAMPLE 9.28. We present some typical values for the block size m, the key-length n, and the number of rounds r for ciphers used in current applications:

- For the DES-algorithm, $m = 64$, $n = 56$, and $r = 16$ (we ignore the first and the last permutations of the bits of the blocks of the texts, since they do not depend on the key and do not carry a cryptographic load).
- For the Russian cipher standard GOST 28147-89, $m = 64$, $n = 256$, and $r = 32$ (here, as is accepted in cryptanalysis, we assume that the s-boxes are known to the cryptanalysts and thus we do not count them in the length of the key n).
- For the new USA standard (Advanced Encryption Standard, AES), the length of the block m is fixed to be 128 and the size of the key n can take one of the following three values: 128, 192, or 256. The number of cipher rounds r is determined according to the values of n (and m); it may be equal to 10, 12, or 14. The AES standard (the ciphers AES-128, AES-192, and AES-256) is a subset of a larger collection of ciphers originally published under the name Rijndael.

Principles of statistical cryptanalysis. The goal of statistical cryptanalysis (as well as of all methods of cryptanalysis) is to develop algorithms for determining the secret key $\boldsymbol{K} \in V_n$. The input of these algorithms is a number of pairs of plaintexts and ciphertexts $(\boldsymbol{X}_i, \boldsymbol{Y}_i)$ obtained by applying the mapping F with a key $\boldsymbol{K}$: $F(\boldsymbol{X}_i, \boldsymbol{K}) = \boldsymbol{Y}_i$, for $i = 1, \dots, N$ (*a cryptographic attack on the key* using known plaintexts and ciphertexts). The set of pairs of plaintexts and ciphertexts $(\boldsymbol{X}_i, \boldsymbol{Y}_i)$, $i = 1, \dots, N$, which is the input of such an algorithm, is called the *material* accessible to the cryptanalyst (the material will be denoted by $\mathcal{M} \subset V_m \times V_m$) and the integer N is called the *volume of the material* ($\#\mathcal{M} = N$).

As a rule, it is supposed that in the material accessible to the cryptanalyst, the plaintexts $\boldsymbol{X}_i$, $i = 1, \dots, N$, are chosen arbitrarily, are equiprobable, independent, and do not comprise the whole space V_m (in some variants of the method, equiprobability and independence are not required; we will always warn the reader when the model of the probability space is different from the one given above).

An important part of the statistical methods used for cryptanalysis are *procedures for statistical classification* (see [**223**]). Any such procedure solves the problem of searching for an unknown parameter based on available observations which are random. The distribution functions of the observations depend on this unknown parameter. If these probability distributions are different, then it is possible, given a large enough number of observations, to distinguish, with a certain probability of error, the distribution law for the observations, and thus to determine the parameter used. This is the essence of the procedure of statistical classification. The number of observations for which, with a given probability of error, it is possible to distinguish the distributions, depends on how "different" the distributions are from each other (a more precise formulation can be found in [**223**]).

In cryptanalysis, the available random observations comprise the material $\mathcal{M}$; the unknown parameter, as a rule, is either part of the key or a linear combination of bits of the key. We denote the set of values of the unknown parameter by $\mathbb{P}$,

$\#\mathbb{P} = s \le 2^n$ (in all variants of the method of statistical cryptanalysis described in the literature, the set $\mathbb{P}$ is a linear space over the field of two elements: $\mathbb{P} = V_{n^*}$, $1 \le n^* \le n$).

Any procedure of statistical classification defines a partition of the space of observations into $T > 1$ disjoint areas: $\mathcal{M} = \mathcal{M}_1 \cup \mathcal{M}_2 \cup \cdots \cup \mathcal{M}_T$; $\mathcal{M}_i \cap \mathcal{M}_j = \varnothing$ for $i \neq j$, $i, j \in \{1, \dots, T\}$. The areas $\mathcal{M}_i$, $i \in \{1, \dots, T\}$, are called *decision areas.* It is always assumed that for any observation $\boldsymbol{m} \in \mathcal{M}$ one can "easily" determine the index $i(m)$ of the decision area to which the observation belongs: $\boldsymbol{m} \in \mathcal{M}_{i(\boldsymbol{m})}$ (in other words, complexity of the algorithm for calculation the number $i(\boldsymbol{m})$ from a given $\boldsymbol{m}$ is small). For any decision area $\mathcal{M}_i$, for $i \in \{1, \dots, T\}$, the procedure of statistical classification also determines an ordered list of length s' $(1 \le s' \le s)$ of the elements of the set $\mathbb{P}$: $\boldsymbol{p}_{i,1}, \boldsymbol{p}_{i,2}, \dots, \boldsymbol{p}_{i,s'}$; $\boldsymbol{p}_{i,j} \in \mathbb{P}$, $i \in \{1, \dots, T\}$, $j \in \{1, \dots, s'\}$, and $\boldsymbol{p}_{i,j_1} \neq \boldsymbol{p}_{i,j_2}$ for $j_1 \neq j_2$.

The procedure for finding the unknown parameter from $\mathbb{P}$ according to the observation $\boldsymbol{m} \in \mathcal{M}$ determines the index $i(\mathbf{m})$ of the decision area to which this observation belongs: $\boldsymbol{m} \in \mathcal{M}_{i(\boldsymbol{m})}$. After that, parameters from $\mathbb{P}$ are examined consecutively according to the list $\boldsymbol{p}_{i(\boldsymbol{m}),1}, \boldsymbol{p}_{i(\boldsymbol{m}),2}, \dots, \boldsymbol{p}_{i(\boldsymbol{m}),s'}$. For any element $\boldsymbol{p}_{i(\boldsymbol{m}),j}$ from the list, an algorithm that checks the stability of the parameter is applied.

In cryptanalysis, the algorithm for checking whether the unknown parameter was correctly determined contains two steps:

(1) determination of the remaining unknown part of the key;
(2) checking whether the complete key is correctly determined.

If the unknown parameter is the complete key (i.e., $\mathbb{P} = V_n$), then the first step is unnecessary. As a rule, in the methods of cryptanalysis which we discuss determination of the remaining unknown part of the key is done by exhaustive search, i.e., all possible unknown bits of the key are systematically tested. Observe that if the parameter is either a part of the key or a nontrivial linear combination of the bits of the key, and if $\mathbb{P} = V_{n^*}$ and $1 \le n^* \le n$, then 2^{n-n^*} variants must be checked.

The algorithm for checking whether the complete key is correctly determined works as follows: it checks consecutively for the pairs $(\boldsymbol{X}_i, \boldsymbol{Y}_i) \in \mathcal{M}$, $\boldsymbol{X}_i, \boldsymbol{Y}_i \in V_m$, $i \in \{1, \dots, N\}$, of plaintexts and ciphertexts from the material $\mathcal{M}$ whether the equalities

$$F(\boldsymbol{X}_i, \boldsymbol{K}^*) = \boldsymbol{Y}_i$$

hold, where $\boldsymbol{K}^* \in V_n$ is the full key being tested.

REMARK 9.29. As a rule, a false key is rejected at the initial steps of the check algorithm. Usually it is enough to check the first k pairs of plaintexts and ciphertexts, where the number k is such that for any pair of k different plaintexts $\boldsymbol{X}_1, \dots, \boldsymbol{X}_k \in V_m$ and for any two different keys $\boldsymbol{K}_1 \neq \boldsymbol{K}_2 \in V_n$, there is a number $j \in \{1, \dots, k\}$ such that $F(\boldsymbol{X}_j, \boldsymbol{K}_1) \neq F(\boldsymbol{X}_j, \boldsymbol{K}_2)$ (the minimum k_0 satisfying this condition is sometimes called the *distance of uniqueness* of the cipher F). Very often the problem of precise computation of the distance of uniqueness of a given cipher is very complicated, and some estimates of it are used instead.

Problem 9.30.[1] Suppose the following hypothesis hold: for an arbitrary equiprobable and independent choice of the keys $\boldsymbol{K}_1$ and $\boldsymbol{K}_2$ $(\boldsymbol{K}_1 \neq \boldsymbol{K}_2)$ from the space V_n and for any k different plaintexts $\boldsymbol{X}_i \in V_m$, $i = 1, \dots, k$, the corresponding

[1]The results in this problem are not used later. It can be skipped it at the first reading.

ciphertexts may coincide only with probability 2^{-mk} (i.e., ciphertexts behave as arbitrarily, equiprobably, and independently chosen vectors from the space V_m):

$$\mathcal{P}\{F(\boldsymbol{X}_i, \boldsymbol{K}_1) = F(\boldsymbol{X}_i, \boldsymbol{K}_2),\ i = 1, \ldots, k\} = 2^{-mk}.$$

Estimate the distance of uniqueness k_0 for such a cipher F. (*Hint*: Find the distribution function for k_0.)

The algorithms for determination of the key are compared with respect to three parameters: N, which is the volume of the material used, Q_0, which is the average complexity of the algorithm, and π_0, which is the reliability of the algorithm. We recall that in the statistical classification procedure we test not all possible values of the parameter in question but only the part included in the list $\boldsymbol{p}_{i(\boldsymbol{m}),1}, \boldsymbol{p}_{i(\boldsymbol{m}),2}, \ldots, \boldsymbol{p}_{i(\boldsymbol{m}),s'}$ (since we allow the inequality $s' \leq s$ to be strict, i.e., $s' < s$). Therefore, the probability that the true value of the parameter $\boldsymbol{p}$ does not belong to the list $\boldsymbol{p}_{i(\boldsymbol{m}),1}, \boldsymbol{p}_{i(\boldsymbol{m}),2}, \ldots, \boldsymbol{p}_{i(\boldsymbol{m}),s'}$ may be positive (i.e., with a nonzero probability we will not find the key). The parameter π_0 characterizes the average probability that the procedure will yield the correct result, i.e., that we will find the key we are searching for.

REMARK 9.31. If the procedure is such that $s' = s$, i.e., if we want to test the whole set $\mathbb{P}$ of possible parameters until we find the true value, then obviously the probability of finding the correct solution is equal to one, i.e., $\pi_0 = 1$. In this case it is not necessary to introduce the notion of reliability of the algorithm and traditionally such a procedure for determining the key is not called statistical.

The complexity of the algorithm for the determination of the key, as well as the probability that it gives the correct result, depends on which plaintexts are arbitrarily chosen and on which key was used for obtaining the material $\mathcal{M}$ (i.e., on the key we search for). In order to compare different methods for statistical cryptanalysis two parameters are used: reliability and complexity. These parameters coincide with the reliability and complexity of the corresponding procedures for statistical classification used in the implementation of the cryptanalysis method.

DEFINITION 9.32. The *average reliability* π_0 *of a statistical classification procedure* (or simply the *reliability*) is equal to the expected value of the probability that the algorithm will find the true key, under the assumption that the key of the cipher algorithm is chosen from the space of keys V_n arbitrarily, equiprobably, and independently of the choice of plaintexts. In other words,

$$\pi_0 = E[\pi(\boldsymbol{K})] = \sum_{\boldsymbol{K} \in V_n} \pi(\boldsymbol{K}) \mathcal{P}\{\boldsymbol{K}\} = \frac{1}{2^n} \sum_{\boldsymbol{K} \in V_n} \pi(\boldsymbol{K}),$$

where

$$\begin{aligned} \pi(\boldsymbol{K}) = \mathcal{P}\{\{\boldsymbol{X}_1, \ldots, \boldsymbol{X}_N\} \in (V_m)^N \mid \\ \boldsymbol{m} = \{(\boldsymbol{X}_1, F(\boldsymbol{X}_1, \boldsymbol{K})), \ldots, (\boldsymbol{X}_N, F(\boldsymbol{X}_N, \boldsymbol{K}))\} \in \mathcal{M}_{i(\boldsymbol{m})}, \\ \boldsymbol{p}(\boldsymbol{K}) \in (\boldsymbol{p}_{i(\boldsymbol{m}),1}, \boldsymbol{p}_{i(\boldsymbol{m}),2}, \ldots, \boldsymbol{p}_{i(\boldsymbol{m}),s'})\} \end{aligned}$$

is the probability to choose N plaintexts $X_1, \ldots, X_N$ such that the observation

$$\boldsymbol{m} = \{(\boldsymbol{X}_1, F(\boldsymbol{X}_1, \boldsymbol{K})), \ldots, (\boldsymbol{X}_N, F(\boldsymbol{X}_N, \boldsymbol{K}))\}$$

that corresponds to these plaintexts and to a fixed key $\boldsymbol{K}$ belongs to the partition $\mathcal{M}_{i(m)}$ such that the parameter list $(\boldsymbol{p}_{i(\boldsymbol{m}),1}, \boldsymbol{p}_{i(\boldsymbol{m}),2}, \ldots, \boldsymbol{p}_{i(\boldsymbol{m}),s'})$ contains the parameter $\boldsymbol{p}(\boldsymbol{K})$ corresponding to $\boldsymbol{K}$.

REMARK 9.33. Another approach to defining the notion of reliability of a statistical classification procedure can be found in the literature. Namely, one can compute the minimum probability that the procedure for determination of the key will work, where the minimum is taken over all keys (in the notation of Definition 9.32):

$$\pi_1 = \min_{\boldsymbol{K}\in V_n} \pi(\boldsymbol{K}).$$

DEFINITION 9.34. The (average) *complexity* Q_0 *of a statistical classification procedure* is the expected value of the number of steps carried out by the algorithm for an arbitrary choice of plaintexts and an arbitrarily, equiprobably, and independently (of the choice of plaintexts) chosen key:

$$\begin{aligned} Q_0 &= E[Q((\boldsymbol{X}_1, \ldots, \boldsymbol{X}_N), \boldsymbol{K})] \\ &= \sum_{\boldsymbol{K}\in V_n;\ \{\boldsymbol{X}_1,\ldots,\boldsymbol{X}_N\}\in (V_m)^N} Q((\boldsymbol{X}_1, \ldots, \boldsymbol{X}_N), \boldsymbol{K})\mathcal{P}\{(\boldsymbol{X}_1, \ldots, \boldsymbol{X}_N)\}\mathcal{P}\{\boldsymbol{K}\} \\ &= \frac{1}{2^n} \sum_{\boldsymbol{K}\in V_n;\ \{\boldsymbol{X}_1,\ldots,\boldsymbol{X}_N\}\in (V_m)^N} Q((\boldsymbol{X}_1, \ldots, \boldsymbol{X}_N), \boldsymbol{K})\mathcal{P}\{(\boldsymbol{X}_1, \ldots, \boldsymbol{X}_N)\}, \end{aligned}$$

where $Q((\boldsymbol{X}_1, \ldots, \boldsymbol{X}_N), \boldsymbol{K})$ is the number of steps carried out by the algorithm for given plain texts $\boldsymbol{X}_1, \ldots, \boldsymbol{X}_N$ and the key $\boldsymbol{K}$.

REMARK 9.35. Similarly to the notion of reliability, complexity is sometimes defined as the maximum number of steps which the procedure performs before it stops (in the notation of Definition 9.34):

$$Q_1 = \max_{\substack{\boldsymbol{K}\in V_n; \\ \{\boldsymbol{X}_1,\ldots,\boldsymbol{X}_N\}\in (V_m)^N}} Q((\boldsymbol{X}_1, \ldots, \boldsymbol{X}_N), \boldsymbol{K}).$$

The complexity, Q_0, and reliability, π_0, of the statistical classification procedure are connected. As a rule, the higher the reliability π_0, the higher the complexity Q_0, and vice versa.

We will state without proof a theorem which gives a complete answer to the question of how, for a given partition of the observation space $\mathcal{M}$ into decision regions $\mathcal{M} = \mathcal{M}_1 \cup \mathcal{M}_2 \cup \cdots \cup \mathcal{M}_T$, can one optimally determine an ordered list $\boldsymbol{p}_{i,1}, \boldsymbol{p}_{i,2}, \ldots, \boldsymbol{p}_{i,s'}$ of elements of the set $\mathbb{P}$ of possible parameters.

THEOREM 9.36. *In the statistical classification procedure described above, for a given partition of the observation space $\mathcal{M}$ into decision regions $\mathcal{M} = \mathcal{M}_1 \cup \mathcal{M}_2 \cup \cdots \cup \mathcal{M}_T$, the average reliability π_0 attains its maximum value*

$$\pi_0 = \max_{\substack{\boldsymbol{p}_{i,1},\boldsymbol{p}_{i,2},\ldots,\boldsymbol{p}_{i,s'}\in\mathbb{P}; \\ i\in\{1,\ldots,T\}}} \pi_0((\boldsymbol{p}_{i,1}, \boldsymbol{p}_{i,2}, \ldots, \boldsymbol{p}_{i,s'}); i \in \{1, \ldots, T\})$$

if and only if for any decision region $\mathcal{M}_i$, the corresponding ordered list $\boldsymbol{p}_{i,1}, \boldsymbol{p}_{i,2}, \ldots, \boldsymbol{p}_{i,s'}$ of elements of the set $\mathbb{P}$ gives rise to the following decreasing *list of posterior probabilities:*

$$\mathcal{P}\{\boldsymbol{p}_{i,1}/\boldsymbol{m} \in \mathcal{M}_i\} \geq \mathcal{P}\{\boldsymbol{p}_{i,2}/\boldsymbol{m} \in \mathcal{M}_i\} \geq \cdots \geq \mathcal{P}\{\boldsymbol{p}_{i,s'}/\boldsymbol{m} \in \mathcal{M}_i\} \geq \max_{\boldsymbol{p}\in\mathbb{P}\setminus\{\boldsymbol{p}_{i,1},\ldots,\boldsymbol{p}_{i,s'}\}} \mathcal{P}\{\boldsymbol{p}/\boldsymbol{m} \in \mathcal{M}_i\},$$

where $\mathcal{P}\{\boldsymbol{p}_*/\boldsymbol{m} \in \mathcal{M}_i\}$ is the conditional probability that the true value of the parameter is equal to $\boldsymbol{p}_*$ if the observed material $\boldsymbol{m}$ lies in the decision region $\mathcal{M}_i$.

In this case the average complexity Q reaches its minimum value

$$Q_0 = \min_{\substack{(\boldsymbol{p}_{i,1},\boldsymbol{p}_{i,2},\ldots,\boldsymbol{p}_{i,s'})\in\mathbb{P};\\ i\in\{1,\ldots,T\}}} Q_0((\boldsymbol{p}_{i,1},\boldsymbol{p}_{i,2},\ldots,\boldsymbol{p}_{i,s'}); i \in \{1,\ldots,T\}).$$

PROBLEM 9.37. Prove Theorem 9.36. (*Hint*: See [**223**].)

The optimal (with respect to Theorem 9.36) statistical classification procedure is called the *Bayes optimal procedure*.

REMARK 9.38. We note that the Bayes optimal procedure is (in the worst case) not optimal with respect to the reliability π_1 and the complexity Q_1.

9.3. Principles of the correlation cryptanalysis method

The correlation analysis method (correlation attack) is widely discussed in the cryptographic literature. As a rule, it is associated with the cryptanalysis of *stream ciphers* (see [**365**]). Very often such ciphers are constructed on the base of pseudo-random generators of sequences. Historically the so-called combining and filtering generators, based on shift registers and Boolean functions, were investigated first. Later, generators based on time clocked shift registers were studied, and also generators whose input functions have memory. The subdivision of generators of pseudo-random sequences given here should not be considered as a classification. However, this subdivision shows that in the analysis of the last two classes of generators the automata-theoretic approach (which is not described in this book) plays a significant role. Therefore, we will focus our attention on combining and filtering generators. The limited scope of this chapter only allows us to discuss basic ideas used in certain variants of the correlation method.

Suppose that a stream cipher (see Figure 9.3) is based on a combining generator with a balanced Boolean function $f \in \mathcal{F}_n$ and n linear feedback shift registers LFSR-i, $i = 1, 2, \ldots, n$.

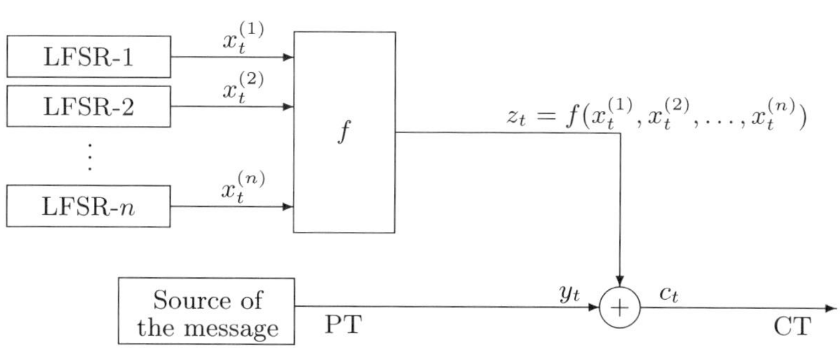

FIGURE 9.3

Denote by s_i, $i = 1, 2, \ldots, n$, the lengths of the corresponding shift registers and by $g_i(\lambda)$, $i = 1, 2, \ldots, n$, the polynomials of the feedbacks of these registers, which are primitive polynomials. The corresponding linear recursive sequences $\boldsymbol{x}^{(i)} = \{x_t^{(i)}\}$, $i = 1, 2, \ldots, n$, have therefore maximal periods $2^{s_i} - 1$, where t is the parameter indexing the steps carried out by the combining generator. At step t the variables $x_t^{(1)}, x_t^{(2)}, \ldots, x_t^{(n)}$ are sent as inputs to the combining function f and the value of the function $z_t = f(x_t^{(1)}, x_t^{(2)}, \ldots, x_t^{(n)})$ is computed, which represents the tth bit of the so-called keystream. Then the plaintext y_t (PT: $\boldsymbol{y} = (y_1, y_2, \ldots, y_N)$) is added modulo 2 to the keystream z_t, giving the ciphertext $c_t = y_t \oplus z_t$ (CT: $\boldsymbol{c} = (c_1, c_2, \ldots, c_N)$). As a key for this cipher we will consider the set K which is the initial state of the shift register. It is clear that the keys in this case can also be polynomials $g_i(\lambda)$, $i = 1, 2, \ldots, m$ (see [**365**]). However, this is not essential for our exposition, and we will focus on the key set given above. Then $\#K = \prod_{i=1}^{n}(2^{s_i} - 1)$ since the initial state of any shift register can be any nonzero vector in the corresponding linear space. Now we will give the cryptographic formulation of the problem we want to solve. It will differ from that in [**365**], where the correlation method was discussed for the first time.

Suppose that a plaintext PT($\boldsymbol{y} = \{y_t\}$) and ciphertext CT($\boldsymbol{c} = \{c_t\}$) pair (see Figure 9.3) is known, and we must find the key $k_{\text{true}} \in K$ that was used to encipher the given plaintext to the corresponding ciphertext. It is clear that since we know PT and CT, we can uniquely calculate the keystream $\boldsymbol{z} = \{z_t\}$, $z_t = y_t \oplus c_t$. Therefore, the solution to this problem can be found using an algorithm for an exhaustive search of the keys (brute force attack). We will further suppose that the cipher does not have a constant key. Testing the elements $k \in K$ in a given order, we get the corresponding sequences $\boldsymbol{z}(k) = z_t(k)$, and comparing them with the known sequence $\boldsymbol{z} = \{z_t\}$, we will get, at some step of the algorithm, the true key $k_{\text{true}} \in K$ such that $\boldsymbol{z}(k_{\text{true}}) = \boldsymbol{z}$. For this we will need $\#K/2$ operations on average (by operation we mean a choice of the consecutive $k \in K$, computation of $\boldsymbol{z}(k)$, and comparison of $\boldsymbol{z}(k)$ with $\boldsymbol{z}$). As a rule, the number $\#K$ is too big, and it is impossible for such an algorithm to finish in a reasonable time.

Now we proceed with an explanation of the correlation method. In order to determine the parameters of the algorithm which implement the correlation method we will need the following statistical model (see Figure 9.4).

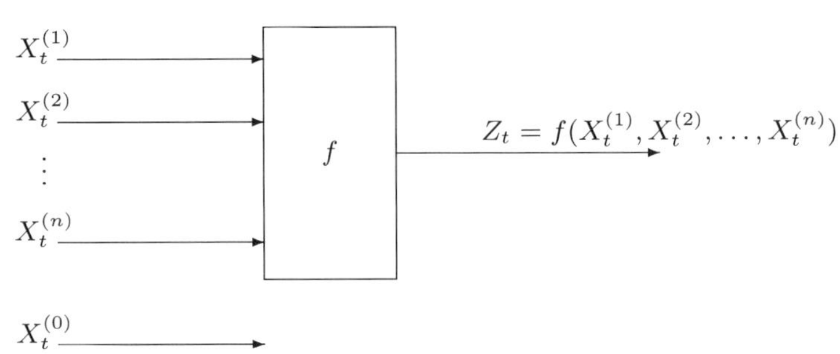

FIGURE 9.4

Let

$$\{X_t^{(0)}\}, \quad \{X_t^{(1)}\}, \quad \ldots, \quad \{X_t^{(n)}\}, \tag{9.17}$$

$t = 1, 2, \dots, N$, be independent uniformly distributed random variables with distribution $\mathcal{P}\{X_t^{(j)} = 0\} = \mathcal{P}\{X_t^{(j)} = 1\}$, for $j = 0, 1, \dots, n$; $t = 1, 2, \dots, N$. Then an arbitrary vector $(X_t^{(1)}, X_t^{(2)}, \dots, X_t^{(n)})$ at any step t has distribution

$$\mathcal{P}\{(X_t^{(1)}, X_t^{(2)}, \dots, X_t^{(n)}) = (a^{(1)}, a^{(2)}, \dots, a^{(n)})\} = 2^{-n}$$

for any $(a^{(1)}, a^{(2)}, \dots, a^{(n)})^\top \in V_n$. As the function f is balanced, the random variables

$$Z_t = f(X_t^{(1)}, X_t^{(2)}, \dots, X_t^{(n)}), \tag{9.18}$$

for $t = 1, 2, \dots, N$ have the same distributions for any t, namely $\mathcal{P}\{Z_t = 0\} = \mathcal{P}\{Z_t = 1\}$, $t = 1, 2, \dots, N$. From our assumption that the random variables (9.17) are independent it follows that $Z_1, Z_2, \dots, Z_N$ are independent. Suppose that for any $t = 1, 2, \dots, N$,

$$\max_{1 \le j \le m} \mathcal{P}\{X_t^{(j)} = Z_t\} > \frac{1}{2}. \tag{9.19}$$

Then we can assume without loss of generality that

$$\max_{1 \le j \le m} \mathcal{P}\{X_t^{(j)} = Z_t\} = \mathcal{P}\{X_t^{(1)} = Z_t\} = q_1. \tag{9.20}$$

As $X_t^{(0)}$ and Z_t are by assumption independent random variables for $t = 1, 2, \dots, N$, we have

$$\mathcal{P}\{X_t^{(0)} = Z_t\} = q_0 = \frac{1}{2}. \tag{9.21}$$

From (9.20) it is clear that $I(X_t^{(1)}, Z_t) \neq 0$ (see §7.1) for any t, and so the random variables $X_t^{(1)}$ and Z_t are dependent random variables (i.e., they coincide with probability q_1 greater than $1/2$). Taking all this into account, we have:

$$\mathcal{P}\{X_t^{(1)} \oplus Z_t \oplus 1 = 1\} = q_1 > \frac{1}{2}, \tag{9.22}$$

$$\mathcal{P}\{X_t^{(0)} \oplus Z_t \oplus 1 = 1\} = q_0 = \frac{1}{2}. \tag{9.23}$$

We now describe an algorithm implementing the correlation method. Let $\boldsymbol{k}_{\text{true}} = (k_{\text{true}}^{(1)}, k_{\text{true}}^{(2)}, \dots, k_{\text{true}}^{(n)}) \in V_{s_1} \times V_{s_2} \times \dots \times V_{s_n}$ be the true key (the initial state), $\boldsymbol{y} = \{y_t\}$ be the plaintext, $\boldsymbol{c} = \{c_t\}$ be the corresponding ciphertext and let the keystream be $\boldsymbol{z} = \{z_t\}$. The initial state of the register with index j, $j = 1, 2, \dots, n$, is $k_{\text{true}}^{(j)}$. In contrast to the full method we will check only the initial state of the first register.

We introduce a random variable ξ which characterizes the correlation between the sequences $Z_1, Z_2, \dots, Z_N$ and $X_1^{(j)}, X_2^{(j)}, \dots, X_N^{(j)}$, where $j = 0$ or 1, as follows:

$$\xi = \sum_{t=1}^{N} (X_t^{(j)} \oplus Z_t \oplus 1). \tag{9.24}$$

The random variable ξ describes the number of successes in the Bernoulli scheme with N independent trials and probability of success q_j (see (9.22), (9.23)). The distribution function of this random variable is (see [**353, 123**])

$$\mathcal{P}_j\{\xi = r\} = \binom{N}{r} q_j^r (1 - q_j)^{N-r},$$

$r = 0, 1, \ldots, N$. Its expectation m_j and variance σ_j^2 are

$$m_j = Nq_j, \quad \sigma_j^2 = Nq_j(1 - q_j).$$

Thus, we may consider two simple hypotheses:

- H_1: the chosen initial state of the first register, $k^{(1)}$, coincides with $k^{(1)}_{\text{true}}$; in this case $x^{(1)}(k^{(1)}) = x^{(1)}(k^{(1)}_{\text{true}})$ and ξ reflects the correlation between $\{X_t^{(1)}\}$ and $\{Z_t\}$;
- H_0: the chosen initial state of the first register $k^{(1)}$ does not coincide with $k^{(1)}_{\text{true}}$. In this case ξ reflects the correlation between $\{X_t^{(0)}\}$ and $\{Z_t\}$.

This can be written briefly as

$$\begin{aligned} \mathcal{P}_{\xi|H_0}(r) &= \mathcal{P}_0(\xi = r) = \binom{N}{r} q_0^r (1 - q_0)^{N-r}, \\ \mathcal{P}_{\xi|H_1}(r) &= \mathcal{P}_1(\xi = r) = \binom{N}{r} q_1^r (1 - q_1)^{N-r}, \end{aligned} \tag{9.25}$$

for $r = 0, 1, \ldots, N$. As is known (see [**353, 223**]), an optimal criterion for checking the hypothesis H_0 against the hypothesis H_1 is constructed on the basis of the following inequality (the Neyman–Pearson lemma):

$$\frac{\mathcal{P}_{\xi|H_1}(r)}{\mathcal{P}_{\xi|H_0}(r)} = \left[\frac{q_1(1 - q_0)}{q_0(1 - q_1)}\right]^r \cdot \left(\frac{1 - q_1}{1 - q_0}\right)^N \geq C, \tag{9.26}$$

where C is some constant. If the inequality (9.26) holds, then the hypothesis H_0 is rejected; otherwise it is accepted. Taking the logarithm of this inequality and performing simple transformations, we can show that inequality (9.26) is equivalent to

$$r \geq C_1 \tag{9.27}$$

for some C_1. To compute the probabilities of first and second type errors, we use the fact that by the central limit theorem (see [**123**]), the number of successes in the Bernoulli scheme with N independent trials and probability of success q_j for a large number N has asymptotically normal distribution with parameters $(Nq_j, \sqrt{Nq_j(1 - q_j)})$. Then

$$\begin{aligned} \alpha &= \mathcal{P}_0\{\xi \geq C_1\} = \mathcal{P}\{\xi \geq C_1 \mid H_0\} \\ &= \mathcal{P}\left\{\frac{\xi - Nq_0}{\sqrt{Nq_0(1 - q_0)}} \geq \frac{C_1 - Nq_0}{\sqrt{Nq_0(1 - q_0)}} \;\middle|\; H_0\right\} \approx 1 - \Phi\left(\frac{C_1 - Nq_0}{\sqrt{Nq_0(1 - q_0)}}\right) \end{aligned} \tag{9.28}$$

is the probability of a first type error, and

$$\begin{aligned} \beta &= \mathcal{P}_1\{\xi < C_1\} = \mathcal{P}\{\xi < C_1 \mid H_1\} \\ &= \mathcal{P}\left\{\frac{\xi - Nq_1}{\sqrt{Nq_1(1 - q_1)}} < \frac{C_1 - Nq_1}{\sqrt{Nq_1(1 - q_1)}} \;\middle|\; H_1\right\} \approx \Phi\left(\frac{C_1 - Nq_1}{\sqrt{Nq_1(1 - q_1)}}\right) \end{aligned} \tag{9.29}$$

is the probability of a second type error, where Φ is the distribution function of the standard normal distribution, i.e.,

$$\Phi(u) = \int_{-\infty}^{u} e^{-\frac{x^2}{2}}\, dx.$$

The probability of a first type error is the probability of rejecting the hypothesis H_0 when it is true; this corresponds to a situation where we take a false key a quantile true key. The probability of a second type error is the probability of accepting the hypothesis H_0 when the opposite hypothesis H_1 is true; this corresponds to a situation where we reject a true key. Choosing first and second type error probabilities (α and β) and using the quantile u_γ of the standard normal distribution for which

$$1 - \Phi(u_\gamma) = \gamma, \tag{9.30}$$

we obtain for the given α and β the bound

$$C_1 \approx Nq_0 + u_\alpha \sqrt{Nq_0(1-q_0)} \approx Nq_1 - u_\beta \sqrt{Nq_1(1-q_1)}. \tag{9.31}$$

The latter relation gives us approximately the necessary volume N_1 of plaintext and ciphertext pairs for implementing the statistical criterion in the algorithm for determining the initial state of the register

$$N_1 \approx \frac{(u_\alpha \sqrt{q_0(1-q_0)} + u_\beta \sqrt{q_1(1-q_1)})^2}{(q_0 - q_1)^2}. \tag{9.32}$$

It is not hard to see the following: (9.32) implies that the necessary number of plaintext and ciphertext pairs is proportional to $\frac{1}{(q_0-q_1)^2}$, i.e., the bigger $|q_0 - q_1|$, the smaller N_1. It is clear that the inequality $N_1 \leq N$ must hold.

Let us choose the probability of a first type error to be $\alpha \approx 2^{-s_1}$ and of a second type to be β. We regard as a single operation the following sequence of actions: choosing the initial state $k^{(1)}$ for the first register, determination of the sequence $\{x_t^{(1)}(k^{(1)})\} = \boldsymbol{x}^{(1)}(k^{(1)})$, and implementation of the statistical criterion for the sequences $\boldsymbol{x}^{(1)}(k^{(1)})$ and $\boldsymbol{z}$. After no more than $2^{s_1} - 1$ single operations, taking into account that $\alpha 2^{s_1} \approx 1$, we will obtain the true part of the key $k_{\mathrm{i}}^{(1)}$ with probability of success $1 - \beta$.

Suppose that similar algorithms to determine the initial state can also be created for the other registers $j = 2, 3, \dots, n$ (see Figure 9.3). Then to find the true key $\boldsymbol{k}_{\text{true}} = (k_{\text{true}}^{(1)}, k_{\text{true}}^{(2)}, \dots, k_{\text{true}}^{(n)})$ we will need no more than $\sum_{j=1}^{n} 2^{s_j} - n$ single operations. The probability of successfully finding the true key will be equal to the product of the probabilities for each of the n stages. But even if similar algorithms to determine $k_{\text{true}}^{(2)}, \dots, k_{\text{true}}^{(n)}$ cannot be created, the initial problem definitely becomes less complicated after finding $k_{\text{true}}^{(1)}$.

REMARK 9.39. The probabilities q_j, $j = 1, 2, \dots, n$, of the correlation sequence $\boldsymbol{x}^{(j)} = \{x_t^{(j)}\}$ (see Figure 9.3) and the key stream $\boldsymbol{z} = \{z_t\}$ as used in the statistical model (see Figure 9.4) are, obviously, parameters defined by the Boolean function f. Furthermore,

$$q_j = 1 - 2^{-s_j}\,\mathrm{wt}(f(x^{(1)}, x^{(2)}, \dots, x^{(n)}) \oplus x^{(j)}).$$

REMARK 9.40. As was noted above, our statement of the problem is different from that in [**365**], where the problem of stream cipher key determination is discussed (see Figure 9.3) for the situation in which only the ciphertexts are known. The correlation between the sequences $\boldsymbol{x}^{(j)} = \{x_t^{(j)}\}$, for $j = 1, 2, \dots, n$, and CT $= \boldsymbol{c} = \{c_t\}$ is considered. It is clear that the knowledge of only statistical characteristics of the plaintext decreases the probabilities of correlation q_j, for $j = 1, 2, \dots, n$ (i.e., they become close to $1/2$). But in principle the problem remains the same. In our opinion, the correlation method of determining the key is a more objective method, as it evaluates the cryptographic qualities of the cipher independently of the properties of the transmitted information and its encoding methods (ASCII, for example).

REMARK 9.41. Now a few words about the term "correlation methods" in cryptanalysis. We think that this notion, reflecting the idea of the "correlation" of different information sequences in the stream cipher, can be understood as a uniting image for various algorithms which determine the key using similar ideas.

In order to construct cryptographic schemes which are resilient with respect to the correlation method it is suggested to use l-resilient functions (see §7.1) as f (see [**362, 365**]). For example, if f is a 1-resilient function, then in our particular case the q_j, $j = 1, 2, \ldots, m$, are equal to $1/2$ (i.e., the $\boldsymbol{x}^{(j)}$ do not correlate with $\boldsymbol{z}$). But if f is not 2-resilient, then a pair of variables whose sum correlates with the value of the function can be found. The sum of linear recursive sequences generated by the corresponding shift registers will also be a linear recursive sequence obtained by the shift register whose feedback polynomial is equal to the least common multiple of the feedback polynomials of the register that generates sequences which occur in the sum. Thus, we return to the initial problem but with longer registers. All these considerations can be repeated for an l-resilient (but not an $(l+1)$-resilient) function f. In general, $l+1$ variables such that the value of the function will correlate with their sum can be found. Using functional terminology, we can speak about the presence of a "good enough" approximation of the function f by a linear function of $l+1$ variables. Is such an approximation better for l-resilient functions? The following theorem holds.

THEOREM 9.42. *Let f be an l-resilient function of $\mathcal{F}_n$ and let $\boldsymbol{u} \in V_n$, $\mathrm{wt}(\boldsymbol{u}) = l+1$, $W_f(\boldsymbol{u}) \neq 0$, and $\mathrm{supp}(\boldsymbol{u}) = M = \{i_1, i_2, \ldots, i_{l+1}\}$.*

Then the minimum possible value among all functions $g \in \mathcal{F}_{l+1}$ (of variables $x^{(i_1)}, x^{(i_2)}, \ldots, x^{(i_{l+1})}$) of the quantity

$$\mathrm{dist}\big(f(x^{(1)}, x^{(2)}, \ldots, x^{(n)}), g(x^{(i_1)}, x^{(i_2)}, \ldots, x^{(i_{l+1})})\big)$$

is attained for the affine function

$$g(x^{(i_1)}, x^{(i_2)}, \ldots, x^{(i_{l+1})}) = \bigoplus_{j=1}^{l+1} x^{(i_j)} \oplus \varepsilon,$$

where $\varepsilon = 0$ if $W_f(\boldsymbol{u}) > 0$, and $\varepsilon = 1$ otherwise. Furthermore,

$$\min_{g \in \mathcal{F}_{l+1}} \mathrm{dist}(f, g) = 2^{n-1} - \frac{1}{2}\left|W_f(\boldsymbol{u})\right|.$$

PROOF. For any vector $\boldsymbol{x} \in V_n$, we denote by $\boldsymbol{x} = (\boldsymbol{y}, \boldsymbol{z})$ its partition, where $\boldsymbol{y} = (x^{(i_1)}, x^{(i_2)}, \ldots, x^{(i_{l+1})})$. Let $d_M(\boldsymbol{v})$ for $\boldsymbol{y} = \boldsymbol{v} \in V_{l+1}$ denote the weight of the subfunction

$$d_M(\boldsymbol{v}) = \mathrm{wt}\Big(f_{i_1, \ldots, i_{l+1}}^{v^{(1)}, \ldots, v^{(l+1)}}\Big).$$

For any function $g \in \mathcal{F}_{l+1}$, we have

$$\mathrm{dist}(f, g) = \mathrm{dist}(f(y, z), g(y)) = \sum_{\boldsymbol{v} \in g^{-1}(0)} d_M(\boldsymbol{v}) + \sum_{\boldsymbol{v} \in g^{-1}(1)} \big(2^{n-l-1} - d_M(\boldsymbol{v})\big).$$

From the last relation it follows that the distance $\operatorname{dist}(f,g)$ is minimum for the function g if and only if

$$(9.33)\qquad g(\boldsymbol{v}) = \begin{cases} 0 & \text{if } d_M(\boldsymbol{v}) < 2^{n-l-2}; \\ 1 & \text{if } d_M(\boldsymbol{v}) > 2^{n-l-2}. \end{cases}$$

In the case $d_M(\boldsymbol{v}) = 2^{n-l-2}$ the value of the function $g(\boldsymbol{v})$ can be chosen arbitrarily. For any $v \in V_{l+1}$ and any $i_s \in M$, where $1 \le s \le l+1$, we have

$$\begin{aligned} d_M(\boldsymbol{v}) + d_M(\boldsymbol{v} \oplus \boldsymbol{e}_{i_s}) &= \operatorname{wt}\big(f_{i_1,\dots,i_s,\dots,i_{l+1}}^{v^{(1)},\dots,0,\dots,v^{(l+1)}}\big) + \operatorname{wt}\big(f_{i_1,\dots,i_s,\dots,i_{l+1}}^{v^{(1)},\dots,1,\dots,v^{(l+1)}}\big) \\ &= \operatorname{wt}\big(f_{i_1,\dots,i_{s-1},i_{s+1},\dots,i_{l+1}}^{v^{(1)},\dots,v^{(s-1)},v^{(s+1)},\dots,v^{(l+1)}}\big) = 2^{n-l-1}, \end{aligned}$$

as the function f is l-resilient (see Lemma 7.8). Let the function $g \in \mathcal{F}_{l+1}$ be such that the distance $\operatorname{dist}(f,g)$ is minimal. As $d_M(\boldsymbol{v}) + d_M(\boldsymbol{v} \oplus \boldsymbol{e}_{i_s}) = 2^{n-l-1}$ for any $\boldsymbol{v} \in V_{l+1}$ and any $i_s \in M$, for $s = 1, 2, \dots, l+1$, it follows from condition (9.33) that

$$(9.34)\qquad g(\boldsymbol{v}) + g(\boldsymbol{v} \oplus \boldsymbol{e}_{i_s}) = 1$$

for any $\boldsymbol{v} \in V_{l+1}$ with $d_M(\boldsymbol{v}) \ne 2^{n-l-2}$. We can assume that relation (9.34) holds for all $\boldsymbol{v} \in V_{l+1}$ since $g(\boldsymbol{v})$ can be arbitrarily determined if $d_M(\boldsymbol{v}) = 2^{n-l-2}$. Therefore, for any $\boldsymbol{v} \in V_{l+1}$, we have

$$g(\boldsymbol{v}) = g(0) \oplus \bigoplus_{j=1}^{l+1} v^{(i_j)}.$$

Since g is an affine function, we have

$$\operatorname{dist}(f,g) = 2^{n-1} - \frac{(-1)^{g(0)}}{2} W_f(\boldsymbol{u}).$$

This distance is minimal if $(-1)^{g(0)}$ and $W_f(\boldsymbol{u})$ have the same sign. □

Therefore, the best approximation of an l-resilient (but not $(l+1)$-resilient) function f by a function of $l+1$ variables will be found among affine functions of the form $g(\boldsymbol{x}) = \langle \boldsymbol{u}, \boldsymbol{x} \rangle \oplus \varepsilon$, where $\operatorname{wt}(\boldsymbol{u}) = l+1$ and $W_f(\boldsymbol{u}) \ne 0$.

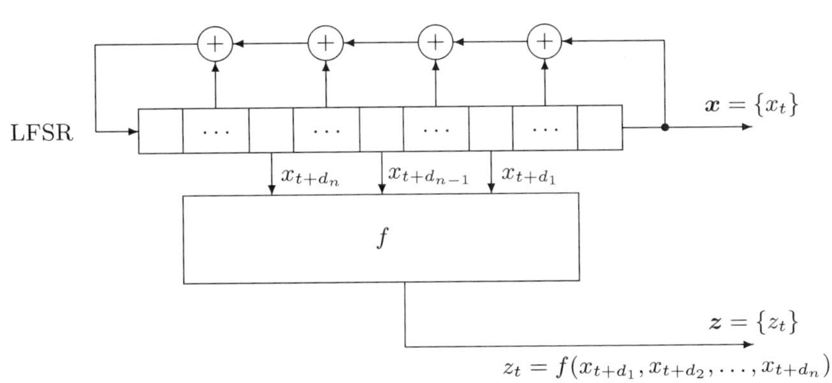

FIGURE 9.5

In place of the filtering generator (see Figure 9.5) constructed using one shift register (LFSR) of length s with a feedback polynomial $g(\lambda)$ and filtering function

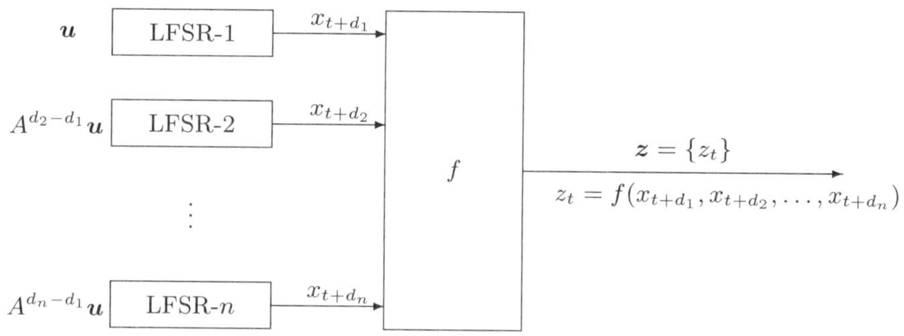

FIGURE 9.6

$f \in \mathcal{F}_n$, [**364**] suggests an equivalent scheme (see Figure 9.6) based on a combining generator which leads the same cipher sequences $\boldsymbol{z} = \{z_t\}$.

A distinctive feature of this generator is that all n registers have the same length s and the same feedback polynomial $g(\lambda)$. In order to get from this generator the sequence $\boldsymbol{z} = \{z_t\}$ obtained earlier by using the filtering generator with initial register state $(x^{(1)}, x^{(2)}, \dots, x^{(n)})$, it is necessary to set the registers of the combining generator to the states $\boldsymbol{u} = A^{d_1}(x^{(1)}, x^{(2)}, \dots, x^{(s)})^\top$, $A^{d_2}(x^{(1)}, x^{(2)}, \dots, x^{(s)})^\top = A^{d_2-d_1}\boldsymbol{u}$, $\dots$, $A^{d_n}(x^{(1)}, x^{(2)}, \dots, x^{(s)})^\top = A^{d_n-d_1}\boldsymbol{u}$, respectively, where A is a linear transformation of the space V_s given by the shift register. In this way we reduce the problem of finding the initial state of the filtering generator resulting in a given output sequence to a similar problem for the combining generator with the additional condition of the presence of certain correlations connecting the initial states of its registers. Details of the application of the correlation method in this case can be found, for example, in [**127, 221, 364**].

Going back to the correlation method described above, it is necessary to make two important remarks. First, to determine the initial state of a particular register from the combining generator, we perform an exhaustive search of all its states. Second, to determine the initial state of the register, we do not use the information that the sequence obtained from the shift register is a linear recursive sequence. These two observations lead to the code-theoretic approach for implementing the correlation method (the fast correlation attack). We will briefly discuss this approach.

Let a linear feedback shift register (LFSR) have length s, number of states 2^s, and feedback polynomial $g(\lambda)$. The sequences of length N obtained from the register with any initial state form an $[N, s]$-linear code $\boldsymbol{C}$. Then the keystream $\boldsymbol{z} = (z_1, z_2, \dots, z_N)$, correlating with the sequence $\boldsymbol{x} = (x_1, x_2, \dots, x_N)$ with probability q, can be represented as a code word sent through a binary symmetric channel (BSC) without memory and with probability of error $p = 1 - q$ (see Figure 9.7).

In this way, the cryptographic problem is formulated as a problem of decoding a linear block code $\boldsymbol{C}$. To send information through a binary symmetric channel with probability of error p, a linear $[N, s]$-code $\boldsymbol{C}$ is used. Let the word $\boldsymbol{z} = (z_1, z_2, \dots, z_N)$ received from the channel be given. The sent code word $\boldsymbol{x} = (x_1, x_2, \dots, x_N)$ has to be recovered.

It is known (see [**122, 131**]) that to be able to uniquely decode the code we need its length N to be at least $N_0 = s/(1 - H(p))$, where $H(p) = -(p \log p + q \log q)$. As a rule, algorithms of this type require the condition $N \gg N_0$ to hold.

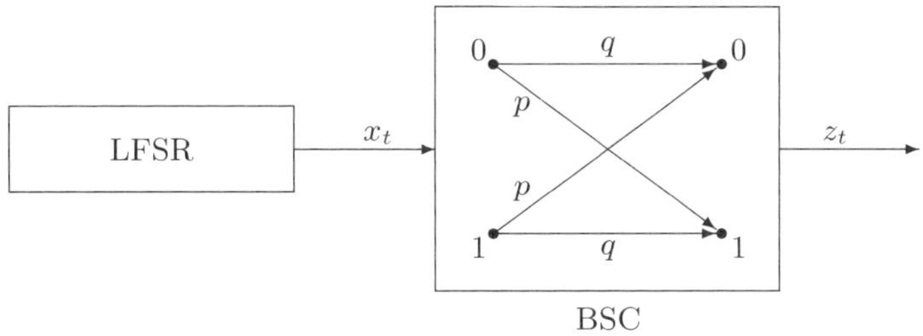

FIGURE 9.7

The basic idea used by many authors for this approach is that the components of the word $\boldsymbol{z}$ must satisfy with high probability certain linear relations which depend on the polynomial $g(\lambda)$. In [**268**] it was shown that under certain conditions, the exhaustive search of the set of states of the register can be avoided. In particular, if the number of feedbacks is small, it is possible to recover the initial state using an iterative procedure whose complexity is smaller than that of the exhaustive search. Using this approach, the so-called syndrome method is suggested in [**425, 426**] for recovering the initial state of the register. This method is undoubtedly similar to the iterative decoding procedure (see [**130**]) for codes with small density of parity checks, which was used in [**84, 85**] to construct an efficient algorithm for recovering the initial state of the register. We should also note the works [**170**]–[**173**], where attempts were made to use ideas of decoding convolutional codes in order to increase the efficiency of the correlation approach.

9.4. Principles of the linear cryptanalysis method

The linear cryptanalysis method is a particular case of the general statistical cryptanalysis method; it is based on expressions of the form

$$\langle \boldsymbol{Y}^{(t_1)}, \boldsymbol{a}\rangle \oplus \langle \boldsymbol{Y}^{(t_2)}, \boldsymbol{b}\rangle = \langle \boldsymbol{K}, \boldsymbol{c}\rangle, \tag{9.35}$$

where $\boldsymbol{a}, \boldsymbol{b} \in V_m$, $\boldsymbol{c} \in V_n$, and $0 \le t_1 < t_2 \le r$ (the pair of vectors $(\boldsymbol{a}, \boldsymbol{b})$ is called a *linear characteristic*). In the linear cryptanalysis method, only those expressions of the form (9.35) which are satisfied with probability different from $\frac{1}{2}$ are used. As will be seen later, the probability with which (9.35) is satisfied, can be most conveniently written in the form

$$\mathcal{P}\{\langle \boldsymbol{Y}^{(t_1)}, \boldsymbol{a}\rangle \oplus \langle \boldsymbol{Y}^{(t_2)}, \boldsymbol{b}\rangle = \langle \boldsymbol{K}, \boldsymbol{c}\rangle\} = \frac{1}{2} + \varepsilon, \tag{9.36}$$

where $-\frac{1}{2} \le \varepsilon \le \frac{1}{2}$. The number ε is called a *deviation* (a bias).

PROBLEM 9.43. Suppose that $\Psi \in \mathcal{F}_{m,n}$ and the vectors of V_m are passed to the input of the mapping Ψ with equal probability. Prove that for any $\boldsymbol{a} \in V_m$ and $\boldsymbol{b} \in V_n$, the deviation

$$\varepsilon_{\boldsymbol{a},\boldsymbol{b}} = \mathcal{P}\{\langle \boldsymbol{x}, \boldsymbol{a}\rangle = \langle \Psi(\boldsymbol{x}), \boldsymbol{b}\rangle\} - \frac{1}{2}$$

is connected with the Walsh–Hadamard coefficients of the mapping Ψ by the following relation:

$$2 \cdot \varepsilon_{\boldsymbol{a},\boldsymbol{b}} = \frac{W_{\langle \Psi(\cdot), \boldsymbol{b}\rangle}(\boldsymbol{a})}{2^n}.$$

The larger the absolute value of the number ε, the higher the reliability of the linear cryptanalysis method. Therefore, when constructing algorithms for finding the key with the aid of linear cryptanalysis, an important problem is to find vectors $\boldsymbol{a}, \boldsymbol{b} \in V_m$ and $\boldsymbol{c} \in V_n$ such that the absolute value of ε is as large as possible: $|\varepsilon| \to \max_{\boldsymbol{a},\boldsymbol{b},\boldsymbol{c}}$.

Later we will describe how to construct algorithms for finding keys based on expressions of the form (9.35). These algorithms use standard techniques involving linear statistical analogs. The main part of the linear cryptanalysis method consists of ways to search for vectors $\boldsymbol{a}, \boldsymbol{b} \in V_m$ and $\boldsymbol{c} \in V_n$ used in (9.35), and also for techniques for estimating the reliability and complexity of the cryptanalysis methods using these expressions (some authors use the phrase "linear cryptanalysis method" to refer to the set of methods used to search for the vectors $\boldsymbol{a}, \boldsymbol{b} \in V_m$, $\boldsymbol{c} \in V_n$ and techniques for estimating the reliability and complexity).

Algorithms used for key determination by the linear cryptanalysis method.

Matsui Algorithm 1. The simplest (but not the most efficient) variant of the linear cryptanalysis method is the algorithm for revealing one bit of the key. This algorithm uses expressions of the form (9.35) for $t_1 = 0$ and $t_2 = r$, i.e., the algorithm uses the expression of the form (9.35), where both the plaintext and the ciphertext are involved:

$$\langle \boldsymbol{X}_i, \boldsymbol{a} \rangle \oplus \langle \boldsymbol{Y}_i, \boldsymbol{b} \rangle = \langle \boldsymbol{K}, \boldsymbol{c} \rangle, \quad i = 1, \dots, N.$$

In the notation of §9.2, the linear cryptanalysis method determines the linear combination $\langle \boldsymbol{K}, \boldsymbol{c} \rangle$ of key bits. The following number is calculated:

$$N_0 = \#\{i \in 1, \dots, N \mid \langle \boldsymbol{X}_i, \boldsymbol{a} \rangle \oplus \langle \boldsymbol{Y}_i, \boldsymbol{b} \rangle = 0\}.$$

A statistical classification procedure splits the whole space of observations into two subsets $\mathcal{M} = \mathcal{M}_0 \cup \mathcal{M}_1$: all observations for which $N_0 \geq N/2$ are placed in $\mathcal{M}_0$, and the subspace $\mathcal{M}_1$ contains the observations for which $N_0 < N/2$. The algorithm (called *Matsui Algorithm 1*) uses the maximum likelihood principle: if $N_0 \geq N/2$ and $\varepsilon > 0$ in equation (9.36), or if $N_0 < N/2$ and, $\varepsilon < 0$ in (9.36), then the algorithm returns the value $\langle \boldsymbol{K}, \boldsymbol{c} \rangle = 0$; otherwise the algorithm returns the value $\langle \boldsymbol{K}, \boldsymbol{c} \rangle = 1$. The remaining key bits are determined by exhaustive search.

Matsui Algorithm 2 (improved). A natural generalization of Matsui Algorithm 1 is to use expressions of the form (9.35) for $t_2 > r$ and/or $t_1 > 0$. In describing this variant of the Matsui algorithm, it is convenient to represent the overall scheme of the block cipher (see Figure 9.2) as in Figure 9.8, separating the first t_1 and the last t_2 rounds.

Put

$$F_{1-t_1} = F_1 \circ \dots \circ F_{t_1}, \qquad F_{(t_2+1)-r} = F_{t_2+1} \circ \dots \circ F_r,$$
$$\boldsymbol{K}^{(1-t_1)} = (\boldsymbol{K}^{(1)}, \dots, \boldsymbol{K}^{(t_1)}), \qquad \boldsymbol{K}^{((t_2+1)-r)} = (\boldsymbol{K}^{(t_2+1)}, \dots, \boldsymbol{K}^{(r)}).$$

In this notation, the intermediate ciphertexts $\boldsymbol{Y}^{(t_1)}$ and $\boldsymbol{Y}^{(t_2)}$ are expressed in terms of the plaintext $\boldsymbol{X}$ and the ciphertext $\boldsymbol{Y}$ as follows:

$$\boldsymbol{Y}^{(t_1)} = F_{1-t_1}(\boldsymbol{X}, \boldsymbol{K}^{(1-t_1)}), \quad \boldsymbol{Y}^{(t_2)} = F^{-1}_{(t_2+1)-r}(\boldsymbol{Y}, \boldsymbol{K}^{((t_2+1)-r)}),$$

where $F^{-1}_{(t_2+1)-r}$ is the permutation (with respect to the first element) which is inverse to the permutation $F_{(t_2+1)-r}$ for a fixed second element. In Matsui Algorithm 2 only the linear combinations $\langle \boldsymbol{Y}^{(t_1)}, \boldsymbol{a}\rangle$ and $\langle \boldsymbol{Y}^{(t_2)}, \boldsymbol{b}\rangle$ of bits of the intermediate ciphertexts are used, but not the intermediate ciphertexts $\boldsymbol{Y}^{(t_1)}$ and $\boldsymbol{Y}^{(t_2)}$ themselves. As a rule, to compute these linear combinations from the plaintext $\boldsymbol{X}$ and the ciphertext $\boldsymbol{Y}$, respectively, it is enough to know only a part of the bits from the sequence of subkeys $\boldsymbol{K}^{(1-t_1)}$ and $\boldsymbol{K}^{((t_2+1)-r)}$. Let us denote by $\widehat{\boldsymbol{K}}$ the "part" of key bits (from the whole key $\boldsymbol{K}$) necessary to compute $\langle \boldsymbol{Y}^{(t_1)}, \boldsymbol{a}\rangle$ and $\langle \boldsymbol{Y}^{(t_2)}, \boldsymbol{b}\rangle$, and by $\widehat{n}$: $\widehat{\boldsymbol{K}} \in V_{\widehat{n}}$ the number of bits in $\widehat{\boldsymbol{K}}$.

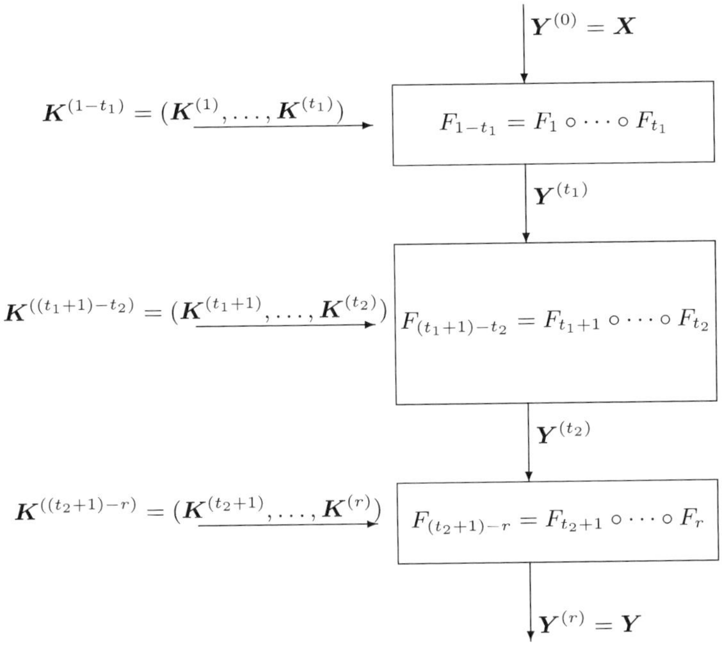

FIGURE 9.8

Matsui Algorithm 2 consists of two phases. In the first phase, for any value $\widehat{\boldsymbol{K}}$ the number

$$N_0(\widehat{\boldsymbol{K}}) = \#\{i \in 1, \dots, N \mid \langle \boldsymbol{Y}_i^{(t_1)}, \boldsymbol{a}\rangle \oplus \langle \boldsymbol{Y}_i^{(t_2)}, \boldsymbol{b}\rangle = 0\}$$

is found. In the second phase, an ordered list $\widehat{\boldsymbol{K}}_1, \dots, \widehat{\boldsymbol{K}}_{2^{\widehat{n}}}$ is first created in decreasing order of the quantities $|N/2 - N_0(\widehat{\boldsymbol{K}}_j)|$, $j = 1, \dots, 2^{\widehat{n}}$:

$$|N/2 - N_0(\widehat{\boldsymbol{K}}_1)| \geq |N/2 - N_0(\widehat{\boldsymbol{K}}_2)| \geq \cdots \geq |N/2 - N_0(\widehat{\boldsymbol{K}}_{2^{\widehat{n}}})|.$$

Then for each $\widehat{\boldsymbol{K}}_j$, $j = 1, \dots, 2^{\widehat{n}}$, a procedure similar to that from Matsui Algorithm 1 is performed: if $N_0(\widehat{\boldsymbol{K}}_j) \geq N/2$ and, in (9.36), $\varepsilon > 0$, or if $N_0(\widehat{\boldsymbol{K}}_j) < N/2$ and, in(9.36), $\varepsilon < 0$, then put $\langle \boldsymbol{K}, \boldsymbol{c}\rangle = 0$; otherwise put $\langle \boldsymbol{K}, \boldsymbol{c}\rangle = 1$. We try to determine the remaining $n - \widehat{n} - 1$ key bits by exhaustive search. If this cannot be done, we go to the next $\widehat{\boldsymbol{K}}_j$ in the list.

REMARK 9.44. In the case where $t_1 = 0$ for some $\boldsymbol{a} \in V_m$, the linear cryptanalysis method can be considered as a cryptanalysis method in the "ciphertext only" attack scenario. For example, if a cryptographic transformation takes ASCII coded symbols as input, then some bits of the input blocks will always be zero and this can be used for an attack on a ciphertext by choosing the vector $\boldsymbol{a} \in V_m$ in a clever way.

Note that the linear cryptanalysis method can be successfully applied not only to blocking cipher cryptanalysis, but also to the analysis of authentication codes and cryptographic hash functions.

Computation of the parameters of a linear code. The main idea of Matsui Algorithm 2 is based on the following assumption.

ASSUMPTION 9.45. *If a linear combination of bits of intermediate ciphertexts and key bits is chosen in such a way that for almost all keys the probability that* (9.35) *is satisfied is different from* 1/2, *then this probability is even more different from* 1/2 *when* (9.35) *is computed with an incorrect "portion" of the key* $\widehat{\boldsymbol{K}}$ *for the plaintext* $\boldsymbol{X}$, *the ciphertext* $\boldsymbol{Y}$, *and intermediate ciphertexts* $\boldsymbol{Y}^{(t_1)}$, $\boldsymbol{Y}^{(t_2)}$.

Formally, this assumption is stated as follows. For all $\boldsymbol{a}, \boldsymbol{b} \in V_m$, $\boldsymbol{c} \in V_n$ *such that the quantity*

$$\left|\mathcal{P}\{\langle \boldsymbol{Y}^{(t_1)}, \boldsymbol{a}\rangle \oplus \langle \boldsymbol{Y}^{(t_2)}, \boldsymbol{b}\rangle = \langle \boldsymbol{K}, \boldsymbol{c}\rangle \mid \boldsymbol{K}\} - \frac{1}{2}\right|$$

is "large" for all keys $\boldsymbol{K}$, *the following is true:*

$$\frac{\left|\mathcal{P}\{\langle \boldsymbol{Y}^{(t_1)}, \boldsymbol{a}\rangle \oplus \langle \boldsymbol{Y}^{(t_2)}, \boldsymbol{b}\rangle = \langle \boldsymbol{K}, \boldsymbol{c}\rangle \mid \widehat{\boldsymbol{K}} = \widehat{\boldsymbol{K}}_*\} - \frac{1}{2}\right|}{\left|\mathcal{P}\{\langle \boldsymbol{Y}^{(t_1)}, \boldsymbol{a}\rangle \oplus \langle \boldsymbol{Y}^{(t_2)}, \boldsymbol{b}\rangle = \langle \boldsymbol{K}, \boldsymbol{c}\rangle \mid \widehat{\boldsymbol{K}} \neq \widehat{\boldsymbol{K}}_*\} - \frac{1}{2}\right|} \gg 1,$$

where $\mathcal{P}\{\langle \boldsymbol{Y}^{(t_1)}, \boldsymbol{a}\rangle \oplus \langle \boldsymbol{Y}^{(t_2)}, \boldsymbol{b}\rangle = \langle \boldsymbol{K}, \boldsymbol{c}\rangle \mid \boldsymbol{K}\}$, $\mathcal{P}\{\langle \boldsymbol{Y}^{(t_1)}, \boldsymbol{a}\rangle \oplus \langle \boldsymbol{Y}^{(t_2)}, \boldsymbol{b}\rangle = \langle \boldsymbol{K}, \boldsymbol{c}\rangle \mid \widehat{\boldsymbol{K}} = \widehat{\boldsymbol{K}}_*\}$, *and* $\mathcal{P}\{\langle \boldsymbol{Y}^{(t_1)}, \boldsymbol{a}\rangle \oplus \langle \boldsymbol{Y}^{(t_2)}, \boldsymbol{b}\rangle = \langle \boldsymbol{K}, \boldsymbol{c}\rangle \mid \widehat{\boldsymbol{K}} \neq \widehat{\boldsymbol{K}}_*\}$ *are the three conditional probabilities that the relation* $\langle \boldsymbol{Y}^{(t_1)}, \boldsymbol{a}\rangle \oplus \langle \boldsymbol{Y}^{(t_2)}, \boldsymbol{b}\rangle = \langle \boldsymbol{K}, \boldsymbol{c}\rangle$ *is satisfied for a fixed key* $\boldsymbol{K}$, *and with intermediate ciphertexts* $\boldsymbol{Y}^{(t_1)}$ *and* $\boldsymbol{Y}^{(t_2)}$ *that are calculated from the plaintext* $\boldsymbol{X}$ *and ciphertext* $\boldsymbol{Y}$ *using true* $\widehat{\boldsymbol{K}}_*$ *and false* $\widehat{\boldsymbol{K}}$ *"portions" of the key, respectively.*

This assumption belongs to the class (often used in cryptanalysis) of so-called *natural cryptographic assumptions*. As a rule, it is impossible to check (i.e., to prove or disprove) these assumptions for real cryptosystems with existing computers (with the increase of the computational power of computers, the parameters of cryptosystems also change!). Besides, usually it is not hard to construct an example of a cryptosystem for which the assumption in question is not satisfied.

PROBLEM 9.46. Construct an example of an iterative mapping $F\colon V_m \times V_n \to V_m$ for which Assumption 9.45 is not satisfied.

For the analysis of real cryptosystems, the use of these assumptions leads to verisimilar conclusions. For Assumption 9.45, the following intuitive argument supports its plausibility: computation of intermediate ciphertexts using an incorrect "portion" of the key $\widehat{\boldsymbol{K}}$ may be regarded as additional rounds of the cryptographic transformation, and in this way the plaintext and the ciphertext become "less dependent", and as a result, the absolute value of the deviation of the linear correlation decreases.

The reliability π_0 of Matsui Algorithm 1 is evaluated by the reliability of distinguishing between the following two simple statistical hypotheses: the hypothesis H_0

that the probability of relation (9.35) is given by (9.36), and the hypothesis H_1 that the probability of relation (9.35) is equal to 1/2 (Assumption 9.45). The standard technique of statistical analysis (see, for example, §9.3) for small enough $|\varepsilon|$ allows us to estimate the reliability π_0 as a quantile of the standard normal distribution $\Phi_{0,1}$ (with expectation 0 and variance 1):

$$\pi_0 \simeq \Phi_{0,1}\big(-2\sqrt{N}\cdot|\varepsilon|\big) = \int_{-2\sqrt{N}\cdot|\varepsilon|}^{\infty} \frac{1}{\sqrt{2\pi}} e^{-x^2/2}\,dx.$$

EXAMPLE 9.47. The following numerical example shows how the value of the reliability π_0, depends on the volume N of the material:

The volume of the material, N	$1/4\cdot|\varepsilon|^{-2}$	$1/2\cdot|\varepsilon|^{-2}$	$|\varepsilon|^{-2}$	$2\cdot|\varepsilon|^{-2}$
Reliability, π_0	84.1%	92.1%	97.7%	99.8%

This example shows that for a reliable determination of the key bit with Matsui Algorithm 1, the volume N of the material must be of order $|\varepsilon|^{-2}$.

To determine the reliability of Matsui Algorithm 2, we need even sharper statistical techniques and additional cryptographic assumptions. But the conclusion remains the same: for a reliable determination of the key with Matsui Algorithm 2, the volume N of the material must be of order $|\varepsilon|^{-2}$.

PROBLEM 9.48. Find an estimate for the reliability of Matsui Algorithm 2.

Construction of linear approximations for iterative mappings. We found that the linear cryptanalysis method can be reduced to a construction of vectors $\boldsymbol{a}, \boldsymbol{b} \in V_m$ and $\boldsymbol{c} \in V_n$ such that the absolute value of ε in expression (9.36) is as large as possible. The reliability and complexity of the method are determined by the value of $|\varepsilon|$, i.e., we must be able to estimate $|\varepsilon|$ in (9.36) when constructing the vectors $\boldsymbol{a}, \boldsymbol{b} \in V_m$, $\boldsymbol{c} \in V_n$.

The linear cryptanalysis method solves this problem using the iterative structure of block ciphers. The first step is the analysis of the nonlinear elements in each round of a ciphering algorithm. For any nonlinear element (and these, as a rule, are Boolean mappings $S \in \mathcal{F}_{k,l}$ with k, l not very large) the probabilities $\mathcal{P}\{\langle \boldsymbol{X}', \boldsymbol{a}'\rangle = \langle S(\boldsymbol{X}'), \boldsymbol{b}\rangle\} = \frac{1}{2} + \varepsilon'$ for all $\boldsymbol{a}' \in V_k$, $\boldsymbol{b}' \in V_l$ are computed for a random equiprobable choice of $\boldsymbol{X}'$ from V_k. To obtain an estimate of the deviation, for the whole algorithm, the sequences (paths) of "coordinated" linear correlations are considered for neighboring rounds (i.e., correlations when the output linear combination of the previous round coincides with the input linear combination of the current round). To do this, the following lemma is used.

LEMMA 9.49 ("Piling-up symbols"; Matsui [**261**]). *Let X_i $(1 \le i \le n)$ be independent random variables, each taking the value* 0 *with probability* $1/2 + \varepsilon_i$ *and value* 1 *with probability* $1/2 - \varepsilon_i$ $(-1/2 \le \varepsilon_i \le 1/2)$. *Then the random variable* $X_1 \oplus X_2 \oplus \cdots \oplus X_n$ *takes the value* 0 *with probability* $1/2 + \varepsilon$, *and the value* 1 *with probability* $1/2 - \varepsilon$, *where*

$$\varepsilon = 2^{n-1}\prod_{i=1}^{n} \varepsilon_i.$$

PROBLEM 9.50. Prove Lemma 9.49. (*Hint*: Use induction on n.)

Since there are many of these "coordinated" sequences, we restrict ourselves to searching for only one "coordinated" path for which the product of deviations obtained according to Lemma 9.49 is large enough. This product is used as an estimate of $|\varepsilon|$.

The number so obtained could be bigger or smaller than the real one. But cryptographers use this approach in practice, and it gives satisfactory results in experimental trials. It is not easy to find at least one coordinated sequence which gives an acceptable answer. Here the "branch and bound" type algorithms are used, as well as ad hoc techniques developed for particular cryptographic algorithms. Thus, as in ancient times, cryptanalysis is still rather an art than a science.

EXAMPLE 9.51. For the DES-algorithm, the following estimates are obtained ([**259**]) for the maximum absolute value of the deviation ε, which depends on the number of rounds (in our notation, on $t_2 - t_1$):

Number of rounds	12	13	14	15	16
$\lvert\varepsilon\rvert$	$1.19 \cdot 2^{-17}$	$1.49 \cdot 2^{-19}$	$1.19 \cdot 2^{-21}$	$1.19 \cdot 2^{-22}$	$1.49 \cdot 2^{-24}$

9.5. Principles of the difference (differential) cryptanalysis method

As the linear cryptanalysis method, the difference cryptanalysis method is a particular case of the general statistical method for cryptanalysis. The difference cryptanalysis method is based on the use of expressions of the form

$$F_{(t_1+1)-t_2}(\boldsymbol{Y}^{(t_1)} \oplus \boldsymbol{a}) \oplus F_{(t_1+1)-t_2}(\boldsymbol{Y}^{(t_1)}) = \boldsymbol{b}, \tag{9.37}$$

where $\boldsymbol{a}, \boldsymbol{b} \in V_m \backslash \{\boldsymbol{0}\}$, $0 \leq t_1 < t_2 \leq r$ (the pair of vectors $(\boldsymbol{a}, \boldsymbol{b})$ is called a *difference characteristic*). Differential cryptanalysis uses the expressions of the form (9.37) that are satisfied with probability much greater than $1/(2^m - 1) \simeq 2^{-m}$:

$$\mathcal{P}\{F_{(t_1+1)-t_2}(\boldsymbol{Y}^{(t_1)} \oplus \boldsymbol{a}) \oplus F_{(t_1+1)-t_2}(\boldsymbol{Y}^{(t_1)}) = \boldsymbol{b}\} = \delta_{\boldsymbol{a},\boldsymbol{b}} \gg 2^{-m}. \tag{9.38}$$

The difference cryptanalysis method specifies algorithms for determining the key that in principle are the same as Matsui Algorithm 2. The general scheme of these algorithms is reduced to checking the keys $\boldsymbol{K}^{(1-t_1)}$, $\boldsymbol{K}^{((t_2+1)-r)}$ of the first t_1 and the last t_2 ciphering rounds (more precisely, to testing "portions" of the keys which are necessary for computation of the vectors included in expression (9.37)). For any checked "portion" of those keys one uses a statistical procedure for distinguishing between two hypotheses:

- H_1: relation (9.37) is satisfied with probability $\delta_{\boldsymbol{a},\boldsymbol{b}} \gg 2^{-m}$;
- H_0: relation (9.37) is satisfied with probability 2^{-m}.

The difference method is based on a cryptographic assumption similar to 9.45.

The volume N of the material necessary for a given reliability π of the method, and thus complexity of the method, is computed in a standard way (see §9.3). All these parameters are determined by the quantity $\delta_{\boldsymbol{a},\boldsymbol{b}}$, and the bigger it is, the more efficient the method. To calculate parameters of this method, some cryptographic assumptions are used, just as in the linear cryptanalysis method.

Similarly to the case of linear cryptanalysis, the main part of the difference cryptanalysis method consists of techniques for finding vectors $\boldsymbol{a}, \boldsymbol{b} \in V_m \setminus \{\boldsymbol{0}\}$ used in (9.37), and also of those for calculating the quantities $\delta_{\boldsymbol{a},\boldsymbol{b}}$.

This problem is solved in the difference cryptanalysis method (similarly to the linear one) by the use of the iterative structure of block ciphers: first the analysis of each round of ciphering algorithm is performed and then sequences (paths) of "coordinated" difference characteristics for neighboring rounds are constructed. The search is limited to only one path. To estimate the probability $\delta_{a,b}$ of a path, the product of the corresponding probabilities in each round is used. Techniques for finding such coordinated paths, as a rule, are similar to those used in the linear cryptanalysis method.

To illustrate the kind of quantities the cryptographers have to work with, we give examples of known values of $\delta_{a,b}$ for the DES-algorithm.

EXAMPLE 9.52. For the DES-algorithm (recall that for this algorithm $m = 64$), the following estimates for the maximum probabilities of difference characteristics depending on the number of rounds (in our notation, on $t_2 - t_1$) are obtained ([**260**]):

Number of rounds	12	13	14	15	16
δ	$1.71 \cdot 2^{-47}$	$1.71 \cdot 2^{-48}$	$1.84 \cdot 2^{-55}$	$1.87 \cdot 2^{-56}$	$1.02 \cdot 2^{-62}$

Comments to Chapter 9

The present chapter plays a special role in this book. As the title suggests, it is essentially via a narrow bridge that the reader is led to the field of cryptanalysis. We tried to choose topics, namely, Boolean functions, which correspond to the main subject of our book. The availability of such a bridge is, we think, necessary for the correct orientation of the interested reader.

Cryptanalysis is an essential branch of cryptology. Cryptanalysis (as a realization of a given combination of cryptanalytic methods), in the long run provides an evaluation for the security of a cipher (a protocol, etc.) in some cases, or allows one to freely obtain (use, change, etc.) information in other cases. This is the most difficult branch of cryptology. Cryptanalysis contains a wide variety of formulations of problems and uses methods from different branches of mathematics to solve them. There have been attempts to classify cryptanalytic methods according to the type of ciphers to which they are applied or to the characteristic features of the algorithms implementing a given method. For the interested reader the authors recommend the books [**10, 15, 150, 151, 269**], in which an overview of modern methods for cryptanalysis can be found. Books on cryptanalysis are usually oriented to specialists in computer security, and not to the mathematically inclined reader.

Together with this, there is a great number of publications on cryptanalysis spread throughout many mathematical journals and proceedings of many cryptology conferences taking place in Europe, USA, Asia, and Australia. Results are published regularly on cryptanalysis of various cryptographic objects (ciphers, protocols, cryptographic hash functions, etc.). References to these scientific publications can be found in the bibliography.

Returning to the essential part of this comment, we note the following. The Berlekamp–Massey algorithm, given here for the Boolean case, naturally allows generalizations to arbitrary finite fields. To get acquainted with them, one can, for example, find them in the books [**10, 22, 225**].

The general scheme of the statistical analysis method for block ciphers is exposed in the papers [**176, 281, 350, 402**].

The idea of linear cryptanalysis was first introduced by the Japanese researchers M. Matsui and A. Yamagishi [**258**] in 1992. One year later Matsui developed the idea into a practical method [**259**] using DES as an example. In the cited work Matsui presents both of his algorithms, and Algorithm 2 is given for $t_1 = 0$ and $t_2 = 15$ (in the notation of §9.4). In this paper Matsui showed how the symmetry of encryption and decryption in DES can be used (all Feistel ciphers have such symmetry). In his next paper [**261**], Matsui showed how two different linear correlations can be used simultaneously for the cryptanalysis of DES; he proposes a method called "branch and bound" to search for "good" linear correlations. In [**175**] the authors show that in practice (in an experimental implementation), the Matsui algorithms work even better than was theoretically expected, but the complexity was not significantly reduced.

In the last few years, several generalizations of the linear cryptanalysis method have been proposed. In [**178**] an idea is suggested of using several linear correlations (this method is called the *method of multiple approximation*), but there is no construction given except for the case where, in all linear correlations, the same combination of bits is included. Recently in [**33**] a statistical evaluation of the multiple approximation method has been presented. In [**194**] the use of nonlinear approximations instead of linear ones is suggested, but it is not discussed how to search for such approximations efficiently and how to evaluate their deviation. In [**195**] the authors show how linear cryptanalysis can be used in chosen plaintext attacks.

The differential cryptanalysis method was developed in 1990 by the Israeli mathematicians Biham and Shamir (the journal version of their talk at a conference is [**31**]). Coppersmith showed [**93**] that this method must have been known to the designers of DES in the early 1970s, but it was classified then. It would be fair to note that an idea close to the differential analysis method was published by Murphy [**280**] in 1990 prior to the work of Biham and Shamir.

Soon after the invention of differential cryptanalysis, it was suggested, when difference characteristics are considered, to replace the coordinate addition of vectors from V_{32} (for DES algorithm) modulo 2 by the sum of these vectors modulo 2^{32}, considering them as binary representations of integers [**25**]. In 1993 the Israeli mathematicians Ben-Aroya and Biham [**20**] proposed to look for difference characteristics assuming that the key takes not all possible values, but only values from some subset. This method is called the *method of partial conditional differentials*. In 1994 the Danish mathematician Lars Knudsen [**193**] proposed the use, by analogy with the differential cryptanalysis method, of the *method of partial differentials*.

Knudsen suggested to "look" at the difference characteristic only for specially chosen bits of the vector, which occur in the correlation. In 1994, Lai [**217**] (see also [**193**]) showed that difference characteristics of higher orders can be used in cryptanalysis: $\bigoplus_{\boldsymbol{a}\in L} F(\boldsymbol{x} \oplus \boldsymbol{a}) = \boldsymbol{d}$, where L is a subspace in V_n, and $\dim L \geq 2$. In 1998, Biham, Biryukov, and Shamir (these results were published one year later in [**30**]) noticed that the so-called *impossible differentials*, i.e., differences that cannot occur, can be used in cryptanalysis. This method of impossible differentials allowed the authors of the cited paper to find a weakness in the cryptographic algorithm Skipjack, the first and (currently) the only algorithm which is officially acknowledged to have been designed by the National Security Agency of the USA. For iteration type ciphers, it is very often impossible to generate differential characteristics which guarantee high probability. In 1999, D. Vagner suggested to use quadruples of texts with a given set of differences (instead of pairs of plaintexts with a given difference). He also suggested a constructive method for generating such quadruples. He named this method the *rectangle method*, which was later developed to the *boomerang method*.

An enormous number of papers is devoted to the application of linear and differential cryptanalysis to particular cryptographic algorithms. The interested reader can find them in the proceedings of conferences such as EUROCRYPT, CRYPTO, ASIACRYPT, Fast Software Encryption, and also in the documents of the competition for the new NIST standard AES.

[TRANSLATOR'S NOTE. Section 9.6 of the Russian original and comments to it are not included in the English edition by the authors' request.]

Bibliography

[1] C. M. Adams, *A formal and practical design procedure for Substitution-Permutation network cryptosystem.* Ph. D. thesis, Department of Electrical Engineering, Queen's University at Kingston, 1990.

[2] C. M. Adams, *On immunity against Biham and Shamir's differential cryptanalysis*, Information Processing Letters, V. 41, 1992, pp. 77–80.

[3] C. M. Adams and S. E. Tavares, *Good S-boxes are Easy to Find*, In Proceedings of Advances in Cryptology: CRYPTO'89, Lect. Notes in Comp. Sci. New York: Springer-Verlag, V. 435, 1990, pp. 612–615.

[4] C. M. Adams and S. E. Tavares, *The structured design of cryptographically good S-boxes*, Journal of Crytology, V. 3, 1990, No. 1. pp. 27–41.

[5] C. M. Adams and S. E. Tavares, *Generating and Counting Binary Bent Sequences*, IEEE Trans. on Inform. Theory, IT 36, No. 5, 1990, pp. 1170–1173.

[6] S. S. Agaian, *Hadamard Matrrices and Their Applications,* Lecture Notes in Mathematics, 1168. Berlin, Heidelberg, New York, Tokyo: Springer-Verlag, 1985.

[7] S. V. Agievich, *On the representation of bent-functions by bent-rectangles*, In Proceedings of the Fifth International Petrozavodsk Conference on Probabilistic Methods in Discrete Mathematics (Petrozavodsk, June 1–6, 2000), Utrecht, Boston: VSP, 2000, pp. 121–135.

[8] E. Akyildiz, I. S. Guloglu, and M. Ikeda, *A Note on Generalized Bent Functions*, Journal of Pure and Applied Algebra, V. 106, No. 1, 1996, pp. 1–9.

[9] A. A. Albert, *Fundamental Concepts of High Algebra,* Chicago: Univ. of Chicago Press, 1956.

[10] A. P. Alferov, A. Yu. Zubov, A. S. Kuzmin, and A. V. Cheremushkin, *Foundations of Cryptography*, Moskva, Helios, Association of Russian Universities, 2001 (in Russian).

[11] A. S. Ambrosimov, *Properties of q-valued logic (bent) functions over finite fields*, Discretnaja matematika, vol. 6, issue 3, 1994, pp. 50–60 (in Russian).

[12] R. Anderson, *Searching for the Optimum Correlation Attack*, Fast Software Encryption, Leuven'94, Lect. Notes in Comp. Sci. New York: Springer-Verlag, V. 1008, 1995, pp. 137–143.

[13] R. Ash, *Information Theory*, New York, London, Sydney: Interscience Publishers, a Division of John Wiley and Sons, 1967.

[14] G. S. Avsarkisyan, *Boolean functions disjunctive decomposition with respect to all variables*, Computernye seti, Riga: Zinatne, vol. 1, pp. 78–94 (in Russian).

[15] A. V. Babash and G. P. Shankin, *Cryptography*, Moskva, Solon-R, 2000 (in Russian).

[16] R. D. Baker, J. H. van Lint, and R. M. Wilson, *On the Preparata and Goethals Codes*, IEEE Trans. on Inform. Theory, IT–29, No. 3, 1983, pp. 342–345.

[17] L. A. Bassalygo, G. V. Zaitsev, and V. A. Zinov'ev, *Uniformly packed codes*, Problems of Information Transmission, vol. 10, issue 1, 1974, pp. 6–9.

[18] L. A. Bassalygo and V. A. Zinoviev, *Remark on Uniformly Packed Codes*, Problems of Information Transmission, vol. 13, issue 3, 1977, pp. 178–180.

[19] *Applied Combinatorial Mathematics*, E. E. Beckenbach (ed.) New York, London, Sydney: John Wiley & Sons, Inc., 1964.

[20] I. Ben-Aroya and E. Biham, *Differential Cryptanalysis of Lucifer*, In Proceedings in Advances of Cryptology: CRYPTO'93, Springer-Verlag, 1993, pp. 187–199.

[21] C. H. Bennet, G. Brassard, and J. M. Robert, *Privacy Amplification by Public Discussion*, SIAM Journal on Computing, V. 17, 1988, pp. 210–229.

[22] E. R. Berlekamp, *Algebraic Coding Theory*, New York, St. Louis, San Francisco, Toronto, London, Sydney: McGrawHill, 1968.

[23] E. R. Berlekamp and L. R. Welch, *Weight Distributions of the Cosets of the* (32, 6) *Reed–Muller Code*, IEEE Trans. on Inform. Theory, IT–18, 1972, pp. 203–207.

[24] S. D. Berman and I. I. Grushko, *On B-Functions Encountered in Modular Codes*, Problems of Information Transmission, vol. 17, issue 2, 1981, pp. 82–88.

[25] T. A. Berson, *Differential Cryptanalysis Mod* 2^{32} *with Applications to MD5*, In Proceedings of Advances in Cryptology: EUROCRYPT'92, Lect. Notes in Comp. Sci., Springer-Verlag, V. 658, 1992, pp. 71–80.

[26] T. Beth and C. Ding, *On Almost Perfect Nonlinear Permutations*, In Proceedings of Advances in Cryptology: EUROCRYPT'93, Lect. Notes in Comp. Sci., New York: Springer-Verlag, V. 765, 1993, pp. 65–76.

[27] J. Bierbrauer, *Bounds on orthogonal arrays and resilient functions*, Journal of Combinatorial Designs, V. 3, 1995, pp. 179–183.

[28] J. Bierbrauer, K. Gopalakrishnan, and D. R. Stinson, *Bounds on Resilient Functions and Orthogonal Arrays*, In Proceedings of Advances in Cryptology: CRYPTO'94, Lect. Notes in Comp. Sci., New York: Springer-Verlag, V. 839, 1994, pp. 247–256.

[29] J. Bierbrauer, K. Gopalakrishnan, and D. R. Stinson, *Orthogonal arrays, resilient functions, error correcting codes and linear programming bounds*, SIAM J. Discr. Math., V. 9, 1996, pp. 424–452.

[30] E. Biham, A. Biryukov, and A. Shamir, *Cryptanalysis of Skipjack Reduced to 31 Rounds Using Impossible Differentials*, In Advances in Cryptology: EUROCRYPT'99, Lect. Notes in Comp. Sci., New York: Springer-Verlag, 1999. V. 1592. pp. 12–23.

[31] E. Biham and A. Shamir, *Differential Cryptanalysis of DES-like Cryptosystems*, Journal of Cryptology, V. 4, No. 1, 1991, pp. 3–72.

[32] A. Biryukov, C. De Cannière, A. Braeken, and B. Preneel, *A Toolbox for Cryptanalysis: Linear and Affine Equivalence Algorithms*, In Advances in Cryptology: EUROCRYPT'2003, Lect. Notes in Comp. Sci., New York: Springer-Verlag, V. 2656, 2003, pp. 33–50.

[33] A. Biryukov, C. De Cannière, and M. Quisquater, *On Multiple Linear Approximations*, `http://www.iacr.eprint-arhiv`, 2004.

[34] R. E. Blahut, *Theory and Practice of Error Control Codes,* Addison-Wesley Publishing Company Inc., Reading, Massachusetts, Menlo Park, California, London, Amsterdam, Don Mills, Ontario, Sydney, 1984.

[35] R. E. Blahut, *Fast Algorithms for Digital Signal Processing,* Addison-Wesley Publishing Company Inc., 1985.

[36] W. Blaser and P. Heinzmann, *New Cruptographic Device with High Security Using Public Key Distribution*, IEEE Student Papers, 1982, pp. 145–153.

[37] V. D. Bliznyuk and M. F. Kholodnyi, *Application of Boolean derivatives to the problem for Boolean function decomposition*, Avtomatica i telemekhanica, issue 5, 1984, pp. 105–112 (in Russian).

[38] D. Bochmann and Ch. Posthoff, *Binäre dynamische Systeme,* Berlin: Akademie-Verlag, 1981.

[39] A. A. Botev, *On the relationship between correlation immunity, nonlinearity and weight for non-balanced Boolean functions*, Matematicheskie voprosy kibernetiki, issue 11, Moskva, Fizmatlit, 2002, pp. 149–162 (in Russian).

[40] A. E. Brouwer and L. M. Tolhuizen, *A Sharpening of the Johnson Bound for Binary Linear Codes*, Design, Codes and Cryptography, V. 3, No. 1, 1993, pp. 95–98.

[41] R. A. Brualdi, N. Cai, and V. S. Pless, *Orphan Structure of the First-Order Reed–Muller codes*, Discrete Mathematics, V. 102, 1992, pp. 239–247.

[42] A. R. Calderbank, G. McGuire, B. Poonen, and M. Rubinstein, *On a Conjecture of Helleseth Regarding Pairs of Binary m-sequences*, IEEE Trans. on Inform. Theory, V. 42, 1996, pp. 988–990.

[43] P. Camion, C. Carlet, P. Charpin and N. Sendrier, *On Correlation Immune Functions*, In Proceedings of Advances in Cryptology: CRYPTO'91, Lect. Notes in Comp. Sci., New York: Springer-Verlag, V. 576, 1992, pp. 86–100.

[44] P. Camion and A. Canteaut, *Construction of t-resilient functions over a finite alphabet*, Advances in Cryptology: Eurocrypt'96, Lect. Notes in Comp. Sci., V. 1070, 1996, pp. 283–293.

[45] P. Camion and A. Canteaut, *Generalization of Siegenthaler Inequality and Schorr–Vaudenay Multipermutations*, Advances in Cryptology: CRYPTO'96, Lect. Notes in Comp. Sci., V. 1109, 1996, pp. 372–386.

[46] P. Camion and A. Canteaut, *Correlation Immune and Resilient Functions Over a Finite Alphabet and Their Applications in Cryptography*, Designs Codes and Cryptography, V. 16, No. 2, 1999, pp. 121–149.

[47] A. Canteaut and E. Filiol, *Ciphertext Only Reconstructing of Stream Ciphers Based on Combination Generators*, Fast Software Encruption'2000, Lect. Notes in Comp. Sci., New York: Springer-Verlag, V. 1978, 2001, pp. 165–180.

[48] A. Canteaut, C. Carlet, P. Charpin, and C. Fontaine, *Propagation Characteristics and Correlation Immunity of Highly Nonlinear Boolean Functions*, In Proceedings of Advances in Cryptology: EUROCRYPT'00, Lect. Notes in Comp. Sci., New York: Springer-Verlag, V. 1807, 2000, pp. 507–522.

[49] A. Canteaut and M. Trabbia, *Improved Fast Correlation Attacks Using Parity-Check of Weight 4 and 5*, In Proceedings of Advances in Cryptology: EUROCRYPT'00, Lect. Notes in Comp. Sci., New York: Springer-Verlag, V. 1807, 2000, pp. 573–588.

[50] A. Canteaut, C. Carlet, P. Charpin, and C. Fontaine, *On Cryptographic Properties of the Cosets of* $RM(1, m)$, IEEE Trans. on Inform. Theory, V. 47, No. 4, 2001, pp. 1494–1513.

[51] A. Canteaut, P. Charpin, and H. Dobbertin, *Binary m-sequences with Three-Valued Cross-Correlation: A Proof of Welch's Conjecture*, IEEE Trans. on Inform. Theory, V. 46, No. 1, 2000, pp. 4–8.

[52] A. Canteaut, P. Charpin, and H. Dobbertin, *Couples de suites binaires de longueur maximale ayant une corrélation croisée à trois valeurs: Conjecture de Welch*, Sér. I Math., Paris: C. R. Acad. Sci., V. 328, 1999, pp. 173–178.

[53] A. Canteaut, P. Charpin, and H. Dobbertin, *Weight Divisibility of Cycle Codes, Highly Nonlinear Functions on* $\mathbb{F}_{2^m}$*, and Crosscorrelation of Maximum Length Sequences*, SIAM Journal of Discrete Mathematics, V. 13, No. 1, 2000, pp. 105–138.

[54] A. Canteaut, M. Daum, H. Dobbertin, and G. Leander, *Normal and Non Normal Bent Functions*, Proceedings of International Workshop on Coding and Cryptography, March, 24–28, Versalles (France), 2003, pp. 91–100.

[55] C. Carlet, *A transformation on Boolean Functions, its Consequences on some Problems Related to Reed–Muller Codes*, EUROCODES'90. Lect. Notes in Comp. Sci., New York: Springer-Verlag, V. 514, 1991, pp. 42–50.

[56] C. Carlet, *The Automorphism Groups of the Kerdock Codes*, Journal of Information & Optimization Sciences, V. 12, No. 3, 1991, pp. 387–400.

[57] C. Carlet, *Partially-bent functions*, In Advances in Cryptology: CRYPTO'92, Lect. Notes in Comp. Sci., Springer-Verlag, V. 740, 1992, pp. 280–291.

[58] C. Carlet, *Partially-bent functions*, Designs Codes and Cryptography, V. 3, 1993, pp. 135–145.

[59] C. Carlet, *Two new classes of bent functions*, In Proceedings of Advances in Cryptology: EUROCRYPT'93, Lect. Notes in Comp. Sci., New York: Springer-Verlag, V. 765, 1994, pp. 77–101.

[60] C. Carlet, *Generalized Partial Spreads*, IEEE Trans. on Inform. Theory, V. 41, No. 5, 1995, pp. 1482–1487.

[61] C. Carlet, *A constraction of bent functions*, Seventh Joint Swedish-Russian International Workshop on Information Theory, St. Petersburg, Russia, 1995, pp. 57–59.

[62] C. Carlet and Ph. Guillot, *A characterization of binary bent functions*, Journal of Combinatorial Theory, Series A, V. 76, No. 2, 1996, pp. 328–335.

[63] C. Carlet and Ph. Guillot, *An alternate characterization of the bentness of binary functions, with uniqueness*, Designs, Codes and Cryptography, V. 14, No. 2, 1998, pp. 33–140.

[64] C. Carlet, P. Charpin, and V. Zinoviev, *Codes, bent functions and permutations suitable for DES-like cryptosystems*, Designs, Codes and Cryptography, V. 15, No. 15, 1998, pp. 125–156.

[65] C. Carlet, *Hyper-bent functions*, PRAGOCRYPT'96, Praga: CTV, GC UCMP, 1996, pp. 145–155.

[66] C. Carlet, J. Seberry, and X. M. Zhang, *Comments on Generating and counting binary bent sequences*, IEEE Trans. on Inform. Theory, V. 40, No. 2, 1994, p. 600.

[67] C. Carlet, *More Correlation-Immune and Resilient Functions over Galois Fields and Galois Rings*, In Proceedings of Advances in Cryptology: EUROCRYPT'97, Lect. Notes in Comp. Sci., New York: Springer-Verlag, V. 1233, 1997, pp. 422–433.

[68] C. Carlet, *On the Propagation Criterion of Degree l and Order k*, In Proceedings of Advances in Cryptology: EUROCRYPT'98, Lect. Notes in Comp. Sci., New York: Springer-Verlag, V. 1403, 1998, pp. 462–474.

[69] C. Carlet, *On the Coset Weight Divisibility and Nonlinearity of Resilient and Correlation Immune Functions*, Sequences and Their Applications: SETA'2001, Discrete Mathematics and Theoretical Computer Science, New York: Springer-Verlag, 2001, pp. 131–144.

[70] C. Carlet and P. Sarcar, *Spectral Domain Analysis of Correlation Immune and Resilient Boolean Functions*, Finite Fields and Its Applications, V. 8, No. 1, 2002, pp. 120–130.

[71] C. Carlet, *A Large Class of Cryptographic Boolean Functions via a Study of the Maiorana–McFarland Constructions*, In Proceedings of Advances in Cryptology: CRYPTO'02, Lect. Notes in Comp. Sci., New York: Springer-Verlag, V. 2442, 2002, pp. 549–564.

[72] C. Carlet and E. Prouff, *On plateaued functions and their constructions*, Fast Software Encryption, 2003.

[73] C. Carlet and Yu. Tarannikov, *Covering sequences of Boolean functions and their cryptographic significance*, Designs, Codes and Cryptography, V. 25, 2002, pp. 263–279.

[74] C. Carlet and A. Klapper, *Upper Bounds on the Numbers of Resilient Functions and of Bent Functions.*

[75] J. M. Carroll and L. E. Robbins, *Using binary derivaties to test an enhancement of DES*, Cryptologia, V. 12, 1988, pp. 193–208.

[76] J. W. S. Cassels, *Rational Quadratic Forms*, London, New York, San Francisko: Academic Press, 1978.

[77] F. Chabaud and S. Vaudenay, *Links between Differential and Linear Cryptanalysis*, In Proceedings of Advances in Cryptology: EUROCRYPT'94, Lect. Notes in Comp. Sci, New York: Springer-Verlag, V. 950, 1995, pp. 356–365.

[78] P. Charpin, A. Tietäväinen, and V. Zinoviev, *On Binary Cyclic Codes with Minimum Distance* $d = 3$, Problems of information transmission, vol. 33, issue 4, 1997, pp. 287–296.

[79] D. Chaum and J. H. Evertse, *Cryptanalysis of DES with a Reduced Number of Rounds; Sequences of Linear Factors in Block Ciphers*, In Proceedings of Advances in Cryptology: CRYPTO'85, Lect. Notes in Comp. Sci., New York: Springer-Verlag, V. 218, 1986, pp. 192–211.

[80] S. Chee, S. Lee, and K. Kim, *Semi-Bent Functions*, In Proceedings of Advances in Cryptology: ASIACRYPT'94, Lect. Notes in Comp. Sci., New York: Springer-Verlag, V. 914, 1995, pp. 107–118.

[81] S. Chee, S. Lee, D. Lee, and S. Sung, *On the Correlation Immune Functions and Their Nonlinearity*, In Proceedings of Advances in Cryptology: ASIACRYPT'96, Lect. Notes in Comp. Sci., New York: Springer-Verlag,V. 1163, 1996, pp. 232–243.

[82] J. H. Cheon and S. Chee, *Elliptic Curves and Resilient Functions*, ICISC'2000, Lect. Notes in Comp. Sci., New York: Springer-Verlag, V. 2015, 2000, pp. 64–72.

[83] J. H. Cheon, *Nonlinear Vector Resilient Functions*, In Proceedings of Advances in Cryptology: CRYPTO'2001, Lect. Notes in Comp. Sci., New York: Springer-Verlag, V. 2139, 2001, pp. 458–479.

[84] V. Chepyzhov and B. Smeets, *On Fast Correlation Attacks on Certain Stream Ciphers*, In Proceedings of Advances in Cryptology: EUROCRYPT'1991, Lect. Notes in Comp. Sci., New York: Springer-Verlag, V. 547, 1991, pp. 176–185.

[85] V. Chepyzhov, T. Johansson, and B. Smeets, *A Simple Algorithm for Fast Correlation Attacks on Stream Ciphers*, Fast Software Encryption'2000, Lect. Notes in Comp. Sci., New York: Springer-Verlag, V. 1978, 2000, pp. 181–195.

[86] A. V. Cheremushkin, *Affine and linear classification methods for binary functions*, Russian Academy of Sciences, Cryptographic Academy of Russian Federation, Trudy po diskretnoi matematike, Moskva, Fizmatlit, vol. 4, 2001, pp. 273–314 (in Russian).

[87] B. Chor, O. Goldreich, J. Hastad, J. Friedman, S. Rudich, and R. Smolensky, *The Bit Extraction Problem for* t*-Resilient Functions*, 26-th Symposium on Foundations of Computer Science, 1985, pp. 396–407.

[88] P. Chose, A. Joux, and M. Mitton, *Fast Correlation Attacks: an Algorithmic Point of View*, In Proceedings of Advances in Cryptology: EUROCRYPT'2002, Lect. Notes in Comp. Sci., New York: Springer-Verlag, V. 2332, 2002, pp. 209–221.

[89] H. Chung and P. V. Kumar, *A New General Construction for Generalized Bent Function*, IEEE Trans. on Inform. Theory, V. 35, No. 1, 1989, pp. 206–209.

[90] J. Clark, J. Jacob, W. Millan, and S. Maitra, *Evolution of Boolean Functions Satisfying Multiple Criteria with Simulated Annealing*, Preprint, 2002.

[91] G. Cohen, I. Honkala, A. Lobstein, and S. Litsyn, *Covering codes,* Elsevier, 1998.

[92] G. D. Cohen, M. G. Karpovsky, H. F. Mattson and, J. Schatz, *Covering radius—survey and recent results*, IEEE Trans. on Inform. Theory, IT-31, No. 3, 1985, pp. 328–343.

[93] D. Coppersmith, *The data encryption standard (DES) and its strength against attacks*, Technical Report RC 18613 (81421), IBM Research Division, December 1992.

[94] H. Cramer, *Mathematical Methods of Statistics*, Princeton University Press, 1946.

[95] C. W. Curtis and I. Reiner, *Representation theory of Finite Groups and Associative Algebras,* New York, London: Interscience Publishers, a division of John Wiley & Sons, 1962.

[96] Th. W. Cusick, *Boolean functions satisfying a higher order strict avalanche criterion*, In Proceedings of Advances in Cryptology: EUROCRYPT'93, Lect. Notes in Comp. Sci., New York: Springer-Verlag, V. 765, 1988, pp. 102–117.

[97] Th. W. Cusick and H. Dobbertin, *Some New 3-valued Crosscorrelation Functions of Binary m-sequences*, IEEE Trans. on Inform. Theory, V. 42, No. 4, 1996, pp. 1238–1240.

[98] Th. W. Cusick, *On constructing balanced correlation immune functions*, Sequences and Their Applications, Proceedings of SETA'98, Springer Discrete Mathematics and Theoretical Computer Science, 1999, pp. 184–190.

[99] M. Daum, H. Dobbertin, and G. Leander, *An Algorithm for Checking Normality of Boolean Function*, Proceedings of International Workshop on Coding and Cryptography, March, 24–28, Versalles (France), 2003, pp. 133–142.

[100] M. H. Dawson and S. E. Tavares, *An expanded set of S-box design criteria based on information theory and its relation to differential-like attacks*, In Proceedings of Advances in Cryptology: EUROCRYPT'91, Lect. Notes in Comp. Sci., New York: Springer-Verlag, V. 547, 1991, pp. 352–367.

[101] P. Delsarte, J. M. Goethals, and F. J. MacWilliams, *On Generalaized Reed–Muller Codes and Their Relatives*, Information and Control, V. 16, 1970, pp. 403–442.

[102] P. Delsarte, *An algebraic approach to the association schemes of coding theory*, Philips Research Reports Supplements, V. 10, 1973.

[103] P. Delsarte, *Four fundamental parameters of a code and their combinatorial significance*, Information and Control, V. 23, No. 5, 1973, pp. 407–438.

[104] O. V. Denisov, *Asymptotic number of the k-order correlation-immune Boolean functions*, Diskretnaya matematika, vol. 3, issue 2, 1991, pp. 25–46 (in Russian).

[105] O. V. Denisov, *Local limit theorem for random binary function part of spectrum distribution*, Diskretnaya matematika, vol. 12, issue 1, 2000, pp. 82–95 (in Russian).

[106] L. E. Dickson, *Linear Groups*, Leipzig: B. G. Teubner, 1901.

[107] J. Dieudonné, *La Geometrie des Groupes classiques*, Springer-Verlag, 1971.

[108] J. F. Dillon, *A survey of bent functions*, The NSA Technical Journal (unclassified), 1972, pp. 191–215.

[109] F. J. Dillon, *Elementary Hadamard Difference sets*, Ph.D. Thesis, University of Maryland, 1974.

[110] C. Ding, G. Xiao, and W. Shan, *The Stability Theory of Stream Ciphers*, Lect. Notes in Comp. Sci., New York: Springer-Verlag, V. 561, 1991.

[111] H. Dobbertin, *One-to-One Highly Nonlinear Power Functions on Finite Field with Characteristic 2*, Appl. Algebra Engr. Comm. Comp., V. 9, 1998, pp. 139–152.

[112] H. Dobbertin, *Almost Perfect Nonlinear Power Functions on $GF(2^n)$*, Preprint.

[113] H. Dobbertin, *Another Proof of Kasami's Theorem*, Preprint.

[114] H. Dobbertin, *Construction of Bent Functions and Balanced Boolean Functions with High Nonlinearity*, Fast Software Encryption—Second International Workshop, Leuven (1994), Lect. Notes in Comp. Sci., New York: Springer-Verlag, V. 1008, 1995, pp. 61–74.

[115] H. Dobbertin, *One-to-One Highly Nonlinear Functions on Finite Field with Characteristic 2*, Appl. Algebra Engrg. Comm. Comp., V. 9, 1998, pp. 139–152.

[116] H. Dobbertin, *Almost Perfect Nonlinear Power Functions on $GF(2^n)$: The Niho case*, Inform. and Comp., V. 151, 1999, pp. 57–72.

[117] H. Dobbertin, *Almost Perfect Nonlinear Power Functions on $GF(2^n)$: The Welch case*, IEEE Trans. Inform. Theory, V. 45, 1999, pp. 1271–1275.

[118] S. M. Dodunekov and V. A. Zinoviev, *Note about Preparata codes*, Trudy 6th Intern. Symp. Information Theory, Moskva–Tashkent, part 2, 1984, pp. 78–80 (in Russian).

[119] J. H. Evertse, *Linear Structures in Blockciphers*, In Proceedings of Advances in Cryptology: EUROCRYPT'87, Lect. Notes in Comp. Sci., New York: Springer-Verlag, V. 304, 1988, pp. 249–266.

[120] B. J. Falkowski, *A Note on the Polynomial Form of Boolean Functions and Related Topics*, IEEE Trans. on Computers, V. 48, No. 8, 1999, pp. 860–864.

[121] M. Fedorova and Y. V. Tarannikov, *On the Constructing of Highly Nonlinear Resilient Boolean functions by Means of Special Matrices*, Progress in Cryptology: INDOCRYPT'2001, Lect. Notes in Comp. Sci., New York: Springer-Verlag, V. 2247, 2001, pp. 254–266.

[122] A. Feinstein, *Foundations of Information Theory*, New York–Toronto–London: McGraw-Hill Book Company Inc., 1958.

[123] W. Feller, *An Introduction to Probability Theory and its Appications*, Third edition, John Wiley & Sons, 1968.

[124] E. Filiol and C. Fontaine, *Highly Nonlinear Balanced Boolean Functions with a Good Correlation-Immunity*, In Proceedings of Advances in Cryptology: EUROCRYPT'98, Lect. Notes in Comp. Sci., New York: Springer-Verlag, V. 1403, 1998, pp. 475–488.

[125] C. Fontaine, *On Some Cosets of the First-Order Reed–Muller Code with High Minimum Weight*, IEEE Trans. on Inform. Theory, V. 45, No. 4, 1999, pp. 1237–1243.

[126] R. Forré, *The Strict Avalanche Criterion: Spectral Properties of Boolean Functions and an Extended Definition*, In Proceedings of Advances in Cryptology: CRYPTO'88, Lect. Notes in Comp. Sci., New York: Springer-Verlag, V. 403, 1989, pp. 450–468.

[127] R. Forré, *A Fast Correlation Attack on Nonlinearly Feed Forward Filter Shift-Register Sequences*, In Proceedings of Advances in Cryptology: EUROCRYPT'89, Lect. Notes in Comp. Sci., New York: Springer-Verlag, V. 434, 1990, pp. 586–595.

[128] R. Forré, *Methods and instruments for designing S-boxes*, Journal of Crytology, V. 3, No. 2, 1990, pp. 115–130.

[129] J. Friedman, *On the Bit Extraction Problem*, 33-rd IEEE Symposium on Foundations of Computer Science, 1982, pp. 314–319.

[130] R. G. Gallager, *Low-density parity-check codes*, MIT Press, Cambridge, MA, 1963.

[131] R. G. Gallager, *Information Theory and Reliable Communication*, New York, London, Sydney, Toronto: John Wiley & Sons Inc., 1968.

[132] M. I. Gelfand, *Lectures on linear algebra*, Moskva, Nauka, 1971; English transl., Dover Publ., New York, 1983.

[133] A. Gill, *Introduction to the Theory of Finite-State Machines*, New York, San Francisco, Toronto, London: McGraw-Hill Book Company Inc., 1962.

[134] J. M. Goethals and S. L. Snover, *Nearly Perfect Codes*, Discrete Mathematics, V. 3, 1972, pp. 64–88.

[135] J. M. Goethals and H. van Tilborg, *Uniformly Packed Codes*, Philips Res. Reports, V. 30, 1975, pp. 9–36.

[136] R. Gold, *Optimal binary sequences for spread-spectrum multiplexing*, IEEE Trans. on Inform. Theory, V. 13, No. 4, 1967, pp. 619–621.

[137] R. Gold, *Maximal recursive sequences with 3-valued recursive crosscorrelation functions*, IEEE Trans. on Inform. Theory, V. 14, 1968, pp. 154–156.

[138] J. D. Golić, *On Security of Nonlinear Filter Generators*, Fast Software Encryption—Cambrige'96, Lect. Notes in Comp. Sci., New York: Springer-Verlag, V. 1039, 1996, pp. 173–188.

[139] J. D. Golić, *Fast Low Order Approximation of Cryptographic Functions*, In Proceedings of Advances in Cryptology: EUROCRYPT'96, Lect. Notes in Comp. Sci., New York: Springer-Verlag, V. 1070, 1996, pp. 268–282.

[140] S. W. Golomb, *On classification of Boolean functions*, IRE Trans. on Circuit Theory, V. 6, 1959, pp. 176–186.

[141] S. V. Golomb, *Theory of transformation groups of polynomials over $GF(2)$ with applications to linear shift register sequences*, Inform. Sci., V. 1, 1968, pp. 209–232.

[142] I. B. Golubov, A. V. Efimov, and V. A. Skvortsov, *Walsh series and transforms. Theory and application*, Math. and Appl. (Soviet Series), 64, Kluwer, Dordrecht, 1991.

[143] G. Gong and S. W. Golomb, *Transform domain analysis of DES*, IEEE Trans. on Inform. Theory, IT-45, No. 6, 1999, pp. 2065–2073.

[144] G. Gong, *Sequence Analysis*, University of Waterloo, Canada, `http://www.cacr.math.uwaterloo`.

[145] K. Gopalakrishnan, D. G. Hoffman, and D. R. Stinson, *A Note on a Conjecture Concerning Symmetric Resilient Functions*, Information Processing Letters, V. 47, No. 3, 1993, pp. 139–143.

[146] K. Gopalakrishnan, *A Study of Correlation-Immune, Resilient and Related Cryptographic Functions*, Ph.D. Thesis, University of Nebraska, 1994.

[147] J. Gordon and H. Retkin, *Are big S-boxes best?*, In Proceedings of Advances in Cryptology: EUROCRYPT'82, Lect. Notes in Comp. Sci., New York: Springer-Verlag, V. 658, 1983, pp. 257–262.

[148] R. R. Green, *A Serial Orthogonal Decoder*, JPL Space Programms Summary, V. 37–39–IV, 1966, pp. 247–253.

[149] R. R. Green, *Analysis of a Serial Orthogonal Decoder*, JPL Space Programms Summary, V. 37–53–III, 1968, pp. 185–187.

[150] A. A. Grusho, E. A. Primenko, and E. E. Timonina, *Analysis and synthesis of cryptographic algorithms. Lectures*, Yoshkar-Ola, 2000 (in Russian).

[151] A. A. Grusho, E. A. Primenko, and E. E. Timonina, *Cryptographic protocols*, Yoshkar-Ola, 2001 (in Russian).

[152] A. E. Gukov and V. P. Chistyakov, *Matrix approach to investigation of preimages of the output sequence of a finite automaton*, Obozrenie prikladnoi i promyshlennoi matematiki, Moskva, vol. 1, issue 1, 1994, pp. 108–117 (in Russian).

[153] X. Guo-Zhen and J. Massey, *A Spectral Characterization of Correlation Immune Combining Functions*, IEEE Trans. on Inform. Theory, V. 34, No. 3, 1988, pp. 569–571.

[154] G. B. Gurevich, *Foundations of the theory of algebraic invaraints*, Moskva, OGIZ, 1948; English transl., Nordhoof, Groningen, 1964.

[155] M. Hall, *The Theory of Groups*, The MacMillan Company, 1959.

[156] M. Hall, Jr., *Combinatorial Theory*, Blaisdell Publishing Company, Waltham (Massachusetts)–Toronto–London, 1967.

[157] Yu. S. Harin, V. I. Bernik, and G. V. Matveev, *Matematical foundations of cryptology*, Minsk, BGU, 1999 (in Russian).

[158] C. Harpes, G. Kramer, and J. L. Massey, *A Generalization of Linear Cryptanalysis and the Applicability of Matsui's Piling-up Lemma*, In Proceedings of Advances in Cryptology: EUROCRYPT'95, Lect. Notes in Comp. Sci., New York: Springer-Verlag, V. 921, 1995, pp. 24–38.

[159] C. Harpes and J. L. Massey, *Partitioning Cryptanalysis*, Proceedings of Fast Software Encryption Workshop'97, pp. 13–27.

[160] M. A. Harrison, *Counting Theorems and Their Applications to Classification of Switching Functions*, In Recent Development in Switching Theory, New York, 1971.

[161] T. Helleseth, *Some Results about the Cross-Correlation Function between Two Maximal Linear Sequences*, Discrete Math., V. 16, 1976, pp. 209–232.

[162] T. Helleseth, T. Klöve, and Mykkeltveit, *On Covering Radius of Binary Codes*, IEEE Trans. on Inform. Theory, IT-24, No. 5, 1978, pp. 627–628.

[163] T. Helleseth, C. Rong, and D. Sandberg, *New Families of Almost Perfect Nonlinear Power Mappings*, IEEE Trans. on Inform. Theory, IT-45, No. 2, 1999, pp. 475–485.

[164] T. Helleseth and V. P. Kumar, *Sequences with Low Correlation*, In: Handbook of Coding Theory, North-Holland, Amsterdam, 1998, pp. 1765–1853.

[165] H. Hollmann and Q. Xiang, *A Proof of the Welch and Niho Conjectures on Cross-correlations of Binary m-sequences*, Preprint, 1998.

[166] D. A. Huffman, *Canonical Forms for Information-Lossless Finite State Logical Machines*, IRE Trans. Circuit Theory, Spec. Suppl, V. 6, 1959, pp. 41–59.

[167] H. Janwa, G. McGuire, and R. M. Wilson, *Double-Error-Correcting Codes and Absolutely Irreducible Polynomials over $GF(2)$*, Journal of Algebra, V. 178, 1995, pp. 665–676.

[168] H. Janwa and R. M. Wilson, *Hyperplane sections of Fermat varieties in P^3 in char 2 and some applications to cyclic codes*, In Proceedings Applied Algebra, Algebraic Algorithms and Error-Correcting Codes, AAECC-10, Lect. Notes in Comp. Sci., Berlin: Springer-Verlag, V. 673, 1993, pp. 180–194.

[169] S. Jiang and G. Gong, *Cryptanalysis of Stream Ciphers—A Survey*, `http://calliope.waterloo.ca/~ggong`, 2002.

[170] T. Johansson and F. Jönsson, *Improved Fast Correlation Attacks on Stream Ciphers via Convolutional Codes*, In Proceedings of Advances in Cryptology: EUROCRYPT'99, Lect. Notes in Comp. Sci., New York: Springer-Verlag, V. 1592, 1999, pp. 347–362.

[171] T. Johansson and F. Jönsson, *Fast Correlation Attacks Based on Turbo Code-Techiques*, In Proceedings of Advances in Cryptology: CRYPTO'99, Lect. Notes in Comp. Sci., New York: Springer-Verlag, V. 1666, 1999, pp. 181–197.

[172] T. Johansson and F. Jönsson, *Fast Correlation Attacks through Reconstruction of Linear Polynomials*, In Proceedings of Advances in Cryptology: CRYPTO'00, Lect. Notes in Comp. Sci., New York: Springer-Verlag, V. 1880, 2000, pp. 300–315.

[173] T. Johansson and F. Jönsson, *Theoretical Analysis of a Correlation Attack Based on Convolutional Codes*, IEEE Trans. on Inform. Theory, V. 48, No. 8, 2002, pp. 2173–2181.

[174] T. Johansson and E. Pasalic, *A Constraction of Resilient Functions with High Nonlinearity*, IEEE International Symposium on Information Theory: ISIT'2000, `http://www.eprint.iacr.org`, No. 2000/053.

[175] P. Junod, *On the complexity of Matsui's attack*, In Proceedings of Selected Areas in Cryptography: SAC'2001 (S. Vadenay, A. M. Youssef, eds.), Lect. Notes in Comp. Sci., Springer-Verlag, V. 2259, 2001, pp. 199–211.

[176] P. Junod, *On the optimality of linear, differential, and sequential distinguishers*, In Proceedings of Advances in Cryptology: EUROCRYPT'2003 (E. Biham, ed.), Lect. Notes in Comp. Sci., V. 2656, 2003, pp. 17–32.

[177] Yu. I. Zhuravlev, *Algorithms for constructing minimal disjunctive normal forms for functions of logic algebra*, Diskretnaya matematika i matematicheskie voprosy kibernetiki, vol. 1, Moskva, Nauka, 1974 (in Russian).

[178] B. S. Kaliski and M. J. B. Robshaw, *Linear cryptanalysis using multiple approximations*, In Proceedings of Advances of Cryptology: CRYPTO'94 (Y. Desmedt, ed.), Lect. Notes in Comp. Sci., Springer-Verlag, V. 950, 1994, pp. 26–39.

[179] M. G. Karpovsky and E. S. Moskalev, *Spectral methods for analysis and synthesis of discrete devices*, Leningrad, Energiya, 1973 (in Russian).

[180] Yu. D. Karyakin, *Fast Correlation Decoding of Reed-Muller Codes*, Problems of Information Transmission, vol. 23, issue 2, 1987, pp. 121–129.

[181] T. Kasami, S. Lin, and W. W. Peterson, *New Generalizations of Reed–Muller Codes. Part I: Primitive Codes*, IEEE Trans. on Inform. Theory, IT-14, No. 2, 1968, pp. 189–199.

[182] T. Kasami, S. Lin, and W. W. Peterson, *Polynomial Codes*, IEEE Trans. on Inform. Theory, IT-14, No. 6, 1968, pp. 807–814.

[183] T. Kasami, *Weight Distributions of Bose–Chaudhuri–Hocquenghem Codes*, In Proceedings of the Conference on Combinatorial Mathematics and Its Applications, Univ. of North Carolina Press, Chapel Hill, NC, 1969, pp. 335–357.

[184] T. Kasami, *Weight Distributions of Bose–Chaudhuri–Hocquenghem Codes*, In: Combinatorial Math. Applications (R. C. Bose, T. A. Dowlings, eds.), Univ. of North Carolina Press, Chapel Hill, NC., 1969, Ch. 9.

[185] T. Kasami, *The Weight Enumerators for Several Classes of Subcodes of the 2-nd Order Binary Reed–Muller Codes*, Information and Control, V. 18, 1971, pp. 369–394.

[186] T. Kasami and N. Tokura, *On the Weight Structure of Reed–Muller Codes*, IEEE Trans. on Inform. Theory, IT-16, No. 6, 1970, pp. 752–825.

[187] T. Kasami, N. Tokura, E. Ivadari, and J. Inagaki, *Coding theory*, Transl. from Japanese, Moskva, Mir, 1978 (in Russian).

[188] A. Kholosha and H. C. A. van Tilborg, *Tensor Transform of Boolean Functions and Related Agebraic and Probabilistic Properties*, `http://www.iacr.org/e-arhiv/2002/`.

[189] K. Kim, *A study on the construction and analysis of substitution boxes for symmetric cryptosystems*, Ph.D. Thesis, Yokohama National Univeristy, Division of Electrical and Computer Engineering, 1990.

[190] K. Kim, T. Matsumoto, and H. Imai, *On generating cryptographically desirable substitutions*, Transactions of the IEICE, V. 73, No. 7, 1990, pp. 1031–1035.

[191] D. P. Kirienko, *Complete description of non-balanced correlation-immune of order* 5 *Boolean functions of* 5 *variables*, Trudy XXIII konferentsii molodykh uchenykh MGU "Sovremennye issledovania v matematike i mekhanike", vol. 2, 9–14 April, 2001, pp. 153–156 (in Russian).

[192] B. M. Kloss and E. N. Nechiporuk, *On the classification of multivalued logic functions*, Problemy kibernetiki, issue 9, 1963 (in Russian).

[193] L. Knudsen, *Truncated and Higher Order Differentials*, In Proceedings of Fast Software Encryption, Second International Workshop, Lect. Notes in Comp. Sci., Springer-Verlag, V. 1008, 1994, pp. 196–211.

[194] L. Knudsen and M. J. B. Robshaw, *Non-Linear Approximation in Linear Cryptanalysis*, In Proceedings of Advances of Cryptology: EUROCRYPT'96 (U. Maurer, ed.), Lect. Notes in Comp. Sci., Springer-Verlag, V. 1070, 1996, pp. 224–236.

[195] L. R. Knudsen and H. E. Mathiassen, *A chosen-plaintext linear attack on DES*, In Proceedings of Fast Software Encryption – FSE'2000 (B. Schneier, ed.), Lect. Notes in Comp. Sci., Springer-Verlag, V. 1978, 2001, pp. 262–272.

[196] S. L. Kolbin, *On some properties of mutually inverse systems of p-valued functions*, Diskretnaya matematika, vol. 6, issue 2, 1994, pp. 145–149 (in Russian).

[197] V. D. Kolesnik and E. T. Mironchik, *Cyclic codes decoding*, Moskva, Svyaz, 1968 (in Russian).

[198] V. D. Kolesnik and G. Poltyrev, *Lectures on Information Theory*, Moskva, Nauka, 1982 (in Russian).

[199] P. S. Korolev, *Quadratic Boolean functions of high order stability*, Matematicheskie voprosy kibernetiki, issue 11, Moskva, Fizmatlit, 2002, pp. 255–261 (in Russian).

[200] P. V. Kumar, R. A. Scholts, and R. L. Welch, *Generalized bent functions and their properties*, Journal of Combinatorial Theory, Series A, V. 40, No. 1, 1985, pp. 90–107.

[201] A. A. Kurmit, *Automata without loss of information of finite order*, Riga: Zinatne, 1972 (in Russian).

[202] K. Kurosawa and T. Satoh, *Generalization of higher order SAC to vector output Boolean Functions*, In Proceedings of Advances in Cryptology: ASIACRYPT'96, Lect. Notes in Comp. Sci., New York: Springer-Verlag, V. 1163, 1996, pp. 218–231.

[203] K. Kurosawa and T. Satoh, *Design of* SAC/PC*(l) of Oder k Boolean Functions and Three Other Cryptographic Criteria*, In Proceedings of Advances in Cryptology: EUROCRYPT'97, Lect. Notes in Comp. Sci., New York: Springer-Verlag, V. 1233, 1998, pp. 434–449.

[204] K. Kurosawa, T. Satoh, and K. Yamamoto, *Highly Nonlinear t-Resilient Functions*, Journal of Universal Computer Science, V. 3, No. 6, 1997, pp. 721–729.

[205] K. Kurosawa, T. Iwata, and T. Yoshiwara, *New covering radius of Reed–Muller codes for t-resilient functions*, SAC'2001, Lect. Notes in Comp. Sci., Springer-Verlag, No. 2259, 2001, pp. 75–86.

[206] K. Kurosawa, T. Johansson, and D. Stinson, *Almost k-wise Independent Sample Spaces and Their Cryptographic Applications*, Journal of Cryptology, V. 14, No. 4, 2001, pp. 231–253.

[207] Yu. V. Kuznetsov, *Classes of Boolean functions invariant with respect to the identification of variables*, Dokl. Akad. Nauk SSSR, vol. 290, issue 4, 1986, pp. 780–785; English transl., Soviet Math. Dokl. **34** (1987), no. 2, 339–344.

[208] Yu. V. Kuznetsov and S. A. Shkarin, *Reed–Muller codes (review)*, Matematicheskie voprosy kibernetiki, Moskva, Nauka, issue 6, 1996, pp. 5–50 (in Russian).

[209] Yu. V. Kuznetsov and V. V. Yashchenko, *On estimation for the degree of nonlinearity of partial Boolean functions*, Vestnik MGU, vol. 119, issue 6, 1993, pp. 36–40; English transl., Moscow Univ. Math. Bull. **48** (1993), no. 6, 32–35.

[210] Yu. V. Kuznetsov and V. V. Yashchenko, *On partial bent functions*, Vestnik MGU, issue 5, 2000, pp. 3–6; English transl., Moscow Univ. Math. Bull. **55** (2000), no. 5, 1–4.

[211] Yu. V. Kuznetsov and V. V. Yashchenko, *On the plateaued Boolean functions*, Proc. XII Intern. School-Semin. "Synthesis and Complexity of Control Systems", Penza, 15–21 October, Part I, Tsenter prikladnykh issledovanii pri mekh.-mat. MGU, 2001, pp. 129–136 (in Russian).

[212] Yu. V. Kuznetsov, *On the number of nondegenerate Boolean forms*, Proc. XI Intern. School-Semin. "Synthesis and Complexity of Control Systems", Nizhnii Novgorod, 20–25 November, Part I, Tsenter prikladnykh issledovanii pri mekh.-mat. MGU, 2001, pp. 105–108 (in Russian).

[213] A. S. Kuzmin, V. L. Kurakin, A. V. Mikhalev, and A. A. Nechaev, *Linear recurring sequences over rings and modules* (Contemporary Math. and its Appl. Surveys, V. 10, Algebra 2, 1994, Moscow), J. of Math. Sciences, V. 76, No. 6, 1995, pp. 2793–2915.

[214] V. G. Labunets and O. P. Sitnikov, *Harmonic analysis of Boolean functions and fuctions of k-valued logic over finite fields*, Tekhnicheskaya kibernetika, issue 1, 1975, pp. 141–148 (in Russian).

[215] G. Lachaud and J. Wolfmann, *The Weights of the Orthogonals of the Extended Quadratic Binary Goppa Codes*, IEEE Trans. on Inform. Theory, V. 36, 1990, pp. 686–692.

[216] X. Lai, J. Massey, and S. Murphy, *Markov Ciphers and Differential Cryptanalysis*, In Proceedings of Advances in Cryptology: EUROCRYPT'91, Lect. Notes in Comp. Sci., New York: Springer-Verlag, V. 547, 1991, pp. 17–38.

[217] X. Lai, *Higher Order Derivatives and Differential Cryptanalysis*, Communications and Cryptography, Kluwer Academic Publishers, 1994, pp. 227–233.

[218] X. Lai, *Additive and Linear Structures of Cryptographic Functions*, Fast Software Encryption, Second International Workshop, Lect. Notes in Comp. Sci., New York: Springer-Verlag, V. 1008, 1995, pp. 75–85.

[219] S. Lang, *Algebra*, Addison-Wesley Publishing Company, Reading, Mass., 1965.

[220] R. J. Lechner, *A Transform Approach to Logic Design*, IEEE Trans. on Computers, C-19, No. 10, 1970, pp. 627–640.

[221] S. Lee, S. Chee, Sa. Park, and Su. Park, *Conditional Correlation Attack on Nonlinear Filter Generators*, In Proceedings of Advances in Cryptology: ASIACRYPT'96, Lect. Notes in Comp. Sci., New York: Springer-Verlag, V. 1163, 1996, pp. 360–367.

[222] W. van Leekwijck and L. Van Linden, *Cryptografische eigenschappen van Boolean functies*, Thesis grad. ESAT Katholieke Universiteit Leuven, 1990.

[223] E. L. Lehmann, *Testing Statistical Hypotheses*, John Wiley, 1959.

[224] V. Levenshtein, *Split orthogonal arrays and maximum independent resilient systems of functions*, Designs, Codes and Cryptography, V. 12, 1997, pp. 131–160.

[225] R. Lidl and H. Niederreiter, *Finite Fields*, Addison-Wesley Publishing Company, Reading, Massachusetts, 1983.

[226] M. Liu, P. Lu, and G. L. Mullen, *Correlation-Immune Functions over Finite Fields*, IEEE Trans. on Inform. Theory, V. 44, No. 3, 1998, pp. 1273–1278.

[227] A. S. Lloyd, *Balance, Uncorrelatedness and the Strict Avalanche Criterion*, Techical Report of Hewlett–Packard Research Laboratories, Bristol, 1989, HPL-ISC-TM-89-012.

[228] A. S. Lloyd, *Characterising and counting functions satisfying Strict Avalanche Criterion of order* $(n-3)$, 2-nd IMA Conference on Cryptography and Coding, 1989.

[229] A. S. Lloyd, *Counting functions satisfying a higher order strict avalanche criterion*, In Proceedings of Advances in Cryptology: EUROCRYPT'89, Lect. Notes in Comp. Sci., New York: Springer-Verlag, V. 434, 1990, pp. 63–74.

[230] A. S. Lloyd, *Properties of Binary Functions*, In Proceedings of Advances in Cryptology: EUROCRYPT'90, Lect. Notes in Comp. Sci., New York: Springer-Verlag, V. 473, 1991, pp. 124–139.

[231] A. S. Lloyd, *Counting binary functions with certain cryptographic properties*, Journal of Cryptology, V. 5, No. 2, 1992, pp. 107–131.

[232] A. Lempel and M. Cohn, *Maximal Families of Bent Sequences*, IEEE Trans. on Inform. Theory, V. 28, No. 6, 1982, pp. 865–868.

[233] A. S. Logachev, *On a recursive decoding algorithm for subsets of first order Reed–Muller codes*, Diskretnaya matematika, vol. 4, issue 2, 1992, pp. 130–135 (in Russian).

[234] O. A. Logachev, A. A. Salnikov, and V. V. Yashchenko, *(Bent) functions over a finite Abelian group*, Diskretnaya matematika, vol. 9, issue 4, 1997, pp. 3–20 (in Russian).

[235] O. A. Logachev, A. A. Salnikov, and V. V. Yashchenko, *Nondegenerate normal form of Boolean functions*, Doklady RAN, vol. 373, issue 2, 2000, pp. 164–167 (in Russian).

[236] O. A. Logachev, A. A. Salnikov, and V. V. Yashchenko, *(Bent) functions and Boolean cube partition*, 12–th International Conference on Formal Power Series and Algebraic Combinatorics FPSAC'00, Supplementary abstracts, Moskva, MSU, 2000, pp. 43–48 (in Russian).

[237] O. A. Logachev, A. A. Salnikov, and V. V. Yashchenko, *Some characteristics of nonlinearity of group mappings*, Diskretnyi analiz i issledovanie operatsii, Series 1, vol. 8, issue 1, 2001, pp. 40–54 (in Russian).

[238] O. A. Logachev, A. A. Salnikov, and V. V. Yashchenko, *Normal form of mappings of finite Abelian groups*, International workshop "Diskretnaya matematika i prilozheniya", 29 January–2 February, 2001, Part III, mekh.-mat. MGU, pp. 315–317 (in Russian).

[239] O. A. Logachev, A. A. Salnikov, and V. V. Yashchenko, *Estimation of some parameters of mappings of finite Abelian groups*, International workshop "Diskretnaya matematika i prilozhenia", 29 January–2 February, 2001, Part III, mekh.-mat. MGU, pp. 318–320 (in Russian).

[240] O. A. Logachev, A. A. Salnikov, and V. V. Yashchenko, *On inheritance of properties under restrictions of Boolean functions*, Diskretnaya matematika, vol. 9, issue 4, 1997, pp. 3–20 (in Russian).

[241] F. J. MacWilliams and N. J. A. Sloane, *The Theory of Error-Correcting Codes*, Amsterdam, New York, Oxford: North-Holland Publishing Company, 1977.

[242] S. Maitra and P. Sarkar, *Enumeration of Correlation Immune Boolean Functions*, 4-th Australasian Conference on Information, Security and Privacy, Lect. Notes in Comp. Sci., New York: Springer-Verlag, V. 1587, 1999, pp. 12–15.

[243] S. Maitra and P. Sarkar, *Highly Nonlinear Resilient Functions Optimizing Siegenthaler's Inequality*, In Proceedings of Advances in Cryptology: CRYPTO'99, Lect. Notes in Comp. Sci, New York: Springer-Verlag, V. 1666, 1999, pp. 198–215.

[244] S. Maitra and P. Sarkar, *Hamming Weights of Correlation Immune Boolean Functions*, Information Processing Letters, V. 71, No. 3–4, 1999, pp. 149–153.

[245] S. Maitra, *Correlation Immune Boolean Functions with Very High Nonlinearity*, `http://www.eprint.iacr.org`, No. 2000/054.

[246] S. Maitra, *Autocorrelation Properties of Correlation Immune Boolean Functions*, In Proceedings of Progress in Cryptology: INDOCRYPT'2001, Lect. Notes in Comp. Sci., New York: Springer-Verlag, V. 2247, 2001, pp. 242–253.

[247] S. Maitra, *Boolean Functions with Important Cryptographic Properties*, Ph. D. Thesis, Indian Statistical Institute, 2001.

[248] S. Maitra and P. Sarkar, *Cryptographically Significant Boolean Functions with Five Valued Walsh Spectra*, Theoretical Computer Science, V. 276, No. 1–2, 2002, pp. 133–146.

[249] S. Maitra and E. Pasalic, *Further Construction of Resilient Boolean Functions with Very High Nonlinearity*, IEEE Trans. on Inform. Theory, V. 48, No. 7, 2002, pp. 1825–1834.

[250] A. I. Maltsev, *Foundations of linear algebra*, W. H. Freeman & Co., San Francisco–London, 1963.

[251] A. A. Malyutin, *Fast correlation decoding of some subsets of first order Reed–Muller codes*, Diskretnaya matematika, vol. 2, issue 2, 1990, pp. 155–158 (in Russian).

[252] Yu. I. Manin, *Cubic forms*, North-Holland, Amsterdam, 1986.

[253] M. Marcus and H. Minc, *A Survey of Matrix Theory and Matrix Inequalities*, Boston: Allyn and Bacon, Inc., 1964.

[254] A. W. Marshall and I. Olkin, *Inequalities: Theory of Majorization and Its Applications*, Academic Press, New York–London–Toronto–Sydney–San-Francisco, 1979.

[255] J. L. Massey, *Threshold Decoding*, MIT Press, Cambridge, Massachusetts, 1963.

[256] J. L. Massey, *Shift-Register syntesis and BCH decoding*, IEEE Trans. on Inform. Theory, IT-17, 1969, pp. 464–466.

[257] S. Matsufuji and K. Imamura, *Real-Valued Bent Function and Its Application to the Design of Balanced Quadriphase Sequences with Optimal Correlation Properties*, Applied Algebra, Algebraic Algorithms and Error-Correcting Codes, 8-th International Conference, AAECC-8, Tokyo, Japan, 1990, Lect. Notes in Comp. Sci., 508, Springer-Verlag, 1991.

[258] M. Matsui and A. Yamagishi, *A new method for known plaintext attack of FEAL cipher*, In Proceedings of Advances in Cryptology: EUROCRYPT'92, Lect. Notes in Comp. Sci., Berlin: Springer-Verlag, V. 658, 1992, pp. 1–91.

[259] M. Matsui, *Linear cryptanalysis method for DES cipher*, In Proceedings of Advances in Cryptology: EUROCRYPT'93, Lect. Notes in Comp. Sci., Springer, V. 765, 1994, pp. 386–397.

[260] M. Matsui, *On Correaltion Between the Order of S-boxes and the Strength of DES*, In Proceedings of Advances in Cryptology: EUROCRYPT'94, Lect. Notes in Comp. Sci., Springer, V. 950, 1994, pp. 366–375.

[261] M. Matsui, *The First Experimental Cryptanalysis of the Data Encryption Standart*, In Advances of Cryptology: CRYPTO'94, Lect. Notes in Comp. Sci., Springer, V. 839, 1995, pp. 1–11.

[262] R. J. McEliece, *On Periodic Sequences from* $GF(q)$, Journal on Combinatorial Theory, Ser. A, V. 10, 1971, pp. 80–91.

[263] R. J. McEliece, *Weight Congruences for p-ary Cyclic Codes*, Discrete Math., V. 3, 1972, pp. 177–192.

[264] R. J. McEliece, *Finite Fields for Computer Scientists and Engineers*, Kluwer Academic Publishers, 2001.

[265] R. L. McFarland, *A Family of Difference Sets in Non-cyclic Groups*, Journal of Combinatorial Theory (A), V. 15, No. 1, 1973, pp. 1–10.

[266] G. McGuire and A. Calderbank, *Proof of Conjecture of Sarwate and Pursley Regarding Pairs of Binary m-sequences*, IEEE Trans. on Information Theory, V. 41, No. 4, 1995, pp. 1153–1155.

[267] W. Meier and O. Staffelbach, *Nonlinearity Criteria for Cryptographic Functions*, In Proceedings of Advances in Cryptology: EUROCRYPT'89, Lect. Notes in Comp. Sci., New York: Springer-Verlag, V. 434, 1990, pp. 549–562.

[268] W. Meier and O. Staffelbach, *Fast Correlation Attacks on certain Stream Ciphers*, Journal of Cryptology, V. 1, 1989, pp. 159–176.

[269] A. Menezes, P. van Oorschot, and S. Vanstone, *Handbook of Applied Cryptography*, CRC Press, 1996.

[270] J. M. Mihaljevic, M. P. C. Fossorier, and H. Imai, *A Low-Complexity and High-Performance Algorithm for the Fast Correlation Attack*, In Proceedings of Fast Software Encryption'2000, Lect. Notes in Comp. Sci., New York: Springer-Verlag, V. 1978, 2001, pp. 196–212.

[271] V. G. Mikhailov, *On the number of preimages of an automaton output sequence*, Obozrenie prikladnoi i promyshlennoi matematiki, Moskva, vol. 1, issue 1, 1994, pp. 118–121 (in Russian).

[272] V. G. Mikhailov, *Generalization of the theorem on the number of preimages of an automaton output sequence*, Obozrenie prikladnoi i promyshlennoi matematiki, Moskva, vol. 1, issue 1, 1994, pp. 122–125 (in Russian).

[273] V. G. Mikhailov, *Asymptotic normality of the number of preimages of an automaton output sequence*, Obozrenie prikladnoi i promyshlennoi matematiki, Moskva, vol. 1, issue 1, 1994, pp. 126–135 (in Russian).

[274] V. G. Mikhailov and V. P. Chistyakov, *On problems of finite automata theory connected with the number of preimages of an output sequence*, Obozrenie prikladnoi i promyshlennoi matematiki, Moskva, vol. 1, issue 1, 1994, pp. 7–31 (in Russian).

[275] W. Millan, A. Clark, and E. Dawson, *Heuristic Design of Cryptographically Strong Balanced Boolean Functions*, In Proceedings of Advances in Cryptology: EUROCRYPT'98, Lect. Notes in Comp. Sci., New York: Springer-Verlag, V. 1403, 1998, pp. 489–499.

[276] K. Miranovich, *Spectral Analysis of Boolean Functions under Non-Uniformity of Arguments.*

[277] C. J. Mitchell, *Enumerating Boolean Functions of Cryptographic Significance*, Journal of Cryptology, V. 2, No. 3, 1990, pp. 155–170.

[278] S. P. Mo, L. Sangjin, and K. Kwangjo, *Improving Bound for the Number of Correlation Immune Boolean Functions*, Information Processing Letters, V. 61, No. 4, 1997, pp. 209–212.

[279] D. E. Muller, *Application of Boolean Algebra to Swithing Circuit Design and to Error Detection*, IRE Transactions on Electronic Computers, V. 3, EC-1954, pp. 6–12.

[280] S. Murphy, *The cryptanalysis of FEAL-4 with 20 chosen plaintexts*, Journal of Cryptology, V. 3, No. 2, 1990, pp. 145–154.

[281] S. Murphy, P. Piper, M. Walker, and P. Wild, *Likelihood estimation for block cipher keys*, Technical report, Information Security Group, Royal Holloway, University of London, 1995.

[282] J. J. Mykkeltveit, *The Covering Radius of the* (128, 8) *Reed–Muller Code is 56*, IEEE Trans. on Inform. Theory, IT-26, No. 3, 1983, pp. 358–362.

[283] Mulan Liu, Peizhong Lu, and G. L. Mullen, *Correlation-Immune Functions over Finite Fields*, IEEE Trans. on Inform. Theory, V. 44, No. 3, 1998, pp. 1273–1278.

[284] P. Naudin and C. Quitté, *Algoritmique Algébrique (avec exercices corrigés)*, Paris, Milan, Barcelone, Bonn: MASSON, 1992.

[285] P. G. Nigmatulin, *Boolean functions complexity*, Moskva, Nauka, 1991 (in Russian).

[286] Y. Niho, *Multi-Valued Cross-Correlation Functions between Two Maximal Linear Recursive Sequences*, Ph.D. Thesis, USCEE Rep., 1972.

[287] V. A. Nosov, *Regularity criterion for a Boolean nonautonomous automaton with divided input*, Intellektualnye sistemy, Moskva, MGU, Russian Technological Science Academy, vol. 3, issue 3-4, 1998, pp. 269–280 (in Russian).

[288] V. A. Nosov, *Construction of classes of latin squares in Boolean data base*, Intellektualnye sistemy, Moskva, MGU, Russian Technological Science Academy, vol. 4, issue 3-4, 1999, pp. 307–320 (in Russian).

[289] K. Nyberg, *Constructions of Bent Functions and Difference Sets*, In Proceedings of Advances in Cryptology: EUROCRYPT'90, Lect. Notes in Comp. Sci., New York: Springer-Verlag, V. 473, 1991, pp. 151–160.

[290] K. Nyberg, *Perfect nonlinear S-boxes*, In Proceedings of Advances in Cryptology: EUROCRYPT'91, Lect. Notes in Comp. Sci., New York: Springer-Verlag, V. 547, 1991, pp. 378–386.

[291] K. Nyberg, *On the Construction of Higly Nonlinear Permutations*, In Proceedings of Advances in Cryptology: EUROCRYPT'92, Lect. Notes in Comp. Sci., New York: Springer-Verlag, V. 658, 1993, pp. 92–98.

[292] K. Nyberg and L. Knudsen, *Provable Security Against Differential Cryptanalysis*, In Proceedings of Advances in Cryptology: CRYPTO'92, Lect. Notes in Comp. Sci., New York: Springer-Verlag, V. 740, 1993, pp. 566–574.

[293] K. Nyberg, *Differentially Uniform Mappings for Cryptography*, In Proceedings of Advances in Cryptology: EUROCRYPT'93, Lect. Notes in Comp. Sci., New York: Springer-Verlag. V. 765. pp. 55–64.

[294] K. Nyberg, *New Bent Mappings Suitable for Fast Implementation*, Fast Software Encryption, Lect. Notes in Comp. Sci., New York: Springer-Verlag, V. 809, 1993, pp. 179–184.

[295] K. Nyberg, *Linear Approximation of Block Ciphers*, In Proceedings of Advances in Cryptology: EUROCRYPT'94, Lect. Notes in Comp. Sci., New York: Springer-Verlag, V. 950, 1994, pp. 439–444.

[296] K. Nyberg, *S-boxes and Round Functions with Controllable Linearity and Differential Uniformity*, Fast Software Encryption Second International Workshop, Leuven, Belgium, 1994, Lect. Notes in Comp. Sci., New York: Springer-Verlag, 1994. V. 1008. pp. 111–130.

[297] K. Nyberg, *S-Boxes and Round Functions with Controllable Linearity and Differential Uniformity*, Fast Software Encryption, Second International Workshop, Lect. Notes in Comp. Sci. New York: Springer-Verlag, V. 1008, 1995, pp. 111–130.

[298] L. J. O'Connor, *Enumeration Nondegenerate Permutations*, In Proceedings of Advances in Cryptology: EUROCRYPT'91, Lect. Notes in Comp. Sci., New York: Springer-Verlag, V. 547, 1992, pp. 368–377.

[299] L. J. O'Connor, *An Analysis of Product Ciphers based on the Properties of Boolean Functions*, Ph.D. Dissertation, University of Waterloo, Waterloo, Ontario, Canada, 1992, p. 171.

[300] J. D. Olsen, R. A. Scholtz, and L. R. Welch, *Bent-Function Sequences*, IEEE Trans. on Inform. Theory, V. 28, No. 6, 1982, pp. 858–864.

[301] S. Palit and K. Roy, *Cryptanalysis of LFSR-Encryption Codes with Unknown Combining Function*, In Proceedings of Advances in Cryptology: ASIACRYPT'99, Lect. Notes in Comp. Sci., New York: Springer-Verlag, V. 1716, 1999, pp. 306–320.

[302] S. Park, S. Lee, S. Sung, and K. Kim, *Improving Bounds for the Number of Correlation-Immune Boolean Functions*, Information Processing Letters, V. 61, 1997, pp. 209–212.

[303] E. Pasalic and T. Johansson, *Further Results on the Relation Between Nonlinearity and Resiliency of Boolean Functions*, IMA Conference on Cryptography and Coding, Lect. Notes in Comp. Sci., New York: Springer-Verlag, V. 1746, 1999, pp. 35–44.

[304] E. Pasalic, S. Maitra, T. Johansson, and P. Sarkar, *New Constructions of Resilient and Correlation Immune Boolean Functions Achieving Upper Bounds on Nonlinearity*, Workshop on Coding and Cryptography: WCC'2001, Paris, Electronic Notes in Discrete Mathematics, New York: Elsevier Science, V. 6, 2001.

[305] E. Pasalic and S. Maitra, *Linear Codes in Constructing Resilient Functions with High Nonlinearity*, Selected Areas in Cryptography: SAC'2001, Lect. Notes in Comp. Sci., New York: Springer-Verlag, V. 2259, 2001, pp. 60–74.

[306] E. Pasalic and S. Maitra, *A Majorana–MacFarland Type Construction for Resilient Boolean Functions on n Variables (n Even) with Nonlinearity* $> 2^{n-1}-2^{n/2}+2^{n/2-2}$, Proceedings of International Workshop on Coding and Cryptography, March, 24–28, Versailles (France), 2003, pp. 365–374.

[307] N. J. Patterson and D. H. Wiedemann, *The Covering Radius of the* $(2^{15}, 16)$ *Reed–Muller Code is at least* 16276, IEEE Trans. on Inform. Theory, IT-29, No. 3, 1983, pp. 354–356.

[308] N. J. Patterson and D. H. Wiedemann, *Correction to "The Covering Radius of the* $(2^{15}, 16)$ *Reed–Muller Code is at least* 16276", IEEE Trans. on Inform. Theory, IT-36, No. 2, 1990, p. 443.

[309] W. W. Peterson and E. J. Jr. Weldon, *Error-Correcting Codes*, MIT Press, Cambridge, Massachusetts, London, England, 1972.

[310] J. P. Pieprzyk, *Error Propagation Property and Application in Cryptography*, IEE Proc., Part E, V. 136, No. 4, 1989, pp. 262–270.

[311] J. P. Pieprzyk, *Nonlinearity of exponent permutations*, In Proceedings of Advances in Cryptology: EUROCRYPT'89, Lect. Notes in Comp. Sci., New York: Springer-Verlag, V. 434, 1990, pp. 80–92.

[312] J. P. Pieprzyk, *On bent permutations*, Technical Report: Department of Computer Science, The University of New South Wales, CS91/11, 1991.

[313] J. P. Pieprzyk and G. Finkelstein, *Towards effective nonlinear cryptosystem design*, IEE Proceedings, part E: Computers and Digital Techniques, November, 6. Department of Computer Science, University of New South Wales, Australian Defence Force Academy, Canberra, ACT 2600, Australia, V. 135, Series E, 1988, pp. 325–335.

[314] V. Pless, *Power Moment Identities on Weight Distributions in Error-Correcting Codes*, Information and Control, V. 6, 1963, pp. 147–152.

[315] *Handbook on Coding Theory*, V. I, II, V. S. Pless and W. C Huffman, Eds., Amsterdam–New York–Oxford–Tokyo: Elsevier, 1998.

[316] G. N. Povarov, *On group invariance of Boolean functions*, Primenenie logiki v tekhnike, Moskva, Akad. Nauk SSSR, 1961, pp. 263–340 (in Russian).

[317] M. M. Postnikov, *Lectures on geometry*, Second semester, Linear algebra, Moskva, Nauka, 1986 (in Russian).

[318] B. Preneel, W. VanLeekwijck, L. Van Linden, R. Govaerts, and J. VanDewalle, *Propagation Characteristics of Boolean Functions*, In Proceedings of Advances in Cryptology: EUROCRYPT'90, Lect. Notes in Comp. Sci., New York: Springer-Verlag, V. 473, 1991, pp. 161–173.

[319] B. Preneel, R. Govaerts, and J. Vandewalle, *Boolean Functions Satisfying Higher Order Propagation Criteria*, In Proceedings of Advances in Cryptology: EUROCRYPT'91, Lect. Notes in Comp. Sci., New York: Springer-Verlag, V. 541, 1991, pp. 141–152.

[320] B. Preneel, *Analysis and Design of Cryptographic Hash Functions*, Ph.D. Thesis, Katholieke Universiteit Leuven, K. Merierlaan 94, 3001 Leuven, Belgium, 1993.

[321] F. P. Preparata, *State-Logic Relations for Autonomous Sequential Networks*, IEEE Trans. Electronic Computers, V. 13, No. 5, 1964, pp. 542–548.

[322] F. P. Preparata, *Convolutional Transformations of Binary Sequences: Boolean Functions and Their Resynchronizing Properties*, IEEE Trans. Electron. Comp., V. 15, No. 6, 1966, pp. 398–409.

[323] I. S. Reed, *A Class of Multiple-Error-Correcting Codes and the Decoding Scheme*, IRE Trans. on Inform. Theory, IT-4, 1954, pp. 38–49.

[324] M. J. B. Robshaw, *Stream Ciphers*, RSA Laboratories, Technical Report TR-701, 1995.

[325] F. Rodier, *On the Nonlinearity of Boolean Functions*, Proceedings of International Workshop on Coding and Cryptography, March, 24–28, Versailles (France), 2003, pp. 397–405.

[326] O. S. Rothaus, *On Bent Functions*, Journal of Combinatorial Theory (A), V. 20, No. 3, 1976, pp. 300–305.

[327] B. Roy, *A brief outline of research on correlation immune functions*, In Information security and privacy: 7-th Australasian conference, ACISP 2002, Melbourne, Australia, July 3–5, 2002, Lect. Notes in Comp. Sci., V. 2384, 2002, pp. 379–394.

[328] R. A. Rueppel, *Analysis and Design of Stream Ciphers*, New York: Springer-Verlag, 1986.

[329] R. A. Rueppel, *Stream Ciphers*, In Contemporary Cryptography: the Science of Information Integrity, Ch. 2, IEEE Press, 1992, pp. 65–134.

[330] B. V. Ryazanov, *On the distribution of spectral complexity of Boolean functions*, Diskretnaya matematika, vol. 6, issue 2, 1994, pp. 111–119 (in Russian).

[331] B. V. Ryazanov and S. I. Checheta, *On the approximation of a random Boolean function by a set of quadratic forms*, Diskretnaya matematika, vol. 7, issue 3, 1995, pp. 129–145 (in Russian).

[332] Yu. L. Sagalovich, *On group invarinace of Boolean functions*, Uspekhi matematicheskikh nauk, vol. 14, issue 6(90), 1959, pp. 191–195 (in Russian).

[333] P. Sarkar and S. Maitra, *Construction of Nonlinear Boolean Functions with Important Cryptographic Properties*, In Proceedings of Advances in Cryptology: EUROCRYPT'2000, Lect. Notes in Comp. Sci., New York: Springer-Verlag, V. 1807, 2000, pp. 485–506.

[334] P. Sarkar and S. Maitra, *Nonlinearity Bounds and Constructions of Resilient Boolean Functions with Important Cryptographic Properties*, In Proceedings of Advances in Cryptology: CRYPTO'2000, Lect. Notes in Comp. Sci., New York: Springer-Verlag, V. 1880, 2000, pp. 515–532.

[335] P. Sarkar, *A Note on the Spectral Characterization of Correlation Immune Boolean Functions*, Information Processing Letters, V. 74, No. 5–6, 2000, pp. 191–195.

[336] P. Sarkar and S. Maitra, *Balancedness and Correlation Immunity of Symmetric Boolean Functions*, Preprint, 2000.

[337] P. Sarkar and S. Maitra, *Cross-Correlation Analysis of Cryptographically Useful Boolean Functions and S-boxes*, Theory of Computing Systems, V. 35, No. 1, 2002, pp. 39–57.

[338] D. Sarwate and M. Pursley, *Crosscorrelation Properties of Pseudorandom and Related Sequences*, Proc. IEEE, V. 68, 1980, pp. 593–619.

[339] P. Savicky, *On the bent Boolean functions that are symmetric*, European Journal of Combinatorics, V. 15, No. 4, 1994, pp. 407–410.

[340] P. Savicky, *Bent functions and random Boolean formulas*, Discrete Mathematics, V. 147, 1995, pp. 1–3.

[341] W. G. Schneeweiss, *On the Polynomial Form of Boolean Functions: Derivations and Applications*, IEEE Trans. on Computers, V. 47, No. 2, 1998, pp. 217–221.

[342] M. Schneider, *Note on the Construction and Upper Bounds of Correlation-Immune Functions*, 6-th IMA Conference, 1997, pp. 295–306.

[343] J. Seberry, X.-M. Zhang, and Y. Zheng, *Nonlinearly Balanced Boolean Functions and Their Propagation Characteristics*, Advances in Cryptology: CRYPTO'93, Lect. Notes in Comp. Sci., New York: Springer-Verlag, V. 773, 1994, pp. 49–60.

[344] J. Seberry, X.-M. Zhang, and Y. Zheng, *On the Constructions and Nonlinearity of Correlation Immune Boolean Functions*, Advances in Cryptology: EUROCRYPT'93, Lect. Notes in Comp. Sci., New York: Springer-Verlag, V. 765, 1994, pp. 181–199.

[345] J. Seberry, X.-M. Zhang, and Y. Zheng, *Relationships Among Nonlinearity Criteria*, In Proceedings of Advances in Cryptology: EUROCRYPT'94, Lect. Notes in Comp. Sci., New York: Springer-Verlag, V. 950, 1995, pp. 376–388.

[346] J. Seberry, X.-M. Zhang, and Y. Zheng, *Improving the Strict Avalanche Characteristics of Cryptographic Functions*, Information Processing Letters, V. 50, 1994, pp. 37–41.

[347] J. Seberry, X.-M. Zhang, and Y. Zheng, *Nonlinearity and propagation characteristics of balanced Boolean functions*, Information and Computation, V. 119, 1995, pp. 1–13.

[348] J. Seberry, X.-M. Zhang, and Y. Zheng, *The relationship Between Propagation Characteristics and Nonlinearity of Cryptographic Functions*, Journal of Universal Computer Science, V. 1, No. 2, 1995, pp. 136–150.

[349] J. Seberry and X.-M. Zhang, *Highly nonlinear* 0-1 *balanced Boolean functions satisfying strict avalanche criterion*, In Proceedings of Advances in Cryptology: AUSCRYPT'92, Lect. Notes in Comp. Sci., New York: Springer-Verlag, V. 718, 1993, pp. 145–155.

[350] A. A. Selçuk, *On probability of success in linear and differential cryptanalysis*, In Proceedings of Security in Communication Networks: Third International Conference, SCN 2002, Amalfi, Italy, September 11–13, 2002 (S. Cimato, C. Galdi, G. Persiano, eds.), Lect. Notes in Comp. Sci., Springer-Verlag, V. 2576, 2002, pp. 174–185.

[351] N. V. Semakov, V. A. Zinoviev, and G. V. Zaitsev, *Uniformly Packed Codes*, Problems of information transmission, vol. 7, issue 1, 1971, pp. 30–39.

[352] J. P. Serre, *Cours D'Arithmétique*, Paris: Presses Universitaires de France, 1970.

[353] B. A. Sevastyanov, *Course on probability theory and matematical statistics*, Moskva, Nauka, 1982 (in Russian).

[354] B. A. Sevastyanov and V. P. Chistyakov, *On the number of input sequences corresponding to the output sequences of a finite automaton*, Obozrenie prikladnoi i promyshlennoi matematiki, Moskva, vol. 1, issue 1, 1994, pp. 96–107 (in Russian).

[355] C. E. Shannon, *Communication theory of secrecy systems*, Bell System Technical Journal, V. 28, 1949, pp. 656–715.

[356] W. Shan, *The Structure and the Construction of Correlation Immune Functions*, MS Thesis, NTE Institute, Xian, 1987.

[357] V. Y. Shen, A. McKellar, and P. Weiner, *A Fast Algorithm for the Disjunctive Decomposition on Switching Functions*, IEEE Trans. on Computers, V. 20., No. 3, 1971, pp. 304–309.

[358] Shestakov, Ed., *Synthesis of electronic computing and controlling schemes*, Moskva, 1954 (in Russian).

[359] V. M. Sidelnikov, *On the mutual correlation of sequences*, Problemy kibernetiki, A. A. Lyapunov, Ed., Moskva, Nauka, issue 24, 1971, pp. 15–42 (in Russian).

[360] V. M. Sidelnikov and A. S. Pershakov, *Decoding of Reed–Muller Codes with a Large Number of Errors*, Problems of information transmission, vol. 28, issue 3, 1992, pp. 269–281.

[361] V. M. Sidelnikov, *Fast algorithms for constructing labeling set for arrays of discrete information*, Russian Academy of Sciences, Cryptographic Academy of Russian Federation, Trudy po diskretnoi matematike, Moskva, vol. 1, 1997, pp. 251–264 (in Russian).

[362] T. Siegenthaler, *Correlation-immunity of Nonlinear Combining Functions for Cryptographic Applications*, IEEE Trans. on Inform. Theory, IT-30, No. 5, 1984, pp. 776–780.

[363] T. Siegenthaler, *Design of Combiners to Prevent Divide and Conquer Attacks*, In Proceedings of Advances in Cryptology: CRYPTO'85, Lect. Notes in Comp. Sci., New York: Springer-Verlag, V. 218, 1986, pp. 273–279.

[364] T. Siegenthaler, *Cryptoanalysis of Nonlinearly Fitered ML-Sequences*, Advances in Cryptology: EUROCRYPT'85, Lect. Notes in Comp. Sci., New York: Springer-Verlag, 1988, pp. 103–110.

[365] T. Siegenthaler, *Decrypting a Class of Stream Ciphers Using Ciphertext Only*, IEEE Trans. on Computers, V. 34, No. 1, C-1985, pp. 81–85.

[366] H.-U. Simon, *A tight $\Omega(\log\log n)$-bound on the time for parallel RAM's to compute nondegenerated boolean functions*, FCT'83, Lect. Notes in Comp. Sci., V. 158, 1984, pp. 439–444.

[367] *Pure algebra. A reference matematical library*, L. A. Skornyakov, Ed., vols. 1, 2, Moskva, Nauka, 1990 (in Russian).

[368] V. I. Solodovnikov, *(Bent)-functions from a finite Abelian group to a finite Abelian group*, Discretnaya matematika, vol. 14, issue 1, 2002, pp. 99–113 (in Russian).

[369] D. R. Stinson, *Resilient Functions and Large Sets of Orthogonal Arrays*, Congressus Numerantium, 1993, V. 92, pp. 105–110.

[370] D. R. Stinson and J. L. Massey, *An Infinite Class of Counterexamples to a Conjecture Concerning Non-linear Resilient Functions*, Journal of Cryptology, V. 8, No. 3, 1995, pp. 167–173.

[371] I. Stradzin, *Invariant groups of Boolean functions of four variables*, Avtomatika i vychislitelnaya tekhnika, number 5, 1968, pp. 18–22 (in Russian).

[372] I. Stradzin, *Affine classification of Boolean functions of five variables*, Avtomatika i vychislitelnaya tekhnika, number 1, 1975, pp. 1–9 (in Russian).

[373] R. L. Stratonovich, *Information theory*, Moskva, Sovetskoe radio, 1975 (in Russian).

[374] T. Sugita, T. Kasami, and T. Fujiwara, *The Weight Distribution of the Third Order Reed–Muller Codes of Length* 512, IEEE Trans. on Inform. Theory, V. 42, No. 5, 1996, pp. 1622–1625.

[375] S. N. Sumarokov, *Prohibitions of binary functions and reversibility for a class of coding devices*, Obozrenie prikladnoi i promyshlennoi matematiki, number 1, 1994, pp. 33–55 (in Russian).

[376] D. A. Suprunenko, *Permutation groups*, Minsk, Nauka i tekhnuka, 1996 (in Russian).

[377] Yu. V. Tarannikov, *On the sets of l-balalnced Boolean sets and functions*, Ph.D. Thesis, Moskva, 1994 (in Russian).

[378] Yu. V. Tarannikov, *On the weight of l-balalnced Boolean functions*, Diskretnyi analiz i issledovanie operatsii, vol. 3, issue 1, 1995, pp. 80–81 (in Russian).

[379] Yu. V. Tarannikov, *On some estimations of the weight of l-balalnced Boolean functions*, Diskretnyi analiz i issledovanie operatsii, vol. 2, issue 4, 1995, pp. 80–96 (in Russian).

[380] Yu. Tarannikov, *On certain bounds for the weight of l-balanced Boolean functions*, Mathematics and Its Applications, A. D. Korshunov (ed.), Operation Research and Discrete Analysis, V. 391, 1997, pp. 285–299. `http://ultra.inria.msu.ru/papers.html`.

[381] Yu. V. Tarannikov, *On the class of Boolean functions uniformly distributed over balls with degree* 1, Vestnik Mosk. Univ., Series 1, number 5, 1997, pp. 17–21; English transl., Moscow Univ. Math. Bull. **52** (1997), no. 5, 18–22.

[382] Yu. Tarannikov, *Limit values for the density of l-balanced k-valued functions defined over the Boolean cube*, International Symposium on Combinatorial Optimization, Bruxelles, April 15–17, 1998, p. 191.

[383] Yu. Tarannikov, *Ramsey-like theorems on the structure and numbers of higher order correlation-immune functions*, Moscow State University, French-Russian Institute of Applied Mathematics and Informatics, Preprint No. 5, Moscow, October 1999, 20 pp. `http://liapunov.inria.msu.ru/PERSONAL/Taran/index.html`.

[384] Yu. Tarannikov, *On a method for the constructing of cryptographically strong Boolean functions*, Moscow State University, French-Russian Institute of Applied Mathematics and Informatics, Preprint No. 6, Moscow, October 1999, 24 p. `http://liapunov.inria.msu.ru/PERSONAL/Taran/index.html`.

[385] Yu. V. Tarannikov, *On the structure and number of correlation-immune functions of the highest orders*, IX Internat. School-Sem. “Synthesis and Complexity of Control Systems”, Nizhnii Novgorod, 16–19 December, 1998, Moskva, Izd. MGU, 1999, pp. 81–92 (in Russian).

[386] Yu. Tarannikov, *On resilient Boolean functions with maximum possible nonlinearity*, Cryptology ePrint Archive (`http://eprint.iacr.org/`), Report 2000/005, March 2000, 18 pp.

[387] Yu. Tarannikov, *On some connections between codes and cryptographic properties of Boolean functions*, Proceedings of Seventh International Workshop on Algebraic and Combinatorial Coding Theory, Bansko, Bulgaria, June 18–24, 2000, pp. 299–304.

[388] Yu. Tarannikov, *On the structure and numbers of higher order correlation-immune functions*, Proceedings of 2000 IEEE International Symposium on Information Theory ISIT2000, Sorrento, Italy, June 25–30, 2000, p. 185.

[389] Yu. Tarannikov, *On resilient Boolean functions with maximal possible nonlinearity*, Proceedings of Indocrypt 2000, Lect. Notes in Comp. Sci., Springer-Verlag, V. 1977, 2000, pp. 19–30.

[390] Yu. Tarannikov and D. Kirienko, *Spectral analysis of high order correlation immune functions*, Cryptology ePrint Archive (`http://eprint.iacr.org/`), Report 2000/050, October 2000, 8 pp.

[391] Yu. Tarannikov, *New constructions of resilient Boolean functions with maximal nonlinearity*, Cryptology ePrint Archive (`http://eprint.iacr.org/`), Report 2000/069, December 2000, 11 pp.

[392] Yu. Tarannikov, *New constructions of resilient Boolean functions with maximal nonlinearity*, 8th Fast Software Encryption Workshop, Preproceedings, Yokohama, Japan, April 2–4, 2001, pp. 70–81.

[393] Yu. Tarannikov and D. Kirienko, *Spectral analysis of high order correlation immune functions*, Proceedings of 2001 IEEE International Symposium on Information Theory ISIT'2001, Washington, DC, USA, June 2001, p. 69.

[394] Yu. Tarannikov, P. Korolev, and A. Botev, *Autocorrelation coefficients and correlation immunity of Boolean functions*, Proceedings of Asiacrypt 2001, Gold Coast, Australia, December 9–13, 2001, Lect. Notes in Comp. Sci., Springer-Verlag, V. 2248, 2001, pp. 460–479.

[395] Yu. V. Tarannikov, *Number characteristics of Boolean functions*, Discrete Math. and Its Appl., Collection of Lectures at Schools for Young Scientists on Discrete Math. and Its Appl., Moskva, Izdat. MGU, Part 1, 2001, pp. 129–144 (in Russian).

[396] Yu. V. Tarannikov, *On the autocorrelation properties of correlation-immune functions*, Proc. VII Intern. Sem. "Discrete Math. and Its Appl.", 29 January–2 February, 2001, Moskva, Izdat. MGU, Part 3, pp. 331–333 (in Russian).

[397] Yu. V. Tarannikov, *On the correlation-immunc and resilient Boolean functions*, Matematicheskie voprosy kibernetiki, Moskva, Fizmatlit, issue 11, 2002, pp. 91–148 (in Russian).

[398] H. C. A van Tilborg, *On Weight in Codes*, Technical Report 71-WSK-03, Department of Mathematics, Technological University of Eindhoven, Netherlands, 1971.

[399] V. D. Tonchev, *Combinatorial Configurations*, New York: Longman, Wiley, 1988.

[400] O. N. Vasilenko, *Number-theoretic algorithms in cryptography*, Moskva, 2003 (in Russian).

[401] L. Yu. Vasiliev and L. Yu. Glagolev, *Metrical properties of disjunctive normal forms*, Diskretnaya matematika i matematicheskie voprosy kibernetiki, V. C. Jablonskii and B. O. Lupanov, Eds., vol. 1, Moskva, Nauka, 1974, pp. 99–148 (in Russian).

[402] S. Vaudenay, *On the weak keys of Blowfish*, In Proceedings of Fast Software Encryption, FSE'96 (D. Gollmann, ed.), Lect. Notes in Comp. Sci., Springer-Verlag, V. 1039, 1996, pp. 27–32.

[403] È. B. Vinberg and A. G. Elashvili, *Classification of three-vectors in the nine-dimensional space*, Trudy seminara po vektornomu i tenzornomu analizu, Moskva, MGU, 1974, pp. 197–233 (in Russian).

[404] I. M. Vinogradov, *Elements of number theory*, Dover, New York, 1954.

[405] F. S. Vinokurov and N. A. Peryazev, *Polynomial decomposition of Boolean functions*, Matem. Zametki, vol. 53, issue 2, 1993, pp. 25–29; English transl., Math. Notes **53** (1993), no. 1–2, 130–133.

[406] B. L. van der Waerden, *Algebra I*, New York: Springer-Verlag, 1991; *Algebra II*, New York: Springer-Verlag, 1991.

[407] A. F. Webster and S. E. Tavares, *On the Design of S-Boxes*, In Proceedings of Advances in Cryptology: CRYPTO'85, Lect. Notes in Comp. Sci., New York: Springer-Verlag, V. 218, 1986, pp. 523–534.

[408] E. J. Weldon, Jr., *New Generalizations of Reed–Muller Codes. Part II: Nonprimitive Codes*, IEEE Trans. on Inform. Theory, IT-14, No. 2, 1968, pp. 199–205.

[409] R. Westwick, *Irreducible Length of Trivectors of Rank Seven and Eight*, Pacific Journal of Mathematics, V. 80, No. 2, 1979, pp. 575–579.

[410] H. Wielandt, *Finite Permutation Groups*, Academic Press, 1964.

[411] S. V. Yablonskii, G. P. Gavrilov, and V. B. Kudryavtsev, *Functions of the algebra of logic and the Post classes*, Moskva, Nauka, 1966 (in Russian).

[412] S. V. Yablonskii, *Introduction to the theory of functions of k-valued logic*, Diskretnaya matematika i matematicheskie voprosy kibernetiki, Moskva, Nauka, vol. 1, 1974.

[413] Y. X. Yang and B. Guo, *Further Enumerating Boolean Functions of Cryptographic Significance*, Journal of Cryptology, V. 8, No. 3, 1995, pp. 115–122.

[414] R. Yarlagadda and J. E. Hershey, *Analysis and synthesis of bent sequences*, Proc. IEE, part E, V. 136, No. 2, 1989, pp. 112–123.

[415] V. V. Yashchenko, *Properties of Boolean mappings that are reducible to their coordinate functions*, Vestnik MGU, Matematika, 1997, no. 4, pp. 11–13; English transl., Moscow Univ. Math. Bull. **52** (1997), no. 4, 11–13.

[416] V. V. Yashchenko, *On the Propagation Criterion for Boolean Functions and on Bent Functions*, Problems of information transmission, vol. 33, issue 1, 1997, pp. 62–71.

[417] V. V. Yashchenko, *On the two characteristics of nonlinearity of Boolean mappings*, Diskretnyi analiz i issledovanie operatsii, Series 1, vol. 5, number 2, 1998, pp. 90–96 (in Russian).

[418] K. Yosida, *Functional Analysis*, Berlin: Springer-Verlag, 1965.

[419] A. M. Youssef and S. E. Tavares, *Spectral Properties and Information Leakage of Multi-Output Boolean Functions*, Proc. 1995 IEEE International Symposium on Information Theory, p. 351.

[420] A. M. Youssef, T. W. Cusick, P. Stănică, and S. E. Tavares, *New bounds on the number of functions satisfying strict avalanche criterion*, Third Annual Workshop on Selected Areas in Cryptography, 1996.

[421] A. Youssef and G. Gong, *Hyper-bent functions*, In Proceedings of Advances in Cryptology: EUROCRYPT'2001, Lect. Notes in Comp. Sci., New York: Springer-Verlag, V. 2045, 2001, pp. 406–419.

[422] L. A. Zalmanzon, *Fourier, Walsh, and Haar transforms and their application in control, communication, and other areas*, Moskva, Nauka, 1989 (in Russian).

[423] O. Zariski and P. Samuel, *Commutative Algebra, Vols. I, II.*, D. Van Nostrand Company Inc., Princeton, 1958.

[424] G. V. Zaitsev, V. A. Zinoviev, and N. V. Semakov, *Fast correlation decoding of block codes*, Kodirovanie i peredacha diskretnykh soobshchenii v sistemakh svyazi, Moskva, Nauka, 1976 (in Russian).

[425] K. Zeng and M. Huang, *On the Linear Syndrome Method in Cryptanalysis*, In Proceedings of Advances in Cryptology: CRYPTO'88, Lect. Notes in Comp. Sci., New York: Springer-Verlag, V. 403, 1990, pp. 469–478.

[426] K. Zeng, C. H. Yang, and T. R. N. Rao, *An Improved Linear Syndrome Algorithm in Cryptanalysis with Applications*, In Proceedings of Advances in Cryptology: CRYPTO'90, Lect. Notes in Comp. Sci., New York: Springer-Verlag, V. 537, 1991, pp. 34–47.

[427] X.-M. Zhang and Y. Zheng, *GAC—the Criterion for Global Avalanche Characteristics of Cryptographic Functions*, Journal of Universal Computer Science, V. 1, No. 5, 1995, pp. 320–337.

[428] X.-M. Zhang and Y. Zheng, *Auto-Correlations and New Bounds on the Nonlinearity of Boolean Functions*, In Proceedings of Advances in Cryptology: EUROCRYPT'96, Lect. Notes in Comp. Sci., New York: Springer-Verlag, V. 1070, 1996, pp. 294–305.

[429] X.-M. Zhang and Y. Zheng, *Characterizing the structures of cryptographic functions satisfying the propagation criterion for almost all vectors*, Designs, Codes and Cryptography, V. 7, 1996, pp. 111–134.

[430] X.-M. Zhang and Y. Zheng, *On the Difficulty of Constructing Cryptographically Strong Substitution Boxes*, Journal of Universal Computer Science, V. 2, No. 3, 1996, pp. 147–162.

[431] X.-M. Zhang and Y. Zheng, *New Lower Bounds on Nonlinearity and a Class of High Nonlinear Functions*, Information Security and Privacy: ACISP'97, Lect. Notes in Comp. Sci., New York: Springer-Verlag, V. 1270, 1997, pp. 147–158.

[432] X.-M. Zhang and Y. Zheng, *Cryptographically Resilient Functions*, IEEE Trans. on Inform. Theory, V. 43, No. 5, 1997, pp. 1740–1747.

[433] X.-M. Zhang, Y. Zheng, and H. Imai, *Connections Between Nonlinearity and Restrictions, Terms and Hypergraphs of Boolean Functions*, ISIT, Cambridge, MA, USA, 1998, p. 439.

[434] J.-Z. Zhang, Z.-S. You, and Z. L. Li, *Enumeration of Binary Orthogonal Arrays of Strength* 1, Discrete Mathematics, to appear.

[435] Y. Zheng and X.-M. Zhang, *Improved Upper Bounds on Nonlinearity of High Order Correlation Immune Functions*, Selected Areas in Cryptography: SAC'2000, Lect. Notes in Comp. Sci., New York: Springer-Verlag, V. 2012, 2000, pp. 264–274.

[436] Y. Zheng and X.-M. Zhang, *On Relationships among Propagation Degree, Nonlinearity, and Correlation Immunity*, In Proceedings of Advances in Cryptology: ASIACRYPT'2000, Lect. Notes in Comp. Sci., New York: Springer-Verlag, V. 1976, 2000, pp. 470–482.

[437] Y. Zheng and X.-M. Zhang, *New Results on Correlation Immune Functions*, International Conference on Information Security and Cryptology: ICISC'2000, Lect. Notes in Comp. Sci., New York: Springer-Verlag, V. 2015, 2001, pp. 49–63.

[438] Y. Zheng and X.-M. Zhang, *On Plateaued Functions*, IEEE Transactions on Information Theory, V. 47, No. 3, 2001, pp. 1215–1223.

[439] Y. Zheng and X.-M. Zhang, *Relationships between Bent Functions and Complementary Plateaued Functions*, In Proceedings of the 2nd International Conference on Information Security and Cryptography, ICISC'99, Lect. Notes in Comp. Sci., Berlin, Heidelbeg, New York: Springer-Verlag, V. 1787, 2000, pp. 60–75.

[440] K. H. Zimmermann, *Beiträge zur algebraischen Codierungstheorie mittels modularer Darstellungstheorie*, Bayreuth. Math. Schr., 1994, No. 48, 278 pp.

Index

Selected Titles in This Series

241 **O. A. Logachev, A. A. Salnikov, and V. V. Yashchenko,** Boolean Functions in Coding Theory and Cryptography, 2012
240 **Kazuya Kato, Nobushige Kurokawa, and Takeshi Saito,** Number Theory 2, 2011
239 **I. Ya. Novikov, V. Yu. Protasov, and M. A. Skopina,** Wavelet Theory, 2011
238 **Leonid L. Vaksman,** Quantum Bounded Symmetric Domains, 2010
237 **Hitoshi Moriyoshi and Toshikazu Natsume,** Operator Algebras and Geometry, 2008
236 **Anatoly A. Goldberg, Iossif V. Ostrovskii, and Iossif V. Ostrovskii,** Value Distribution of Meromorphic Functions, 2008
235 **Mikio Furuta,** Index Theorem. 1, 2007
234 **G. A. Chechkin, G. A. Chechkin, A. L. Piatnitski, A. L. Piatnitski, A. S. Shamaev, and A. S. Shamaev,** Homogenization, 2007
233 **A. Ya. Helemskii,** Lectures and Exercises on Functional Analysis, 2006
232 **O. N. Vasilenko,** Number-Theoretic Algorithms in Cryptography, 2007
231 **Kiyosi Itô,** Essentials of Stochastic Processes, 2006
230 **Akira Kono and Dai Tamaki,** Generalized Cohomology, 2006
229 **Yu. N. Lin′kov,** Lectures in Mathematical Statistics, 2005
228 **D. Zhelobenko,** Principal Structures and Methods of Representation Theory, 2006
227 **Takahiro Kawai and Yoshitsugu Takei,** Algebraic Analysis of Singular Perturbation Theory, 2005
226 **V. M. Manuilov and E. V. Troitsky,** Hilbert C^*-Modules, 2005
225 **S. M. Natanzon and S. M. Natanzon,** Moduli of Riemann Surfaces, Real Algebraic Curves, and Their Superanalogs, 2004
224 **Ichiro Shigekawa,** Stochastic Analysis, 2004
223 **Masatoshi Noumi,** Painlevé Equations through Symmetry, 2004
222 **G. G. Magaril-Il'yaev and V. M. Tikhomirov,** Convex Analysis: Theory and Applications, 2003
221 **Katsuei Kenmotsu,** Surfaces with Constant Mean Curvature, 2003
220 **I. M. Gelfand, S. G. Gindikin, and M. I. Graev,** Selected Topics in Integral Geometry, 2003
219 **S. V. Kerov,** Asymptotic Representation Theory of the Symmetric Group and its Applications in Analysis, 2003
218 **Kenji Ueno,** Algebraic Geometry 3, 2003
217 **Masaki Kashiwara,** D-modules and Microlocal Calculus, 2003
216 **G. V. Badalyan,** Quasipower Series and Quasianalytic Classes of Functions, 2002
215 **Tatsuo Kimura,** Introduction to Prehomogeneous Vector Spaces, 2002
214 **L. Š. Grinblat,** Algebras of Sets and Combinatorics, 2002
213 **V. N. Sachkov and V. E. Tarakanov,** Combinatorics of Nonnegative Matrices, 2002
212 **A. V. Mel′nikov, S. N. Volkov, and M. L. Nechaev,** Mathematics of Financial Obligations, 2002
211 **Takeo Ohsawa,** Analysis of Several Complex Variables, 2002
210 **Toshitake Kohno,** Conformal Field Theory and Topology, 2002
209 **Yasumasa Nishiura,** Far-from-Equilibrium Dynamics, 2002
208 **Yukio Matsumoto,** An Introduction to Morse Theory, 2002
207 **Ken'ichi Ohshika,** Discrete Groups, 2002
206 **Yuji Shimizu and Kenji Ueno,** Advances in Moduli Theory, 2002
205 **Seiki Nishikawa,** Variational Problems in Geometry, 2002
204 **A. M. Vinogradov,** Cohomological Analysis of Partial Differential Equations and Secondary Calculus, 2001
203 **Te Sun Han and Kingo Kobayashi,** Mathematics of Information and Coding, 2002
202 **V. P. Maslov and G. A. Omel'yanov,** Geometric Asymptotics for Nonlinear PDE. I, 2001

For a complete list of titles in this series, visit the
AMS Bookstore at **www.ams.org/bookstore/**.